CONTENTS

TECHNICAL GRAPHICS

EDWIN T. BOYER

FREDERICK D. MEYERS

FRANK M. CROFT, JR.

MICHAEL J. MILLER

JOHN T. DEMEL

ALL OF THE OHIO STATE UNIVERSITY

WILEY

JOHN WILEY & SONS, INC.

NEW YORK   CHICHESTER   BRISBANE   TORONTO   SINGAPORE

Text design by Karin Gerdes Kincheloe
Cover design by Lee Goldstein
Cover art by Marjorie Dressler
Production supervised by Lucille Buonocore
Copyediting supervised by Marjorie Shustak
Acquisitions Editor—Charity Robey

Library of Congress Cataloging in Publication Data:
Technical graphics / Edwin T. Boyer . . . [et al.].
        p.   cm.
    Includes bibliographical references and index.
    ISBN 0-471-85689-4
    1. Engineering graphics.   I. Boyer, Edwin T., 1937–
T353.T284   1991
604.2—dc20   90-39861
                                                             CIP

Printed in the United States of America

10 9 8 7 6 5 4 3 2 1

ABOUT THE AUTHORS

Edwin T. Boyer has been an Associate Professor of Engineering Graphics at The Ohio State University since 1984. He earned his B.Ed. degree at Keene State College in 1962 and his M.S. degree at Indiana State University in 1963. In 1976 he earned the D.Ed. degree at Texas A&M University.

Dr. Boyer taught drafting and graphic arts at the high school level and then founded his own business, Visual Grafix. His company produced engineering, architectural, and land contour scale models for industry, as well as other graphics and illustrations for specific client needs. He taught for a year at Texas A&M University before teaching Industrial Design and Drafting at Northern Illinois University for eight years.

Dr. Boyer is an active member of professional organizations including the Engineering Design Graphics division of the American Society for Engineering Education. The International Technology Education Association, and he is a Life Member of The Council on Technology Teacher Education. Dr. Boyer organized and supervised the team that prepared all the camera-ready graphics for this book.

Frederick D. Meyers, P.E., has been an Associate Professor of Engineering Graphics at The Ohio State University since 1982. Prior to his appointment there he served as a project engineer, engineering manager and technical director with Owens-Corning FIBERGLAS Corporation over a period of 30 years. He also served as an engineering officer in the U.S. Air Force. His early teaching experience was as an Instructor in Engineering at the A&M College of Texas. He holds B.S. and M.B.A. degrees from Ohio State.

Professor Meyers is an active member of the American Society for Engineering Education and the American Society of Mechanical Engineers. He was named Ohio Engineer of the

Year in 1972. He is a member of Pi Tau Sigma and Tau Beta Pi, engineering honorary societies.

Frank M. Croft, Jr. P.E., is an Associate Professor in the Department of Engineering Graphics at The Ohio State University, where he has been since 1984. Before joining the faculty at Ohio State, Dr. Croft had taught 11 years on the college level. He started his academic career at West Virginia Institute of Technology, where he spent three years, and then moved on to the Speed Scientific School, the Engineering College of the University of Louisville, where he spent eight years. He taught engineering graphics at West Virginia Tech and was involved in graphics and transportation at the University of Louisville. During this period with the University of Louisville, he completed a Ph.D. in engineering at Clemson University in South Carolina. In addition to his Ph.D., Dr. Croft has a M.S. in Engineering from West Virginia College of Graduate Studies and a B.S. in Aerospace Engineering from Indiana Institute of Technology.

Dr. Croft has been very active with the Engineering Design Graphics Division (EDGD) of the American Society for Engineering Education (ASEE) since 1974. He has served on the publications staff of the *EDGD Journal* as Associate Editor and as Advertising Manager. He has served as Program Chairman for both the Mid-Year Meeting of the Division and the Annual Conference of ASEE. He has served the division as Vice Chairman and as Chairman.

Michael J. Miller, P.E., began teaching full-time in 1985. Prior to that he had served as an Adjunct Professor while working full-time as an engineering research administrator at The Ohio State University. He also spent five years in the aerospace and electronics industries as a practicing engineer. Professor Miller holds B.S. and M.B.A. degrees from The University of Michigan and The Ohio State University, respectively, and is an Assistant Professor of Engineering Graphics at Ohio State. He is a member of the American Society for Engineering Education and the American Institute of Industrial Engineers.

John T. Demel, P.E., is Professor, Chairman of the Engineering Graphics Department, and Associate Dean, College of Engineering, at The Ohio State University. He holds the degrees of Ph.D. and M.S. in Metallurgy from Iowa State University and a B.S. in Mechanical Engineering from The University of Nebraska-Lincoln. Previously he taught at Texas A&M University and Savannah State College. He also served as an officer in the Army Ordnance Corps.

Professor Demel is a national leader in implementing computer graphics in engineering. He has co-authored texts in computer graphics with Professor Miller and authored or co-authored over 20 publications on engineering graphics. He is active in the American Society for Engineering Education, American Society of Mechanical Engineers, and National Computer Graphics Association.

ACKNOWLEDGMENTS

The authors wish to thank all the members of their immediate families for their patience and understanding during the many hours required to prepare this text. Their support during this period was greatly appreciated.

The effort and support of the editorial and production staff of John Wiley & Sons is gratefully acknowledged. We would especially like to thank Charity Robey, Editor, for her support and encouragement during this project.

Most of the finished illustrations in this book were produced by a team of illustrators. There were many people involved with the illustrations for this project, and the authors would like to acknowledge their efforts. Those illustrators who made the most significant contributions were: Casey D. Doyle, Jeffrey S. Wallace, Marilyn R. Bahney, and Wayne J. Kleman, Jr. Other contributors were: K.J. Bomlitz, R.E. Spica, D.P. O'Donovan, R.L. Long, T.L. Myers, and M.E. Edwards. Their dedication to this project was greatly appreciated.

Faculty members of the Department of Engineering Graphics of The Ohio State University who contributed to this text were: Professors L.O. Nasman, R.I. Hang, R.A. Wilke, and G. Bertoline. One graduate student who made a significant contribution in problem development was M.S. Stephenoff.

We appreciate the assistance of the American Society of Mechanical Engineers, particularly Frida G. Yeghiazarian, for their help in determining the latest versions of the standards quoted in this text.

Finally, the authors wish to express their gratitude and sincere appreciation to the reviewers of the manuscript. Without their input this text would not have been possible. Reviewers for this book were: Terry Brown, Cincinnati

Technical College; James D. Forman, Rochester Institute of Technology; Peter A. Giambalvo, California Polytechnic State University; Randy Shih, Oregon Institute of Technology; Donald L. Simon, Trus Joist Corp.; Leonard Weiss, New York City Technical College; and Charles E. Wiser, SUNY–College of Technology at Alfred.

Additonal reviewers were: Ronald Barr, University of Texas at Austin; Homer L. Bosserman, Monterey Peninsula College; W. George Devens, Virginia Tech; George W. Eggeman, Kansas State University; Jon E. Freckleton, Rochester Institute of Technology; Larry Genalo, Iowa State University; J. David Holloway, Ohio Northern University; Klaus Kroner, University of Massachusetts; Jacques Simoneau, The Royal Military College of Canada; and Joe Tuholsky, West Virginia Tech.

E.T.B.
F.D.M.
F.M.C.
M.J.M.
J.T.D.

DEDICATION

WE DEDICATE THIS BOOK TO THE MEMORY OF **THOMAS E. FRENCH**, ONE OF THE GREAT PIONEERS OF ENGINEERING GRAPHICS. HIS DEDICATION, LEADERSHIP, AND CONTRIBUTIONS TO THE FIELD DURING HIS YEARS AS CHAIRMAN OF THE ENGINEERING DRAWING DEPARTMENT (1906–1942) AT THE OHIO STATE UNIVERSITY FORM A CLASSIC MODEL OF ACADEMIC LEADERSHIP AND EXCELLENCE.

PREFACE

This book is designed to meet the needs of modern technical graphics curricula and employs the current American National Standards Institute (ANSI) graphic standards. Instructors today have less time available to teach, and students have less time available to learn more material than ever before. The number of credit hours allocated for teaching technical graphics has steadily shrunk, while the field of design has steadily become more complicated. In recent years we have seen traditional steel and cast iron technology replaced by a wealth of new materials. These new materials often require radically different fabrication methods than the older materials. At the same time, a great variety of new design, drafting, and analytical tools have become available; many of these tools are computer-based. There is much more material to cover, and fewer hours to do the job.

This textbook meets the needs of today's technical graphics programs by addressing the new technologies. The traditional graphics topics have been streamlined so that they can be presented in a shorter period of time. We have eliminated obsolete material. Our approach to this text has been to critically examine every paragraph of material to assure that it is up-to-date, essential to the text, and presented in a forthright manner. All material is written in clear, uncomplicated language. The examples used throughout this book are of objects and assemblies that should be familiar to the student. For instance, a bicycle chain and sprocket are used to illustrate a line tangent to a circle.

A table saw fixture is used throughout the book as a "common problem." The fixture is introduced in Chapter 3 as a sketch. The sketch is converted to a drawing in Chapter 4. The drawing is then refined and modified in subsequent chapters to illustrate the material presented in each of those chapters.

Computer-aided design and drafting (CADD) methods are included where appropriate. If you are using a computer graphics system to do the drawing exercises, you will be able to complete the exercises without having to refer to an appendix or a special chapter. Instructions for CADD are set in a special typeface so that you will be able to tell which portions of the text are dealing with manual methods and which are for the computer-aided methods.

In each chapter, care is taken to provide instruction on the latest technologies, including CADD and drawing reproduction techniques. Old techniques that are seldom used are not presented unless they will help you understand the new techniques. Since there are many CADD software packages available, the CADD discussion is generic and the instructor will need to provide specific information on hardware and software systems available to you. This text is available packaged with a CADD program called GraphiCad. This package is designed specifically for educational applications; it is easy to learn and use, and it has many of the functions of commercial CADD packages. You will own the software if it is purchased with this text; this ownership gives you the right to use the software on any available computer. When you complete your education and enter the work force, you will probably have access to or be required to use graphics workstations.

Students who are going to use a computer and CADD software must know or learn how to perform basic computer operations. These include booting the hardware system, starting the CADD program, disk initialization, file saving and retrieving, making backup copies of disks, and recovering in case the program is halted by an error.

Typical CADD computer system components are identified in Chapter 1. It is desirable, but not necessary; that you have some understanding of the inner workings of computer system components. This will give you more confidence when using the system and may help you to know what to do if an error occurs. Programming experience is not essential.

The book does not assume that you have had previous graphics experience. It begins with the development of basic skills that you will need. Geometric constructions, visualization and sketching techniques, orthographic projection, and pictorial drawing are described in detail. Illustrations and photographs are used every step of the way to illustrate the points being made by the written material.

The material proceeds logically to develop your ability to describe the shape and size of objects. Important drawing conventions are described and illustrated. We describe the use of sectional views and show the insides of objects. Conventional methods of drawing sections are presented along with illustrations of each convention. These conventions are based on the American National Standards Institute's (ANSI) latest published standards.

We cover size description, adding dimensions to the drawing, in detail. The specification of tolerances, the allowed variations in dimensions, is presented as you begin to understand methods of making objects and reasons why variations in size must be allowed.

Production is the ultimate use of a working drawing. You will find a short description of the production processes and information on tolerances for metric and English measuring systems. You will also be introduced to surface control, datums, and geometric dimensioning and tolerancing. What is presented here will not make you an expert; you will be aware of advanced dimensioning methods and have a good foundation upon which you can build more expertise.

As you gain in ability to draw a single object, the idea of joining two or more objects to create an assembly is introduced. We discuss fastening and joining methods and describe various types of fasteners. Welding and use of adhesives, two very popular methods of joining objects, are also covered.

Design and presentation tools that are available to you are covered next. Three dimensional problems are solved with traditional hand tools and the computer. The 3-D space geometry (including descriptive geometry) techniques are presented using real world examples whenever possible. The pattern for a three-dimensional model is included inside the back cover of this book. When assembled, the model can provide tactile as well as visual input for thinking in three dimensions. Construction methods for shade and shadow are developed to provide better presentation of ideas. Presentation graphics combining text, drawings, and graphs are used to sell ideas. This integration of communications is presented just before the design process so that you will be ready to include presentation as part of the design process.

The design process is described so that you will be aware of the many steps that are necessary before an idea becomes a completed design. With this description, we show how graphics is used to document the steps of the design process. Comprehension of this material will help ''pull together'' all the other information presented in the text.

The instructor's manual contains suggestions for organizing

courses or course sequences of various types and lengths, as well as problem solutions and transparency masters. Workbooks of additional problems are also available from the publisher, as well as solutions for the workbook problems. The book is available with or without the software packaged inside.

We hope that both students and teachers will enjoy learning with this text. Please write to us in care of the publisher if you have comments on how the book could be made more useful.

EDWIN T. BOYER
FREDERICK D. MEYERS
FRANK M. CROFT, JR.
MICHAEL J. MILLER
JOHN T. DEMEL

CONTENTS

TERMS YOU WILL SEE IN THIS CHAPTER

DESIGN   Process of planning and creating a solution to a problem.

SIMULATION   To create a mathematical description or a pictorial representation of an object or a process.

SOFTWARE   Programs that make a computer run. System software provides the computer's essential general functions. Application software performs specific tasks such as CADD.

MODELING   Design technique where the characteristics of an object or process are stored in a computer and used to produce and work with an image of the design on the screen.

CADD   Computer-aided design and drafting. Refers to computer software and hardware for creating engineering drawings on a computer screen.

DIMENSIONS   Information added to a drawing to show the sizes of the objects the drawing represents.

.25

ORTHOGRAPHIC PROJECTION   Method for describing objects using multiple views. Each view is taken at right angles to the adjacent views.

SECTION VIEWS   Cutaway views of objects showing internal features.

1

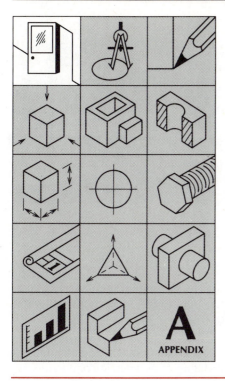

APPENDIX A

GETTING STARTED

1.1 INTRODUCTION

This book is designed for you, the designer graduate of the
1990s. It will provide you with the instruction in **design**
graphics skills that you will need in your profession. Design is
the essence of engineering, and graphics is the language of
design. Similar to English, or any other language, it is the
means we use to communicate our ideas to others. We have all
had the experience of trying to describe something, such as
how to get to our house, to someone else. We may have ended
up drawing them a map. It has been said that a picture is worth
a thousand words. That saying may seem trite, but it is very
true. A single picture can often illustrate an idea that cannot be
communicated in any other way.

  As a graduate designer, you will enter the design profession
at a very exciting time. Historically, new designs have been
created at the drafting table, proven out by constructing
physical models, and tested in the laboratory and in the field.
When the test program was complete, tooling was designed and
fabricated, and production was started. Rapid technological

Figure 1.1 Computer-aided design and drafting workstations are used by many designers. (Courtesy of Ohio State University Engineering Publications)

advances are changing the field to one where the design is created at a computer-aided design (CAD) workstation (Figure 1.1). During the design phase, the design team has access to computer-aided engineering (CAE) design analysis tools that can be used to test the design in **simulation.** This means that with **software** that permits **modeling** of geometric solids, a mathematical model of the object can be created on the computer. The mathematical model can then be tested on the computer just as a real physical model might be tested in a laboratory, and design errors can be corrected without having to construct a physical model (Figure 1.2). All of the data that

Figure 1.2 Editing drawing on CADD workstation. (Courtesy of Ohio State University Engineering Publications)

Figure 1.3   Computer numerical control milling machine. (Courtesy of Ohio State University Engineering Publications)

are used in the design are stored in the computer. When the product designer is satisfied that the design is correct, the tool designers can use the same data base to design the needed tooling without ever having seen an actual physical model of the object. When it is time to begin production of the object, the manufacturing personnel can use the same design data to program the computer-controlled machines that will be used in its production (Figure 1.3).

Because of the ease with which a design can be created, tested, and modified without ever having to create an actual drawing on paper or a model in the shop, designers can try out many different designs quickly and economically. They are free to be very creative.

You will need a variety of skills in order to make effective use of the power of the new computer-based tools available for design. The ability to visualize and describe objects will be increasingly important as more use is made of computer-generated images instead of physical models. It will be necessary for you to be able to visualize what an object under design will look like and to be able to create both manually drawn sketches of the object and computer-drawn images.

1.2   *WHAT YOU WILL LEARN FROM THIS TEXT*

This text is written for design students. It will teach the graphics skills that they must master. Each chapter will introduce a specific topic area and provide instruction in both manual methods and computer methods, so that you may become proficient in sketching, use of drawing instruments, and use of **computer-aided design and drafting (CADD)** systems. You will also have learned how to create a drawing or a sketch that communicates the details of your design to others.

1.2.1 Lettering

Lettering is used on drawings to identify the drawing, the scale (size) at which it is drawn, who drew it, who approved it, revision dates, **dimensions,** notes, and so on. In this chapter, you will learn techniques for lettering freehand, with the use of lettering instruments, and on a computer-aided drafting system.

1.2.2 Geometric Constructions

A drawing is made up of geometric shapes, lettered dimensions, and notes. In Chapter 2 you will learn effective methods for performing the most often used geometric constructions. These will include drawing lines parallel or perpendicular to a given line, dividing lines into several parts, bisecting angles, and constructing polygons, circles, and arcs, plus several other constructions.

1.2.3 Sketching Techniques

Designers need a quick way to present and study their ideas. They can create a drawing with instruments, but that often takes a lot of time and may require instruments that they do not have with them. Sketching, which is creating freehand drawings, is a powerful tool since it requires only a pencil and some paper. It is a quick way to present an idea. Furthermore, a sketch can be modified quickly and easily. When the idea has been worked out to the designer's satisfaction, the sketch can be used as the basis for detailed drawings. Sketches are also an important part of the records in a designer's lab book. They document the designer's work and provide a record for patent applications and other legal purposes. A drafter working from a good sketch can create the drawing without asking many questions.

Designers increasingly rely on making sketches of their ideas and then completing the design by computer. Because of this, drawing with instruments (triangles, T-squares, compasses) will be done infrequently. However, the ability to create drawings with instruments will still be important. A very large number of manually created drawings are in existence and must be maintained and modified.

1.2.4   Drawing with Instruments

A well-made sketch can contain as much detail as a final drawing done with instruments. However, a sketch is seldom acceptable for use in manufacturing. Therefore, it is necessary to make a final drawing from the sketch. This drawing might be made on a CADD system or manually with instruments. Basic instrument use will be presented in this chapter. Additional instruction on use of instruments as well as use of a computer drawing system will be presented as needed throughout the text.

1.2.5   Shape Description

Orthographic Projection

A drawing of an object is made so that the graphic image appears to be the same shape as the object that is being represented. Thus, the drawing provides a shape description of the object.

When the drawing is to be used for creation of the object, it will generally provide three different views of the object, most often the front, top, and a side view. The views are placed in a standard arrangement (Figure 1.4). This drawing technique is known as **orthographic projection.** It is covered in Chapter 4.

Pictorial Drawings

Pictorial drawings are generally used to describe an object or assembly, how it goes together, and how it is used (Figure 1.5). There are usually no dimensions or other notes that would tell how to make the object since that information would be on the detail drawings. A person assembling a toaster, for example, does not need to know the exact dimensions of each part, just where each part goes. Techniques for producing the various types of pictorial drawings will be presented in Chapter 5.

Perspective Drawings

Perspective drawings are used to illustrate how an object will look when it is completed (Figure 1.6). They provide a three-dimensional (3D) view of the object. Unlike axonometric drawings, parallel lines are not drawn parallel in perspectives but converge to a point, called a vanishing point. As a result, parts of the object that are close to the observer are drawn larger than those farther away. This technique adds realism since distant real-life objects appear smaller to us than closer ones. Perspective can be applied to just one axis (one-point perspective), to two axes (two-point perspective), or to all three axes (three-point perspective). You will learn more about perspective drawings in Chapter 5.

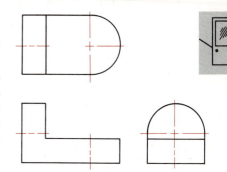

Figure 1.4 Orthographic projection uses multiple views to describe object.

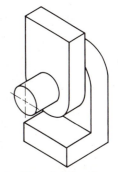

Figure 1.5 Pictorial drawings are created to resemble picture of object being represented.

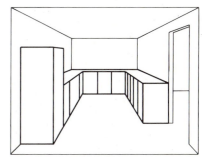

Figure 1.6 Linework of kitchen layout. (Courtesy of Utley Company)

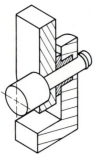

Figure 1.7 Sectional views are drawn so that object appears to have been cut open to display its interior details.

1.2.6 Sections and Conventions

Designers often create drawings that show a part or assembly that has had a portion cut away to show interior details (Figure 1.7). Such drawings are known as **section views.** They make it easier to understand how an object is constructed or how an assembly functions. You will learn to make sectional views of single objects and of assemblies.

Sometimes a portion of a drawing would be difficult to understand if it were drawn in a technically correct manner. A number of techniques have been devised to make the drawings more understandable. These special techniques are called conventional practices. They will be covered in detail in Chapter 6.

1.2.7 Size Description

It is necessary in most instances to indicate the size of the object as well as its shape. Size description is provided in two ways. One is to specify a scale, such as half size, meaning that one unit on the paper represents two units on the actual object. The second method is to mark dimensions on the drawing (Figure 1.8). In most cases, both methods are used. Drawing to a certain scale permits the drafter to lay out the object on the paper using a scale that is graduated in the proper increments. However, determining an object's size by measuring the drawing does not provide accurate dimensions. Therefore, design drawings include both dimensions and tolerances. Dimensions show the desired size. Tolerances tell the person making the object how much the size can vary from the dimensions shown (Figure 1.9). The basic concepts of dimensioning and tolerancing are covered in Chapter 7.

Chapter 8 builds on the dimensioning basics that are introduced in Chapter 7. It covers advanced dimensioning concepts and techniques, including tolerancing for metric dimensions, geometric tolerancing, and an introduction to manufacturing processes.

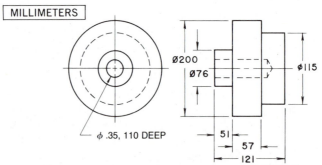

Figure 1.8 Dimensions are added to orthographic drawing to show size of object being represented.

1.2.8   Fastening and Joining

Most of the devices we use are made of two or more parts fastened together into an assembly using fasteners, as shown in Figure 1.10. There are many methods of fastening. Bolts, screws, rivets, welds, and adhesives are just a few of the many devices and materials that are used in fastening. In some cases, the fastener does not permit any movement between the objects, while in other cases it is important to permit such movement. You will learn more about fasteners and fastening techniques in Chapter 8.

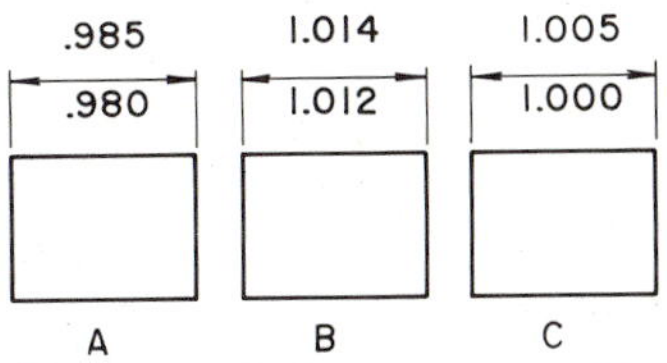

1.2.9   Drawing Notes

Notes are often added to a drawing. Notes are used to indicate the kind of material the object is to be made from, how many units of the object will be needed, and any special requirements (Figure 1.11). For instance, a note on a drawing of a table leg might say that it will be made from oak and that four will be needed for each table. Drawing notes are covered in Chapters 7 and 9.

1.2.10   Production Drawings

Production drawings, often called working drawings, are used to guide the workers who create the object or assembly (Figure 1.12). They include all the information needed to produce the object: shape description, size description, quantities, notes, and assembly instructions. Designers have working drawings of mechanical objects such as gears and of assemblies such as transmissions. Civil designers have working drawings showing how a portion of a highway is to be built. Electrical designers may have working drawings showing the layout of printed circuit boards or of electrical wiring in a building. Piping diagrams used by chemical engineers are another example of working drawings. Working drawings of airplane wings and other aircraft components are used by aeronautical designers. Production drawings are covered in Chapter 10.

1.2.11   Descriptive Geometry

Many design tasks, such as piping layouts and duct work, often require the techniques of descriptive geometry for their solution. Descriptive geometry is the application of the principles of spatial geometry to the solution of size, shape, and space relationships of objects and parts of objects. For instance, descriptive geometry can be used to determine the necessary size and shape of a flat piece of metal that is to be formed into a transition piece. (Figure 1.13).

Figure 1.9   Tolerances are added to dimensions to show size of object being represented.

.985 .980   1.014 1.012   1.005 1.000

A   B   C

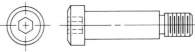

Figure 1.10   Fasteners such as these are used in many assemblies.

In Chapter 11 you will learn the basic concepts of descriptive geometry. You will learn about lines, planes, and solid objects and the relationships among them. You will also learn how to determine the intersections of one-dimensional, two-dimensional, and three-dimensional objects.

Chapter 12 will give you practice in applying the concepts from Chapter 11 to practical, real-world problems. You will learn to create flat patterns for making objects such as boxes and cans. You will also learn to create maps and contour plots. You will work with vectors and learn to combine vector quantities. Finally, you will learn to add shades and shadows to your drawing to create a more realistic appearance when drawing pictorial views.

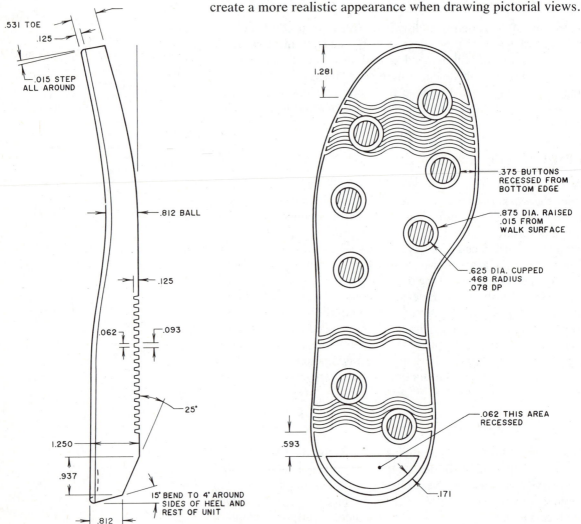

Figure 1.11   Notes are an important part of a drawing. (Courtesy of Jones and Vining, Inc.)

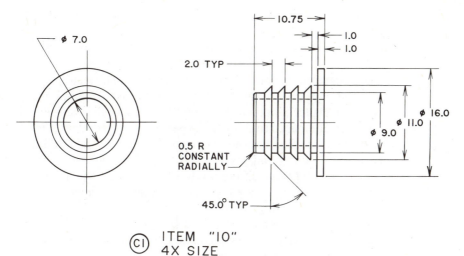

Figure 1.12 Production (working) drawings provide details needed to make object. (Courtesy of Ford Motor Company)

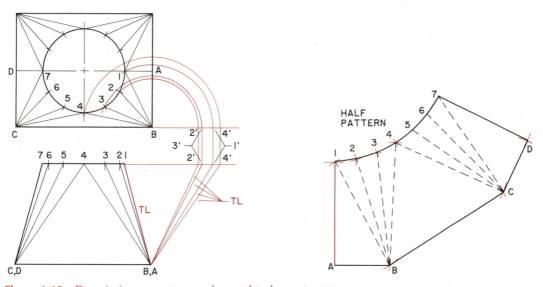

Figure 1.13 Descriptive geometry can be used to lay out pattern.

1.2.12 Presentation Graphics

The term *presentation graphics* refers to the design of material for reports and presentations. It includes charts, graphs, illustrations, reports, and audiovisual materials. The work can include producing charts and graphs, illustrations, diplomas, letterheads, logos, presentation quality reports, and even complete publica-

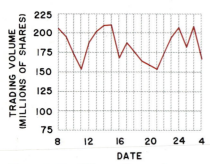

A FAMILY BUDGET EXPENDITURES BY CATEGORY

Figure 1.14 Pie charts provide wealth of information in small space.

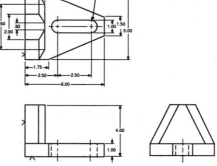

Figure 1.15 *XY*-coordinate charts show relationship between two variable values.

tions such as magazines and books. You will learn about this field in Chapter 13.

Charts and graphs are very valuable tools for presenting information. The information they provide is not intended to take the place of detailed numerical data but rather to supplement it. Charts and graphs are used to show all sorts of relationships. Sales charts show at a glance whether sales are increasing or decreasing. Pie charts are often used to show the expenditures of a budget (Figure 1.14). Relationships between economic data and time are easily shown with charts and graphs (Figure 1.15). You will learn to select graph styles for different types of data. You will also learn step-by-step how to create the graph manually and how to create it on a computer using chart and graph software.

The term *desktop publishing* has been coined to describe the process of merging text and graphics by computer to produce both text and pictures on the same page (Figure 1.16). Use of desktop publishing software is covered in Chapter 13.

TABLE SAW FIXTURE
Model GCF–145A
DESIGN DRAWING AND PRODUCTION SCHEDULE

Table saw fixture, Model GCF–145A, is standard on Contractor Saw, Model GCTS–90357 and available for Models GCTS–90356 and GCTS–90357. Fixture production schedules for the first half-year are expected to increase inventory by 20,000 units in anticipation of construction upturns and resulting brisk mid-year demand.

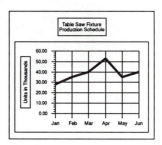

Figure 1.16 One-page document produced with desktop publishing software.

Figure 1.17 CADD workstations are powerful design tools. (Courtesy of Ohio State University Engineering Publications)

1.2.13 Computer Graphics

Computer graphics provides an exciting set of new capabilities that are much more powerful than anything the designer has ever had available before. Computer graphics includes graphing, desktop publishing, and includes both two-dimensional (2D) and three-dimensional (3D) CADD systems (Figure 1.17). Modeling software gives the designer the ability to create a three-dimensional computer model of an object. The model can then be moved or rotated, its size or shape can be changed, or it can be combined with other objects. The designer can examine it from many different angles. Structural analysis software permits the designer to test the design of an object to determine whether it meets the requirements for strength, resistance to temperature changes, resistance to bending under load, and resistance to shock. These tests can be conducted in simulation, meaning that they are done by the computer on a computer-generated model meeting the designer's specifications.

Computer-aided Design

Many design firms now use computer systems to store their design data and create their drawings. The CAD systems or CADD systems are in use in most large companies. As microcomputers become more powerful, an increasing number of smaller companies are getting these systems. The main difference between the two types of systems is that CAD systems provide more tools to aid in the design process while CADD systems provide more tools to aid in the drafting process.

The CADD systems can produce drawings that are equivalent to those produced in the "conventional" way with pencil and paper (Figure 1.18). However, with a computer-based design system, the drawing is created on the screen of the computer display terminal, and the information (data) needed to reproduce the drawing is stored in a data file on a computer data disk or on magnetic tape. When the drawing is complete, it can be drawn on paper or other drawing media using a plotter that is connected to the computer (Figure 1.19). If the drawing is small and a "rough" copy is all that is needed, the drawing may be reproduced on a graphics printer.

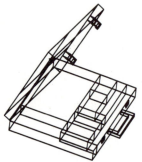

Figure 1.18 Drawing produced on CADD system. (*Computer-generated image*)

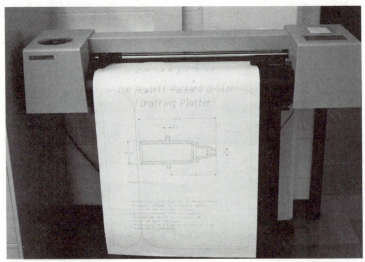

Figure 1.19 Pen plotters producing precision drawings. (Courtesy of Ohio State University Engineering Publications)

It is no longer enough for the designer to be able to produce a set of drawings showing how an object is to be made. Historically, that is what has been done. It was then necessary for a tool designer to determine what tools would be needed to produce the object and to design any special tools that were required. Then the drawings of the object and the necessary tools to make it were provided to the manufacturing people. They would have to study the drawings and decide how to make it. Sometimes the tooling would not work. In that case, the manufacturing people and the tool design people would need to work together to modify the tooling or to develop new methods for making the object.

Now, it is becoming very common for tool design people to use computer files of design data created by the product designer to design the needed tools. Just as the product designer can test the product design in simulation, the tool designer can test the tooling by having the computer simulate the necessary steps to create the object from the raw material. The best method of manufacturing the object can be determined, and the special tools can be designed correctly before the manufacturing process is started.

Computer-aided Manufacturing

Computer-aided manufacturing (CAM) involves using the computer to control the machines that are making the object. For several years, machines have been controlled by creating a tape specifying the required operations and then having the machine follow the instructions on the tape. These machines are known as numerical control (NC) machines. Newer versions of this type of control are computer numerical control (CNC) and direct numerical control (DNC) (Figure 1.20). With CNC or DNC a computer is connected to the NC machine and is programmed to direct its actions.

Computer-integrated Manufacturing

In many of the preceding cases, it is still necessary for a human to study the design drawings and create the necessary tape or computer program to guide the operation of the machines that will produce the object. However, the CAD and CAM systems are now combined in many companies. The result is known as CAD/CAM or computer-integrated manufacturing (CIM). With these systems the same data file is used throughout the design process and the manufacturing process. The file is first created by the designer using the design and analysis capabilities. If new tools are needed, the tool designer uses the file and the same design and analysis software to design the tools. Then the file is used with the manufacturing software to create the instructions for the NC machines to produce the parts. It is possible to complete the entire process without ever producing a drawing on paper. The ability to use the actual design data through all phases of the design-and-manufacturing cycle eliminates the errors that are caused when drawings have to be read and interpreted by humans and the information has to be transferred manually from system to system.

Flexible Manufacturing Systems

Flexible manufacturing systems (FMSs) permit the concepts and techniques of CIM to be used by small companies and others that do not produce enough objects of any one design to make it profitable to set up a complete production line for the object. Flexible manufacturing systems permit rapid and inexpensive changeover from production of one part to production of another, often by simply reprogramming the machine(s) involved (Figure 1.21).

Figure 1.20 Computer numerical control machines provide automated manufacturing. (Courtesy of Ohio State University Engineering Publications)

Figure 1.21 Flexible manufacturing center. (Courtesy of Ohio State University Engineering Publications)

1.3 *TOOLS YOU WILL NEED*

The tools you will need will depend on whether you are going to produce sketches, final drawings using instruments, or drawings using a computer system. This section will discuss the tools required for each of these processes.

1.3.1 Tools for Sketching

The tools you will need to produce acceptable sketches will include paper, pencils, and an eraser. An erasing shield is also useful. The paper may be ruled with a rectangular or an isometric grid (Figure 1.22). Thus, you can follow the lines and will find it easier to produce a neat sketch. Some people use unruled paper laid over a ruled sheet. This provides both the advantages of a ruled sheet and the better final appearance of the unruled sheet.

Many types of pencils are designed especially for creating sketches, drawings, and illustrations. The type of pencil you select will have a strong effect on your ability to produce good work. The thickness and hardness of the pencil lead are also important. Lead hardnesses vary from 6B (very soft) to 9H (very hard). The softer leads (HB to F) are generally used for sketching. The harder leads are better for detail drawing. A more detailed discussion of pencils and leads will be found in Section 1.4.

The type of eraser you use is also very important. Some will erase cleanly, others will not. Some are designed to erase small

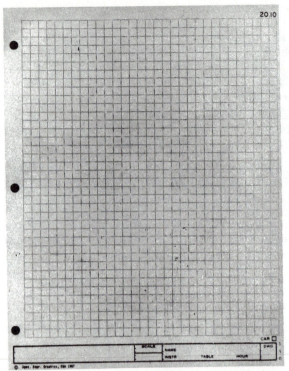

Figure 1.22 Ruled paper is excellent for sketching.

Figure 1.23 Erasers used by drafters.

areas. Others are designed for bigger jobs. The three types of eraser material most often used for drafting are gum, pink rubber, and white plastic (Figure 1.23).

1.3.2. Tools for Instrument Drawing

Instrument drawing requires many more tools than sketching. Figure 1.24 shows a typical set of tools for a beginning student. It is important that you be familiar with the tools and their uses.

This chapter will describe the media, that is, the material, on which drawings are generally made. Paper, vellum, and polyester film are the most commonly used graphics media. Much work in computer graphics is done on the screen of a video display (Figure 1.25).

Whether you are sketching, doing detail drawing, or drawing illustrations, it is important that you be able to measure distances and be able to draw lines perpendicular or parallel to each other. You will need to be able to draw ellipses, irregular curves, and other special shapes. A number of special drafting tools are available to help you. They include scales for measuring distances (Figure 1.26); T-squares; triangles; templates for circles, ellipses, and other shapes; irregular curves (Figure 1.27); drafting ma-

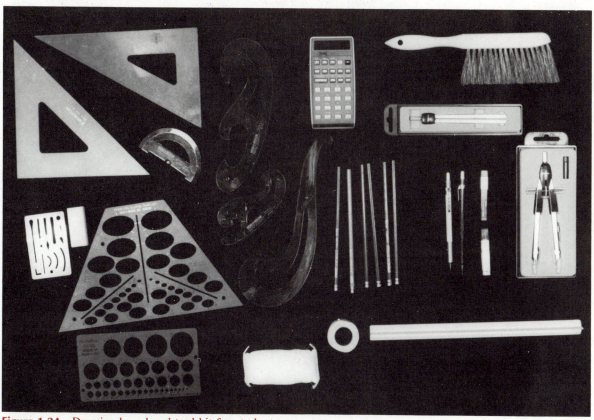

Figure 1.24   Drawing board and tool kit for student use.

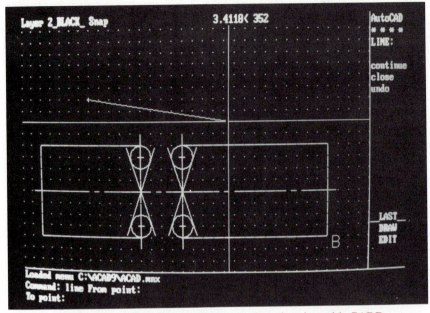

Figure 1.25 CRT displays are used when drawing with CADD systems. (*Computer-generated image*)

Figure 1.26 Designers use variety of scales.

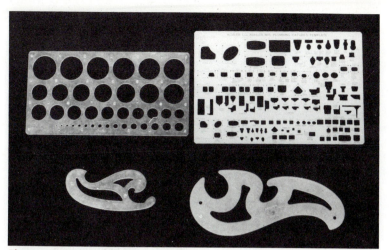

Figure 1.27 Templates and irregular curves make drafting faster and easier.

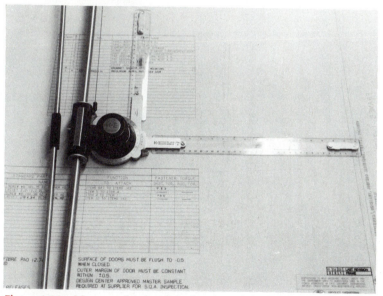

Figure 1.28 Many professional designers and drafters use drafting machines. (Courtesy of Ohio State University Engineering Publications)

chines (Figure 1.28); parallel bars (Figure 1.29); and drawing instruments such as compasses and dividers (Figure 1.30).

Scales may be English or metric. The English scales are graduated and calibrated in inches. The metric scales are graduated and calibrated in millimeters. Graduations are the divisions on

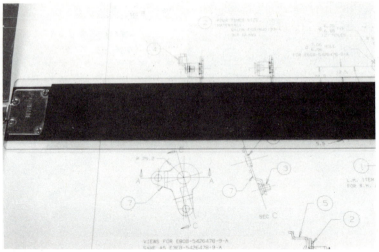

Figure 1.29 Parallel bars are used by some drafters. (Courtesy of Ohio State University Engineering Publications)

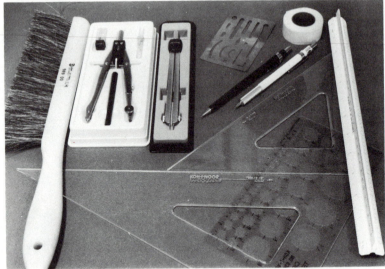

Figure 1.30 Drafting instruments are available as boxed sets. (Courtesy of Ohio State University Engineering Publications)

the scale, and calibrations are the numbers indicating the values those divisions represent. Within these two measuring systems (English and metric), there are a number of scales that are designed to meet the needs of certain users.

The English scales may be graduated in decimal units, as is the civil engineer's scale, or in fractional units, as are the mechanical engineer's scale and the architect's scale. Some scales are full divided (Figure 1.26*a*), and some are open divided. Metric scales are coming into greater use as more designs are being created in the metric system (Figure 1.26). For a further discussion of scales and scaling, refer to Section 7.2.

1.3.3 Tools for Drawing by Computer

Central Processing Unit

The heart of a computer graphics system is the central processor (CPU or system unit). It may be a very large mainframe computer, a medium-sized minicomputer, or even a small microcomputer. The larger computers can usually support more simultaneous users and jobs. However, the power of computers is increasing so rapidly that today's microcomputers are often more powerful than the mainframes of just a few years ago. Workstations designed around powerful microcomputers can support multiple simultaneous tasks and even multiple simultaneous users.

Input Devices

A computer system must have more than just a CPU. In order to communicate with users and other devices, it must have input devices and output devices. It must also have data storage devices.

There are several input devices in common use on CADD systems. Certainly the most common is the keyboard. Nearly all computer programs expect some input from the keyboard, even if it is only the command to run the program.

A joystick can also be used as an input device (Figure 1.31). Most people are familiar with the use of joysticks, which are very popular for computer games. They provide very fast response and thus fast action.

Most CADD systems use a mouse as a primary input device (Figure 1.32). Mouse input is similar in many respects to joystick input but gives the user better control. The mouse is generally preferred to the joystick for CADD systems.

Digitizing tablets are found on many systems (Figure 1.33). They are often used as menu boards for selecting tasks to be done. Another use for digitizing tablets is to copy a drawing from paper to the computer.

Scanners are another type of input device. They are used to automate the process of copying a drawing from paper to the computer. They can often interpret and copy text and symbols as well as graphics.

Still another method of inputting data is imaging. This process

Figure 1.31 Joysticks are found on some systems. (Courtesy of Ohio State University Engineering Publications)

Figure 1.32 Mouse is a popular CADD system input device. (Courtesy of Ohio State University Engineering Publications)

uses a video camera and a computer with appropriate software to process and store the data. Laser-based systems can provide three-dimensional (*X, Y, Z*-coordinate) data at high speeds.

Output Devices

The primary output devices that computer systems use are cathode-ray tube (CRT) displays. The name refers to their television-like display tube. There are two main types of CRTs. An alphanumeric CRT can display letters, numbers, and punctuation marks but cannot display graphics images. A graphics CRT is designed to display lines, circles, curves, and other drawing features, often in color. In most cases, it can also display alphanumeric information. Some systems use only a graphics display. Others use both an alphanumeric display and a graphics display (Figure 1.34).

As was stated earlier, it is often necessary to produce a copy of the drawing on paper or other material. The two types of devices that are most often used for this purpose are plotters (Figure 1.35) and graphics printers (Figure 1.36). Two types of plotters in general use are pen plotters and electrostatic plotters. The pen plotters use special drafting pens and create the drawing one line at a time on paper or other drawing media somewhat as you would do with your drawing pencil. Electrostatic plotters produce the image by placing an electrical charge on the media. Plotters can produce very high quality drawings.

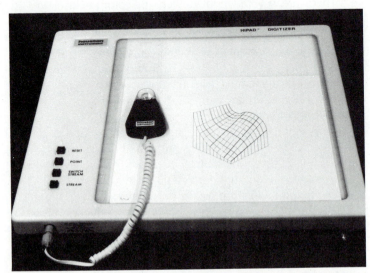

Figure 1.33 Digitizing tablets are versatile CADD system input devices. (Courtesy of Ohio State University Engineering Publications)

Figure 1.34 Some CADD workstations use two displays (Courtesy of Ohio State University Engineering Publications)

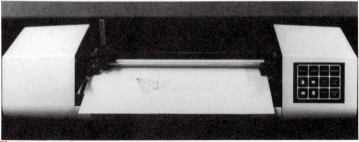

Figure 1.35 Pen plotter for small drawings. (Courtesy of Ohio State University Engineering Publications)

Graphics printers are capable of printing a complex page as a pattern (matrix) of dots such as on the screen. Some of them use pins to strike an inked ribbon. Others spray a "jet" of ink on the paper. Still other printers use a laser beam. The laser printers can combine graphic and nongraphic (text) material and produce very high quality output. The quality of drawing a printer produces depends on the capabilities of both the printer and the software, so it will vary from system to system, but it is generally not as good as drawings produced by plotters.

A third type of copy-producing device works somewhat like an office copier. It produces on paper an image of what is on the display screen.

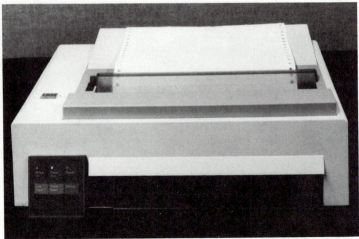

Figure 1.36 Dot matrix graphics printer. (Courtesy of Ohio State University Engineering Publications)

Data Storage Devices

The two types of data storage devices most commonly found on computers are disk drives (Figure 1.37), which store data on disks that have been coated with a magnetic material, and tape drives (Figure 1.38), which store data on tape that has been coated. The magnetic tape is similar to the audio tape used in tape recorders and players.

Disk drives are found on virtually every computer system. It is not possible to keep all the needed programs, data, and other

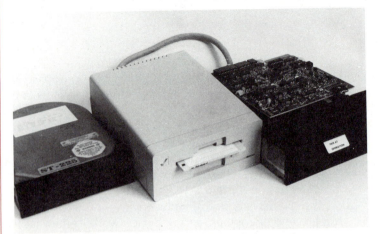

Figure 1.37 Disk drives come in variety of sizes and types. (Courtesy of Ohio State University Engineering Publications)

information in the main memory of the computer. Even if it were, the information would all be lost if the system failed due to a power failure or other problem. Disk drives provide permanent storage and fast access to data. They are used both as input devices to load programs and data into the computer and as output devices to store programs and data.

Tape drives are used on many systems. They provide backup information in case something happens to the disk drive. It is standard procedure to copy the files from the disk to the backup tape on a regular basis. Tape drives are not generally used to replace disk drives. They lack the random-access capability of disk drives, so it usually takes much more time to get information from a tape than it does from a disk.

Software

Of course, no computer system can be complete without software. The term *software* refers to the programs that provide instructions to the computer. There are two types of software. The first is system software. It provides the basic instructions to the computer that tells it how to read from the disk drives, how to display information on the screen, how to send information to the printer, and so on. The second type is application software. It includes the programs and data files that provide the specific instructions and information for the job that is to be done. Computers generally have one set of system software. They would have several sets of application software, one set for each type of task to be performed. Word processing would require one set, computer-aided design would need a different set, and so on.

Figure 1.38   Tape drives are used to provide backup for disk drives. (Courtesy of Ohio State University Engineering Publications)

1.4   *USING YOUR DRAWING TOOLS*

1.4.1.   Sketching

Sketches are made without the aid of instruments. Grid paper is often used to aid in sketching straight lines (Figure 1.39). An erasing shield will help you to correct mistakes. The erasing shield is a small piece of flat metal containing a number of different-shaped holes (Figure 1.40). To use it, place it on your drawing so that all or part of the marks to be removed show through one of the holes but parts to be saved do not. Then erase the marks through the hole. This prevents you from accidentally erasing a line you want to keep.

The pencil should be relatively soft, such as HB, and should be sharpened so as to produce a clean dark line. Your normal pocket pencil is somewhat harder but is generally acceptable.

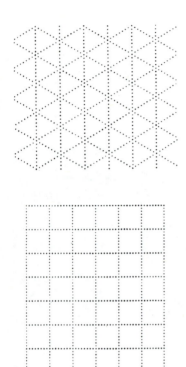

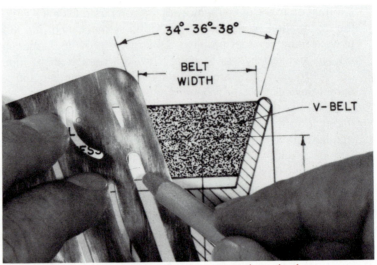

Figure 1.40 Erasing shields help when correcting mistakes.

Figure 1.39 Rectangular and isometric grid paper.

Correct pencil movements and other techniques to help you to produce good sketches will be covered in Chapter 3. You should start by sketching lightly so that you can erase mistakes easily. When the sketch is the way you want it, go back and darken the lines, making the visible lines the darkest.

1.4.2 Alphabet of Lines

The weight, or thickness, of a line provides valuable information. Dark, wide lines are used to show the outlines and other visible lines that form a part of the object. Slender, lightweight lines are used if they are not to be part of the final drawing. Dark, thin lines are used to mark the centers of circles and arcs and for other purposes that will be covered in later chapters. Similarly, the style in which the line is drawn is important. Visible lines are drawn solid. Hidden lines, that is, lines that describe edges that are hidden behind other surfaces, are shown dashed. Still other line styles are used to show the locations of circular features, a cutting plane for sectional views, and so on. Drafters use a standard set of styles and widths of lines in their work. Each type of line is designed for a specific purpose. The set of line styles and widths is referred to as the alphabet of lines (Figure 1.41).

1.4.3 Instrument Drawing

Straight Lines
The T-square historically has been a basic tool in instrument drawing. It is used to draw horizontal lines on the paper and is also used to provide a reference for lines at other angles. The T-

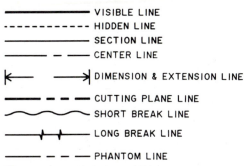

VISIBLE LINE
HIDDEN LINE
SECTION LINE
CENTER LINE
DIMENSION & EXTENSION LINE
CUTTING PLANE LINE
SHORT BREAK LINE
LONG BREAK LINE
PHANTOM LINE

Figure 1.41 Alphabet of lines.

square is placed on the drawing board with the blade going from left to right and the top of the T pressing firmly on one side of the board (Figure 1.42). The paper is attached to the board so that its edge or reference line is parallel to the blade of the T-square. The T-square may be slid up or down on the board as needed. It may also be used as a straight edge to draw a line between any two points (Figure 1.43). It should not be used as shown in Figure 1.44.

The T-square is gradually falling into disuse. Many student exercises are designed so they are not needed. Professional drafters use drafting machines or parallel bars (Figure 1.45). And of course, CADD systems do not use instruments at all.

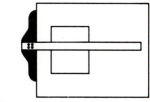

Figure 1.42 Using T-square.

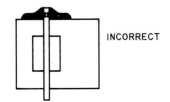

Figure 1.43 T-squares can be used at an angle.

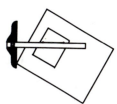

INCORRECT

Figure 1.44 Incorrect use of T-square.

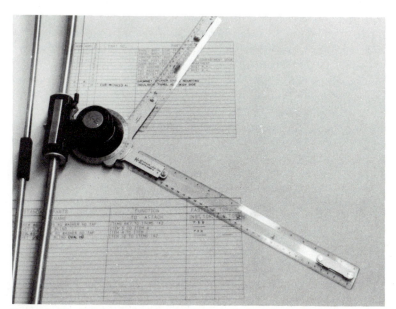

Figure 1.45 Drafting machine blades are designed to aid drafter. (Courtesy of Ohio State University Engineering Publications)

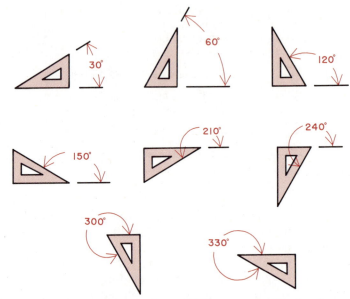

Figure 1.46 Constructing angles at 30° increments with triangle.

Your triangles will permit you to accurately draw lines at 15° increments from any reference line. The 30°–60° triangle can be used with the T-square, parallel bar, or another triangle to draw lines at 30° increments (Figure 1.46). Similarly, the 45° triangle can be used to draw lines at 45° increments (Figure 1.47). The two triangles can be used together to provide the 15°-increment capability (Figure 1.48). You can also use your triangles to draw lines parallel to another line. To do that, place two triangles in position, as shown in Figure 1.49, with one edge of one triangle along the line. Then hold the other triangle to keep it from moving and slide the first one along it. The edge of the triangle will move away from the line but will remain parallel to it. When the edge of the triangle is the desired distance from the line, draw the new line.

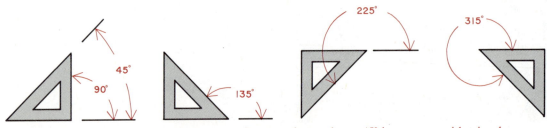

Figure 1.47 Constructing angles at 45° increments with triangle.

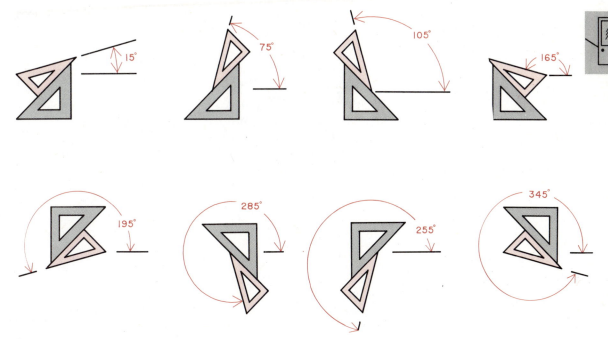

Figure 1.48 Using two triangles to construct angles at 15° increments.

The two triangles can also be used to draw lines that are perpendicular to other lines, as illustrated in Figure 1.50. Two of the edges of each triangle are perpendicular to each other, so when one of these edges is along or parallel to a line, the other edge is perpendicular to that line.

The protractor is shaped similar to a half circle or a circle with a flat bar through it (Figure 1.51). The flat bar contains a straight

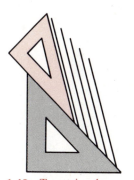

Figure 1.49 Two triangles can be used to draw parallel lines.

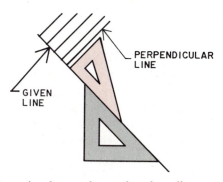

Figure 1.50 Two triangles can be used to draw lines perpendicular to other lines.

Figure 1.51   Using protractor to measure angles. (Courtesy of Ohio State University Engineering Publications)

edge with a mark in the center to show the location of the center of the circle or arc. To use the protractor to measure the angle between two lines, place the straight edge along one of the lines forming the angle, with the mark on the intersection of the two lines. Read the angle on the circular portion of the protractor where the other line crosses it.

The drafting machine combines the capabilities of a T-square, triangles, and a protractor. The head of the machine has two blades fixed at right angles to each other. They can be rotated to fixed settings selected by markings on the head (Figure 1.52). The arms of the drafting machine are constructed so that the angle of the blades does not change as the head is moved from position to position.

Drawing Straight Lines with Your CADD System

In order for your CADD system to draw a line, it must know the location of each end of the line. Two values are used to locate each end of the line. The first value is the X coordinate, which is the horizontal position of the point on the drawing, and the second value is the Y coordinate, which is the vertical position of the point. If you were drawing with a three-dimensional drawing system, you would also have a Z coordinate (Figure 1.53). Your system will provide more than one way for you to specify the location of the points.

One way to specify the location of a point is to determine what the *X* and *Y* coordinates are and then enter the values from the keyboard. That is usually the most accurate way to do it but not always the easiest. A more commonly used method is to use the graphics cursor, which is a cross-hair-like pair of lines that can be moved around the screen with the keyboard arrow keys, a mouse, or other input device. When the intersection of the cursor lines is on the desired spot on the screen, the space bar or mouse button is pressed to select the point. This is less accurate since the cursor does not move smoothly. The screen surface is made up of phosphor dots called picture elements, or pixels. The cursor jumps from pixel to pixel on the screen. The number of pixels, and thus the distance between them, will vary from system to system. To overcome this problem, CADD systems provide dot grids and the ability to snap the cursor to the grid or to another known location such as the end of an existing line.

To draw a line on your system at a specified angle, you can use the graphics cursor to locate one end of the line and then move the cursor until the angle and line length displays on the screen show the desired values. Then press the space bar or mouse button to select the point. A line will be drawn between the two points. On some systems, you will need to type in the angle and line length from the keyboard.

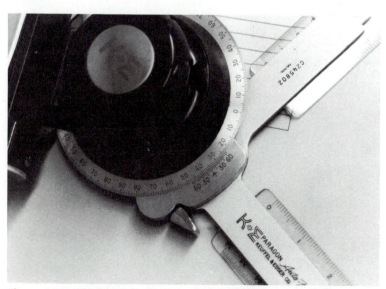

Figure 1.52 Drafting machine's protractor head is used to draw lines at set angles. (Courtesy of Ohio State University Engineering Publications)

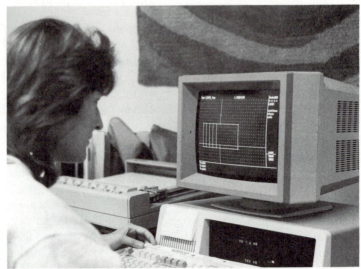

Figure 1.53 Drawing straight lines on CADD system. (Courtesy of Ohio State University Engineering Publications)

Circles and Arcs

Your instrument set should contain at least one and perhaps two compasses (Figure 1.54). They are called bow compasses. Look at your bow compass. One leg has a metal point. The other leg has a sharpened pencil lead. To draw a circle, the metal point is pressed into the paper at the center of the desired circle, and the compass is swung about it with the lead pressing on the paper (Figure 1.55). If an arc of a circle is desired, the compass is swung only through the range of the arc (Figure 1.56).

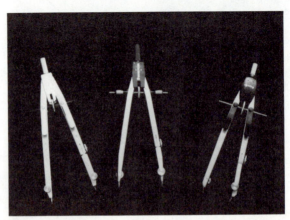

Figure 1.54 Compasses are available in variety of styles and sizes. (Courtesy of Ohio State University Engineering Publications)

Circle templates (Figure 1.57) are often used to draw circles and arcs of common sizes. They are somewhat easier to use than compasses but provide only a limited range of common circle diameters. They do not have a mark for the center of the circle but have four marks at 90° increments around the edge of the circle. The drafter first draws horizontal and vertical reference lines through the center point and then lines up the marks on the template with these lines to locate where the circle is to go.

Drawing Circles and Arcs on Your CADD System

Similar to lines, circles are described by two points. They are the center and one point on the circle. Your system may offer you several alternate ways of selecting a circle, such as three points on the circle, two points on opposite sides of the circle, or its center and radius. If you select one of the alternate ways, the software will use the information you provide to calculate the locations of the points it needs.

To draw a circle, select one of the methods from the menu and follow the prompts. Each system is slightly different, but most give you the option of selecting each point with the cursor

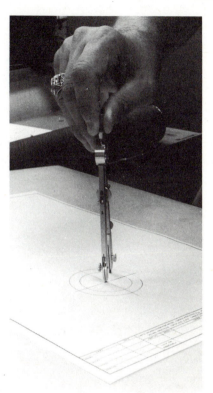

Figure 1.55 Using compass to draw circles. (Courtesy of Ohio State University Engineering Publications)

Figure 1.56 Using compass to draw arcs. (Courtesy of Ohio State University Engineering Publications)

Figure 1.57 Drawing with circle template. (Courtesy of Ohio State University Engineering Publications)

or typing its coordinates from the keyboard. If you are using the center and radius, you may be required to type in the value of the radius.

Arcs are drawn in the same way as circles except that the system needs to know three points, the center and the two end-points of the arc. Again, your system may offer several alternatives for entering the information and will calculate the needed values if they are not provided directly.

Drawing Ellipses

Ellipses are much more difficult to draw than circles. They are typically drawn in one of three ways. The first is to construct the ellipse by plotting several points on its surface and then connecting them with a smooth curve. This is a slow process. Reasonably good ellipses can be made by drawing four arcs of circles (four-center method) as described in Chapter 2.

Many ellipse templates are available and are widely used by drafters. Due to the wide range of sizes and shapes of ellipses, it may be impractical to have a full set of ellipse templates. If so, either the construction method or the four-center approximation method just mentioned can be used.

Drawing Ellipses on Your CADD System

Not all CADD systems provide ellipse-drawing capability. If yours does not, you may need to approximate the ellipse using the same techniques described for manual methods. With some three-dimensional CADD systems you can draw the ellipse in an orientation where it appears as a circle and then rotate it into the desired position.

Some CADD systems allow you to create ellipses in the following way: Draw a square by specifying one corner (vertex) and then the opposite one. A circle will be drawn in the square. Now, move the cursor (which is "attached" to the second vertex) around the screen. The square will be distorted into a rhombus and the circle into an ellipse. When the ellipse appears as you want it, press a key to fix it in that position.

Drawing Irregular Curves

Irregular curves, often called splines, are smooth curves that are made up of various shapes and thus cannot be drawn as combinations of arcs and straight lines. Irregular curves are drawn by plotting several points on the curve and then fitting a smoothly curved line to the points. An irregular curve, often called a French curve (Figure 1.58) generally is used for this construction. There are many different shapes and sizes of irregular curves. However, none is likely to fit the entire set of points, so the curve is drawn in sections, by fitting the curve to as many points as possible, drawing a portion of the curve, fitting the curve to more points, drawing another portion of the curve, and so on, until the curve is complete (Figure 1.59).

Drawing Irregular Curves with Your CADD System

Not all CADD systems offer spline drawing capability. For those that do, the input techniques vary somewhat. A typical system

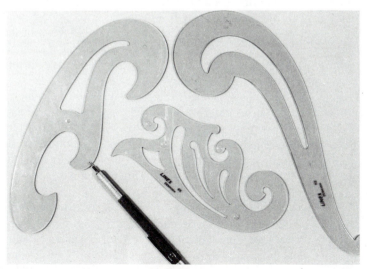

Figure 1.58 Drafters use wide variety of irregular curves. (Courtesy of Ohio State University Engineering Publications)

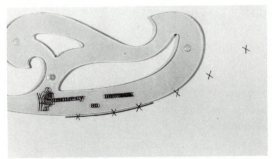 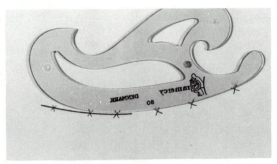

Figure 1.59 Many shapes can be created by using irregular curve.

will ask you to locate each point for your spline either by typing its coordinates from the keyboard or by locating it with the graphics cursor. When you have entered all of the points, the software will perform calculations to select locations for many additional points between the ones you entered. It will then connect all of the points with straight lines, but the lines will be so short that it will appear to be a smooth curve. Other systems permit you to "sketch" or draw a line freehand by moving the graphics cursor about the screen. In this case, the quality of the curve is dependent on your ability to move the cursor in a smooth manner.

Measuring

Most drawings are created to a scale, that is, a certain size. If the drawing is the same size as the object, the scale is full size. If each line on the drawing is half as long as the corresponding edge on the object, the scale is half size. Architects often use a scale of 1 in. on the drawing equals 8 ft on the object. Designers work in fractional or decimal inches. They may also work in the metric system. Civil designers and others working on large projects may work in feet, miles, or kilometers. Figure 1.26 shows a variety of scales that are used by designers and architects.

Drawings should not be scaled, that is, you should not place a scale on the drawing to determine the size of some feature. The size should be shown as a dimension. However, it is important to draw the object accurately to a selected scale. Failure to do so will result in the shape of the drawing being different than the shape of the object being represented. This can cause confusion for anyone using the drawing.

Correct use of the scale is important if you are to maintain accuracy in your drawing. Place your scale on the drawing and then use a sharp pencil held absolutely vertical to carefully mark the points with a short dash or "tic" mark (Figure 1.60). When possible, measure distances from a single point (Figure 1.61) to avoid cumulative error in measurements.

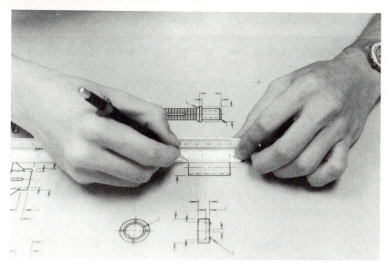

Figure 1.60 Scale is placed on drawing and distance required is carefully marked by points. (Courtesy of Ohio State University Engineering Publications)

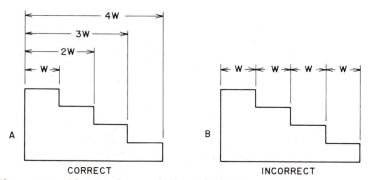

Figure 1.61 Measure from a single surface as in (*a*), avoid error buildup as in (*b*).

When using an open divided scale, place your scale so that the divided section can be used to measure the fractional or decimal portion of the distance. That portion is then added to the number of whole units to provide the complete dimension.

Measuring on Your CADD System

The CADD system allows you to measure distances by several means. In a typical case, the units in which you are working are selected from a listing. Then the system establishes an "origin" whose X, Y- or X, Y, Z-coordinate values are all zero (0). You can elect to have the system print out the cursor coordinates as it is moved about the screen. On some systems, once you enter the first point for a line, circle, or other entity, the distance and angle of the current cursor location from that starting point can be displayed. Often, you can key in the location of a point in terms of either (1) its coordinates relative to the origin or (2) its distance and angle from another point.

Inking Drawings

Drawings that are going to be placed in a permanent file as master drawings are generally traced on vellum or polyester film from the original drawing. Copies of the drawing are then made on a reproduction machine, and the original is placed in the master file. Technical ink pens are used for this purpose (Figure 1.62).

Drawings made on a CADD system are stored on disk or tape. It is not necessary to maintain a permanent file of tracings. In fact, it is undesirable since the copy on disk or tape is really the master. However, if desired, high-quality plots can be made on vellum or polyester film.

1.5 *LETTERING*

1.5.1 Lettering Instruments

Common lettering instruments for technical drawings are pencils and pens. There are three types of pencils: the wooden pencil,

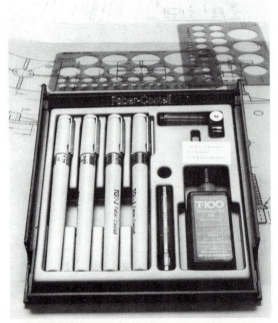

Figure 1.62 Technical ink pen set. (Courtesy of Ohio State University Engineering Publications)

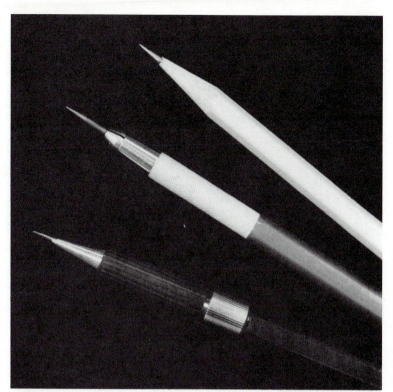

Figure 1.63   Types of pencils used for lettering on engineering drawings.

(*a*)

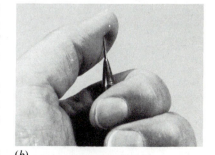

(*b*)

Figure 1.64   Correctly (*a*) and incorrectly (*b*) retracting lead in fine-line pencil.

the lead holder with a clutch, and the fine-line pencil (Figure 1.63). Of the three, the fine-line pencil is by far the most popular today.

Fine-line pencil lead is retracted by pressing the button on the top of the pencil and pressing the lead against a hard surface (Figure 1.64*a*). Care must be taken to never push the lead in using the thumb and finger (Figure 1.64*b*).

The lead holder is a type of mechanical pencil that holds long pieces of lead of varying degrees of hardness. The leads range from the very hard 9H to the very soft 6B. The working end of the lead must be pointed using a pencil pointer (Figure 1.65). The pointer puts a very sharp conical point on the end of the lead. The point should be slightly dulled before use. If the sharp point is not dulled, it may snap off during the initial strokes made with the lead. Not only will this lead only make a rough line, but also the point could snap and cause injury (Figure 1.66).

While lettering with this style of pencil, you should rotate the pencil in your hand after making several letters or numbers. Rotate the pencil between each word. Some people find that they must rotate the pencil between letters in a single word. This ro-

Figure 1.65 Pointers used to prepare working end of lead held in lead holder.

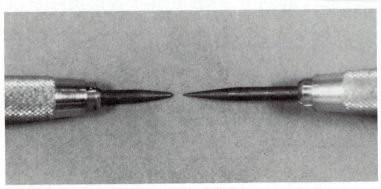

Figure 1.66 Point should be slightly dulled.

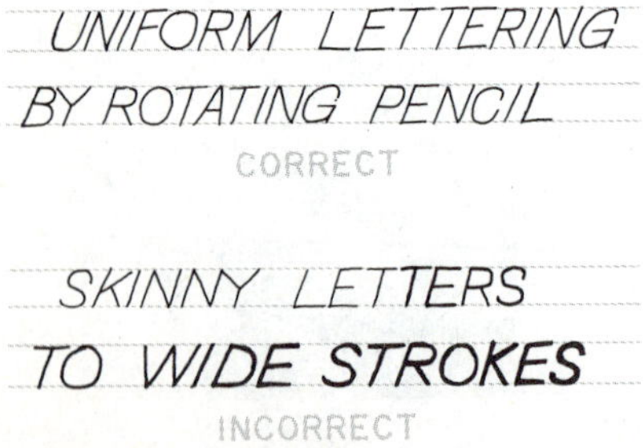

UNIFORM LETTERING BY ROTATING PENCIL
CORRECT

SKINNY LETTERS
TO WIDE STROKES
INCORRECT

Figure 1.67 Rotating pencil produces equal line widths.

tation process helps keep the point on the pencil uniform and prevents the lead from developing a flat spot. If the lead point becomes dull, the width of the strokes will be too wide; then the lettering will not be uniform and may not be clear (Figure 1.67).

The newest style of mechanical pencil is the fine-line pencil. The fine-line pencil is available in four metric sizes: 0.3, 0.5, 0.7, and 0.9 mm (Figure 1.68). These fine-line pencils are not available in inch sizes. Note that the lead for each of these pencils is a

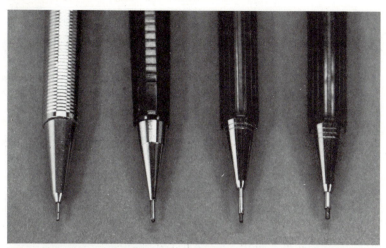

Figure 1.68 Fine-line pencils are manufactured to hold leads of different diameters.

different diameter. A 0.5-mm pencil should be used for lettering. The other widths are used for drawing lines of different widths.

A softer lead (such as HB lead) is used in fine-line pencils in order to produce black lines. The lines will be more uniform because the lines become as wide as the diameter of the lead (Figure 1.69). Another advantage of using fine-line pencils is the elimination of the need for a pointer; this saves time and reduces graphite dust. For students as well as others who carry their

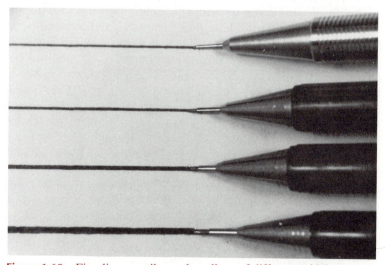

Figure 1.69 Fine-line pencils produce lines of different widths.

Figure 1.70   Erasers for erasing pencil lines from vellum or film.

drafting instruments from location to location, this is an advantage because the graphite shavings from the previously required pointer will not spill and smear all over the other equipment.

Many of the commercial leads available for fine-line pencils have a plastic base instead of a graphite base. This lead is designed for use on drafting film, a clear medium, as well as vellum, a type of drawing paper that is translucent. The soft plastic erasers are very effective when erasing lines from either the film or vellum and will not scratch the surface of the film such as some other erasers (Figure 1.70).

Technical pens have replaced the ruling pen for drawing or tracing ink lines and ink lettering (Figure 1.71). Technical pens have drawing tips of different diameters that produce lines of different widths (Figure 1.72). The technical pens need not be cleaned after every use. If you will not be using your pens for a period of time, take the time to draw a few lines with each pen every few weeks or so to keep the ink from drying in the point. If this is done, you can go for several years without having to clean the pens and can just add more ink occasionally.

Erasing ink lines from vellum requires a little more work and an abrasive eraser. It is better to pass the eraser over the ink line to be removed 100 times lightly than 10 times pressing so hard that the surface of the vellum is damaged (Figure 1.73). Erasing ink lines from film is easy and clean when using one of the plastic, chemically treated erasers that have been manufactured and treated for removing a particular manufacturer's ink (Figure 1.74).

1.5.2 Drawing Surfaces

There are the three types of drawing surfaces upon which the designer will draw lettering: drafting film, vellum, and opaque paper.

Vellum is a special type of paper that is translucent. Both lead and ink produce dark opaque lines on this type of surface and are easily erased.

Drafting film is more expensive but much more durable and dimensionally stable; that is, it does not expand or shrink when

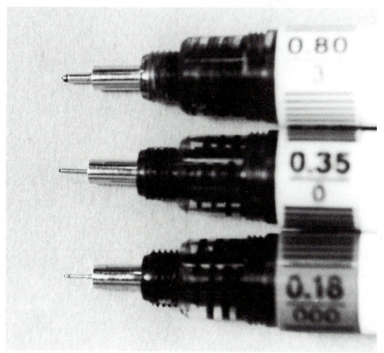

Figure 1.71 Technical pens for drawing ink lines.

.005 IN.	.007 IN.	.010 IN.	.012 IN.	.014 IN.	.020 IN.	.024 IN.	.028 IN.	.031 IN.	.039 IN.	.047 IN.	.055 IN.	.079 IN.
6x0 .13 mm	4x0 .18 mm	3x0 .25 mm	00 .30 mm	0 .35 mm	1 .50 mm	2 .60 mm	2.5 .70 mm	3 .80 mm	3.5 1.00 mm	4 1.20 mm	6 1.40 mm	7 2.00 mm

Figure 1.72 Technical pens are used to draw ink lines of different widths.

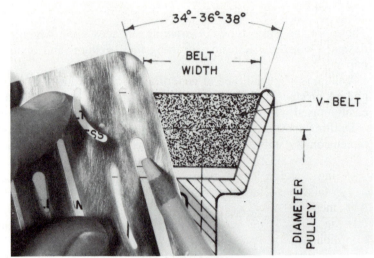

Figure 1.73 Erasing ink lines from vellum.

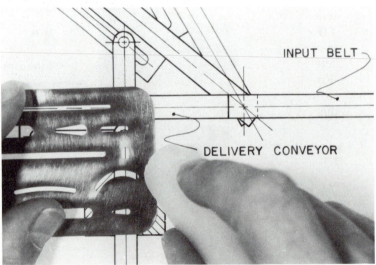

Figure 1.74 Erasing ink lines from film.

temperature or humidity changes. It is the best surface to use if a great number of copies must be made from the same master. It will also last much longer than vellum. A master is a final drawing that is the result of design work and revisions. This drawing is used as the master for reproduction purposes. The master drawing usually is stored in a safe place such as a fireproof file in order to protect it. The master drawing is never used in the shop; it is protected so clean copies can be made for shop use.

Opaque paper is used for smaller applications, such as common $8\frac{1}{2} \times 11$-in. (216×280-mm) standard paper. This paper is commercially available in a variety of grades and is more absorbent than vellum. Erasures on paper are not as clean as on either film or vellum.

1.5.3 Lettering Guidelines

Lettering guidelines are drawn for all numbering and all lettering on the face of all drawings. Lettering guidelines are thin, very light lines that only serve as a guide to the person doing the lettering. These lines should not be dark enough to interfere with the visibility of the lettering or even dark enough to be seen when copies are made (Figure 1.75).

Lettering guidelines are usually $\frac{1}{8}$ in. (3 mm) apart. Lettering is never placed in the space adjoining a line such as a border line (Figure 1.76). A space of $\frac{1}{8}$ in. (3 mm) is usually left between lines of lettering. Lettering is never placed in adjacent $\frac{1}{8}$-in. (3-mm) spaces (Figure 1.77).

Time-saving devices are commercially available to help draw these lettering guidelines. The Ames lettering guide and the Bradock–Rowe triangle are examples (Figure 1.78). To use these devices, position the row of holes and simply draw the guidelines by sliding the device along a straight edge (Figure 1.79).

Figure 1.75 Lettering guidelines should be drawn before lettering freehand.

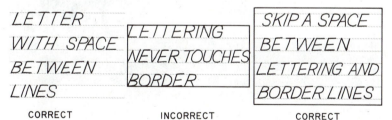

CORRECT   INCORRECT   CORRECT
Figure 1.76 Lettering is never placed adjacent to border line.

1.5.4 Lettering Styles

Nearly all pencil lettering drawn by the designer is drawn freehand. The style of letters used is single-stroke Gothic style (Figure 1.80). Note the common errors made while lettering; they are illustrated in color (Figure 1.81).

With rare exception, lettering is drawn using uppercase letters

INCORRECT
Figure 1.77 Do not letter in adjacent line spaces.

ALVIN
L666

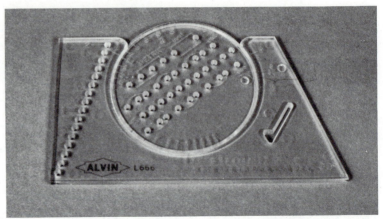

Figure 1.78 Lettering guidelines are made more uniformly with lettering devices.

ALVIN
L666

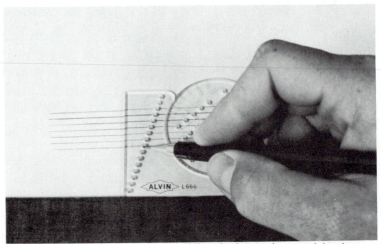

Figure 1.79 Draw guidelines by sliding device against straightedge.

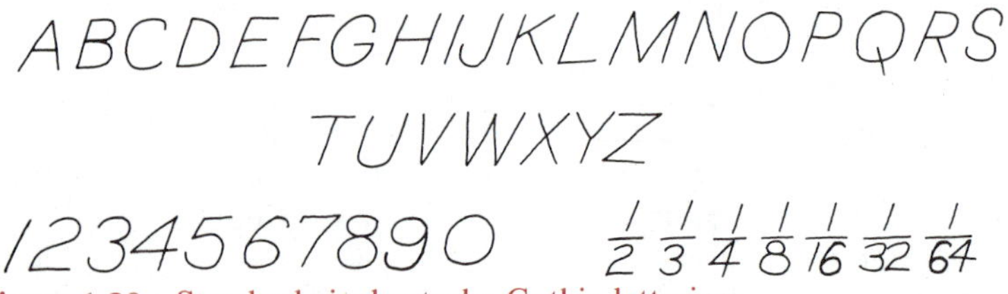

ABCDEFGHIJKLMNOPQRS
TUVWXYZ
1234567890
1/2 1/3 1/4 1/8 1/16 1/32 1/64

Figure 1.80 Standard single-stroke Gothic lettering.

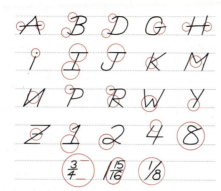

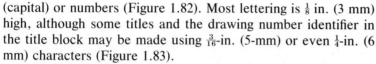

Figure 1.81 Common errors are often made when forming letters and numbers.

(capital) or numbers (Figure 1.82). Most lettering is $\frac{1}{8}$ in. (3 mm) high, although some titles and the drawing number identifier in the title block may be made using $\frac{3}{16}$-in. (5-mm) or even $\frac{1}{4}$-in. (6 mm) characters (Figure 1.83).

Upper and lowercase letters are used in civil engineering when making plats. These are usually made in ink with a mechanical lettering device of some sort (Figure 1.84).

The lettering may be drawn with either vertical or inclined strokes (Figure 1.85). With the vertical characters, the stems of the letters (a vertical line extending from guideline to guideline) are vertical. With inclined characters, these stems are drawn at a uniform incline. The standard incline is 68°. One edge of the Ames lettering guide is manufactured at this angle so that you may use it as a guide (Figure 1.86). Whether you elect to use vertical or inclined letters and numbers makes little difference, but you must be consistent and use only one style on any one drawing.

Whole numbers are drawn $\frac{1}{8}$ in. (3 mm) high, as are the capital letters. However, fractions must be treated differently. Fractional numbers are nearly as large as whole numbers but extend above and below the lettering guidelines. Note that the fraction line is a horizontal line (Figure 1.87). Also, the height of the fraction is twice the height of the whole numbers.

1.5.5 Character Uniformity

Good lettering is characterized by uniformity. The appearance of a well-lettered drawing will be directly related to how uniform it appears. The uniformity of line widths has already been addressed.

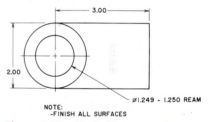

Figure 1.82 Capital letters and numbers are generally used when lettering engineering drawings.

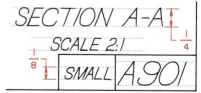

Figure 1.83 Letters and numbers larger than $\frac{1}{8}$ in., are used for titles and drawing identifiers.

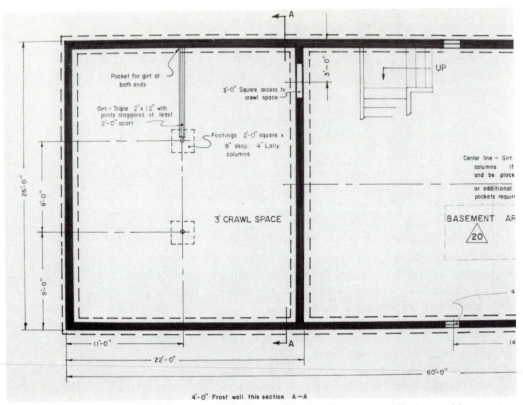

Figure 1.84   Civil engineering drawings are usually lettered in upper- and lowercase letters. (Courtesy of Timothy C. Whiting)

Uniform blackness is very important. All of the characters must be lettered *black,* not gray. This is very important because characters not dark enough may not copy well and may cause errors in reading the drawing. The characters must be lettered uniformly and black enough to copy under any conditions.

A uniform slant to the vertical letters will enhance the appearance of the drawing. Uniformity when drawing the curved portion of the characters having curves is also important (Figure 1.88).

Stability in the appearance of lettering is shown with example characters such as B, P, and the number 8. If these characters are drawn with the middle stroke drawn at the horizontal center, they will appear below center. For this reason, the top part of these letters is drawn slightly smaller to provide space for drawing the middle stroke above the horizontal center of the letter. If there is any question about this, turn the lettering upside down and observe the illusion (Figure 1.89).

1.5.6 Spacing

Both letter spacing and word spacing must be addressed. Letter spacing is the spacing between individual letters. This space should be very small if the words are to appear as words and not just as random letters (Figure 1.90).

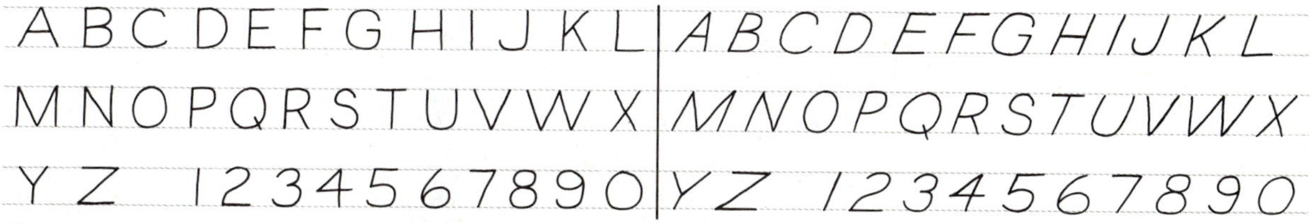

A B C D E F G H I J K L | *A B C D E F G H I J K L*
M N O P Q R S T U V W X | *M N O P Q R S T U V W X*
Y Z 1 2 3 4 5 6 7 8 9 0 | *Y Z 1 2 3 4 5 6 7 8 9 0*

Figure 1.85   Either vertical or inclined letters or numbers are acceptable.

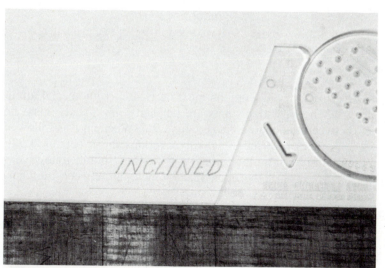

Figure 1.86   One edge of Ames lettering device is manufactured at 68° angle to be used as guide for inclined letters.

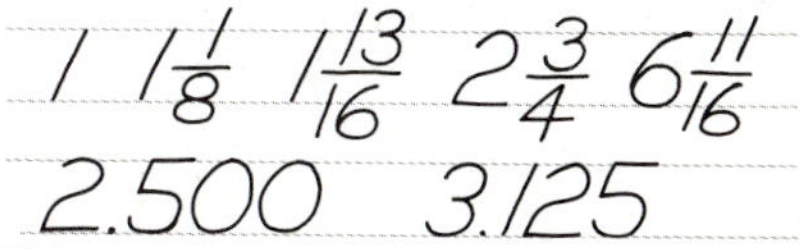

Figure 1.87   Fraction numbers extend above and below $\frac{1}{8}$-in. lettering guidelines.

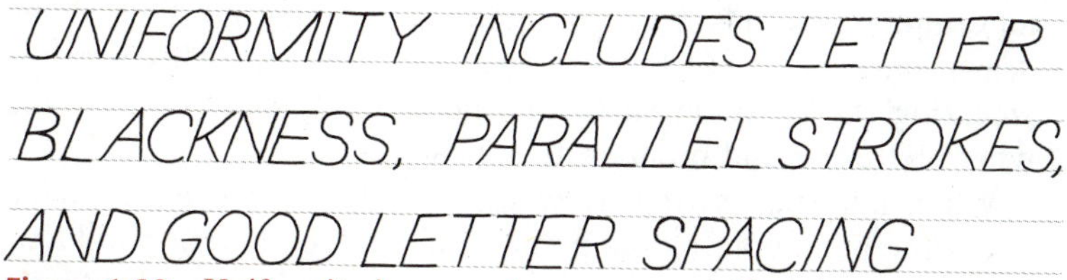

UNIFORMITY INCLUDES LETTER
BLACKNESS, PARALLEL STROKES,
AND GOOD LETTER SPACING

Figure 1.88   Uniformity is a very important consideration.

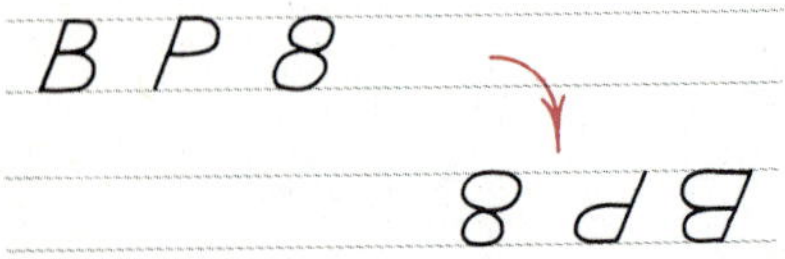

B P 8

8 d 8

Figure 1.89 ''Rule of stability'' requires lower part of some letters and numbers to be larger than top.

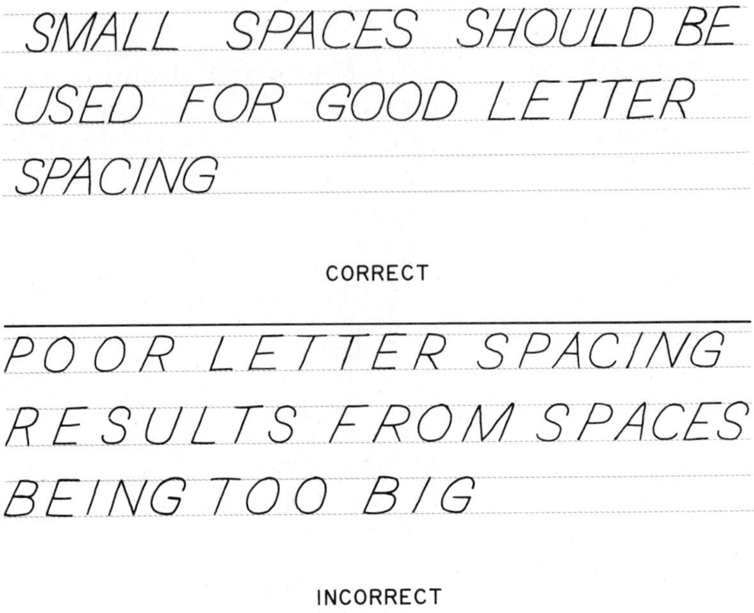

SMALL SPACES SHOULD BE USED FOR GOOD LETTER SPACING

CORRECT

POOR LETTER SPACING RESULTS FROM SPACES BEING TOO BIG

INCORRECT

Figure 1.90 Letter spacing should be very small to make letters visually appear as words, not letters.

HIGH *HI*

OHIO

COLUMN

Figure 1.91 Letters with adjoining stems require more letter spacing.

TOTAL *AL*

LATE

LT *ALTER*

Figure 1.92 Certain letter combinations require overlapping of adjacent letters.

A few letter combinations must be treated with special attention. When two letter stems are adjacent, more space must be left between the characters (Figure 1.91). When an L is followed by a T, the top of the T actually overlaps the L (Figure 1.92). When letters follow the letter A or L, they should follow very closely because of the open space appearing between the characters (Figure 1.92).

Word spacing should be wider than letter spacing. Letter spacing should be so small that the characters nearly touch. However, word spacing should be approximately the width of a medium-wide character such as the letter E (Figure 1.93).

1.5.7 Suggestions

There are several techniques that you can use to improve your lettering. When making an engineering drawing, all the lines are drawn dark first, and the lettering is drawn last. This prevents smearing the lettering since the most lead is put on the paper where lettering and numbering appear.

When lettering on the drawing, place a clean sheet of paper under the hand holding the pencil. This prevents your hand from ever touching the drawing and prevents sweat, hand oil, and other foreign materials from being transferred to the surface of the drawing (Figure 1.94). The use of eraser chips from a dry cleaning pad can aid significantly in keeping the surface of the drawing clean.

If you are working on a translucent medium, such as vellum, and a significant amount of lettering will appear on the drawing surface, such as on a specification sheet, it may be a good idea to draw the border lines and lettering guidelines on the *back* of the drawing sheet. This allows for many erasures or changes to be made on the front side of the sheet without damaging the guidelines. Remember, the copy machine will not know on which side of the vellum or film you are drawing (Figure 1.95).

After the drawing is complete, the tape holding it to the board may be removed. You may then find it much more comfortable to rotate the sheet slightly on the board before starting to letter (Figure 1.96).

You will find that your lettering will be be much more uniform if your entire forearm is resting on the surface of the drawing table. This simply offers more support from your arm and drawing hand (Figure 1.97).

1.5.8 Mechanical Lettering

There are many mechanical devices and aids available on the commercial market for use while making letters by mechanical means. This mechanically produced lettering is almost exclusively used when drawing with ink for presentation purposes. The Leroy type of lettering equipment is one type used most often (Figure

LETTERS AND
NUMBERS ARE
USED WITH PICTURES
ON A DRAWING

Figure 1.93 Word spacing requires more space than letter spacing.

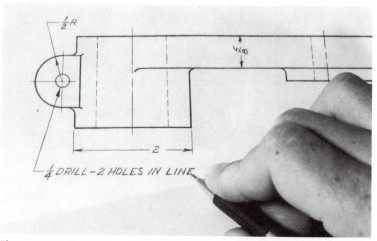

Figure 1.94 Protective sheet under lettering hand helps keep surface of sheet receptive to ink or pencil.

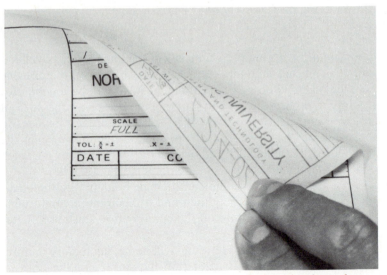

Figure 1.95 When much erasing is anticipated, draw border and guidelines on one side of sheet and lettering on other side.

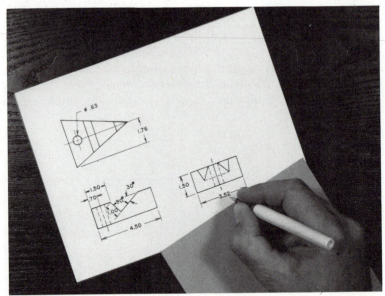

Figure 1.96 Rotate paper to most comfortable position for lettering.

1.98). Technical pens are inserted into a scribe having three points extending from its lower side (Figure 1.99). The point at the end of the handle (1) slides in the wide groove at the bottom of the scale. The thin point (2) (called the stylus) is then inserted in the groove representing the desired letter in the scale. The third point

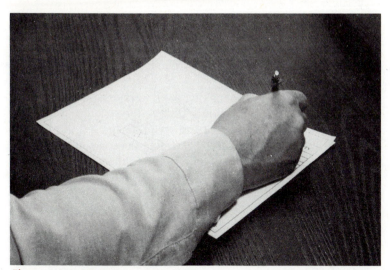

Figure 1.97   Place paper on table in such a position as to allow your entire forearm to rest on table.

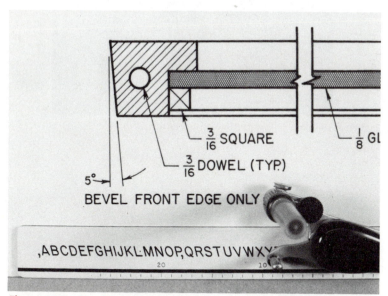

$\frac{3}{16}$ SQUARE

$\frac{3}{16}$ DOWEL (TYP.)

$\frac{1}{8}$ GL

5°

BEVEL FRONT EDGE ONLY

,ABCDEFGHIJKLMNOP,QRSTUVWXY

20   10

Figure 1.98   Hand lettering equipment is used to produce mechanical lettering.

is near the pen point (3) and acts as a support for the pen point (Figure 1.100).

To form a letter, move the stylus within the groove of the desired letter with the pen point resting on the drawing surface. Then tilt the scribe back and place the stylus in the next letter.

Slide the scale until it is in the correct position. Trace that letter and repeat this step as many times as necessary. Once the origin point for the new letter has been established, keep your eye on the stylus and engraved template while drawing the letter or number (Figure 1.101).

The skill that the person making the letters must develop is the spacing of the letters and words. For letter spacing, first letter the first character. Then place the stylus in the left extreme of the new letter on the template. Then slide the scale until you line up by eye with the right extreme of the old letter. Remember

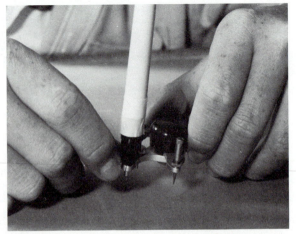

Figure 1.99 Pen is secured in arm of lettering device.

Figure 1.100 Adjusting screw keeps pen point at correct distance from paper.

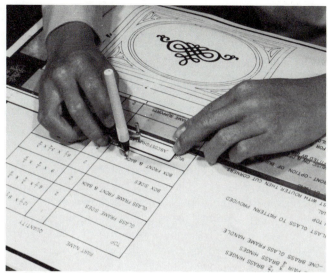

Figure 1.101   Forming letters and spaces using lettering device.

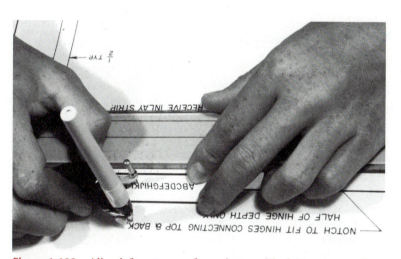

Figure 1.102   Align left extreme of new letter with right extreme of previously drawn letter.

to align the left extreme of the new letter with the right extreme of the previously drawn letter (Figure 1.102).

Those of you who must draw a lot of mechanical lettering and can afford the expense of an automatic lettering machine will benefit greatly from the features available on such an automated device. Options such as proportional spacing, degrees of slant

Figure 1.103 Automatic lettering device is used by professionals. (Courtesy of MUTOH America, Inc.)

and memory, as well as predrawn symbols for many disciplines are available (Figure 1.103).

Some designers use ink lettering and numbering that are mechanically generated. They often use upper- and lowercase letters (Figure 1.104).

Ink drawings can be produced easily using CAD systems, and today many industries are using CAD for documentation.

1.5.9 Lettering by Computer

Lettering can be created quickly and easily on a CADD system. Each letter is a graphic shape just as circles, rectangles, and polygons are graphic shapes. They are drawn on the computer screen one line at a time just as other graphic shapes are drawn.

Since each letter is a graphic shape, most CADD systems offer a variety of options for drawing them. The options generally include the height and width of the letters, the angle at which

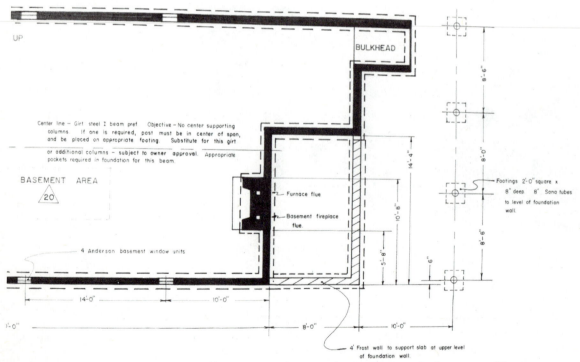

Figure 1.104 Civil engineers often use both upper and lowercase letters and also make presentation drawings using mechanical lettering. (Courtesy of Timothy C. Whiting)

the letters will be drawn, whether they are upright or slanted, and in many cases, a choice of letter styles.

The height and width of the letters can be specified separately on most CADD systems. Sometimes you select each individually. With other systems, you select the height and then the ratio of height to width (aspect ratio).

The lettering styles that are available will depend on the computer hardware and software system that is being used. Most CADD systems offer a choice of fonts (lettering styles) for text on the screen. Sometimes additional fonts can be purchased to meet special needs. Fonts are stored in computer files that are read by the CADD system.

When you plot a drawing, the lettering styles available will depend on whether the letters are being plotted as they are drawn on the screen, using the CADD system fonts, or the plotter's built-in character set is being used.

Your CADD software will ask you to enter (type in) the text that is to be drawn and to mark the location where you want it on your drawing. You may enter one letter, a line of text, or even several lines of text at one time. The amount of control you have over the lettering will depend on how much text you are entering at one time. If you enter one letter at a time, each letter can be sized and placed as you want it. If you enter a line of text, the letter size and style you have selected will be used for the entire line. Letters will be drawn using the starting point you select for the line and the preset spacing between letters and words. If you enter several lines of text at once, then the spacing between lines will also be determined by the CADD software.

Text for dimensions is determined and placed automatically if you use automatic dimensioning. You can specify the text (letter and number) size to use and where the dimension is to be placed, but the system calculates the dimension and draws it in.

The exact procedure for lettering with a CADD system will depend on your software, but in a typical case, you will create notes by making the proper menu selection, typing in the text, and using the graphics cursor to locate the starting point for the text string. If you are dimensioning, you will typically select the type of dimension from the menu. Then with the graphics cursor, you will locate the starting and ending points for the dimension and where the dimension text is to be placed. The system will calculate the value to be placed there.

Some CADD systems include the option of entering text from a file instead of the keyboard. This is convenient for entering large amounts of text such as lengthy notes, bills of material, and similar items. The text can be created with a text editor or word processor and stored in the file for access by the CADD system.

1.6 *SUMMARY*

As a student who will be practicing in your chosen field well into the twenty-first century, you will be in the work force at a very exciting time. Technical design graphics will become increasingly important as automation continues to mechanize the workplace. Machines controlled by computer rely very heavily on geometric descriptions of the objects to be produced. More and more frequently, these descriptions are created during the design process by a designer using a CADD workstation. While the designer may be drawing by moving a mouse rather than triangles and a pencil, the graphics principles are the same. The designer must be able to visualize and graphically describe three-dimensional objects. Sketching will also continue to be very important since designers continue to create designs on paper even if they modify them on a CADD system.

Throughout this book, we shall present graphics techniques using both manual and computer-based methods. We will try to keep the description of computer-based techniques general enough so that they will be applicable to whatever CADD system you are using.

PROBLEMS

Problem 1.1
What is a data base? How is it used?

Problem 1.2
Why do engineers make sketches? What tools and materials are needed to make a sketch?

Problem 1.3
What is meant by shape description?

Problem 1.4
What is meant by size description?

Problem 1.5
What does tolerance mean?

Problem 1.6
Why is a drawing drawn to scale?

Problem 1.7
If a 6-in. long object is drawn to half scale, how long will it be on paper?

Problem 1.8
For what are section views used?

Problem 1.9
What are conventional practices?

Problem 1.10
Why are notes added to a drawing? Name three uses for drawing notes.

Problem 1.11
What is a working drawing? For what are working drawings used?

Problem 1.12
What are illustrations? How do illustrations differ from working drawings?

Problem 1.13
For what are illustrations used?

Problem 1.14
What is meant by fastening and joining? Name four types of devices, materials, or processes that are used for fastening or joining.

Problem 1.15
What is descriptive geometry? How is it used by the designer?

Problem 1.16
What is a pie chart? Name a use for pie charts.

Problem 1.17
Draw a series of $\frac{1}{8}$-in. spaces on your paper using very light horizontal lettering guidelines. Using capital single-stroke Gothic lettering

(inclined or vertical is your choice), letter the following paragraph (remember to only letter in every other space):

> Dimensions are added to detailed engineering drawings to specify the size or location of all features. Notes may also be included to explain additional details about the object. Dimensions and notes become the communication link between the designer and the craftsman.

Problem 1.18

Draw the following sets of numbers and labels. Use $\frac{1}{8}$-in. high whole numbers, but let the fractions extend a little above and a little below the guidelines:

Whole Numbers	1, 234, 567, 890
Fractions	$\frac{1}{4}$, $\frac{3}{8}$, $\frac{5}{16}$, $\frac{7}{32}$, $\frac{9}{64}$
Mixed numbers	$6\frac{5}{16}$, $4\frac{1}{2}$, $22\frac{7}{32}$, $9\frac{7}{8}$

Problem 1.19

Letter the following in every other space of the $\frac{1}{8}$-in. lettering guidelines.

> When dimensioning, I will form the letters, numbers, and mixed numbers correctly. All letters and numbers will be black and crisp so they will copy well. Clear uniform lettering is essential for accurate communication.

Problem 1.20

Letter the following notes within the $\frac{1}{8}$-in. lettering guidelines. Letter in every other space.

> Note: all unspecified radii $R\,\frac{1}{8}$
>
> Ø 4.13, 5 holes, equally spaced
>
> $\frac{1}{2}$-20UNF-2B
>
> Chamfer 3 × 45° both ends
>
> 13.5 drill, spotface 024 × 1.5 deep, 2 holes
>
> Top security, do not copy
>
> For in-house price quote only, not for production

Problem 1.21

What is meant by modeling? For what are physical models used? For what are computer-generated models used?

Problem 1.22

You are asked to sketch three orthographic views of an object. What equipment and materials will you need for the job?

Problem 1.23

You are creating a drawing with instruments and must add several arcs and circles to it. What tool(s) could you use to draw the arcs and circles?

Problem 1.24

What is simulation? How is it used in computer-aided design? Is it used in manual design?

Problem 1.25
What is meant by numerical control? How does computer numerical control differ from numerical control?

Problem 1.26
What is computer-integrated manufacturing?

Problem 1.27
What are some similarities and differences between computer-integrated manufacturing and flexible manufacturing?

Problem 1.28
How is a T-square used? Can you use it to draw horizontal lines? Vertical lines? Lines at various angles?

Problem 1.29
You have just purchased a drawing instrument kit. It contains two triangles. What size angles can you draw with one triangle? What size angles can you draw with the other?

Problem 1.30
Using just your T-square, the two triangles, and a pencil, how many different angle lines can you draw? What angles are they?

Problem 1.31
You are to draw a line at 75° to the horizontal using your T-square and your two triangles. Sketch a rectangle representing the drawing board and show the positions of the T-square and each triangle.

Problem 1.32
Your drawing instrument kit contains a pair of dividers. For what are they used?

Problem 1.33
You must lay a distance of 3.6 in. on your drawing. Name the tools from your kit that you will use for this task and how you will use them.

Problem 1.34
Name two types of erasers commonly used when creating drawings.

Problem 1.35
Your drawing contains two lines that are very close together. What tool do you have in your kit that will permit you to erase one line without erasing the other?

Problem 1.36
Name three types of drawing media (material) on which drawings are commonly made.

Problem 1.37
You have plotted several points on your drawing paper and must connect them with a smooth curve. The curve will be of irregular shape. What tools do you have in your drawing kit for this purpose?

Problem 1.38
Drawing pencils are made with several different hardnesses. Name three hardnesses that are available.

Problem 1.39
You are going to create a sketch. What lead hardness will you choose? Why?

Problem 1.40
Name the principal components of a computer-aided drawing system.

Problem 1.41
Name three input devices that are commonly used with a computer.

Problem 1.42
Name three output devices that are commonly used with a computer.

Problem 1.43
You have created a drawing on your computer. Name two types of output devices that can be used to make a copy of it.

Problem 1.44
Name two important differences between disk drives and tape drives.

Problem 1.45
What are tape drives generally used for on modern computer systems?

Problem 1.46
Name two types of computer software that might be used by a designer.

Problem 1.47
How is a digitizing tablet used?

Problem 1.48
How is a mouse used?

Problem 1.49
Name a task that could be performed with either a digitizing tablet or a mouse.

Problem 1.50
Name a task that could be performed with a digitizing tablet but not with a mouse.

Problems 1.51–1.70
Draw the figures shown using either manual instruments or a CADD
system as directed by your instructor. Use the dimensions shown or
other dimensions as directed.

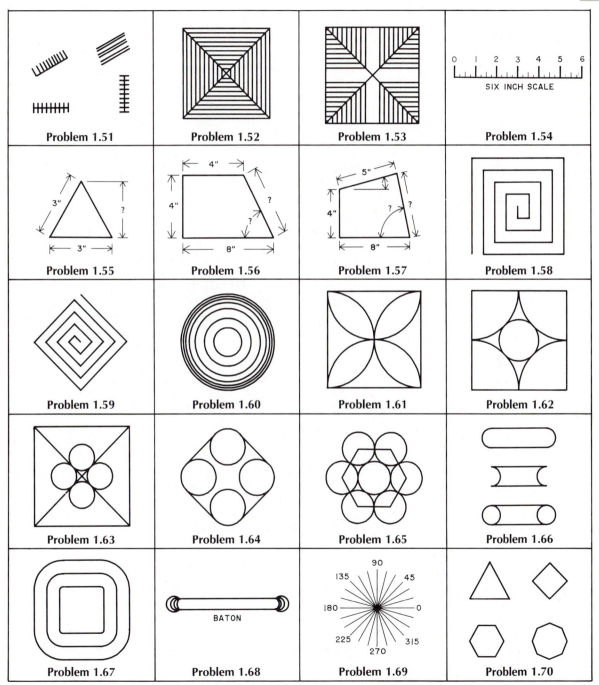

| Problem 1.51 | Problem 1.52 | Problem 1.53 | Problem 1.54 |

| Problem 1.55 | Problem 1.56 | Problem 1.57 | Problem 1.58 |

| Problem 1.59 | Problem 1.60 | Problem 1.61 | Problem 1.62 |

| Problem 1.63 | Problem 1.64 | Problem 1.65 | Problem 1.66 |

| Problem 1.67 | Problem 1.68 | Problem 1.69 | Problem 1.70 |

TERMS YOU WILL SEE IN
THIS CHAPTER

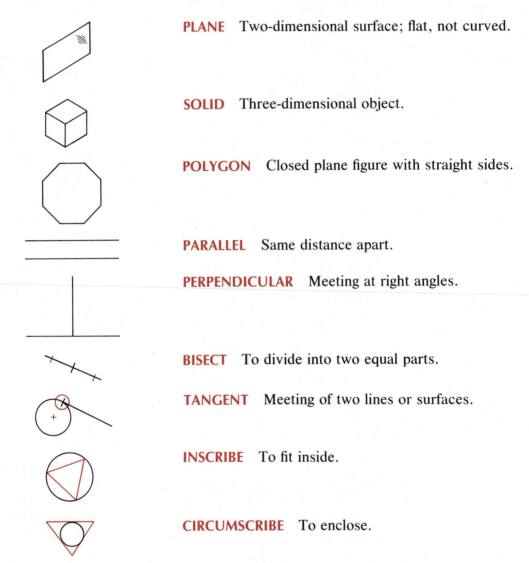

PLANE   Two-dimensional surface; flat, not curved.

SOLID   Three-dimensional object.

POLYGON   Closed plane figure with straight sides.

PARALLEL   Same distance apart.

PERPENDICULAR   Meeting at right angles.

BISECT   To divide into two equal parts.

TANGENT   Meeting of two lines or surfaces.

INSCRIBE   To fit inside.

CIRCUMSCRIBE   To enclose.

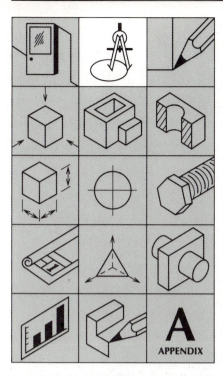

2

GEOMETRIC CONSTRUCTIONS

A
APPENDIX

2.1 INTRODUCTION

Before you can make a complete technical drawing, you need to refresh your knowledge of basic geometry. Many of the constructions we need in graphics and for building or producing useful devices are the same ones learned in plane geometry.

You will use some of the basic geometric constructions you learned. You can build on this knowledge and learn new uses for geometry. Several geometric figures and shapes you may know by name are shown for reference in Figure 2.1. If you do not know these figures and shapes and their names, you should learn them.

When you finish the work in this chapter, you will know some simple geometric shapes and constructions that are useful for solving graphics problems. You may have seen many of these constructions before and will remember them; others may be new to you. The constructions you will need include **parallels, perpendiculars,** regular **polygons,** triangles, **tangents,**

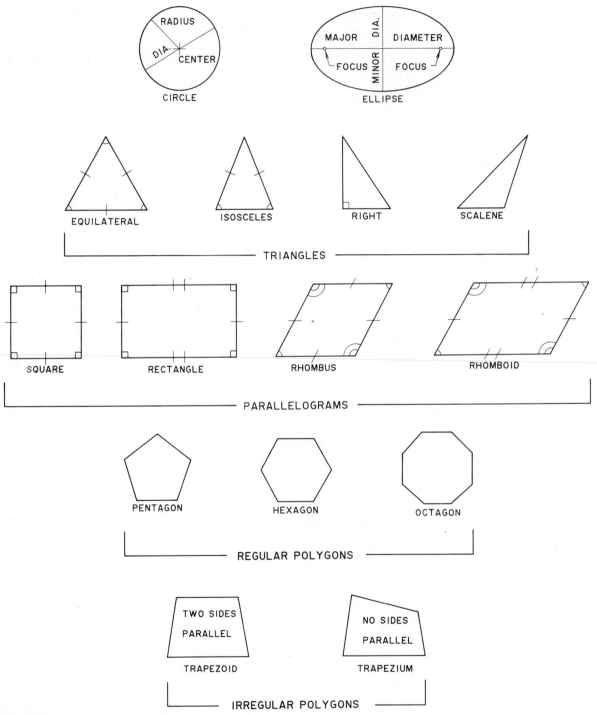

Figure 2.1 Common plane figures.

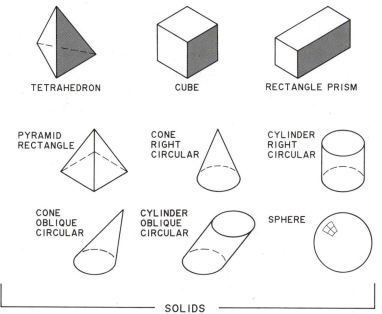

TETRAHEDRON   CUBE   RECTANGLE PRISM

PYRAMID
RECTANGLE

CONE
RIGHT
CIRCULAR

CYLINDER
RIGHT
CIRCULAR

CONE
OBLIQUE
CIRCULAR

CYLINDER
OBLIQUE
CIRCULAR

SPHERE

SOLIDS

Figure 2.2   Some common solids.

Figure 2.3   Door with parallel sides.
(Photo by Howard Keith Williams)

ellipses, and involutes. There are many other construction methods that may be of interest; the ones presented here are those needed to understand the theory and solve the problems presented later in this book. The **plane** figures and **solid** shapes you will need to know are shown in Figures 2.1 and 2.2.

2.2   *TO DRAW A LINE PARALLEL TO A GIVEN LINE*

A line parallel to a given line is a construction you will use very often. Many objects around us include parallel lines. Consider a door—the sides are usually parallel. If they are not, the door may not close or it may have large cracks that create drafts (Figure 2.3).

A parallel bar or a T-square makes it easy to draw parallel lines. A T-square provides parallel horizontal lines as you slide it up and down on the drawing board, keeping the head of the T-square firmly against the edge of the board (Figure 2.4). When you put a triangle on the T-square and slide the triangle, it will generate a series of parallel lines. These parallel lines may be vertical or at the angle of the triangle (Figure 2.5). If you are not using a T-

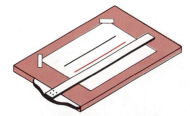

Figure 2.4   To draw horizontal parallel lines using T-square, (1) align T-square with given horizontal line on paper, (2) tape paper on top two corners with T-square in place, and (3) move T-square to desired position and draw new line.

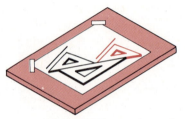

Figure 2.5 To draw vertical parallel lines using T-square and triangles (tape paper down before starting), (1) slide triangle along upper edge of T-square with 90° angle on T-square until vertical edge of triangle is aligned with given line and (2) slide triangle, still on T-square, to desired position and draw new line.

Figure 2.6 To draw parallel lines using only triangles, (1) place two triangles on drawing surface hypotenuse to hypotenuse; (2) align one edge of upper triangle with given line; and (3) hold lower triangle firmly in place, slide upper triangle to desired position, and draw line.

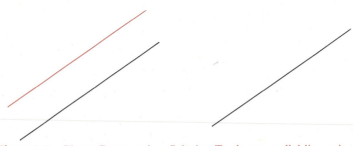

Figure 2.7 **Given–Construction–Solution** To draw parallel line using computer graphics, (1) create line on screen using LINE task and (2) move cursor to desired position for one end of new line and use DUPLICATE, COPY, or PARALLEL function to create parallel line. Alternate method: Note angle of line on screen and create new line at same angle. *(Computer-generated image)*

square, you can make parallel lines by holding one triangle solidly in place and sliding another triangle along the edge of the first one (Figure 2.6). Figure 2.7 demonstrates parallel lines with computer graphics.

2.3 TO DRAW A LINE PERPENDICULAR TO A GIVEN LINE

Perpendicular lines are important features in drawings for making buildings, machine parts, and most other man-made objects. We are used to walls being plumb (perpendicular to floors) and are uncomfortable if they appear to lean. Carpenters, masons, and ironworkers all learn to plan and build structures that have square corners (Figure 2.8). As designers, we need to be able to draw perpendicular lines quickly and accurately (Figures 2.9–2.12). Later

Figure 2.8 Masons must learn to make square corners (perpendicular) when building walls. (Photo by Howard Keith Williams)

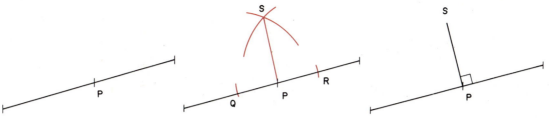

Figure 2.9 Given–Construction–Solution To draw line perpendicular to given line at given point using a compass, (1) set compass to any convenient radius, (2) place compass point on given point *P* and draw two arcs cutting given line on each side of given point, (3) increase radius of your compass and draw two arcs from cut points on line (*Q* and *R*) that intersect on same side of line *S*, and (4) draw line from given point *P* to intersection *S*. This line (*PS*) is perpendicular.

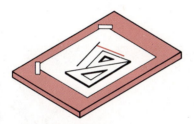

Figure 2.10 To draw line perpendicular to given line at given point using triangles, (1) place two triangles on drawing surface hypotenuse to hypotenuse, (2) slide pair of triangles until side of one triangle is parallel to given line, (3) hold second triangle firmly in place and slide first triangle along hypotenuse until side perpendicular to given line lies on given point, and (4) draw perpendicular along edge of triangle.

in this book you will learn to make orthographic drawings and use descriptive geometry. Both of these subjects require accurate perpendiculars.

2.4 *TO DIVIDE A LINE INTO A NUMBER OF PARTS*

We often need to divide a line into a number of parts, and very often the number of parts is not easily divisible into the length of the line. Dividing a 4-in.-long line into two parts is easy; dividing a line $4\frac{3}{16}$ in. long into exactly five parts does not appear to be as easy. If you were planning a graph, you might have only so much space on the paper and need to divide the axes into a number of parts that are not whole numbers or simple fractions (Figure 2.11). Highways usually have a stripe down the center. To be sure that

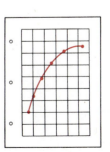

Figure 2.11 To design graph, axes must be divided evenly.

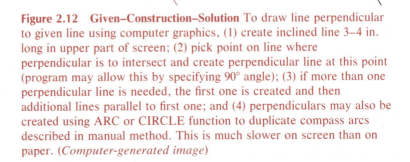

Figure 2.12 **Given–Construction–Solution** To draw line perpendicular to given line using computer graphics, (1) create inclined line 3–4 in. long in upper part of screen; (2) pick point on line where perpendicular is to intersect and create perpendicular line at this point (program may allow this by specifying 90° angle); (3) if more than one perpendicular line is needed, the first one is created and then additional lines parallel to first one; and (4) perpendiculars may also be created using ARC or CIRCLE function to duplicate compass arcs described in manual method. This is much slower on screen than on paper. (*Computer-generated image*)

Figure 2.13 Center stripe on highway should bisect road. (Photo by Howard Keith Williams)

the stripe is in the middle of the road, the highway crew has to **bisect** the highway, that is, divide it into two equal parts (Figure 2.13).

The method for dividing a line into any number of parts is based on the principle of similar triangles, which you may remember from plane geometry (Figures 2.14–2.16). The method for bisecting a line into exactly two equal parts is one you may also remember (Figure 2.17).

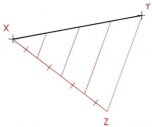

Figure 2.14   Given–Construction–Solution To divide line into any number of parts: (1) Draw line of known length (*XZ*) from one end of given line (*XY*) at any convenient angle. Line of known length (*XZ*) chosen should contain easy multiple of number of parts needed. (If four parts are needed, line could be 2 in. long, each part $\frac{1}{2}$ in. long.) (2) Connect open end of given line to open end of line of known length (*Y* to *Z*). (3) Draw lines parallel to connector line from each of divisions on known line. Where these lines cross given line, you have divided it into the required number of parts.

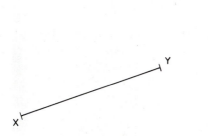
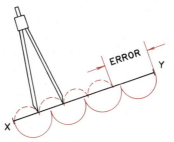

Figure 2.15   Given–Construction–Solution Dividing a line into any number of parts by successive approximations also works for dividing a circular arc: (1) Estimate how long one part would be. Set dividers to this length. (2) Step off number of parts on line by "walking" dividers along line. Distance will probably be too long or too short. (3) Change setting of dividers by estimate of total error divided by number of parts (error/*n*). If error is $\frac{1}{2}$ in. short and four parts are needed, lengthen dividers by about $\frac{1}{8}$ in. (4) Step off parts again on line. If still not exactly right, change divider setting by (new error)/*n*. Three tries will usually give a good fit.

2.5   *TO BISECT AN ANGLE*

Whenever there are two lines or two pieces of material intersecting in space, there is an angle between them. For neatness and strength we often make a mitered joint in which we split the angle and have the parts meet on a line that cuts the angle exactly in two. Steel framing for buildings and other structures can be cut this way (Figure 2.18). From plane geometry, you may remember the construction for the bisector of an angle (Figures 2.19–2.21). You may also remember that the bisector of an angle is the locus of all points equidistant from the sides of the angle. We shall use this fact when we solve some descriptive geometry problems.

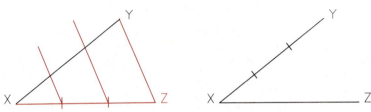

Figure 2.16 **Given–Construction–Solution** To divide line into any number of parts using computer graphics: (1) Create line on screen about 3–4 in. long using line task. It should be near middle of screen and inclined. (2) Name line *XY*. (3) Create horizontal line *XZ* exactly 6 in. long. (Line at any angle can be used; however, most graphics programs give distances more readily if line is horizontal or vertical.) (4) Name endpoint of line *Z*. (5) Create line from *Z* to *Y* (start at *Z*). (6) Move cursor to point on *XY* exactly 2 in. from *Z* (use distance counter) and duplicate line *ZY* or create parallel to it, creating line parallel to *ZY*. (7) Move cursor to point on *XY* exactly 4 in. from *Z* and repeat step 6. Parallel lines cut *XY* into three equal pieces. (*Computer-generated image*)

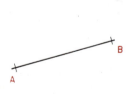

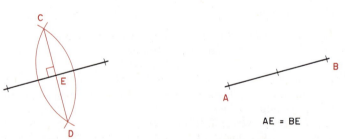

Figure 2.17 **Given–Construction–Solution** To divide line into two parts (bisect line), (1) set compass to radius greater than one-half length of line *AB*, (2) swing intersecting arcs with this radius from ends of the line, and (3) join intersections of arcs with straight line (*CD*) (this line is perpendicular bisector of given line; midpoint is at *E*).

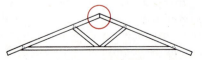

Figure 2.18 Mitered joint in roof truss.

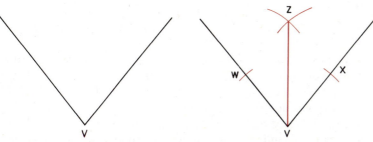

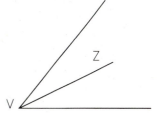

Figure 2.19 Given–Construction–Solution To bisect angle using compass, (1) set compass to any convenient radius less than length of the sides of angle, (2) place compass point on vertex *V* of angle and swing arcs cutting each side of angle, (3) place compass on each of cut points (*W* and *X*) and swing arcs intersecting with angle (*Z*), and (4) draw line from vertex *V* to intersection *Z* (this is bisector).

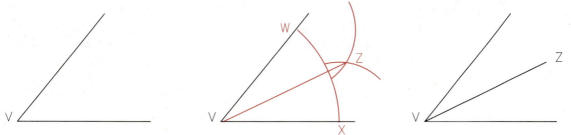

Figure 2.20 Given–Construction–Solution To bisect angle using computer graphics: (1) Create two intersecting lines 3–4 in. long near middle of screen. (2) Name intersection vertex *V* using TEXT. (3) Create arc with center at *V* and radius a little shorter than shortest line; make arc cross both lines. (4) Name cut points on lines *W* and *X*. (5) Create two more arcs of same radii with centers at *W* and *X*; make them long enough to intersect inside angle. (6) Name intersection point *Z*. (7) Create line *ZV*. This is bisector. (*Computer-generated image*)

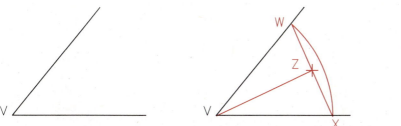

Figure 2.21 Given–Construction–Solution Alternate method to bisect angle using computer graphics: Follow steps 1–4 of Figure 2.20. Then create line *WX*, find midpoint of line *WX* using midpoint or center function and name point *Z*; create line *ZV*, which is bisector. (*Computer-generated image*)

2.6   TO CONSTRUCT A TRIANGLE GIVEN TWO SIDES AND THE INCLUDED ANGLE

Sail makers and other technical people apply graphic trigonometry in their work (Figure 2.22). A triangle has three sides and three angles. If we know the three sides, two sides and the included angle, or two angles and the included side, we can draw the triangle. We may find the information in different forms—two sides and the included angle is a fairly common way of describing a triangle. Our sail maker may know the length of the mast, the length of the boom, and the angle between them. The sail maker would then lay out the triangular pattern for the sail as described here (Figures 2.23 and 2.24).

2.7   TO CONSTRUCT A REGULAR POLYGON

Regular polygons are closed plane figures having sides of equal length and equal internal angles. You will see polygons in nature and in man-made objects. Consider the honeycomb built by bees; it is a collection of hexagons (six-sided figures) attached to each other to provide maximum efficiency in the use of space (Figure

Figure 2.22   Sailboat with triangular sail. (Photo by Howard Keith Williams)

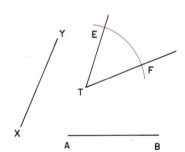

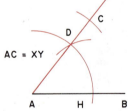

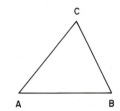

Figure 2.23   Given–Construction–Solution To construct triangle given two sides and included angle using manual graphics: (1) Draw line equal in length to one of given lines (*AB*). (2) Set compass to radius shorter than sides of angle (and shorter than line *AB*) and swing arc from vertex *T*, which cuts both sides *E* and *F*. (3) Do *not* change compass setting; now swing same-size arc from *A* on one side of line *AB*, cutting *AB* at *H*. (4) Place compass point on *E* and *F* and set compass to distance *EF*. (Dividers may be used for this operation.) (5) Do *not* change compass setting; now swing arc (or transfer divider distance) along arc from point *H*, cutting arc at *D*. (6) Draw line from point *A* through point *D*. (7) Set compass to length of second given line and swing arc from *B*, which cuts line *AD* at some point *C*. Triangle *ABC* is desired triangle containing two given lines and given angle.

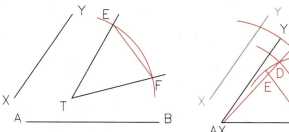
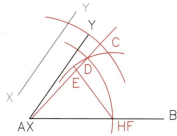
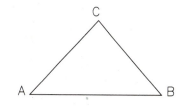

Figure 2.24 Given–Construction–Solution To construct triangle using computer graphics: (1) Create one line horizontally on lower part of screen and use counter to measure length. (2) Name line *AB*. (3) Create arc with center at *A* and length equal to second line and use counter to measure arc length. Make angle of arc greater than given angle. (4) Create line starting at *A* that crosses arc and makes given angle with line *AB*. Use angle counter to measure angle. (5) Name intersection point *C*. (6) Create line *CB*. Triangle *ABC* is desired triangle containing two lines of given length and given angle. (*Computer-generated image*)

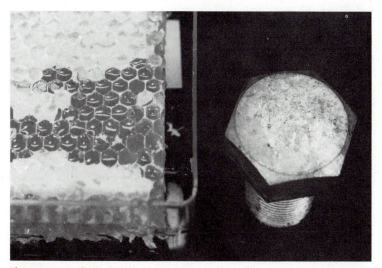

Figure 2.25 Cells in honeycomb are series of hexagons; many bolt heads are hexagons also.

2.25). Each cell in the honeycomb provides food storage for a new insect. A large polygon built by man is the Pentagon outside Washington, DC, which houses the Department of Defense. Construction methods are illustrated in Figures 2.26–2.28.

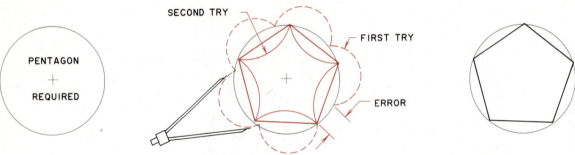

Figure 2.26 **Given–Construction–Solution** To construct polygon of any number of sides using manual graphics, (1) draw circle that contains polygon, (2) divide circumference of circle into as many parts as there are sides in polygon using method of successive approximations, and (3) connect points on circumference.

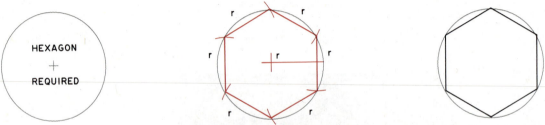

Figure 2.27 **Given–Construction–Solution** To construct polygon of six sides using manual graphics (special construction for just six sides), (1) draw circle that contains (circumscribes) polygon, (2) step off radius of circle along circumference (it will fit exactly six times, think about angles and number of sides in terms of plane geometry), and (3) connect points on circumference.

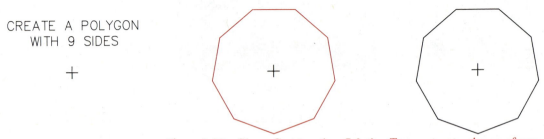

Figure 2.28 **Given–Construction–Solution** To construct polygon of any number of sides using computer graphics, use a POLYGON task (available with several programs) and create polygon with any number of sides by following prompts on screen. Typically, you would be asked for the number of sides, the location of the center, and the location of one vertex. (*Computer-generated image*)

2.8   TO CONSTRUCT A CIRCLE THROUGH THREE GIVEN POINTS NOT IN A STRAIGHT LINE

The circle is common in nature and in man-made devices. When we cut through a tree trunk or a flower stem, we see a circle. Every time a bicycle or an automobile passes, we see circles in the wheels. The circle rolls, and it is the most efficient way of enclosing an area. (The ratio of the circumference of a circle to the area inside the circle is a minimum; no other plane figure can enclose the same area with a smaller perimeter.)

The construction shown here will help you build a circle that passes through three given points or to find the center of a circle if three points on the circumference are known (Figures 2.29 and 2.30). This is useful when we need to repair or replace a broken wheel or gear (Figure 2.31).

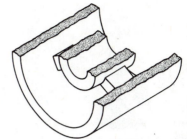

Figure 2.29   Center and diameter of broken wheel can be found by using three points on circumference.

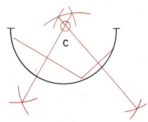
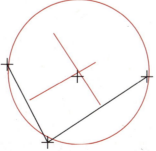

Figure 2.30   Given–Construction–Solution To construct circle using manual graphics: (1) Connect two points on circle to make straight line; repeat to form another straight line. (2) Construct perpendicular bisector to each line. (Intersection of bisectors is center of circle.) (3) Set compass to length from center (intersection) to end of any of lines and swing circle.

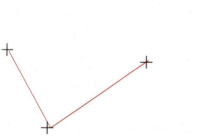
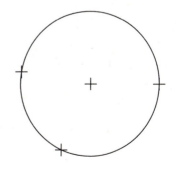

Figure 2.31   Given–Construction–Solution To construct circle through three points using computer graphics: (1) Create points on screen and connect them to create two straight lines. (2) Construct perpendicular bisectors of the lines (Figure 2.12). (Intersection of bisectors is center of circle.) (3) Use CIRCLE task to create required circle. *Note:* Some programs have routine that creates circle directly from three points. (*Computer-generated image*)

Figure 2.32 Bicycle chain is tangent to sprockets. (Photo by Howard Keith Williams)

2.9 TO CONSTRUCT A LINE TANGENT TO A CIRCLE THROUGH A GIVEN POINT

A line tangent to a circle is visible when you look under the hood of a car and see the fan belt going around a pulley. You also see a tangent when you look at the chain on a bicycle going around the sprockets (Figure 2.32). The most useful devices have contours that are a combination of straight sides and curves that are tangent to the straight-line contours.

The word *tangent* means to touch, so the point of tangency is where the circle just touches the straight line. A radius drawn from the center of the circle to this point will always be perpendicular to the tangent line. You will use this fact in your constructions (Figures 2.33–2.35).

2.10 TO CONSTRUCT A LINE TANGENT TO TWO CIRCULAR ARCS

If you examine the fan belt in a car, you will find that it is not only tangent to one circle (pulley) but is also tangent to one or more other pulleys (Figure 2.36). The next time you are looking at machine tools—lathe, milling machine, or big drill press—note that the straight-sided contours are rounded wherever they join

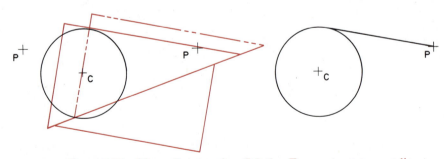

Figure 2.33 Given–Construction–Solution To construct tangent line using manual graphics (two-triangle method): (1) Place triangles on drawing surface hypotenuse to hypotenuse (as for drawing parallel or perpendicular lines). (2) Slide triangles so that one edge of upper triangle passes through given point and touches circle. (3) Hold lower triangle firmly in place and slide upper triangle until other edge passes through center of circle. (4) Mark point where this edge cuts circle arc using short dash. This is point of tangency. (5) Draw line from point of tangency to given point. (Slide upper triangle back to its original position; now you know where to stop tangent line.)

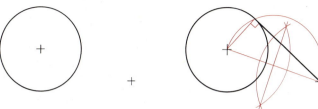

Figure 2.34   Given–Construction–Solution To construct tangent through point on circle: (1) Draw line from given point to center of circle. (2) Bisect line (Figure 2.16). (3) Set compass to radius one-half length of line and swing half circle with line as diameter and that cuts given circle. Point of intersection of two arcs is point of tangency. (4) Draw line from given point to point of tangency. (5) Draw line from center of circle to point of tangency. This line makes right angle with tangent line and confirms point of tangency. *Note:* Theorems of plane geometry teach that any triangle inscribed in semicircle is right triangle. This construction gives perpendicular at point of tangency.

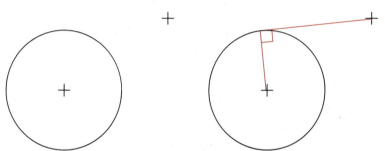

Figure 2.35   Given–Construction–Solution To construct tangent using computer graphics: (1) Create circle on screen using CIRCLE task (record coordinates of center). (2) Create point on screen at least 2 in. away from circle. (3) Through point create line that touches circle. Make construction line longer than just touching circle. Use zoom function to check that line just touches circle. This is tangent line. (4) Create line through center of circle perpendicular to tangent line. Intersection with circle (and tangent line) is point of tangency. (*Computer-generated image*)

another surface. Here you have a straight line tangent to two circle arcs (Figures 2.37 and 2.38).

2.11 *TO CONSTRUCT AN ARC TANGENT TO A LINE (OR LINES)*

Nature does not produce plants or animals with sharp corners. Sharp corners are easily broken. When you study the mechanics

Figure 2.36 Fan belt on automobile engine is tangent to all pulleys it touches. (Courtesy of Ohio State University Engineering Publications)

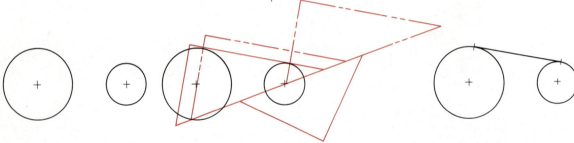

Figure 2.37 **Given–Construction–Solution** To construct tangent line using manual graphics: (1) Place triangles on drawing surface hypotenuse to hypotenuse. (2) Align one edge of upper triangle so that it touches both arcs. Draw light line connecting two arcs. (3) Hold lower triangle firmly in place and slide upper triangle so that other edge passes through center of circle; mark tangent point with short dash. (4) Slide upper triangle so that it passes through center of other circle and mark this tangent point. (5) Connect two points.

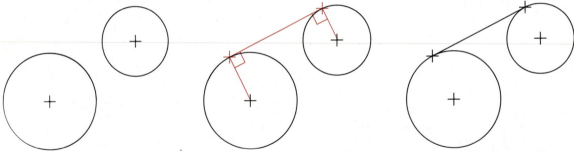

Figure 2.38 **Given–Construction–Solution** To construct tangent line using computer graphics: (1) Create two circles on screen, preferably of different diameters and not touching each other. (2) Create line that touches both circles. Check accuracy with ZOOM task. Delete line and re-create it if it does not just touch both circles. (3) Create line through center of each circle perpendicular to tangent line. Points of tangency occur where these lines touch tangent line. (Some programs have TANGENT task that locates tangent.) (*Computer-generated image*)

of materials, you will learn that destructive forces increase at internal sharp corners (stress concentration). When we design mechanisms that will be heavily loaded, we avoid sharp corners, just as nature does. We do this by rounding outside corners and filling in interior corners with a curve (a fillet) (Figures 2.39–2.43).

Figure 2.39 Exterior and interior corners on machinery are rounded with arcs tangent to straight surfaces. (Courtesy of Ohio State University Engineering Publications)

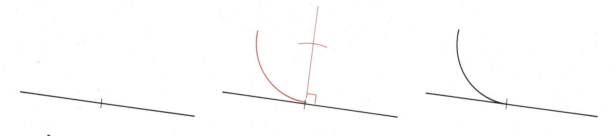

Figure 2.40   Given–Construction–Solution To construct arc tangent to one line using manual graphics, (1) draw perpendicular to given line that is longer than radius of arc, (2) set compass to desired radius, (3) swing arc from intersection of two lines that cuts perpendicular line, and (4) set compass point on cut point and draw tangent arc.

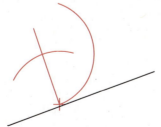

Figure 2.41   Given–Construction–Solution To construct arc tangent to one line using computer graphics if line is horizontal or vertical, read coordinates on screen and move cursor distance of radius away from line to locate center for arc. Alternate method: If line is not horizontal or vertical, use manual graphics method to locate center; then create arc. (*Computer-generated image*)

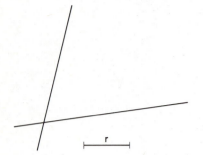

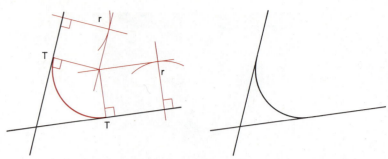

Figure 2.42 Given–Construction–Solution To construct arc tangent to two intersecting lines using manual graphics: (1) Follow steps 1–3 in Figure 2.40 to find point that is radius distance away from each line. (2) Draw line parallel to each given line through cut points. These lines will intersect at point equidistant from both given lines. (3) Draw perpendicular from this point to each given line. (These are tangent points.) (4) Swing arc with its center at intersection that touches each given line where perpendicular cuts it.

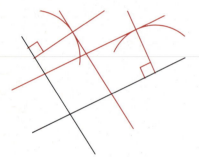

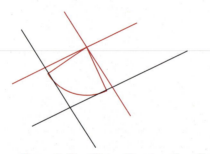

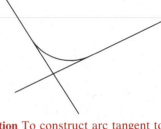

Figure 2.43 Given–Construction–Solution To construct arc tangent to two intersecting lines using computer graphics if lines are horizontal and vertical, read coordinates on screen to locate center of arc at radius distance from each line. Create arc. (Some programs have FILLET function that locates arc.) Alternate method: If lines are not horizontal and vertical, uses manual graphics method to locate center and then create arc. (*Computer-generated image*)

2.12 *TO CONSTRUCT AN ARC TANGENT TO A LINE AND AN ARC*

Not all roads are straight. Some roads have curves, and some have curves leading into other curves. We need to know how to lay out such patterns. If you ever need to design cams (such as those that move the valves in an automobile engine), you will also need to know how to fit an arc tangent to a line and another arc (Figures 2.44–2.46).

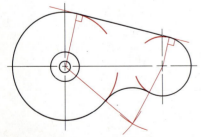

Figure 2.44 Cam development may require arcs tangent to straight lines and other arcs.

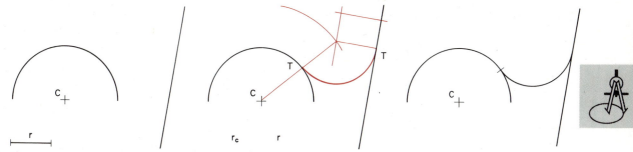

Figure 2.45   Given–Construction–Solution To construct arc tangent to line and arc using manual graphics: (1) Set compass to radius of existing arc. (2) Increase (or decrease) radius by length of radius of arc to be drawn. (3) Swing new arc with increased (decreased) radius from original center; be sure it is in area of line. (4) Draw line parallel to existing line and exactly radius distance away from existing line; this line should cut new arc. Cut point is center for desired arc. (5) Align triangle with center of desired arc and center of existing arc and mark where this line crosses existing arc. This is point of tangency. (6) Draw perpendicular from center of desired arc to given line. Cut point (intersection) is point of tangency. (7) Set compass to radius of desired arc and swing it from center, drawing from tangent point to tangent point.

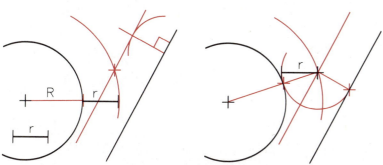

Figure 2.46   Given–Construction–Solution To construct arc tangent to line and arc using computer graphics: Follow the method using manual graphics in Figure 2.45 and (a) record coordinates of circle center and ends of lines and (b) use ZOOM function to increase accuracy when creating desired arc. You may be able to use MOVE, COPY, PARALLEL, or PERPENDICULAR functions available in some programs. (*Computer-generated image*)

2.13   *TO CONSTRUCT AN ARC TANGENT TO TWO ARCS*

The design of some gears requires that we locate the pitch circle of one tangent to the pitch circles of two other gears (Figure 2.47).

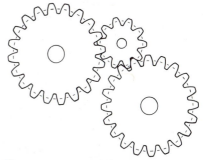

Figure 2.47   Two idler gears driven by third gear.

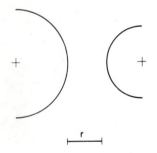

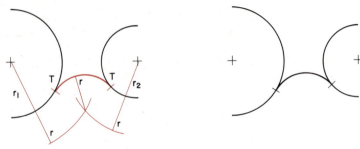

Figure 2.48 **Given–Construction–Solution** To construct arc tangent to two given arcs using manual graphics: (1) Increase or decrease radius of each given arc by length equal to radius of desired arc. (2) Swing concentric arcs around each of given arcs with new radii. Intersection of these arcs locates center of desired arc. (3) Align triangle with new center and each of given centers. Make short dash where these lines of centers cross given arcs. These are tangent points. (4) Set compass to radius of desired arc and swing arc from tangent point to tangent point.

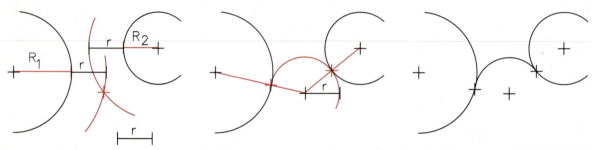

Figure 2.49 **Given–Construction–Solution** To construct arc tangent to two given arcs using computer graphics, follow the method using manual graphics in Figure 2.48. *(Computer-generated image)*

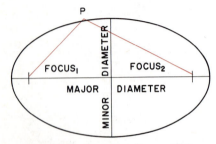

$\overline{F_1P} + \overline{F_2P}$ = K = MAJOR DIAMETER

Figure 2.50 Ellipse showing its foci and major and minor diameters.

(The pitch circles of circular gears are the lines along which the gears transmit motion.) We may also use this construction to avoid the sharp corners that would occur if we had a straight-line contour connecting two circular shapes (Figures 2.48 and 2.49).

2.14 TO CONSTRUCT AN ELLIPSE

An ellipse is a closed plane curve. Every point on its curve is located so that the sum of the distances from the point on the curve to two points (the *foci*) on the long axis (the major diameter) is constant (Figure 2.50). The ellipse is also a conic section, the

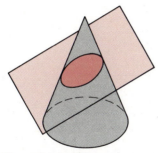

Figure 2.51 Ellipse as conic section.

curve produced when you cut a right-circular cone with a plane not perpendicular to the axis but making a greater angle with the axis than the elements of the cone (Figure 2.51).

We often use ellipses in making drawings because every view of a circle except the one looking straight at it (normal) is an ellipse (Figure 2.52) or an edge view. Ellipses are an attractive shape and make a pleasing design element. Some gears are also made with an elliptical shape. Several constructions for drawing ellipses are given in Figures 2.53–2.57. Commercial design rooms use ellipse templates for drawing ellipses. A common template is one used for drawing circles in isometric pictorial drawings (Figure 2.58).

Figure 2.52 Coin is round when viewed normally; from any other angle it is seen as ellipse or edge.

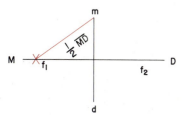
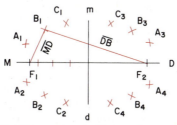
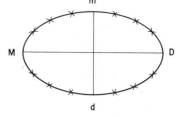

Figure 2.53 **Given–Construction–Solution** To construct ellipse using foci method: (1) Set compass to one-half length of major diameter ($\frac{1}{2}$MD). (2) Swing arcs from ends of minor diameter (points m and d) intersecting major diameter at points f_1 and f_2, the foci. (3) On one end of major diameter mark points between focus and intersection of diameters. (Three points are shown here; more points provide greater accuracy.) (4) With foci as centers, set compass to lengths f_1A and f_2A and swing arcs intersecting in each quadrant at points APV_1PVPY, A_2, A_3, and A_4. Repeat for points B and C so that point is plotted in each quadrant for each selected point. (5) Connect intersection points with irregular curve.

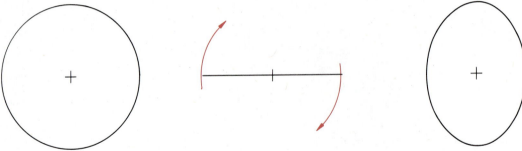

Figure 2.54 Given–Construction–Solution To construct ellipse using computer graphics, use foci method given in Figure 2.53 and connect points using curve-plotting task. (This method is very time consuming; if alternate is available, try it.) Alternate method: Some graphics programs create round and edge views of circle and then view edge from any angle to create true ellipse by projection. This method is shown in Chapter 11. Some programs also have an Ellipse command. (*Computer-generated image*)

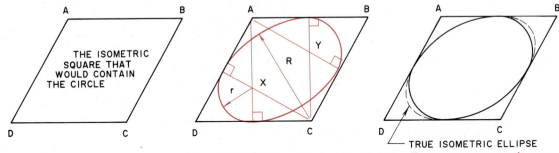

Figure 2.55 Given–Construction–Solution To construct approximate ellipse using four-center method when making isometric pictorial drawings (discussed in Chapters 3 and 5): (1) Lightly draw isometric square that contains circle of given diameter; each side of square is same length as diameter and interior angles are 60° and 120°, as shown in square *ABCD*. (2) From each of large-angle vertices (*A* and *C*) draw lines to midpoints of each side of square. These lines are perpendicular bisectors of sides. (3) Set compass to length of bisector lines and swing arcs from vertices *A* and *C* to feet of perpendiculars (points of tangency). (4) From points where bisectors intersect each other (*X* and *Y*) as centers, swing small arcs to points of tangency. Approximate four-center ellipse results. Note how well it matches true ellipse; it is good approximation and faster for this application than any other method, except template.

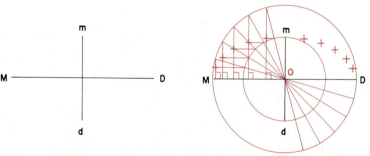
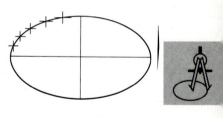

Figure 2.56   Given–Construction–Solution To construct ellipse using concentric circle method: (1) Set compass to one-half major diameter ($\frac{1}{2}$MD) and draw circle with center at 0, intersection of diameters. (2) Set compass to one-half minor diameter ($\frac{1}{2}$md) and draw circle with center at 0. (3) Use triangles to divide circle into equal parts, drawing radial lines through center intersecting both large and small circles (30° increments give 12 equal parts; 15° increments, using triangles in combination, give 24 equal parts.) (4) From intersection of each radial line with large circle draw line perpendicular to major diameter. (5) From intersection of each radial line with small circle draw line perpendicular to minor diameter. Where two perpendiculars for any pair of points cross is the point on ellipse. (You may want to construct one pair of perpendiculars at a time and find point where they cross in order to reduce confusion.) (6) Connect points using irregular curves.

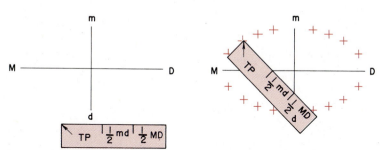
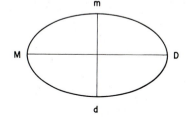

Figure 2.57   Given–Construction–Solution To construct ellipse using trammel method: (1) Select small card or piece of paper with straightedge longer than one-half major diameter for building tool (trammel is used for transferring measurements). (2) Mark one end of trammel as tracing point with arrow. This point will locate points on ellipse. (3) From tracing point measure one-half major diameter and mark it with dash and MD. (4) From tracing point measure one-half minor diameter and mark it with dash and md. (5) Lay trammel across diameters so that MD point is on the minor diameter and md point is on major diameter; tracing point now gives point on curve. Mark it with a short dash. (6) Move trammel around axes keeping MD point on minor diameter and md point on major diameter and mark points on ellipse; mark three or more in each quadrant. (7) Connect points with irregular curves.

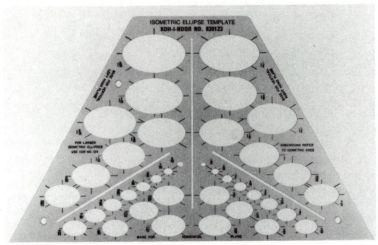

Figure 2.58 Ellipse template for drawing circles as ellipses in isometric pictorial drawings.

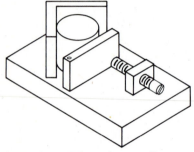

Figure 2.59 Triangular set of surfaces encloses cylindrical pin on this test fixture.

2.15 *TO INSCRIBE A CIRCLE IN A TRIANGLE*

A classic construction from Euclidean geometry is the circle **inscribed** in a triangle. Inscribed means that the circle is tangent to each of the sides of the triangle.

You may need to inscribe a circle in a triangle if you are designing tools to hold circular parts in a shop (Figure 2.59). A common shipping container for rolled drawings is a triangular

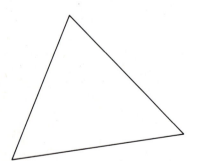
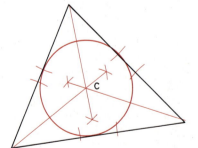
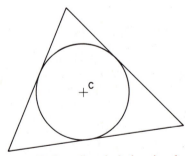

Figure 2.60 Given–Construction–Solution To inscribe circle in triangle using manual graphics: (1) Construct bisector of each angle of triangle (Figure 2.14). (2) Swing circle from intersection of bisectors tangent to sides of triangle. *Note:* Theorems of plane geometry teach that bisector of angle is locus of all points equidistant from sides of angle. Point where bisectors intersect is then equidistant from all sides of angles. This is proper location for center of circle with radius distance that is the same to all sides.

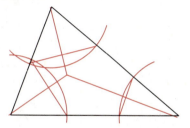
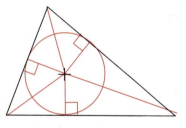
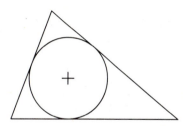

Figure 2.61 Given–Construction–Solution To inscribe circle in triangle using computer graphics: (1) Bisect angles of triangle. (2) Create circle with center at intersection of bisectors and that just touches the sides. You can locate points of tangency by creating perpendiculars from center to sides of triangle. Use ZOOM function to improve accuracy. (*Computer-generated image*)

carton. How large a roll of drawings could you ship in a triangular carton shaped similar to an equilateral triangle with 2-in. sides? The constructions for this problem are given in Figures 2.60 and 2.62.

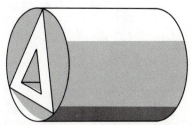

Figure 2.62 Round shipping container protects triangular part.

2.16 *TO CIRCUMSCRIBE A CIRCLE ABOUT A TRIANGLE*

Another construction from Euclidean geometry is fitting a circle around a triangle so that the vertices of the triangle all lie on the circle. This construction is also useful for designing tools and packaging (Figures 2.61–2.64).

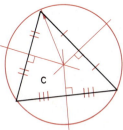
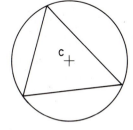

Figure 2.63 Given–Construction–Solution To **circumscribe** circle around triangle using manual graphics: (1) Construct perpendicular bisectors of sides of triangle (Figure 2.16). (2) Set compass to distance from intersection of perpendiculars to one vertex of triangle (distance should be same to all vertices). (3) Construct circle passing through all three vertices. *Note:* Theorems of plane geometry teach that perpendicular bisector of line segment is locus of all points equidistant from ends of line segment. Point where bisectors intersect is equidistant from ends of all sides. This is proper location for center of circle with radius distance the same from all vertices.

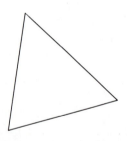

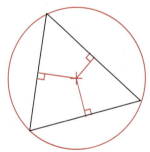

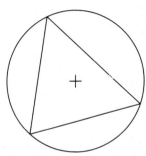

Figure 2.64 Given–Construction–Solution To circumscribe circle around triangle using computer graphics: (1) Find midpoints of sides. (2) Create perpendiculars at these midpoints. (3) Create circle with point of intersection of perpendiculars as center and radius that just touches sides. Use ZOOM function to improve accuracy. (*Computer-generated image*)

Figure 2.65 Involute of circle is path traced by string unwound from it.

2.17 *TO CONSTRUCT AN INVOLUTE OF A CIRCLE*

The involute of a circle is the path traced by unwinding a string wrapped around the circle (Figure 2.65). This shape is used for gear teeth to assure true rolling motion between circular gears.

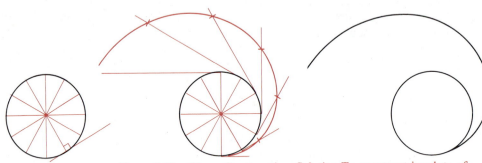

Figure 2.66 Given–Construction–Solution To construct involute of circle by manual graphics: (1) Divide circle into number of parts using radial lines or using method of successive approximations on circumference (Figure 2.15). (2) Construct tangents (perpendiculars) at each division point (Figure 2.33). (3) Set dividers to length of arc of one part of circle, and beginning at any point on circle measure this distance on tangent line. Mark distance with short dash. (4) Go to next tangent line and measure two parts of length of arc. Mark it with short dash. (5) Continue around circle adding one more part to number measured on tangent lines, completing as much of the circle as directed by instructor. (6) Connect points with irregular curves (curve plotted may get quite large; more than one irregular curve may be needed to get smooth fit among the points).

You can develop involutes for lines, triangles, and other shapes, but the involute of a circle has the greatest application for engineers. Methods for constructing involutes are shown in Figures 2.66 and 2.67.

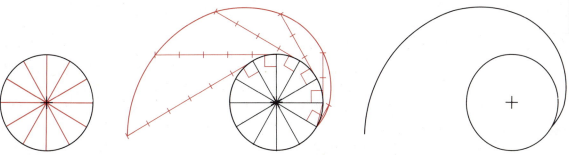

Figure 2.67 Given–Construction–Solution To construct involute of circle by computer graphics, follow the method for manual graphics in Figure 2.66 and connect plotted points with curve-fitting tasks. (*Computer-generated image*)

2.18 *SUMMARY*

After practicing the material presented in this chapter you will be able to use basic geometric constructions when needed for making useful technical drawings. The constructions will be used often when you apply descriptive geometry to solving problems. You will find use for parallels, perpendiculars, regular polygons, triangles, tangents of all sorts, and ellipses. Some of these constructions may have application to solving problems later in your professional career when you may be designing gears, tooling for manufacturing, machine parts, or highways. Everything you learn is a building block for future learning: These constructions are not just an end in themselves but foundation blocks for new knowledge.

PROBLEMS

Problem 2.1
(Solve these problems on $8\frac{1}{2} \times 11$ drawing paper with the long dimension placed vertically.)
   (a) Draw a line 4.2 in. long in the upper third of your paper; divide it into five equal parts using the method of similar triangles.
   (b) Draw a line $2\frac{15}{16}$ in. long in the middle third of your paper; bisect it using your compass. Erect a perpendicular bisector $1\frac{9}{16}$ in. long on one side of the original line and extend it $\frac{7}{8}$ in. on the other side.
   (c) Draw a line 100 mm long in the lower third of your paper; divide it into seven equal parts by the method of successive approximations.

Problem 2.2
   (a) Use your triangles to draw a 45° and a 30° angle on your paper. Make the legs about 50 mm long and bisect each angle with your compass.
   (b) Draw a 60° angle with your triangles. Make the legs about $1\frac{3}{4}$ in. long and bisect it. Check the results with the 30° angle on your triangle.

Problem 2.3
   (a) Draw a triangle with legs 45 and 72 mm long with an included angle of 45°.
   (b) Think of a new way to build a triangle; draw a triangle with *angles* of 45° and 60° and an included *side* of 64 mm.
   (c) Here is another one to try. Build a triangle with three sides each 64 mm long (an equilateral triangle). What is the size of each angle?

Problem 2.4
   (a) Construct a hexagon with sides 1.30 in. long.
   (b) Construct a square that just fits (inscribed) in a circle 3.00 in. in diameter.
   (c) Construct a square with sides 1.40 in. long. (All the sides are equal; and all the internal angles are 90°. What a great way to practice perpendiculars!)

Problem 2.5
Draw two circles that have 4 in. between their centers. One circle should have a $1\frac{1}{2}$-in. radius; the other should have a 1-in. radius. Construct four tangent lines touching both circles; one pair crosses between the circles; the other pair does not. Mark all tangent points with red dashes.

Problem 2.6
Draw a reduced-size copy of your 30°–60°–90° triangle; the longest leg making the right angle should be 5.00 in. long. Round the 30° angle

with a 0.50-in. radius; round the 60° angle with a 0.40-in. radius. Mark the tangent points with red dashes. Bisect the 90° angle.

Problem 2.7
Construct an ellipse with a major diameter of 165 mm and a minor diameter of 95 mm. Use the trammel method. Draw a copy of your trammel, in one position, on your ellipse. Find the foci of this ellipse and mark them with green crosses.

Problem 2.8
    (a) Draw a reduced-sized copy of your 45° triangle; each of the legs making the 90° should be 4.50 in. long. Inscribe a circle in the triangle. Mark the tangent points where the circle touches the sides of the triangle with blue dashes; draw a small (0.13 diameter) blue circle around the center of the inscribed circle.
    (b) Now circumscribe a circle around your triangle. Draw a small red circle around the center of the circumscribed circle.

Problem 2.9
    (a) Locate the center for a circle 4 in. down from the top of your paper and 3 in. from the left side. Draw a circle of $1\frac{1}{2}$ in. diameter using this center. Starting at the bottom of this circle, draw the involute formed when you unwind the circumference; do not go lower on the sheet than the bottom of the circle. Use your irregular curves to connect the points.
    (b) Draw an equilateral triangle with 1-in. sides that is $2\frac{1}{2}$ in. from the left side of the paper and 2 in. above the title block or border. Start at the lower right corner and unwind the involute.

Problem 2.10
Prepare a graph that converts inches to millimeters (1 in. = 2.54 mm). Make the x axis 7 in. long; divide it into 10 parts. Make the y axis 9 in. long; divide it into 254 parts, but mark only every tenth part.

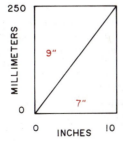

Turn your drawing paper with the long dimension horizontal for problems 2.11 and 2.12.

Problem 2.11
Draw a triangular roof truss as shown. (Scale: 1 in. = 2 ft.) Bisect the angles for miter cuts of the truss members.

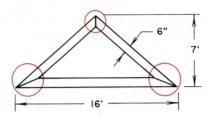

Problem 2.12
Lay out the pattern for the side plate of a link of roller chain. Draw it double size. Mark tangent points with a red dash.

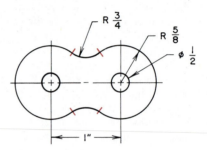

Solve problems 2.13–2.20 on your computer screen using a graphics program.

Problem 2.13
Create a horizontal line 4.50 in. long near the top of your computer screen. Erect a line 2.75 in. long perpendicular to the first line at its midpoint.

Problem 2.14
Create a line 5.20 in. long near the bottom of your screen that is neither horizontal nor vertical. Divide this line into six equal parts.

Problem 2.15
Create a circle 4 in. in diameter near the middle of your screen. Inscribe a hexagon in the circle; connect alternate vertices of the hexagon to form another closed plane figure. What is this figure?

Problem 2.16
Construct a triangle with sides $1\frac{1}{2}$, 2, and $2\frac{1}{2}$ in. long. Circumscribe a circle around this triangle. What is the diameter of the circle? Is there a right angle in the triangle? If your answer is yes, prove it!

Problem 2.17
Construct a pentagon of any convenient size. Join alternate vertices to form a star. Delete lines in the central portion of the star so that the center is open.

Problem 2.18

Two small pulleys are 76.1 mm apart on their line of centers. The smaller is 25.4 mm in diameter; the larger is 50.7 mm. Show the pulleys and the path of the belt that goes around them.

Problem 2.19

Create a block letter T on your screen 4.32 in. high and 4.32 in. wide. The horizontal and vertical members should be 1.00 in. Round all exterior and interior sharp corners with a radius of .25 in. (Rounded exterior corners on machine parts are called *rounds;* rounded interior corners are called *fillets.*)

Problem 2.20

Construct two circles with 1.50-in. diameters 2.50 in. apart on a horizontal line of centers near the top of your screen. Install a circle with a diameter of 3.00 in. tangent to both of the original circles. Does this remind you of a famous amusement park?

Solve Problems 2.21–2.27 using pencil and paper or on a computer screen. If you do not have a color monitor or colored pens in your plotter, change any color instructions to black.

Problem 2.21

Lay out the center lines for a portion of a highway and fit the needed curves to the drawing. Mark the tangent points with a red dash (Scale: 1 in. = 100 ft.)

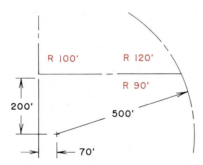

Problem 2.22

Draw a view of the trammel base. Round all external corners with a 6-mm radius and all internal corners with an 8-mm radius. Show tangent points with a red dash.

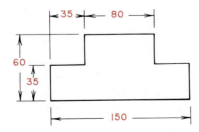

Problem 2.23

Make an outline drawing for a cam latch. Show tangent points in red. (Scale: full size.)

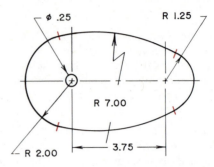

Problem 2.24

What is the diameter of the largest circular pipe that can be put through area A of the truss? (Scale: 1 in. = 10 ft.) For more practice check areas B and C.

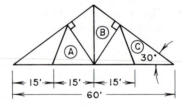

Problem 2.25

A special triangular beam is to be shipped in a round protective tube. What is the inside diameter of the needed tube? (Scale: full size.)

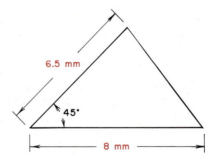

Problem 2.26

Your Aunt Eleanor wants to have the largest possible elliptical flower bed in her front yard. The space available is shown in the accompanying figure. Design a trammel you can use to determine points on the elliptical curve. Also make a scale drawing of the bed

using the concentric circle method (edge view method on a computer).
(Scale: 1 in. = 10 ft.)

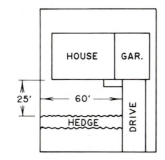

Problem 2.27

Your little sister sends you a broken wheel from her red wagon and
asks you to tell her how big a wheel she needs to replace it. You
measure two chords on the wheel as shown in the accompanying
figure. What is the diameter?

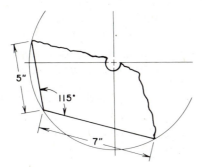

TERMS YOU WILL SEE IN THIS CHAPTER

"BOXING"   Process of making marks on established center lines to act as guides when sketching circles or arcs.

IRREGULAR CURVE   Curve that is not a function of a radius.

CONSTRUCTION LINES   Very light lines used as guides when sketching. If drawn correctly, they will be so light that the darkened final sketch lines will render the construction lines nearly invisible.

ORTHOGRAPHIC SKETCH   Sketch made in orthographic projection, that is, a sketch where the direction of sight is perpendicular to the principal plane of projection. Most often, the top, front, and right side views are shown.

ISOMETRIC SKETCH   Sketch drawn as an isometric drawing; that is, the height axis is vertical, but the width and depth axes are drawn 30° from the horizontal. This produces an easy-to-read-and-sketch pictorial view.

WIDTH   Size of an object measured from side to side.

DEPTH   Size of an object measured from front to back.

HEIGHT   Size of an object measured from top to bottom.

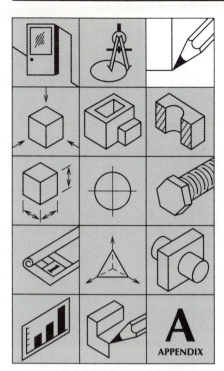

3

TECHNICAL SKETCHING

3.1 INTRODUCTION

The graphic language of the engineering world is used to communicate ideas and concepts through a series of pictures and symbols. After these ideas, concepts, and details for a project have been finalized, precise technical drawings are produced so that parts may be manufactured. These drawings are usually made from sketches generated during the project-planning stage.

The objective of making a sketch is to make a quick, graphic representation of an idea that can be easily communicated to a co-worker, a drafter, or a client. (This sketch is generally made without the aid or use of technical drawing instruments.) The primary need for making a sketch is to represent what you see rather than to give technical information. Sketches are an integral part of analytical calculations where they are used to define the physical objects being analyzed. For this reason, sketches tend to show an object as it appears, with little emphasis on hidden or internal surfaces or features, although they may be included if necessary. The sketch also serves as

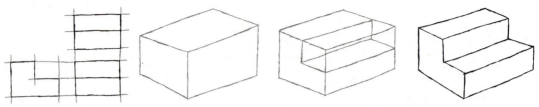

Figure 3.1 Sample of four steps for making sketch.

an *immediate* graphic communication, often drawn under varying conditions such as on the job site or in a business meeting (Figure 3.1).

Sketching, similar to writing, is a skill that can be developed through practice. This chapter will show you how to sketch effectively, and it will address materials and show engineering applications. Techniques and suggestions will be given so your sketches will look more realistic. You can practice sketching anywhere and at any time because the only items needed to make a sketch are a writing instrument, a drawing surface, and possibly an eraser.

Sketching is a process that generally precedes the action of creating a drawing on a computer or on a drawing board. This chapter offers you sketching techniques with only brief explanations concerning drawing theory. Detailed theory of projections will be given in Chapter 4.

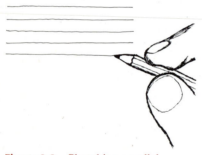

Figure 3.2 Sketching parallel horizontal lines.

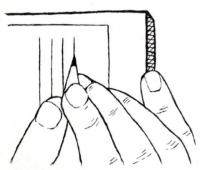

Figure 3.3 Using side of board as guide to sketch parallel lines.

3.2 SKETCHING HORIZONTAL LINES

Lines commonly used for sketches include straight lines and curved lines. Straight lines can be divided into horizontal, vertical, or inclined lines. The normal way to draw parallel horizontal lines is from left to right for right-handed people or right to left for left-handed people. You should be able to develop enough skill to draw many evenly spaced horizontal lines while leaving uniform spaces between the lines (Figure 3.2).

Care should be used to avoid drawing arcs instead of horizontal lines. This occurs when the elbow becomes stationary. You must move the entire arm while drawing horizontal lines. This is particularly important when sketching long horizontal lines. Long lines may be sketched more accurately by sketching a series of short horizontal lines end to end. An effective technique for drawing parallel lines is to draw the lines while guiding your finger along the edge of the drawing board. Another important tip is to put the pencil on the starting point and look at and draw to the finishing point (Figure 3.3).

3.3 *SKETCHING VERTICAL LINES*

Vertical lines are usually drawn from top to bottom. You should be able to develop enough skill to draw uniformly spaced parallel vertical lines. The edge of the drawing surface can be used for this as well.

3.4 *SKETCHING ANGULAR LINES*

Angular lines are drawn from lower left to upper right. This is a comfortable position for the right-handed person. The most uncomfortable and difficult lines to sketch are angular lines extending from upper left to lower right. A helpful hint would be to rotate the paper to a position where it would be more comfortable to sketch these lines (Figure 3.4).

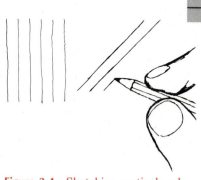

3.5 *SKETCHING CURVED LINES*

Curved lines are identified as either circles, arcs, or irregular curves. Circles are sketched most accurately by first sketching center lines. The radius is then marked on these center lines as guides. These marks form a box within which the circle can easily be sketched. This process is often called **"boxing"** a figure. Lastly, the circle is sketched in two motions. First, the upper left side of the circle is sketched in a counterclockwise direction, and then the lower right side of the circle is sketched in a clockwise direction (Figure 3.5).

Large circles may require a second set of center lines rotated at a 45° angle in order to increase the number of guides from four to eight. Also, four arcs may be required. If so, the paper should

Figure 3.4 Sketching vertical and angular lines.

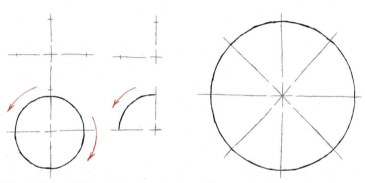

Figure 3.5 Guide marks for circles and arcs.

be rotated for completion of the last two quadrants. Arcs are sketched similar to circles. The orientation of the arc and the size of the arc will govern how the person sketches the arc. For medium and large arcs, it is recommended that center lines and guides be drawn.

3.6   SKETCHING IRREGULAR CURVES

Irregular curves (those without a given radius) should be sketched using pencil strokes that are most comfortable for you. Very light lines should be sketched first to check for the correct shape. Draw dark lines over the light lines to complete the curve.

3.7   DRAWING MEDIA

After becoming familiar with the most comfortable sketching strokes, identify an appropriate media for the sketch. Drawing media are the materials upon which you will make your sketch. We all know that sketches are made on everything from match book covers to expensive papers or films. In Chapter 1 we learned about films and vellums. However, the interest here is to identify those paper or film materials that are commonly used in the engineering field.

Technical sketches are most often made on $8\frac{1}{2} \times 11$-in. opaque paper. For copying purposes, the original sketch is referred to as the master. Sketches made on this opaque paper may be copied on a normal copy machine. Normal copy machines generally accept $8\frac{1}{2} \times 11$-in. masters; however, some office copy machines will copy sheets $8\frac{1}{2} \times 14$ or even 11×17 in.

Some paper may be printed with a grid of thin, lightly colored lines, which are a significant aid to the person doing the sketching. Common grid patterns are square or isometric (Figure 3.6). The printed grids may be printed on opaque paper or on a translucent paper or film. Again, sketches made on opaque paper may be copied on a standard office copy machine, and sketches made on translucent stock can be copied on copy machines or on Diazo equipment.

3.8   GRIDS FOR SKETCHING

A special printed grid is the "dropout" grid. This consists of a light blue grid printed on translucent paper or film. The grid serves

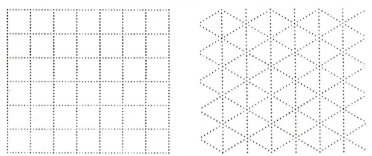

Figure 3.6 Sample grid patterns that are useful for making refined sketches.

as a guide to the person sketching, but when the copies are made, the grid will not copy and the sketch stands by itself. This is a great aid to making better appearing sketches.

Translucent paper or film may be placed over a printed grid as an overlay during the sketching process. Then any printed grid may be used. Special large printed grids are available for this purpose, such as perspective grids for interior or exterior views at various angles. This allows you to have the advantage of using a selected grid system without the grid appearing on the master or on the copies (Figure 3.7).

Sometimes a special grid may be printed or drawn on the back side of the paper. Engineering problem paper is a typical example. In this case, the grid is printed on the back of the sheet, and you sketch on the front of the sheet. One such example might be when a number of lettering guidelines are used on a large specification sheet. If many erasures or changes must be made, the guidelines will not be disturbed when the erasures are made (Figures 3.8 and 3.9).

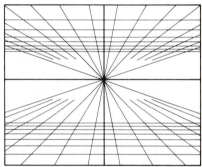

Figure 3.7 Grid patterns are used to make more accurate perspective sketches.

3.9 *LINE WEIGHTS*

Line weight refers to the line blackness or width of a line. The very light lines used to sketch for size, proportion, and position are called layout lines, or **construction lines.** They are so light that they would not appear if copied using a standard copy process. Construction lines are the first lines drawn. After the shape of the object has been sketched, the lines are refined. Then the visible lines are darkened with black solid lines. Hidden lines represent nonvisible edges or surfaces and are drawn with black dashed lines. If the original construction lines are drawn light enough, none should have to be erased after the sketch has been darkened and refined (Figure 3.10).

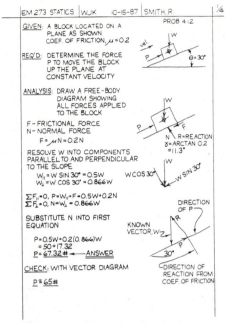

Figure 3.8 Example of typical engineering problem.

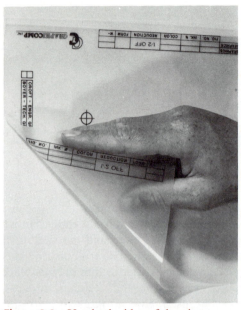

Figure 3.9 Use both sides of drawing media to protect guidelines if many erasures are anticipated.

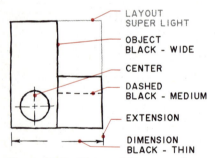

Figure 3.10 Line weight conventions used for finished drawings are also used for sketching.

3.10 SKETCHING INSTRUMENTS

Sketches are generally made with black lines from a lead pencil. The common wood pencil has a number 2 lead, which may be used to draw the layout lines simply by allowing the weight of the pencil to pass over the paper. Then during the darkening process, the pencil must be pushed harder against the paper to gain the proper blackness of the final sketch.

Sketches made using a lead holder are commonly drawn with 2H lead. Pencils with a harder lead may be used for the layout lines, and pencils with a softer lead may be used for shading or illustrating after the sketch has been outlined. Similar to wooden pencils, these must be sharpened with a pointer. The more contemporary lead pencils used in technical graphics are the fine-line mechanical pencils. The 0.3-mm fine-line pencil may be used for the construction lines, and a 0.5- or 0.7-mm pencil will likely be used to make the dark refined lines. Fine-line pencils do not require a pointer because the lines take the width of the diameter of the lead (Figures 3.11 and 3.12).

Colored pencils are not commonly used for sketching. One reason for not using colored pencils is that the purpose of the sketch is to quickly convey information, not to make a finished illustration. Also, some copy machines are "color blind" to cer-

Figure 3.11 Wood pencil and lead holders both require special pointers.

tain colors such as blue or red. However, if color is important for communication, such as for comments or corrections, the use of color should be considered on analytical sketches created by the designer. As the lines are darkened, remember that all the final lines should be dark so they are clear and copy well. The width of the lines on detail sketches should follow the accepted conventions for line widths. This means the object lines will be the widest, hidden lines will be of medium width, and lines such as center lines, dimension lines, extension lines, and so on, will be thin lines (Figure 3.10).

3.11 ENLARGING AND REDUCING USING A GRID

Several methods may be used to enlarge or reduce the size of a sketched image. One of the easiest methods of increasing or decreasing the size of the image, especially if the original was made on a grid, is to construct a larger or smaller grid and simply plot enough points to reconstruct the original image (Figure 3.13).

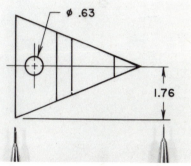

Figure 3.12 Fine-line pencils are manufactured in different sizes to produce given line widths and require no pointers.

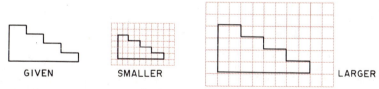

Figure 3.13 Given image presented on grid may be enlarged or reduced by simply altering size of grid.

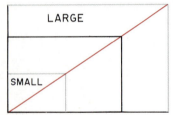

Figure 3.14 Rectangle may be enlarged or reduced by changing size of rectangle by using diagonal that is drawn through opposite corners of given rectangle.

3.12 *ENLARGING AND REDUCING WITH A RECTANGULAR DIAGONAL*

The rectangular diagonal is another useful aid for sketching. If an extended diagonal is drawn across opposite corners of a given rectangle, a proportionally larger or smaller rectangle can easily be constructed (Figure 3.14).

3.13 *ENLARGING AND REDUCING BY CHANGING THE FOCAL DISTANCE FROM THE OBJECT TO THE PLANE*

If you are sketching large buildings from a distance, many proportions may be transferred by holding the pencil at arms distance from the eye. This will help estimate the proportion of heights of buildings, bridges, or other similar large objects (Figure 3.15). The greater the focal distance between the object and the picture plane, the shorter the height will appear on the picture plane (Figure 3.16).

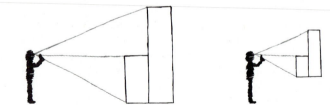

Figure 3.15 Proportion of large object may be proportionately reduced by holding pencil at arms length and sketching shortened heights accordingly.

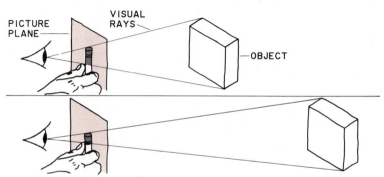

Figure 3.16 The further object is located from picture plane, the shorter the height will appear on picture plane.

3.14 SKETCHING SMALL OBJECTS WITH A HAND FOR PROPORTION

A very common sketching technique employed when showing small objects is to show them in relation to a hand. For instance, an object pictured by itself may not be representative of its actual size. However, the association of size may easily be illustrated by showing a hand holding the object. Other familiar objects may be used in place of the hand such as a pencil or a scale. For large objects, for example, tractors or buildings, it may be helpful to sketch a human figure next to the object to indicate proportion (Figures 3.17 and 3.18).

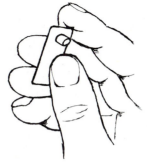

Figure 3.17 Relative proportion of small objects may be represented by showing them with a hand.

Figure 3.18 Proportion of large objects may be represented by showing them with a person.

3.15 SKETCHING SOLIDS

You must learn to sketch two- and three-dimensional shapes accurately. One approach to sketching a manufactured product is to assume that all the components making up the shape of the assembly are simply a collection of outlines of various basic shapes. These shapes are comprised of cubes, rectangular solids, cylinders, cones, and pyramids.

The major concern when drawing cubes or rectangular solids is that the sides must appear parallel in the sketch (vertical lines are always vertical even though the angular lines may vary) (Figure 3.19). The cylinder also requires lines to be drawn parallel to each other. This is most easily accomplished by sketching layout lines that define the two ends of the cylinder and then connecting the end surfaces (Figure 3.20). For the cone and pyramid, the center lines of the base should be sketched first, then the altitude,

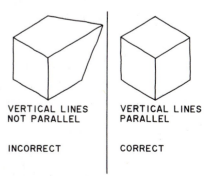

Figure 3.19 All vertical lines must be parallel or distorted view will result.

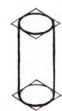

Figure 3.20 Define ends of cylinder, and connect ellipses with parallel lines to complete cylinder.

and lastly the sides (Figure 3.21). Using these sketching layout procedures, it is easy to sketch such objects as a sander and drill press (Figures 3.22 and 3.23).

A good exercise to improve sketching techniques is to sketch basic shapes over and over until you are comfortable with the steps involved in the process. You will also develop speed through this practice.

Figure 3.21 When sketching cone or pyramid, first construct center lines, then base and altitude, and finally sides of cone or pyramid.

Figure 3.22 Simple sketch of electric sander.

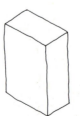
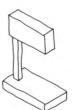
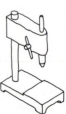

Figure 3.23 Making sketch: (1) block out outline, (2) block out major features, (3) refine shapes, and (4) add details and darken.

3.16 *FOUR STEPS TO SKETCHING AN OBJECT*

There are four major steps to follow when making a sketch of most objects or manufactured products. The first step is to *block out the outline* of the object or assembly to be sketched with light construction lines. This includes orienting the view. For instance, you may outline the object in isometric or perspective before drawing a detailed sketch. Proper orientation is very important in order to show the most descriptive view of the object in the front view.

The second step when sketching a product is to *block out the major features* of the product. This is generally accomplished by sketching light construction lines to outline the major shapes in the form of solids (cubes, rectangular solids, cylinders, cones, or pyramids) or variations of these and other solid shapes.

The third step is to *refine the shapes* already sketched. This step takes into account particular details such as rounded edges, sharp edges, finished and cast surfaces, and so on.

The fourth and last step is to *darken* only the layout lines you want to appear in the final sketch. This should be done with black lines, in accordance with the accepted line widths (Figures 3.23 and 3.24).

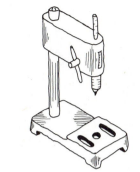

Figure 3.24 Sketch of drill press with shading.

3.17 *WHERE DO IDEAS COME FROM THAT REQUIRE SKETCHING?*

There are many reasons for making technical sketches. The major reason is to produce a quick graphic picture of an idea or object to convey to another person or group. One might ask: From where does the original idea or concept come? There are many sources from which you should be able to make a sketch of an object or product. The normal classroom procedure of sketching one or more views from a view given in a textbook is totally unrealistic when compared to the real world. After a sketch has been made using some of the proportional techniques discussed, additional information in the form of dimensions or notes may be added (Figure 3.25).

One source for the original image is from your mind. Another source is from a picture or technical drawing. This happens frequently in the classroom. Product modification is a good example of a real-world experience requiring a need for sketching. A third source is from a photograph of the object. Recall that translucent paper or film can be placed over the surface of the photo in order to make a refined sketch. Alterations can easily be made to the sketch of a photograph. Another source is a model or prototype

NOTES:

NO. REQ'D: 15,000
MATERIAL: PLASTIC
COLOR: RED

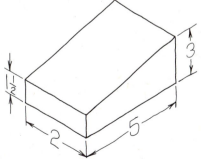

Figure 3.25 Dimensions and notes may be added to sketch.

Figure 3.26 Sketches are often made from verbal information given over phone.

of the object. Now you are asked to sketch from a three-dimensional object, which means you must visualize and be able to draw what you see. Making a sketch while looking at the object is essential. Sketches needed to make product modifications or alterations to products are very important. You should be able to make a better graphic image if the actual object is within sight during the sketching process.

Another method of receiving information from which a sketch is to be made is from a written description. This means you must visualize from a written description without the aid of any visuals such as the product, photos, or drawings.

Often the information required for a sketch is given by a verbal description. In the real world a person receives information over the telephone daily and often has to make a sketch representing this information. Unfortunately, this skill is rarely developed in the classroom. Some helpful tips include first orienting either top, bottom, left, or right (or north, south, east, or west). Also, have the person first briefly describe the entire object to be sketched before you even try to draw a line. This enables you to visualize the general concept of the object. Then have the person describe the parts by major components before refining the details. After you have sketched the item, take the time to describe your sketch back to the originator in order to check that you correctly sketched the features required. It would also be appropriate to send a copy of the sketch in the mail so that you and your correspondent are working from the same picture (Figure 3.26).

There are several descriptions provided at the end of the chapter for practice. These exercises require you to make sketches using many of the methods described in this chapter.

3.18 *BASIC SHADING OF SOLIDS*

Shading is the process of indicating surface texture, shape description, or depth by adding certain sketching patterns. Seldom will the drafter take the time to shade various parts being sketched. However, if such basic shading is necessary, the rectangular shapes should be shaded showing a light, medium, and dark surface. A cylinder will be darker on the upper and lower edges of the length of the axis view. A similar treatment is applied to the cones, except the shading gets narrower toward the point. On the pyramid, the two adjacent surfaces should be shown by the contrast between medium and dark shading. A suggestion for making the shading look more uniform is to dull the point of the pencil before

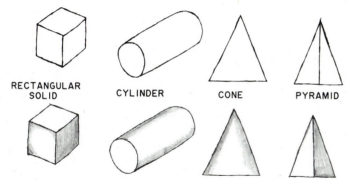

Figure 3.27   Shading techniques applied to basic solids.

trying to shade large surfaces. The development of shading technique, as all parts of the sketch, requires practice. Daily obser-
vation of objects in the sunlight or in light areas can aid in this
process (Figure 3.27).

3.19   *QUICK ORTHOGRAPHIC PROJECTION*

Sketches made in **orthographic** projection are generally the top,
front, and right side views of an object. However, the back, bottom, and left side views may have to be sketched occasionally.
The front view is the most critical. It should be the most characteristic view of the object. The front view indicates the **height**
and **width** dimensions. The right side view is projected directly
to the right of the front view and indicates the height and **depth**
of the object. The top view is projected directly above the front
view and indicates the dimensions for width and depth (Figure
3.28).

   Before sketching an object's front, top, and right side views,
first determine the best direction to view the front view. Then
sketch the front view (Figure 3.29).

   This is a sketch, so follow all four steps of the sketching process
already described. First, outline the extremes of the box, which
will include the entire front view, with light layout lines. Then
sketch a layout of the major features. Next sketch a layout of the
minor details. Lastly, darken the lines that should remain on the
final sketch remembering to employ the correct line widths. You
must develop these lines for all the required views. Both visible
and hidden surfaces and edges are represented when sketching
orthographic views.

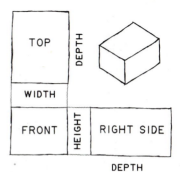

Figure 3.28   Position of commonly
shown views in orthographic
projection.

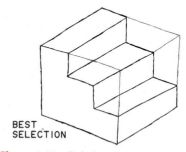

Figure 3.29   It is important to select
best position for front view.

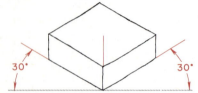

Figure 3.30 Isometric axes are vertical and 30° above horizontal (to left and right).

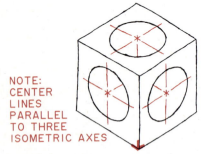

NOTE:
CENTER
LINES
PARALLEL
TO THREE
ISOMETRIC AXES

Figure 3.31 When sketching ellipses (circles shown on isometric planes), it is important to have ellipses oriented correctly. This is done by orienting center lines for each ellipse parallel to axes of that isometric surface.

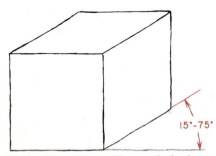

Figure 3.32 Depending on desired representation, receding axis of oblique sketch may vary from 15° to 75°. Most common receding angle for oblique sketching is 45°.

3.20 *QUICK ISOMETRIC PICTORIAL SKETCH*

The most common pictorial view to sketch is an **isometric** view. This is a sketch having a vertical axis and two axes at 30° from the horizontal. The axes shown are located at the near bottom corner of the object. When sketching, height dimensions are sketched along or parallel to the vertical axis, and the width and depth dimensions are sketched on or parallel to each of the two angular axes. Remember, the four steps still apply to making the sketch.

There are a few differences and similarities when drawing isometric views. Note that the parallel lines of rectangular surfaces are still parallel, but they are not 90° to each other. However, they are parallel to one of the three original axes lines (Figure 3.30).

The center lines for circles drawn in isometric are parallel to one of the three isometric axes. This means the circles will appear as ellipses. Note orientation of the long and short distances across the ellipses when they are drawn on the three isometric planes (Figure 3.31).

The four steps for making a sketch are still in effect. Sketch the outline box first using the extreme height, width, and depth. Then the major solids and features are sketched with layout lines using height, width, and depth. The details are then added. Lastly, darken (to black) the lines you want to appear in the final sketch. Only the visible lines are darkened when making an isometric pictorial sketch.

3.21 *QUICK OBLIQUE PICTORIAL SKETCH*

An oblique pictorial view is quickly drawn and shows one of the principal views (front, top, or side view) in the plane of the paper. This means two of the principal dimensions (height and width if the front view is drawn in the plane of the paper) will be sketched on the horizontal and vertical axes. The depth dimension will be sketched on an angular axis slanting away from the object at a selected angle (often 45°). When most of the detail is on a single surface, an oblique sketch can be effective, especially if most of the arcs or circles fall on this plane. In this case, the front surface is sketched as a front view; then the depth axis is added, at 45°. Examples of objects that could be appropriately sketched in oblique would be a set of kitchen cabinets, a desk, a speedometer, or a gauge. When sketching in oblique, the same four steps are applicable. It is common practice to darken only visible lines when sketching oblique pictorials. The oblique sketch may be easier when only rectangular grid paper is available. The normal surface can be sketched accurately, as can any selected receding axis (Figures 3.32–3.34).

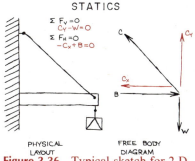

Figure 3.33 Oblique sketch of desk.

3.22 *SPECIFIC TECHNOLOGY APPLICATIONS*

A typical engineering problem follows. The problem is sketched on engineering problem paper, and the information on top of the page gives the class, the student's name, the date, and the problem number. The problem itself contains the given information, including a sketch, a description of what is required, the analysis with equations, some sketches and the solution (Figure 3.35).

Some examples of analytical sketches are illustrated. These represent the type of sketches you might be asked to produce in industry or on the job site (Figures 3.36–3.43).

Figure 3.34 Oblique sketches of some gauges.

Figure 3.35 Common layout used for stating and solving technical problems.

Figure 3.36 Typical sketch for 2-D statics problem.

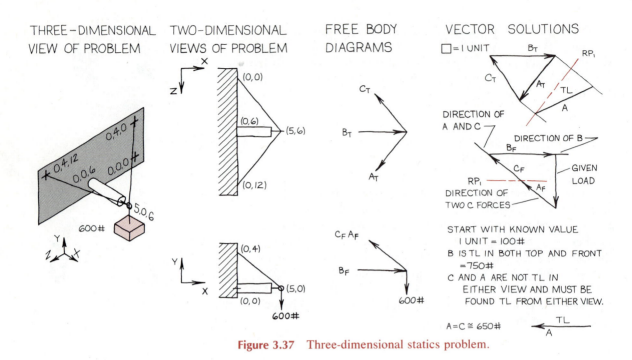

THREE-DIMENSIONAL VIEW OF PROBLEM

TWO-DIMENSIONAL VIEWS OF PROBLEM

FREE BODY DIAGRAMS

VECTOR SOLUTIONS

Figure 3.37 Three-dimensional statics problem.

STRENGTH OF MATERIALS

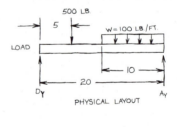

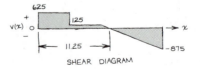

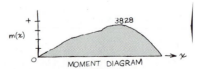

Figure 3.38 Sketch of typical technology application.

MACHINE DESIGN

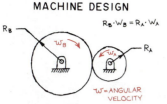

Figure 3.39 Sketch of typical mechanical technology application.

FLUID MECHANICS

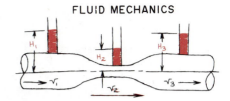

Figure 3.40 Mechanical, chemical or civil engineering technologists might use this type of sketch.

TYPICAL MICROSTRUCTURE

Figure 3.41 Metallurgical engineers sketch microstructures of materials.

CIRCUITS

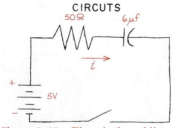

Figure 3.42 Electrical, welding, and industrial engineering technologists often sketch circuits.

DYNAMICS

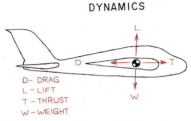

Figure 3.43 Aeronautical engineers typically sketch effects of forces acting on aircraft.

3.23 SAMPLE PROBLEM: TABLE SAW FIXTURE

The authors have developed a common example problem that will be used to apply the principles presented in this chapter through Chapter 13. With the exception of this chapter, all of the illustrations and applications will be computer generated.

These illustrations show how the computer may be used to generate different views of an object. Different software packages will be used to generate these views. These applications will be located just before the summary of each chapter.

In this chapter you learned to sketch graphic representations of ideas you developed. In this case, a fixture designed for use on a table saw when cutting equal-length pieces has been conceived. This fixture has a right angle, is strong, and has an elongated hole that will be used when fastening the fixture to the table of the saw. Figure 3.44 shows how the fixture will be used on the table of the saw, and Figure 3.45 is a sketch of the fixture. Sketching is the starting place for most computer drawings.

Figure 3.44 Table saw fixture in use.

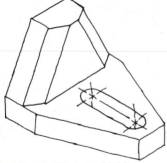

Figure 3.45 Sketch of table saw fixture prior to generating computer images.

3.24 *SUMMARY*

Sketched images generally show what the viewer visualizes rather than technical details. Often the engineer does not have time to make detailed engineering drawings so the sketches are given to a drafter who will make the technical drawings. Therefore, the designer must be able to sketch clearly and accurately. Also, the designer may be required to sketch in the field, on the job, or in a meeting in order to better communicate ideas or concepts to others. Sketching is an extremely useful skill when used to make quick graphic images to help enhance communication.

PROBLEMS

Problems 3.1–3.20
These problems are drawn in isometric on an isometric grid. You are
to sketch these objects in orthographic (draw the top, front, and right
side views) using ¼-in. square grid paper. Each space on the isometric
grid equals one square on your grid paper. Leave two spaces between
the views. *Note:* Problems 3.1–3.5 have only normal surfaces,
problems 3.6–3.10 also have inclined surfaces, problems 3.11–3.15
have oblique surfaces, and problems 3.16–3.20 have curved surfaces.

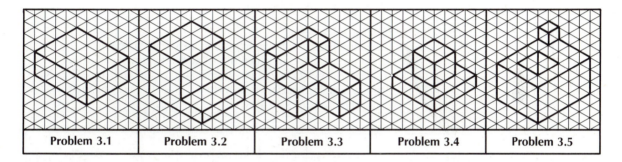

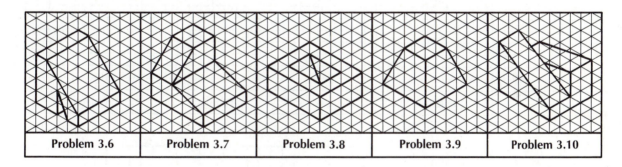

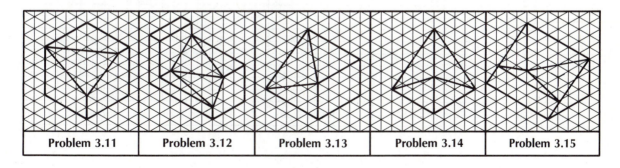

Problems 3.1–3.20 (continued)

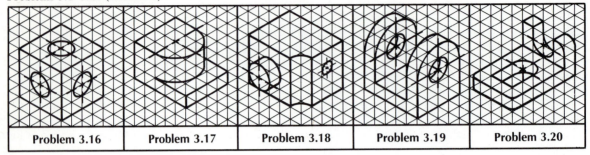

| Problem 3.16 | Problem 3.17 | Problem 3.18 | Problem 3.19 | Problem 3.20 |

Problems 3.21–3.40

These problems are drawn in orthographic on square grids. You are to
sketch these objects in isometric on isometric grid paper. *Note:*
Problems 3.21–3.25 have only normal surfaces, problems 3.26–3.30
also have inclined surfaces, problems 3.31–3.35 have oblique surfaces,
and problems 3.36–3.40 have curved surfaces.

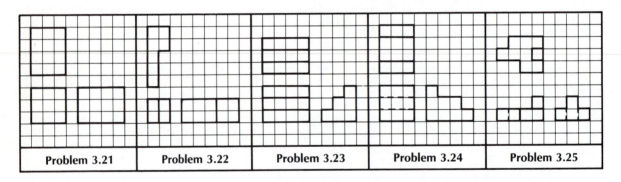

| Problem 3.21 | Problem 3.22 | Problem 3.23 | Problem 3.24 | Problem 3.25 |

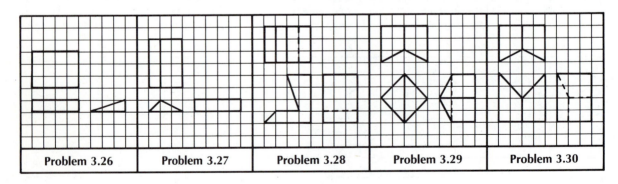

| Problem 3.26 | Problem 3.27 | Problem 3.28 | Problem 3.29 | Problem 3.30 |

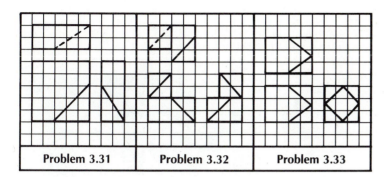

Problem 3.31   Problem 3.32   Problem 3.33

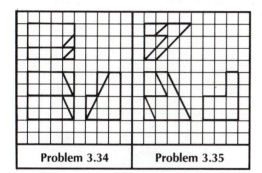

Problem 3.34   Problem 3.35

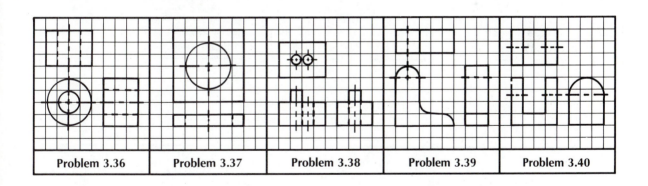

Problem 3.36   Problem 3.37   Problem 3.38   Problem 3.39   Problem 3.40

Problems 3.41–3.45
Sketch these photographed objects in isometric and add simple shading to the appropriate surfaces of each object.

| Problem 3.41 | Problem 3.42 | Problem 3.43 | Problem 3.44 | Problem 3.45 |

Problems 3.46–3.50
Sketch these photographed objects in isometric and add a sketch of your hand holding each of these objects.

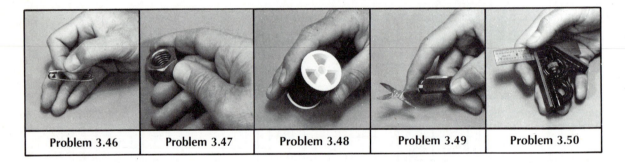

| Problem 3.46 | Problem 3.47 | Problem 3.48 | Problem 3.49 | Problem 3.50 |

Problems 3.51–3.55
Sketch these photographed objects in isometric.

| Problem 3.51 | Problem 3.52 | Problem 3.53 | Problem 3.54 | Problem 3.55 |

Problem 3.56

Sketch in isometric a cube (length of sides X) that has a cylinder extending from the center of each of the six faces of the cube. The axis of each of the cylinders is $2X$, and the diameter is $\frac{1}{2}X$.

Problem 3.57

Sketch in isometric a hand-held remote control channel selector for a television or VCR. This is to be a rectangular solid with a width of $4X$, depth of $2X$, and height of X. A rectangular window appears at the top, and three rows of three keys each are slightly raised from the top surface of the rectangular solid.

Problem 3.58

Sketch a cylindrical salt and pepper shaker in isometric. They are both the same diameter and have glass on the lower two-thirds. A shiny metal cap covers the upper third of each shaker. Sketch some holes in the top of each shaker. Both shakers are one-third full of salt or pepper. Sketch the salt shaker upright and the pepper shaker on its side.

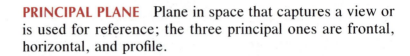

TERMS YOU WILL SEE IN
THIS CHAPTER

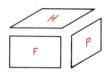

PRINCIPAL PLANE   Plane in space that captures a view or is used for reference; the three principal ones are frontal, horizontal, and profile.

LIMITING ELEMENT   Boundary line of a curved surface shown in the noncircular view.

INCLINED SURFACE   Plane surface not parallel to any of the principal planes but perpendicular to one of them, where it appears as an edge.

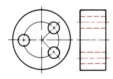

OBLIQUE SURFACE   Plane surface not parallel or perpendicular to any of the principal planes.

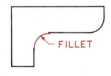

CONVENTION   Accepted practice that violates the strict rules of graphics in order to improve clarity.

FILLET   Interior rounded corner on an object.

ROUND   Exterior rounded corner on an object.

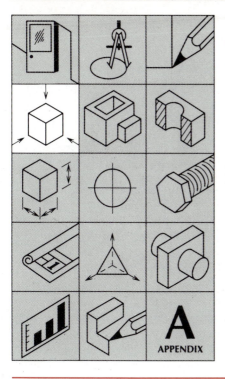

APPENDIX

4

ORTHOGRAPHIC PROJECTION

4.1 INTRODUCTION

Engineering graphics is the "language" of engineering and designing. It is through graphics that designers communicate ideas to engineers and other technical people. Just as English or Spanish has rules of grammar that must be followed, the graphics language also has rules. English and Spanish have exceptions to some rules to avoid confusion. This is also true of the graphics language.

4.1.1 Communicating in Three Dimensions

The world we live in is a three-dimensional world. All the objects we see every day have width, depth, and height. The world and all the objects in it are three-dimensional. The designer must know how to work and think in three-dimensional space.

How can we represent the width, height, and depth of objects on a two-dimensional medium such a sheet of paper or the computer screen? Over the centuries, designers have

Figure 4.1 Two common letters shown as pictorial sketches and in orthographic projection.

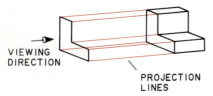

Figure 4.2 Orthographic projectors, straight and parallel.

developed a way in which objects can be represented on paper and now on the computer screen. This method is called orthographic projection. It is the international language used to communicate thoughts and ideas in the design of real-world objects to enhance all of our lives. Orthographic projection is truly international in that a designer in the United States can understand the shape description provided with orthographic projection methods by an Italian designer. They may not understand each other in normal conversation, but the drawing "speaks" a common language understood by all who have been trained in the techniques and procedures of orthographic projection (Figure 4.1).

The term *orthographic* comes from the Greek words *ortho* meaning straight and *graphos* meaning written. Orthographic projection means that the object is described using straight parallel projectors (Figure 4.2).

4.2 THEORY OF ORTHOGRAPHIC PROJECTION

There are several methods of projection. They all require an object, an observer, a picture plane, and projectors from the object to the observer.

Figure 4.3 shows an object in space with an observer at a known

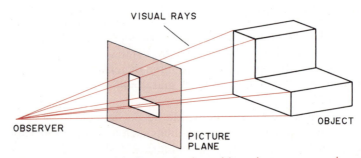

Figure 4.3 True perspective projection with projectors converging at observer's eye.

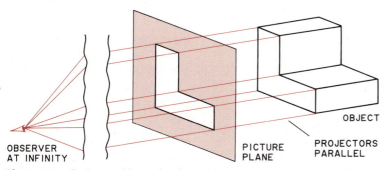

Figure 4.4 Orthographic projection with parallel projectors and observer at infinity.

distance from the object. There is also a picture plane between the observer and the object. Projection lines extend from the object through the picture plane to the observer. A projected image is formed on the picture plane. Since the observer is at a known or finite distance from the object, the projection lines meet at the eye of the observer, and the projected image on the picture plane is smaller than the actual object. This projected image is called a true perspective projection. The image on the picture plane is similar to a photograph of the object. The exact size and shape of the image are a function of

the distance between the object and the observer,

the angle between the object and the observer, and

the location of the picture plane.

A change in any one of the these factors will change the image. Forty people in a classroom looking at an object in the front of the room will each see it differently, for none of them will have exactly the same point of view.

Figure 4.4 shows the same object, picture plane, and observer with the observer located at an infinite distance from the object. The projection lines from the object through the picture plane going to the observer are now parallel. The image shown on the picture plane is not a perspective projection, but it is an orthographic projection of the object. The projectors, or "visual rays," as they are sometimes called, intersect the picture plane at right angles (perpendicular). The image on the picture plane is a one-to-one mapping of the points projected from the object. Every observer viewing this object from the front will be considered to be an infinite distance away. The image is now the same for all observers:

All are at the same distance from the object,

the angle of viewing is the same for all, and

the location of the picture plane does not affect the image.

Every observer viewing the object from the front will now have exactly the same image.

4.2.1 Third-Angle and First-Angle Projection

Orthographic projection uses a horizontal picture plane and a vertical picture plane that are perpendicular to each other to capture images. Figure 4.5 shows these planes in their proper orientation. Think of the planes as being infinitely large and intersecting each other so as to divide space into four areas. Each of these four areas is referred to as a quadrant. When the object is placed in the first quadrant, it is between the observer and the picture plane. Projectors from the observer pass to the object and then to the picture plane to create an image. This projection method is referred to as first-angle orthographic projection. It is used in Europe.

When the object is placed in the third quadrant, the picture plane is between the observer and the object. Projectors from the object pass through the picture plane to the observer and show an image on the picture plane. This projection method is referred to as third-angle orthographic projection. Third-angle orthographic projection is the standard in the United States and Canada (Figure 4.6).

Third-angle orthographic projection will be used throughout this book. The top view will always be above the front view, right side views will be to the right of the front view, and left side views will be to the left of the front view. A first-angle drawing will have the top view *below* the front view, the right side to the *left* of the front view, and the left side view to the *right* of the front view. If you encounter a drawing that appears to be first-

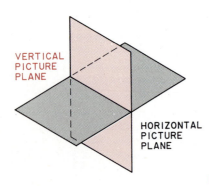

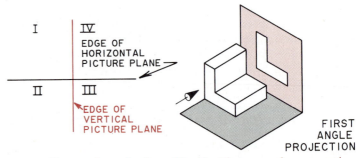

Figure 4.5 First-angle projection with object between observer and picture plane.

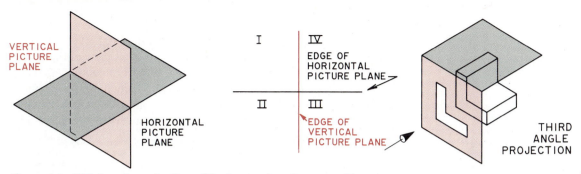

Figure 4.6 Third-angle projection with picture plane between object and observer.

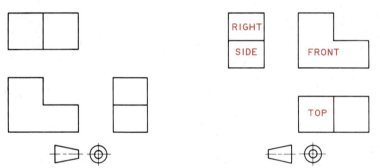

Figure 4.7 International symbol that identifies method of projection.

angle projection, check for the international symbol for projection (near the title block). Study Figure 4.7 and note the little symbol at the bottom of the drawing. When the small end of the cone frustum is to the right, it indicates third-angle projection; when the small end is to the left, it indicates first-angle projection.

4.3 *PRINCIPAL VIEWS AND PLANES*

When using third-angle orthographic projection, you should imagine that the picture plane is always between you and the object and the projectors pass through the picture plane from the object to you. The projectors "map" the points of the object on the picture plane.

4.3.1 Front View and the Frontal Plane

Figure 4.8 shows a picture plane in a position that gives an image of the object showing its characteristic shape. This view of the object is called the front view, and the picture plane on which the view is projected is called the frontal plane.

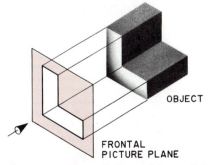

Figure 4.8 Frontal plane with parallel projectors capturing front view.

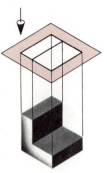

Figure 4.9   Horizontal plane with parallel projectors capturing top view.

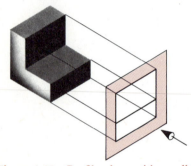

Figure 4.10   Profile plane with parallel projectors capturing side view.

4.3.2   Top View and the Horizontal Plane

Figure 4.9 shows a picture plane in a position that gives an image showing a "bird's eye" view of the object. This view of the object is called the horizontal view, and the picture plane is called the horizontal plane. Other names for this view include the top view and the plan view. Regardless of the name, the view is always looking down on the object; that is, the observer is imagined to be overhead with the horizontal plane between the object and the observer.

4.3.3   Side View and the Profile Plane

Figure 4.10 shows the picture plane in a position that gives an image of the object showing its depth. This view is called the side view, and the picture plane is called the profile plane. A side view may be drawn of the right side of the object or the left side of the object. Both views are considered to be side views.

4.3.4   Other Principal Views and Planes

Figures 4.11 and 4.12 show the picture plane in a position that gives an image of the object as viewed from below and from the rear, respectively. These views are called the bottom view, and the rear view and the picture planes are called the bottom plane and the rear plane. Notice that the bottom view and the rear view of the object are nearly the same as the top view and front view, respectively. The bottom and rear views are used only when there is interesting information on the bottom or rear of the object, which would be difficult to understand without these views.

The six views discussed in the preceding are called principal views. An object has six principal views: front, top, right side,

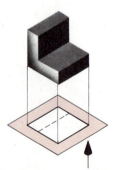

Figure 4.11   Another position for horizontal plane; underneath object to capture bottom view.

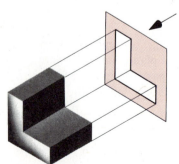

Figure 4.12   Rear view with picture plane and observer behind object.

left side, bottom, and rear. Only the views that best describe the object are used to make an orthographic drawing or sketch of the object.

4.3.5 Relation of the Principal Views

When the picture planes of the principal views are assembled in their proper position around the object, a projection box, or "glass box," is created. Each of the principal views is projected onto each of the **principal planes** (Figure 4.13). When the glass box is unfolded and laid flat on the paper, the orientation of the orthographic views can be seen (Figure 4.14). *This orientation is the only acceptable way to describe shapes using orthographic projection. Placing views of an object anywhere on the paper and labeling them is unacceptable,* as shown in Figure 4.15. The relationship of the views and the orientation of the views in proper orthographic position is as important to the person "reading" the drawing as the actual shape of each view.

Later in this chapter you will be introduced to views that are not principal views. They are called auxiliary views; they also follow all the principles of orthographic projection.

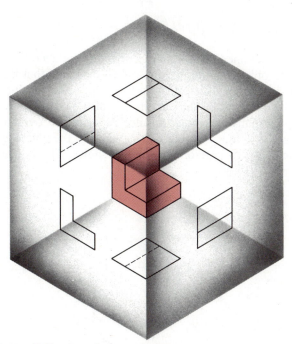

Figure 4.13 Collection of planes that forms projection box, or glass box.

Figure 4.14 Projection box opened to show relationship of orthographic views.

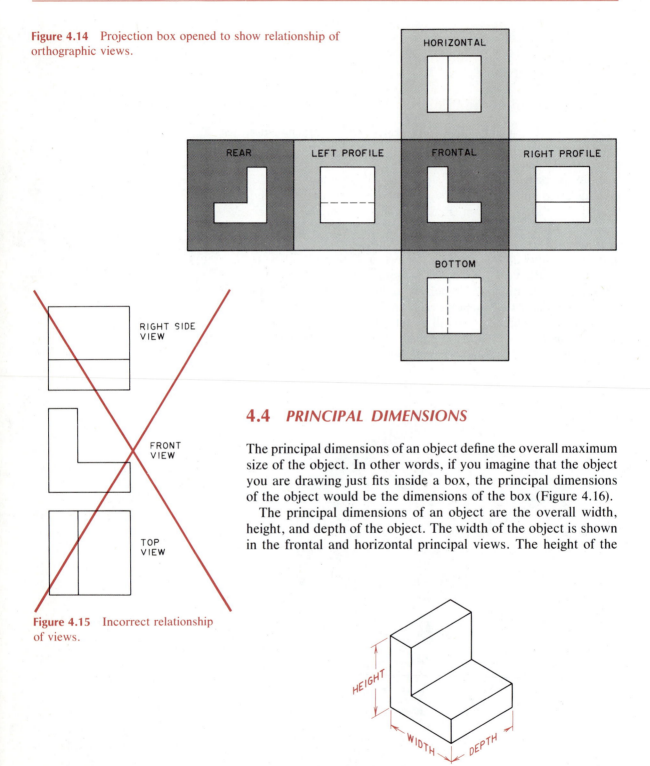

Figure 4.15 Incorrect relationship of views.

4.4 *PRINCIPAL DIMENSIONS*

The principal dimensions of an object define the overall maximum size of the object. In other words, if you imagine that the object you are drawing just fits inside a box, the principal dimensions of the object would be the dimensions of the box (Figure 4.16).

The principal dimensions of an object are the overall width, height, and depth of the object. The width of the object is shown in the frontal and horizontal principal views. The height of the

Figure 4.16 Principal dimensions that describe overall maximum size.

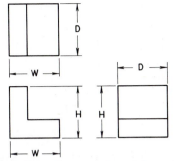

Figure 4.17 Principal dimensions on views where they appear.

object is shown in the frontal and profile views. The depth is shown in the horizontal and profile views. Each principal view shows two of the principal dimensions. All of the principal views together describe the three-dimensional shape of the object (Figure 4.17).

4.5 NONLINEAR FEATURES

Complex objects often have nonlinear features described in their shape. Circles and arcs are nonlinear features that need special attention when drawing them orthographically. Figure 4.18 shows the orthographic views of a cylinder. The top view shows the cylinder as a circle, and the front view shows the cylinder as a rectangle. Simply looking at the front view does not describe the object. Both views together define the object as a cylinder.

4.5.1 Center Lines

A significant feature shown in the orthographic views of a cylinder or any circular object is the center line. Figure 4.18 shows the correct way to show a center line in the circular view and in the longitudinal view (the view where the cylinder resembles a rectangle). The center line is a thin black line and is broken as shown in Figure 4.18. The view in which the cylinder appears as a circle has the center lines intersecting to form a cross. This cross is the point where the end of a compass would be positioned in order to draw the circle. The center line shown in the longitudinal view of the cylinder is simply the axis of the cylinder; it may be called the axial center line. A cross is not shown in this view.

Parts of cylinders or arcs are shown in a similar way. Center lines are required if half a circle or more is shown. A quarter of a circle does not require a center line. The center lines used with arcs are the same as the center lines used with cylinders (Figure 4.19).

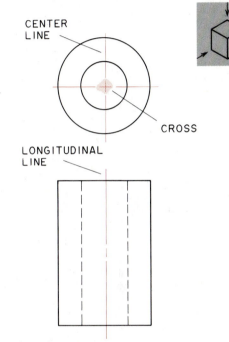

Figure 4.18 Center lines for cylindrical objects.

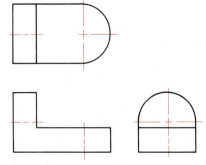

Figure 4.19 Center lines are needed for half a cylinder or more.

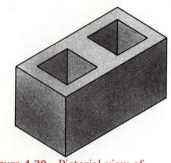

Figure 4.20 Pictorial view of concrete block.

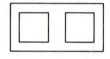

Figure 4.21 Orthographic views of concrete block.

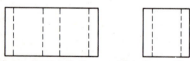

Figure 4.22 Orthographic views of concrete block showing hidden details.

4.6 *VISIBILITY*

Visibility refers to what you can see when you look at the object. You will be able to see some parts of the object; other parts will be hidden from your eyes. When you change your location, some parts of the object that were hidden will become visible and some parts that were visible will be hidden. All features that can be seen in a view are shown with solid lines. The outline or boundaries of the view will always be shown with solid lines. Other features inside the outline that are visible in that view will also be shown with solid lines. Look at a concrete block (Figure 4.20). It has two rectangular holes in it. You can see the shape of these holes in the top view. You cannot see the holes in the front or side views (Figure 4.21). The holes are still there; this set of views does not tell you everything about this concrete block.

4.6.1 Describing Hidden Features

The next set of views gives you more information (Figure 4.22). Here you see dashed lines in the front and side views, which represent the holes. Dashed lines are used to show features of an object that the observer should know about but are not visible in a given view. In this case, they give you more information about the block. You now know that the holes go all the way through the block; that information was not available from the top view. This example demonstrates an old saying in graphics: In orthographic projection, when you have one view of something, you have nothing. We need two or more views to describe a three-dimensional object completely in orthographic projection. Always remember that you are using a two-dimensional sheet of paper or computer screen to describe a three-dimensional object.

Now look at Figure 4.23. There is a hole shown in the top of the cylinder in the pictorial view. You do not know from this view if the hole goes all the way through or only part way through. (If it only goes part way through, you cannot see through the

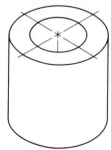

Figure 4.23 Pictorial view of hollow cylinder.

block. A hole that does not go all the way through is called a blind hole. A hole that does go all the way through is called a through hole.) The orthographic views without hidden lines (Figure 4.24) do not tell whether you have a blind hole or a through hole. The next figure (Figure 4.25) shows clearly that the hole goes only part way through the cylinder. The dashed lines here complete the description of the object.

4.7 THE ALPHABET OF LINES

All languages have alphabets. The English language has 26 letters in its alphabet. Russian, Japanese, and Arabic have much different alphabets. The graphic language of designers also has its own alphabet, to which you were introduced in Chapter 1. At this time you will need to use only four parts of this alphabet. These lines are shown in Figure 4.26. Another kind of line will be described when you learn about section views.

4.7.1 Visible Lines

Visible lines are the most important lines in our alphabet. They are drawn as solid black lines. Drawing standards specify that the lines be 0.032 in., or 0.7 mm, wide. Mechanical drawing pencils and pens are specified in metric sizes and are available in the 0.7-mm width.

4.7.2 Dashed Lines for Hidden Features

Dashed lines are used to represent hidden features. They are drawn as $\frac{1}{8}$-in.- (3-mm-) long dashes separated by $\frac{1}{16}$-in.- (1.5-mm-) long spaces. They may be shortened a little on short lines and lengthened a trifle on long lines. They should be solid black lines 0.022 in., or 0.5 mm, wide. Standard mechanical pencils and pens are available in a 0.5-mm size.

4.7.3 Junctions of Lines

There are established practices for the junctions of solid lines and dashed lines. Solid lines should always meet completely with no gaps. Dashed lines should meet other dashed lines at the dashes;

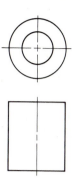

Figure 4.24 Orthographic views of cylinder.

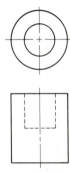

Figure 4.25 Orthographic views of cylinder showing hidden details.

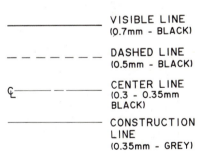

Figure 4.26 Partial alphabet of lines.

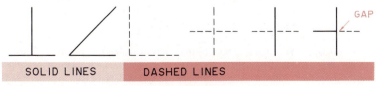

Figure 4.27 Line junctions for visible and dashed lines.

the junction should not be at an open space. Dashed lines should cross solid lines at a space if possible. When a dashed line is a continuation of a solid line, a gap should be left to show that these are different lines (Figure 4.27).

4.7.4 Center Lines

Center lines are used to show the center of circles and the axes of cylindrical objects. They are composed of $\frac{3}{4}$-in. (19-mm) dashes and $\frac{1}{8}$-in. (3-mm) dashes with a $\frac{1}{32}$-in. (1-mm) space between them. Center lines should be solid black lines .016 in., or .35 mm, wide.* Standard pens are available that are .35 mm wide; standard pencils are available with leads measuring .3 mm. The pencils require care when you use them; it is easy to break the fine leads. The symbol for center line is also shown on the alphabet of lines. Sometimes it is placed on a center line to call attention to it. The symbol may also be used as a shorthand way of writing center line in a document ($\mathcal{C}$).

4.7.5 Construction Lines

Construction lines are the lines you will use when you start a drawing and are planning the views graphically. They are made very light gray (just dark enough to be seen by the person making the drawing). They should be thin. You might use a 0.5- or a 0.3-mm mechanical pencil to draw them. The original engineering documents never leave the design office; only copies are sent to a shop or construction site. If your construction lines are properly drawn, they will never show on a copy of the drawing. There is no need to erase construction lines on an engineering drawing unless it is a presentation drawing that would be shown directly to a client to explain a project.

4.7.6 Precedence of Lines

The graphics alphabet has a feature we call precedence. Some lines are considered more important than others. If they occur in the same place, one type of line can cover up another. Visible lines are the most important and take precedence over, or cover up, hidden lines or center lines. Hidden lines take precedence

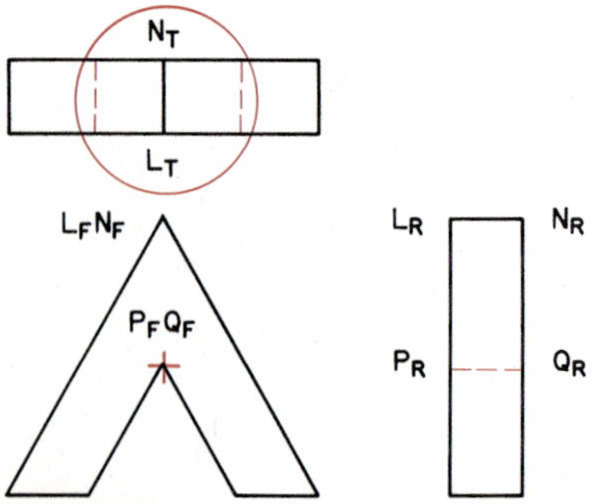

Figure 4.28 Inverted V-block; in top view visible line *NL* covers hidden line *PQ*.

---

*This text recommends the .022-in. width for hidden lines. The current American National Standards Institute (ANSI) standards specify only two widths of lines: thick, .032 in. (0.7 mm), and thin, .016 in. (.35 mm). Note that these widths are not direct conversions between measurement systems but are specified separately in the standards. Your authors believe that pens available for computer drawing and the pens and pencils available for manual graphics make it relatively easy to provide more line widths. The larger alphabet of lines makes it easier to understand the drawings.

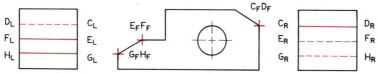

Figure 4.29   Visible lines cover hidden lines and center lines; hidden lines also cover center lines.

over center lines but not over visible lines. The views of the "inverted V" block in Figure 4.28 contain a line PQ that is the intersection of the two sloping sides inside the block. This line appears in the side view but not in the top view. It is coincident with line NL in the top view, and line NL, being visible, covers up line PQ.

The views in Figure 4.29 show some more examples of precedence. The horizontal center line of the circular hole is at the same elevation as line EF, so in the side views line EF is visible and covers up the center line of the circle. Note that it is desirable to show the ends of the center line extending beyond the boundaries of the side view in line with the side view of EF. Line GH is the intersection of two surfaces in the front view; it appears as a solid line in the left side view. The boundary of the circular hole is in line with GH and would appear as a dashed line in the side view. The solid line GH takes precedence, and you see the solid line GH but not the hidden boundary line of the circular hole. Line CD is coincident with the upper boundary line of the circular hole. Both would appear as dashed lines in the left side view, so one dashed line is shown in that view. In the right side view line CD would be visible, and the upper boundary line would be covered by the solid line $C_R D_R$. Line EF is dashed in the right side view to represent the hidden line; it takes precedence over the center line in this view.

Construction lines are always covered by visible, dashed, or center lines for they are on the drawing only for your convenience in building the views of the object. Hopefully, there will be no conflict on your drawings, and the construction lines will all disappear when the drawings are copied.

4.8   USING SUBSCRIPTS

The last time line CD was described, it was shown with subscripts $(C_R D_R)$. This is standard practice when describing points on an engineering drawing. Points in the front view are labeled F, in the top view T, in the right side view R, and in the left side view L. When you label points, add the subscripts for each view in which the point is shown.

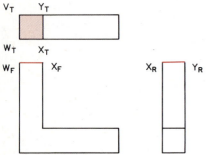

Figure 4.30 Line in space connecting two points.

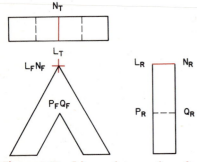

Figure 4.31 Line as edge view of plane surface.

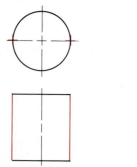

Figure 4.32 Line as intersection of plane surfaces.

4.9 WHAT A LINE REPRESENTS

As you look at a drawing, you will learn to ''read'' it just as you learned to read written language. A line on a drawing can represent several different features in our three-dimensional world. Look at Figure 4.30, and you will see points *A* and *B* connected by a line. This line would be called line *AB*. A line is the shortest distance between two points. You will use lines to connect points when you learn about descriptive geometry later in this book.

4.9.1 A Line as the Edge View of a Surface

A line can represent the edge view of a plane surface. Look at Figure 4.31, and you will see the flat, horizontal surface on the top of the L-shaped block. The corners are named *V*, *W*, *X*, and *Y*. In the front view there is a visible line $W_F X_F$. This line describes the flat, horizontal surface in the front view. The visible line $X_R R_R$ describes the flat, horizontal surface in the right side view.

4.9.2 A Line as an Intersection

A line can also represent the intersection of two surfaces. Look at Figure 4.32, and you will see the two **inclined surfaces** on the outside of the block meeting at the top. They meet in line *LN*. This line is the intersection of the two plane surfaces and appears as line $L_T N_T$ in the top view and as line $L_R N_R$ in the right side view. Sometimes a line can represent more than one feature. Look at Figure 4.31 again; the line $X_R Y_R$ represents the edge view of the plane *VWXY* in the right side view, and it also represents the intersection of plane *VWXY* with the vertical plane surface on the rising leg of the L-block.

4.9.3 A Line as a Limiting Element

A line can also represent the **limiting element** of a curved surface. Look at the cylinder in Figure 4.33. Trace the outline of the circular top view with your finger or a pencil point. As you go from the front of the cylinder to the rear, you go farther to the right (or left) until you reach a maximum distance from the center and then your pointer starts going back in the other direction. The point of change is the outermost point on the boundary of the circle. In the three-dimensional cylinder the outermost vertical line on the surface is called the limiting element. It represents the outermost vertical line, or element, that can be seen in the noncircular view. The colored vertical lines are the limiting elements in the front view. They are the outermost boundaries of this view of the cylinder.

Figure 4.33 Line as limiting element of curved surface.

4.9.4   What Can a Line Represent?

You have seen that a line on a drawing can represent several different ideas. It can be a line (a two-dimensional object). It can be the edge view of a plane surface. It can be the intersection of two surfaces, or it can be the limiting element of a curved surface. These possibilities are summarized in Figure 4.34. When you see a line on an orthographic drawing, it has to mean one or more of the following:

1. a line in space,
2. the edge view of a plane surface,
3. the intersection of two surfaces, and/or
4. the limiting element of a curved surface.

These are the only things a line can represent!

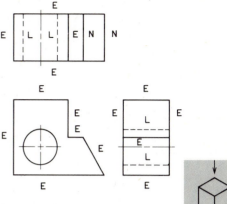

E - EDGE VIEW OF A SURFACE

N - INTERSECTION OF TWO SURFACES

L - LIMITING ELEMENT OF A CURVED
    SURFACE

Figure 4.34   What can a line represent?

4.10   *WHAT A POINT REPRESENTS*

A point on a drawing can represent a single point in space (Figure 4.35). It can also represent the end view of a line. Look at Figure 4.36. The right side view of line AB is $B_R A_R$, a point that is the end view of the line. An end view of a line is sometimes called the "point view." You can see this on a three-dimensional object in Figure 4.37. The lines LN and PQ, which appear as lines in the top and side views, are points in the front view. Notice that line PQ is not visible in the top view. It is covered by line $L_T N_T$, which is visible in this view, and by the rules of precedence, the visible line covers the hidden line. Look at $P_R Q_R$ in the side view. It is shown as a dashed line for it is hidden. The dashed line can be seen for there is no visible line to cover it.

Figure 4.35   Point in space shown by small cross.

Figure 4.36   End view or point view of line.

4.10.1   A Point as the Intersection of Lines

A point can also represent the intersection of lines. Look at Figure 4.38. Lines RS and ST intersect at point S. This point appears in all three views. If you look carefully, you will see that a vertical line on the object also passes through point S. So a point can represent the intersection of two or more lines.

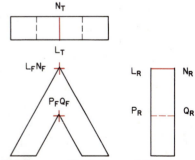

Figure 4.37   End views of lines on object.

4.10.2   What Can a Point Represent?

A point can represent three things: a point alone, the end view of a line, or the intersection of two or more lines. Every point on an orthographic drawing represents one or more of these three things.

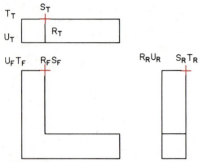

Figure 4.38 Intersection points on object.

4.11 *CREATING A MULTIVIEW ORTHOGRAPHIC DRAWING*

Engineering drawings should have enough views to provide all the information needed for fabrication or construction. For three-dimensional objects this means that more than one view is required. For example, when making a bolt, the designer must specify the length of the bolt and the shape of the bolt head (Figure 4.39). In a construction project there are normally a plan view and elevation views of a building to be built (Figure 4.40). Whenever you look at an object, you must think about which view of the object will provide the most information. An automobile is normally viewed from the side. This is the view that gives the

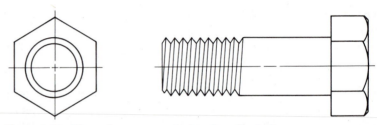

Figure 4.39 Two views are needed to describe bolt.

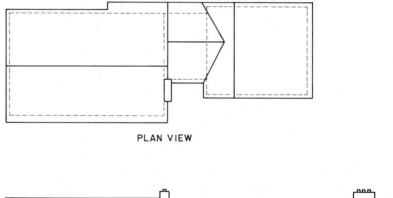

PLAN VIEW

FRONT
ELEVATION

SIDE
ELEVATION

Figure 4.40 Three views describe outside of building. (Courtesy of Timothy C. Whiting)

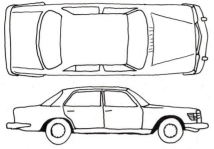

Figure 4.41   Three views describe automobile.

most information about the intended use of the vehicle. It is easy to see the number of doors and windows and determine whether it is a two-door sedan, four-door sedan, station wagon, or convertible. If all of these vehicles were viewed from the front, they would all appear to be the same. A view from the rear or the top of the car would help distinguish between the convertible, the station wagon, and the sedan but would not identify them as completely as the view from the side does (Figure 4.41).

4.11.1   Choice of Views

You should choose the most descriptive or characteristic view and use it as the front view. It is also good practice to select the front view so that the longest dimension of the object appears as the *width* in the front view. The other views required should then be chosen. These will usually be right or left side views and top or bottom views. A simple cylindrical object such as a can requires two views: one to show the circular shape and the other to show the height (Figure 4.42).

Figure 4.42   Cylinder can be described with two views.

4.11.2   Planning a Drawing

Before you start a drawing, you should plan how it is going to look and how it will fit on the paper. This is best done by sketching the views to be drawn and deciding how they will fit on the paper (Figure 4.43). This sketch may be made on scrap paper or the back of an old drawing. The object in Figure 4.43 is best described in a set of views that are wider than they are high, so the paper would be turned with the long dimension horizontal. If the dimensions of the views were as shown, the views alone would require 14 in. in width and 9 in. in height, and space is required between the views. What size paper do you have on which to draw? If it is 11 × 17 in., you can probably fit the drawing on the

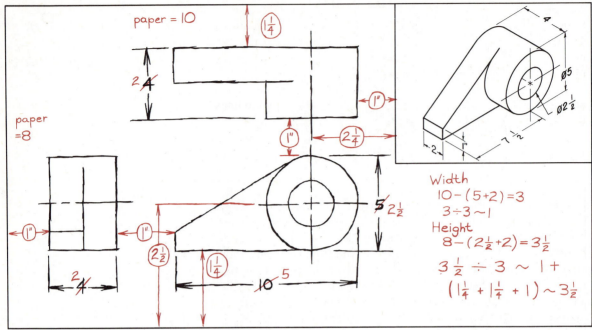

Figure 4.43 Planning drawing takes quick sketch of views and some simple calculations.

paper. What do you do if your paper is only $8\frac{1}{2} \times 11$ in.? The drawing will not fit the paper so it must be drawn less than full size (drawn to scale). Often-used scales are half size and quarter size—which is the best choice? Drawn half, the width becomes 5 in. on the paper, the height becomes $2\frac{1}{2}$ in., and the depth would be 2 in. (note the numbers in color). Now will it fit on $8\frac{1}{2} \times 11$-in. paper? It appears that it will, and now we need to plan for the space between the views. An $8\frac{1}{2} \times 11$-in. sheet will have about 8×10 in. of usable space. Subtracting the space required for views from the total available space leaves 3 in. of width and $3\frac{1}{2}$ in. of height. You could distribute this space as shown by the circled numbers on the sketch. *Only now are you ready to make a drawing; the planning or layout step is important.*

4.11.3 Developing the Drawing

Tape your paper to the drawing surface and set up your straight edges for horizontal and vertical lines. If you plan to use a border line, draw it lightly first. Now use your circled dimensions to locate the principal dimensions of the object. Draw the box that will contain each view with light construction lines and put in the center lines for circular features (Figure 4.44*a*). Add light construction lines for major features on each view; some features cannot be drawn directly on a view but must be projected from another view (e.g., tangent points in noncircular views must be projected from the tangent found in the circular view).

Using your compass or circle template, draw the circles and arcs that appear in each view (Figure 4.44*b*). Then draw the lines tangent to the circular arcs. Find the tangent points by construct-

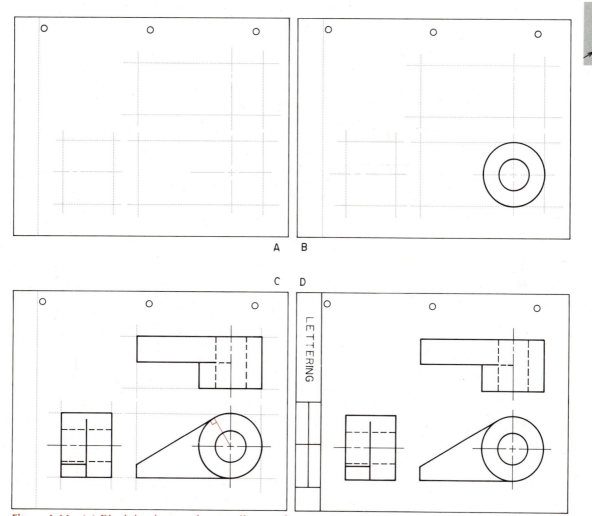

Figure 4.44 (*a*) Block in views and center lines as first step in making drawing. (*b*) Construct circles and arcs before adding straight lines. (*c*) Add visible and hidden lines. (*d*) Finish drawing with lettering and notes.

ing a perpendicular from the center of the arc to the tangent line. (It is much easier to fit a line to an arc than to try to adjust the center of an arc to fit a line.) Now check projections for any features not shown with construction lines. Add them. Start down the paper making thick horizontal lines for visible lines; then make thinner dashed lines for hidden features. Work across the paper making thick vertical lines and thin dashed lines. Draw any lines that are not horizontal or vertical. Check your center lines and draw them as thin lines with long and short dashes (Figure 4.44*c*). If you have a border line, now is the time to make it a wide, black line. The last task is to complete your title block and notes with neat lettering (Figure 4.44*d*).

4.11.4 Checking Your Work

Your drawing should be complete. Check it carefully by locating each feature in every view and making sure the features are in projection. In a work environment you would make a copy of the drawing to be checked by a more experienced drafter; this is usually called a "check print." The experienced checker examines the copy of your drawing and notes in red any needed corrections. When those changes are made, the drawing is rechecked before any copies leave the design room.

4.12 CREATING A DRAWING ON THE COMPUTER SCREEN

You will need a computer terminal or personal computer that has graphics capability and a software package, known as a computer-aided design and drafting (CADD) program designed to produce technical graphics. There are many software programs available for engineering graphics. Some are intended for student use; these programs are fairly simple and will lead you through the steps in making a drawing much as you would draw with pencil and paper. Other programs were designed for professional drafters; these programs have many options and great flexibility. They take longer to learn to use but offer advantages in speed and features when you take the time to master them. In order to print your work, you will need a printer capable of graphics or a plotter primarily intended for graphics but that can also plot words or numbers.

4.12.1 Selecting a Grid

When using a CADD program to create a three-view drawing, first determine the grid size to be used. If the dimensions are to be in inches, normally a .125-, .25-, or .50-in. grid is used. If you are working in millimeters, you would usually select a 10-mm grid. Then select one of the grid dots on the screen as the starting point, or origin (0.0,0.0). This is usually the first visible dot in the lower left corner of the screen. When you have chosen the dot and reset the origin, it is a good idea to mark that dot so you can easily find the origin again. This can be done with two crossed lines or a small polygon.

4.12.2 Selecting the Layers

Once the grid has been selected, it is helpful to decide how many layers will be needed for the drawing. A CADD program allows you to draw on many different layers. You can view a single layer or many layers at one time. Simple drawings may be plotted on one layer. A simple plan for more complex drawings is to put construction lines on one layer and all the visible lines on a second layer; hidden lines may be drawn on a third layer. This allows you to keep the construction lines but gives you the option of omitting them from the printed or plotted drawing.

4.12.3 Selecting Pen Widths and Colors for a Plotter

You or your instructor will need to select the plotting colors, line styles, and line weights that will be used for a drawing. Normally a single color, such as dark blue or black, would be chosen to be used with the various line weights. This primary color can be represented on the screen by white if you are working on a monochrome CADD system. If colors are available, you should choose those that can be copied on a color copier.

Examples of colors, assuming a black background, that might be used and their accompanying line weights are:

Monochrome (Optional Weight)	Color	Use Application	Thickness (mm)
Solid/pattern (medium)	Green	Reference planes/lines, hidden lines, lettering	0.5
Solid (thin)	Red	Center, dimension, and extension lines	0.3
Solid (thick)	Blue	Visible lines	0.7
Dot	Yellow	Construction lines	0.3

4.12.4 Developing the Drawing on the Screen

After you have selected the origin, layers, line weights, and colors, you can start with the construction lines that define the location and overall size of the view on your construction layer. First, draw one horizontal line all the way across the screen to establish a Y coordinate for the lower limit of the front and profile views. Then repeat this line at the appropriate Y coordinates to establish the height of these views and the front and back of the top or bottom views.

Construction Lines

Construction is best done with a line type or color that will not be used for the finished drawing. You might choose a dot line pattern or the color yellow as outlined in the preceding. Yellow shows up well on a black monitor screen but would not be easy to read on white plotting paper. Save the high-visibility colors (black, blue, red, and green) for lines that must appear on the finished drawing.

It is efficient to locate the centers of circles and arcs on your construction layer (Figure 4.45a). Block in all of your views and then switch to your visible line layer for circles, arcs, and lines. When you locate tangent points, it is a good idea to bring them back to the construction layer. They are construction information and should not appear on the finished drawing.

Visible Lines

Circles and arcs should be drawn before the straight lines that are tangent to them (Figure 4.45b). This allows lines of the correct length to be added to the drawing. Otherwise, you may find yourself putting in lines followed by arcs and circles and then having to replace the original lines. Most CADD programs provide a continuous readout of the cursor position and display the length of a line or radius once the initial point or center has been selected. These features will help you locate the tangent points and draw the tangent lines. If the program you are using has a zoom feature, you can increase the scale of the tangent areas and locate the tangents accurately. At this time you should add any detailed construction that you could not determine before (Figure 4.45c) and carry it back to your construction layer.

Finishing the Drawing

The next step is to add remaining visible and hidden lines. Your last operation is to switch to a lettering or text mode and add the title block information and notes to your drawing (Figure 4.45d). You may choose to put text on the same layer as visible lines or reserve a layer just for text and notes.

Alternate Methods for Layering

If you decide to use more than one layer for the finished drawing, the following plan may be used. The first, or zero, layer can be

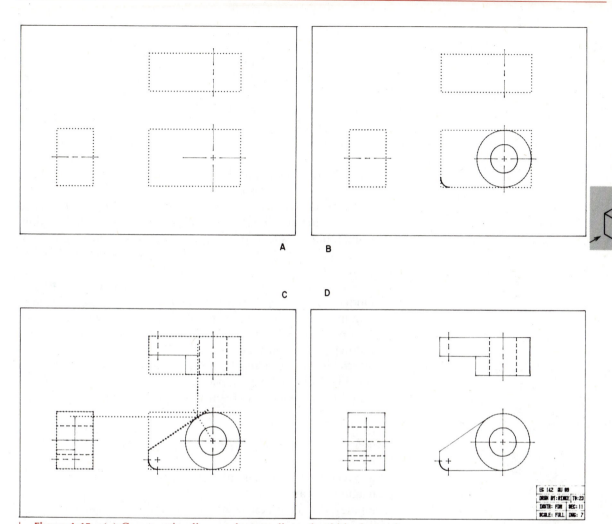

A B

C D

Figure 4.45 (a) Construction lines and center lines should be located on construction layer. (b) Construct circles and arcs on visible layer before adding straight lines. (c) Add visible and hidden lines; locate any tangent points. (d) Lettering and text complete drawing, usually on separate layer. Drawing can be displayed or printed without construction lines. (*Computer-generated image*)

used for construction. The second can be used for the visible lines. The third can be used for hidden lines and text material, and the fourth can be used for center lines, dimension lines, leaders, and extension lines. This plan allows a single pen per layer for the finished drawing exclusive of the construction, but it does assume that one plotting color with multiple line widths is available.

Another plan you may find efficient puts the construction on the first layer, the visible and hidden lines on the second layer, the center lines on the third layer, and the leaders, dimension, extension lines, and text on the fourth layer. This allows you to concentrate on the actual drawing on layers 2 and 3 and to add the size description (dimensioning) later or as needed.

If the steps outlined in the preceding have been followed, you should be able to tell the CADD system to plot the second and third layers or second, third, and fourth layers. This omits the first, or zeroth, layer where you did the construction so there is very little cleanup required. If lines have been drawn an incorrect length, CADD systems normally require you to erase the original line and replace it with a new line of the correct length. Some systems allow the line to be "shortened" or "lengthened." When using multiple layers, you should check your drawing a layer at a time to ensure that you have plotted the correct lines on the correct layer.

Use of layers is important for many types of construction or fabrication drawings. For example, an architectural/engineering firm may plan a building by putting the basic plan for each floor on one or two layers. They can add a layer for each of the following functions: heating and air conditioning, plumbing, electrical, and furniture layout. This technique allows them to provide the basic building plan to each specialty contractor with only the information that contractor needs.

4.13   *READING SURFACES OF OBJECTS*

You have read about what lines and points represent on a drawing and about creating an orthographic drawing manually or on a computer. As you practice with this information, you will become skilled at making drawings. You will also be called upon to read or interpret drawings in classes and in your professional career. One important part of reading orthographic projections is being able to recognize various types of surfaces.

4.13.1   Surfaces That Are True Size and Shape

The object in the next series should be an old friend. It is the L-shaped block you have seen before. Now all of the points on the surface of the object are labeled in all views. You can name every point, every line, and every plane surface on the object. Look at planes *VWXY* and *EFGH* in the top view (Figure 4.46). Do they appear in true size? Yes, they do because they are parallel to the horizontal plane. All the lines in the plane are horizontal lines. They are all true length in the top view, and the whole surface is

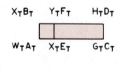

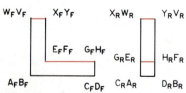

Figure 4.46 Horizontal surfaces in true size on top view.

true size. How do these horizontal surfaces appear in the front and side views? They appear as edges. You see a single line that is the edge view of the entire plane surface. *If a plane surface appears in true size in one view, it is parallel to the picture plane of that view and it will appear as an edge in adjacent views.*

Now examine the L-shaped plane *WXEGCA* (Figure 4.47). It is true size in the front view; it appears as an edge in the top view and in the right side view. Planes *GHCD* and *XYFE* are true size in the side view (Figure 4.48); they appear as edges in the top and front views. As you read drawings in the future, look for the surfaces that are parallel to the principal planes. These surfaces will be in true size in one view and will appear as edges in adjacent views.

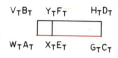

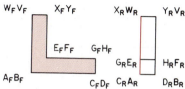

Figure 4.47 Frontal surface in true size on front view.

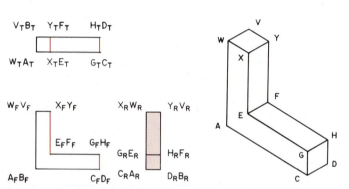

Figure 4.48 Profile surfaces in true size on side view.

4.13.2 Surfaces That Are Inclined

Figure 4.49 is another familiar shape, the inverted V. This time all of its points are named, so you can identify every point, line, and surface on the object. Surface *LNYX* appears as an edge in the front view, but it is neither an edge nor true size in adjacent views. It does appear as a rectangle in both the top and right side views. Plane surface *LYNX* is an inclined surface. It appears as an edge in one view and shows its characteristic shape in adjacent views. *It is not in true size in these views.* Now examine the views of the F-block (Figure 4.50). The inclined surface appears as an edge in the side view. You see the F shape in the front view and again in the top view. The characteristic shape is not in true size in either view, but you can identify the F in both views.

If a line represents the edge view of a plane surface, that surface will appear in true size in an adjacent view and as an edge again in the other principal view or that surface will appear in its characteristic shape in the other principal views.

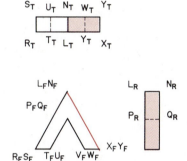

Figure 4.49 Inclined surface that is not true size in any principal view.

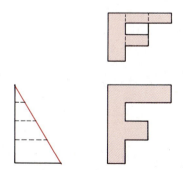

Figure 4.50 An F-block with inclined surface that shows its characteristic shape in two views and appears as edge in other view.

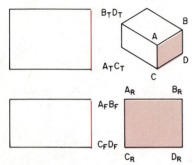

Figure 4.51 Rectangular prism with colored surface parallel to principal plane. Which plane?

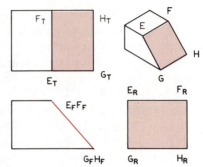

Figure 4.52 Prism with inclined colored surface.

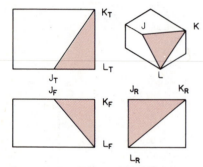

Figure 4.53 Prism with oblique colored surface.

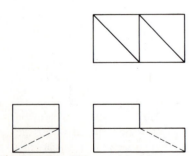

Figure 4.54 Object to sketch.

4.13.3 Surfaces That Are Oblique

Look at Figure 4.51. First, you see a rectangular prism that has a vertical plane on its right end. This surface, named *ABCD*, shows up as an edge in the top and front views. Because it is parallel to the profile plane, it shows up in true size in the right side view. In Figure 4.52 the right end of the block has been cut. Plane *EFGH* shows up as an edge in the front view and as a rectangle, but not true size, in the top and front views. Plane *EFGH* is an inclined surface. In Figure 4.53 the right end has again been cut. The cut is made so that no edge appears. Instead, we see the cut face in each of the principal views. Plane *JKL* is an **oblique surface** for it does not appear as an edge in any of the principal views. It does show up as a triangular shape in all views—none are in true size.

4.13.4 Reading by Sketching

The small pictorial sketches should help you to understand what has happened to the block each time. When you are reading orthographic drawings, it will help you to understand the object if you also make sketches of it. Start with the box that would contain the object and add lines to the sketch as you read each line of the drawing. You may try several positions for lines on the sketch before you get agreement with all the orthographic views. This is a part of the reading and visualizing process. Your eraser becomes as useful or more useful than your pencil when reading and sketching. Make a pictorial sketch of the object in Figure 4.54. If you start sketching it as the views are oriented in orthographic projection, you will find that the features at the right end of the object will not be visible in the sketch (Figure 4.55). At this time you would start a new sketch oriented similar to Figure 4.56. In this pictorial you can see what has happened to the right end of the block and still see most of the left end. When you brighten the lines of the sketch, you get Figure 4.57, which shows more of the object than the sketch started in Figure 4.55. Study this object. Can you find surfaces that are true size in the orthographic views? Can you find an inclined surface? An oblique surface?

4.13.5   Curved and Tangent Surfaces

Some plane surfaces can also be identified as tangent to curved surfaces. Look at the wedge in Figure 4.58. The upper and lower surfaces can be described as being inclined and horizontal. They can also be described as being tangent to the cylindrical portion of the wedge. When a plane surface is tangent to a curved surface, you can locate the point of tangency in the circular view by drawing a perpendicular to the edge view of the plane from the center of the curve. This method will locate the intersection or tangent point for you, but *no line exists at this point.* The plane surface joins smoothly with the curved surface without a line of intersection. Look again at the top and side views of the wedge. There are no lines in these views in projection with the tangent point shown in the front view.

4.13.6   Conventional Practices

Figure 4.59 shows another kind of intersection with a curved surface. The vertical cylindrical part of the object meets a horizontal rectangular prism. The prism is as wide as the diameter of the cylinder, and the points of tangency can be seen in the top view where the vertical center line touches the visible edges of the prism. In the front view the edges of the prism end at the points of tangency. There are no lines of intersection; the contours blend smoothly. Now look at the small triangular part that joins the cylinder and the prism. It is not tangent, and you can see a visible line of intersection, $C_R D_R$, in the side view. You also see the intersection of the inclined face of the triangular block with the horizontal prism in the top view, $E_T F_T$. Note the colored line on the front view. Think about what you have learned so far about orthographic projection. If the line CB were projected carefully from the top view to the front view, it would appear in the position of the colored line. It is drawn in black in projection with the limiting element of the cylinder. Is this a mistake? No, it is a **convention.** A convention is an irregular practice in the graphic language just as some verb forms are irregular in languages. Nothing would be gained by drawing the line in true projection in the front view, and it would take extra time of the drafter. Everyone

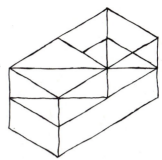

Figure 4.55   First try: box that contains object and some of the lines illustrated.

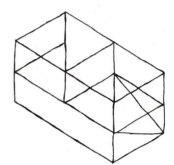

Figure 4.56   Second try: Turned around, it is easier to read.

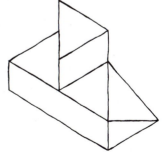

Figure 4.57   Completed sketch with all features visible.

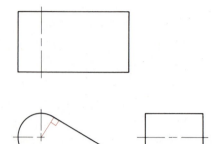

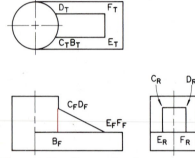

Figure 4.58 Wedge with curved and tangent surfaces.

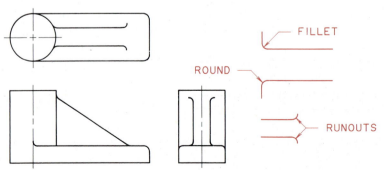

Figure 4.59 Conventional treatment of rib intersecting cylinder.

familiar with the graphic language understands this convention, so you should use it when you make drawings.

4.13.7 Fillets, Rounds, and Runouts

The small triangular piece on the object in Figure 4.59 is called a rib. It is used to add strength to the object and to help maintain the vertical position of the cylinder. Sharp intersections on real objects are avoided when possible because forces that can cause fracture build up at sharp corners. If you were going to use this object in a machine, you could make it stronger by rounding the intersections of the rib with the cylinder and the prism. You would add **fillets** to these junctions—additional material that is rounded to make a smooth transition from one surface to another. When you do this, the views will change. Study Figure 4.60 and compare it with Figure 4.59. The sharp corners *CD* and *DE* are gone. Instead, you find smooth curves, fillets, joining the surfaces. The lines of intersection disappear in the top and side views. Instead, the parallel edges of the rib now end in runouts, which show that you have added fillets in the other view where you can see the curves. Another common practice used to reduce breakage is to round sharp outside corners. When you round these corners, the result is called a **round.** Note that in the new, improved version of the object one corner of the prism has been rounded. (The other corners have been left square because they will need to match with the surfaces of some other parts.)

4.14 *AUXILIARY VIEWS*

Many objects have surfaces that are not parallel to any of the principal planes. When you describe the shape of these objects

Figure 4.60 Fillets, rounds, and runouts found on real objects make them stronger.

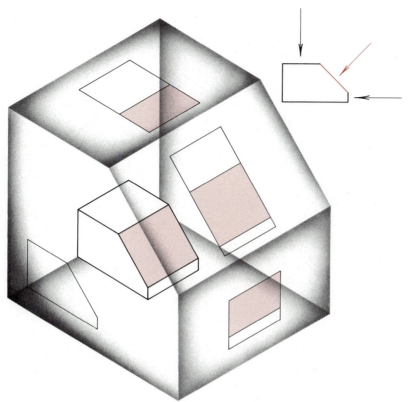

Figure 4.61 Projection box with inclined plane that captures image of inclined face of object in true size.

with principal views, the surfaces not parallel to the principal planes will be distorted. They will not be in true size or shape. Some of the dimensions of the surface will appear shorter than they really are; they will not be in true length.

We can obtain a true-size view of such surfaces by assuming a point of view so that our line of sight and projectors are perpendicular or normal to the surface. The image can be captured on a picture plane, which is not one of the principal planes. It will be parallel to the surface and perpendicular to the projectors. Such planes are called auxiliary planes. An infinite number of auxiliary planes are possible; in orthographic projection new planes must always be installed perpendicular to one of the existing planes.

Figure 4.61 shows an object with an inclined face in a projection box. Three of the planes are familiar to you. They are the principal planes: frontal, horizontal, and profile. One new plane has been added to the collection of planes, an auxiliary plane parallel to

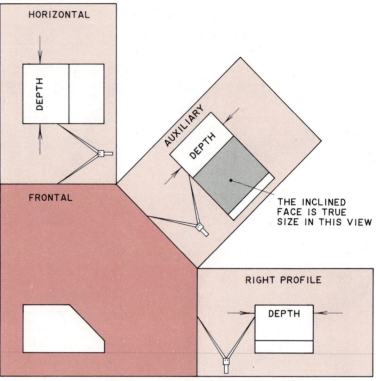

HORIZONTAL

DEPTH

AUXILIARY

DEPTH

THE INCLINED
FACE IS TRUE
SIZE IN THIS VIEW

FRONTAL

RIGHT PROFILE

DEPTH

Figure 4.62   Projection box opened to show relationship of auxiliary view to other orthographic views.

the inclined surface and perpendicular to the frontal plane. This plane is perpendicular to the line of sight and to the projectors that are normal to the inclined face. In the front view of the object you see the inclined face as an edge. You will also see the auxiliary plane as an edge in the front view, *parallel to the inclined face of the object.* Compare this picture to Figure 4.13, which introduced the projection box.

When the collection of planes in three-dimensional space is opened up and laid flat on a two-dimensional surface, you see an arrangement such as Figure 4.62. The top view is still above and in projection with the front view. The side view is to the side of and in projection with the front view. The new auxiliary view is projected perpendicular to the edge view as the projection box opens up along the hinge line with the front view and the auxiliary plane lays out flat. Note that the front edge of the object is the same distance away from the hinge line in the auxiliary view as it is in the top and side views. The depth of the object is the same in all views. Compare the relationships in this view to Figure 4.14.

4.14.1 Preparing an Auxiliary View Drawing

Drawings with auxiliary views follow the same principles as orthographic drawings with principal views. The projectors are perpendicular to the picture planes, and the observer is assumed to be at an infinite distance, so the projectors are all parallel for any one view. The picture planes for adjacent views (those views in projection with each other) are perpendicular.

Figure 4.63 shows the construction of an auxiliary view. You need at least two views to establish the size and location in space. In Figure 4.63 the surface we want to describe in true size appears as an edge in the front view, so we assume a line of sight per-

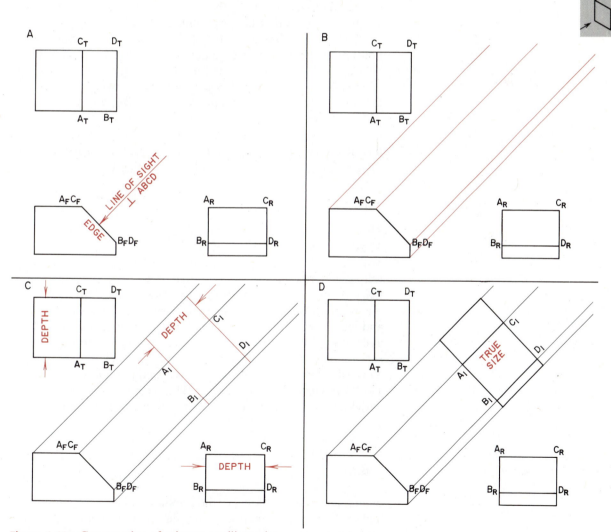

Figure 4.63   Construction of primary auxiliary view.

pendicular to this edge view. All the projectors for this auxiliary view will be parallel to this line of sight. The picture plane for the auxiliary view will be perpendicular to these projectors and parallel to the edge view of the surface we want to describe.

We select a convenient location for the front of the object in the auxiliary view just as you select a location for the top or side views at some convenient distance from the front view. Perpendicular to the projectors, we establish the line that represents the front of the object. Using dividers, we establish the rear of the object the same distance behind the front as it measured in the top or side views. This is the depth of the object. All points are located, and the visible lines are drawn as thick, black lines.

The relationships among the views are as shown in Figure 4.62, where the projection box was unfolded. When you prepare an auxiliary view, allow enough distance from the view from which you are projecting so that you do not overlap an existing view, such as the top or side view in this problem. A complete auxiliary view is not always needed; a partial view, showing only the surfaces or features that now appear in true size, will be adequate if other information is complete in the principal views. Auxiliary views may be projected from any existing view. You construct the view by projecting from an existing view and measuring from a view already in projection with the existing view.

4.14.2 Other Auxiliary Views

The views discussed in the preceding are primary auxiliary views. They are projected directly from principal views. We can also project auxiliary views from other auxiliary views as long as we follow the principles of orthographic projection. Successive auxiliary views are named in order as they are removed from principal views: primary, secondary, and tertiary. We can construct more and more views further removed from the principal views. However, it is difficult to maintain accuracy as we try to keep many sets of projectors parallel and to make perfect measurements. You will learn more about successive auxiliary views in Chapter 11, "3-D Geometry Concepts." This chapter will provide you with knowledge for solving problems in three-dimensional space.

4.15 *KEYS TO READING THREE-DIMENSIONAL OBJECTS*

Seeing a two-dimensional orthographic drawing with your eyes and reading this information in your mind as the three-dimensional object is a skill. People begin the study of graphics with different levels of this skill. Some people seem to see a three-dimensional

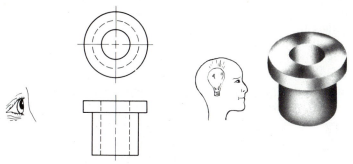

Figure 4.64 Visualization. Eye collects information; mind processes it to form three-dimensional image.

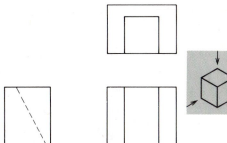

object immediately when they look at an orthographic drawing (Figure 4.64). Others (most of us) have to practice this skill as they would any other, by starting with simple tasks and progressing to more difficult ones. The goal is to look at the multiview drawing and visualize the three-dimensional object in the mind. If this visualization does not occur immediately, there are methods and steps for organizing and interpreting the information that will lead you to the correct visualization of the object.

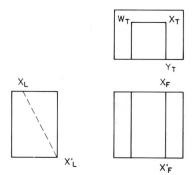

Figure 4.65 Can this object be visualized in three dimensions?

4.15.1 Identify and Locate Points on the Object

The first method is to identify *points* on the object and locate those points in all views. Look at the object in Figure 4.65. Perhaps you can visualize it right away, perhaps not. In either case, go through the steps of organizing the information in the given views. Start by naming points and locating them in all views. In Figure 4.66 point X is named in the top view. It is the intersection of lines WX and XY. Point X in the front view must be in projection with point X_T. It could be at X_F or at X'_F. Now look at the left side view. If you project from the front view, point X could be at X_L or at X'_L. Check the top view: X_T is not on the front of the object. It is about two-thirds of the distance from the front toward the rear. So X' cannot be the correct location for point X in the front or side views. You can name several points on the object and locate them by projection and measurement in the other views. Any locations that do not agree in all three views cannot be correct.

Figure 4.66 Read object by naming points and locating them in all views.

4.15.2 Identify and Locate Lines on the Object

A second method is to identify *lines* on the object and locate those lines in all views. Check Figure 4.67 and locate line XY in the top view. Line XY in the front view could be the point view $X_F Y_F$ or the line $X_F Y'_F$. Now look at the side view. Point Y is on the front

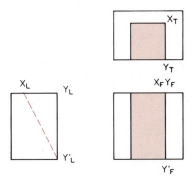

Figure 4.67 Read object by naming lines and locating them in all views.

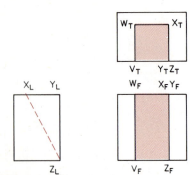

Figure 4.68 Read object by identifying planes (coloring them works well) and locating them in all views.

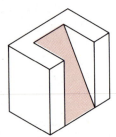

Figure 4.69 Object in three-dimensional sketch with inclined plane in color.

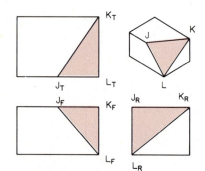

Figure 4.70 Shape repetition can help identify oblique surface.

of the object and line XY could be the horizontal line $X_L Y_L$ or the hidden, inclined line $X_L Y'_L$. Which is it? Either line could be the solution, for both points Y and Y' are in projection. How can that be? There are *two* lines that meet the requirements of the solution—there are *two* lines on the drawing, XY and XY'. In this example you followed a line on the object and found that you were locating two lines.

4.15.3 Identify and Locate Surfaces on the Object

A third method is to identify *planes* on the object and locate those planes in all views. Check Figures 4.67 and 4.68: point Y' has been renamed point Z and two other points on the colored plane named W and V, similar to points X and Z. These identify the surface you see in color in each view. Start with this surface in the top view. It resembles a small rectangle within the visible outline of the top view. Project this surface to the front view. It must lie between the two vertical lines inside the view, and it must include points below the top, horizontal surface of the object. If the colored plane were all on the top, horizontal surface, it would be one big surface because line XY and the other visible lines in the small rectangle would not exist. A line on a view of a three-dimensional object can represent only the edge view of a plane surface, the intersection of two surfaces, or the limiting element of a curved surface. Since the colored plane dips below the top plane, it must go all the way to the bottom. There are no breaks or lines between the top and the bottom. When you look at the side view, the only plane surface that could include points $VWXZ$ is the line XZ. This must be the edge view of the plane surface. (If you include point Y, you would have more than one surface or a warped, nonplanar surface.) In this drawing, line XZ in the side view represents the edge view of an inclined plane surface. This surface appears in its characteristic rectangular shape in both the top and front views. You can now identify more points in the side view, W_L and V_L.

You have seen Figure 4.70 before. It is an object with an oblique surface JKL. Note that JKL is a triangle in every view; look for such shape repetition to locate oblique as well as inclined surfaces on an object. Following plane surfaces on the views is a powerful method for reading a drawing. As you identify planes, you identify several points and lines at the same time.

4.15.4 Curved Surfaces and Limiting Elements

When you have curved objects, there is another type of line you can identify. This is the limiting element of the curved surface.

Examine Figure 4.71. The colored line, line *AB*, appears in the front view. It is a limiting element of the cylinder. This line appears as a point in the top view. The line segment $A_F B_F$ continues, and you see a line $C_F D_F$ in the front view that resembles an extension of *AB*. It is not. These are two different lines. Line $C_F D_F$ is the edge view of plane surface *CDEF*. Now check line *GH*. It is a limiting element of the cylinder in the side view and appears as a point in the top view. Where is *GH* in the front view? Where is *AB* in the side view? These lines do not appear in all views; they are elements of the cylinder and become visible lines only when they are the *limiting* elements or boundary lines of a view. Be careful when working with curved objects: The limiting elements are visible lines only in views where they describe the boundaries of the object.

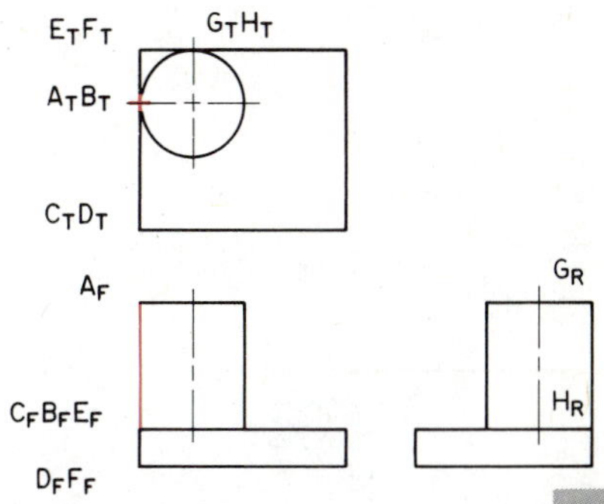

Figure 4.71 Limiting elements of cylinder; not same lines in every view.

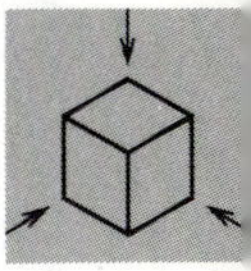

4.15.5 Reading an Object by Sketching It

Another way to help visualize an object is to sketch a pictorial view of it. This method was described earlier in this chapter. It takes very little time and almost always helps you to understand a drawing. Begin by sketching lightly the pictorial "box" that would contain the principal dimensions of the object: width, height, and depth. Then sketch in parts of the object and remove those areas that are not a part of the finished object.

4.15.6 Reading by Modeling the Object

If you have used all of the methods described in the preceding and still have difficulty visualizing an object, try modeling it in three dimensions. You can use plastic foam, soft wood, or modeling clay. The clay is recommended because you can use it over and over again. The principles are the same as sketching. Form a three-dimensional box that would contain the object, lightly scribe lines on the surfaces of the box to match lines on the drawing, and remove the pieces that are not a part of the object. If the object is mostly cylindrical, you can start with a cylindrical shape or join a cylindrical shape to another box. A pocket knife or putty knife and a pointed stick are the only tools you need to make readable models (Figure 4.72). Do not worry if your corners are not sharp and your plane surfaces are not perfectly flat. These models should be aids in understanding; they need not be works of art (Figure 4.73).

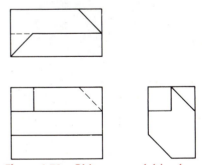

Figure 4.72 Object to model in clay, soap, or foam.

Modeling on the Computer

You can also model the object on a computer. When you have learned to use your computer to make pictorial drawings, you can "sketch" on the computer screen. Some programs will pro-

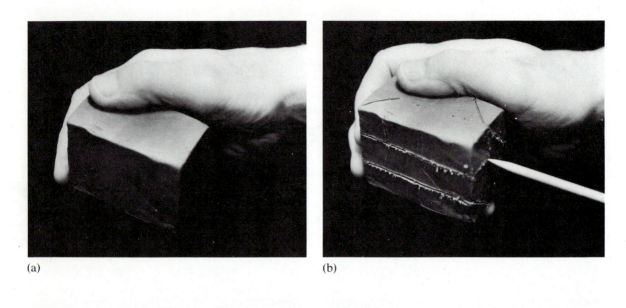

(a)   (b)

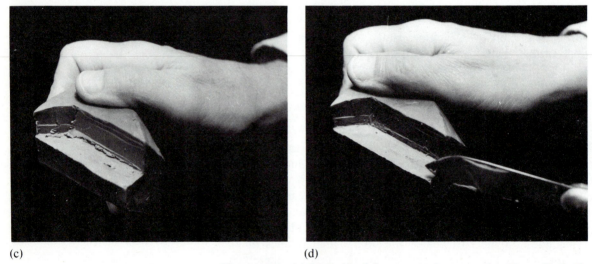

(c)   (d)

Figure 4.73 (*a*) Make block large enough to contain object. (*b*) Scribe lines from views on faces of block. (*c*) Cut along lines so that shapes on all faces remain same. (*d*) Clean up corners and surfaces with knife or pointed stick.

vide a pictorial grid so that you can easily construct the box that will contain the dimensions of width, height, and depth. As in manual sketching, you will want to place those lines on the pictorial drawing that match the multiview drawing and agree with all of the given views. You can erase on the computer screen and try lines in a new position. It is a good idea to layer your pictorial drawing. The first layer could be the pictorial grid, the

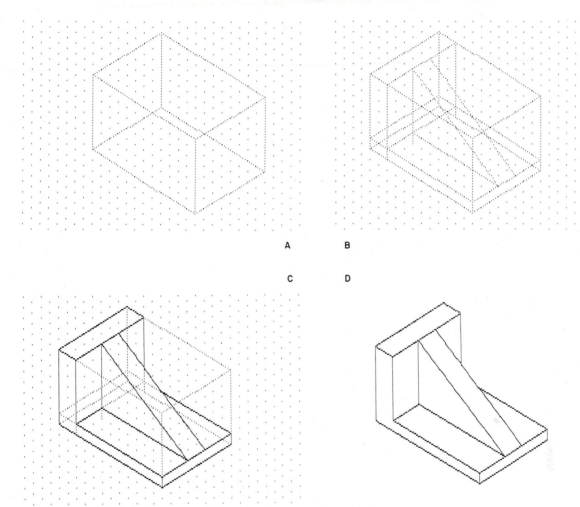

Figure 4.74 (*a*) Select grid for first layer and build box that contains object. (*b*) Use construction-type lines to rough out shape. (*c*) Switch to visible layer and add solid lines where should be; check agreement with views. (*d*) Turn off grid layer. You now have model of object on computer screen. (*Computer-generated image*)

second layer the box that would contain the object and the trial lines, and the third layer the finished object (Figure 4.74).

4.16 SAMPLE PROBLEM: TABLE SAW FIXTURE

Most CADD software packages allow you to develop orthographic views. Therefore, drawing the three views of this object in orthographic projection should be easy using the computer.

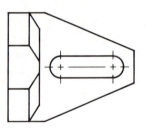

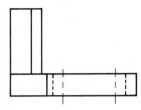
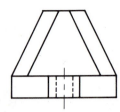

Figure 4.75 Orthographic views of the table saw fixture produced using a CADD package.

Figure 4.75 illustrates what the three-view orthographic drawings of the table saw fixture look like when produced using a CADD package. When using simple software packages, all lines in each view must be drawn individually. Projections from view to view are easier if GRID and/or SNAP features are available. A large modeling package may be able to produce the views directly in one step.

Some of the more powerful CADD software packages will automatically develop a data base that records the coordinates of each point. This data base can be used to draw the third view after the first two have been drawn. Other CADD packages allow you to put in features in three-dimensional coordinates and then choose the orientation for the orthographic views. The example in this chapter was created in this mode.

As you will see in following chapters, this data base can be used to produce many different types of drawings. The ability to manipulate the data base is one of the great advantages of using the more powerful CADD software packages.

4.17 *SUMMARY*

Orthographic projection is a fundamental part of the graphic language that all engineers and design professionals need to know. It is used to communicate with other technical people and with production and construction people. It is truly an international language.

Remember that the observer is an infinite distance away from the object; this makes all lines of sight (projectors) parallel. Projectors are all parallel to each other and perpendicular to the picture plane. The principal views include the front, top, and side; they may also include bottom and rear views. Auxiliary views may be needed to describe objects with surfaces not parallel to principal planes. Auxiliary views are always made with a line of sight perpendicular to the plane of the surface to be described. The new auxiliary picture plane will always be perpendicular to the line of sight and to one of the existing picture planes. It will be parallel to the surface to be described.

Visible features are shown with solid lines and hidden features with dashed lines. Center lines are shown for cylindrical objects if half a circle or more appears in one view. You should learn to read lines just as you learned to read the alphabet.

Creating a multiview drawing requires planning which views will best describe the object and how they will be positioned on the screen or paper. The overall dimensions of width, height, and depth should be blocked in as you start the drawing. All views should be developed together. Often, one view cannot be completed until intersections and other important information are plotted in another view.

Both reading drawings and making them require practice. Keys to understanding include visualization in three dimensions; identifying points, lines, and surfaces; and sketching in three dimensions with a pencil or on a computer screen. Modeling in clay or other material is also helpful. The problems at the end of this chapter will give you an opportunity to practice drawing and reading in orthographic projection.

PROBLEMS

Problems 4.1–4.12

Sketch the given views in each drawing. Sketch the lines that are missing in any of the views.

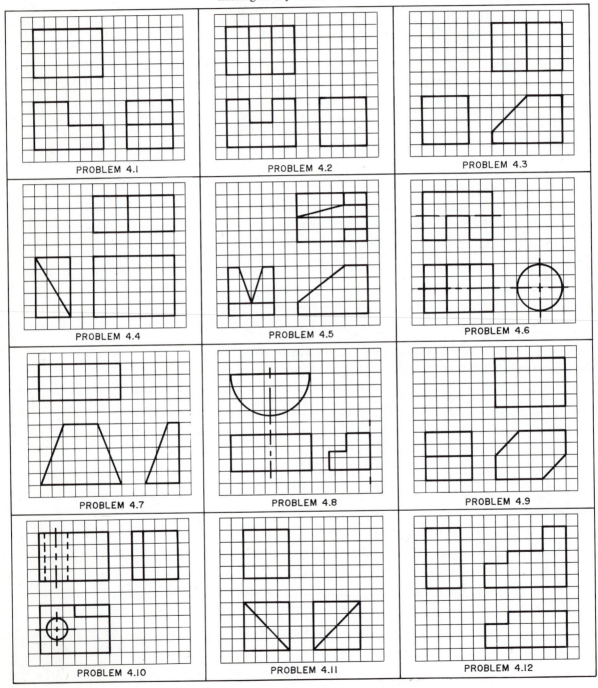

PROBLEM 4.1

PROBLEM 4.2

PROBLEM 4.3

PROBLEM 4.4

PROBLEM 4.5

PROBLEM 4.6

PROBLEM 4.7

PROBLEM 4.8

PROBLEM 4.9

PROBLEM 4.10

PROBLEM 4.11

PROBLEM 4.12

Problems 4.13–4.24
Sketch the given views in each drawing. Sketch the missing view as it
relates to the given views.

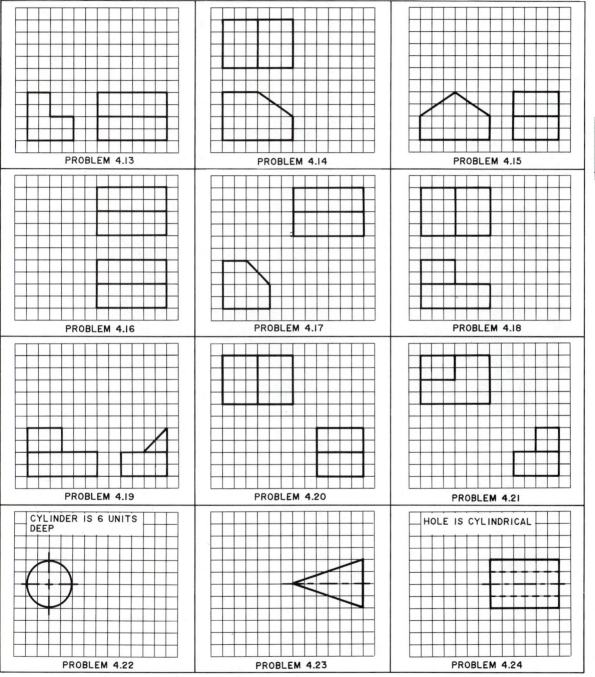

PROBLEM 4.13

PROBLEM 4.14

PROBLEM 4.15

PROBLEM 4.16

PROBLEM 4.17

PROBLEM 4.18

PROBLEM 4.19

PROBLEM 4.20

PROBLEM 4.21

CYLINDER IS 6 UNITS DEEP

PROBLEM 4.22

PROBLEM 4.23

HOLE IS CYLINDRICAL

PROBLEM 4.24

Problems 4.25–4.27
Read the orthographic drawings and sketch a pictorial of the object
defined.

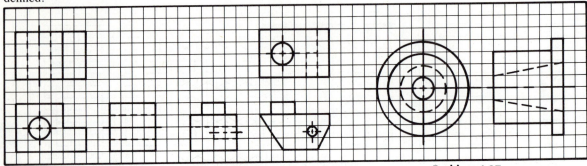

Problem 4.25   Problem 4.26   Problem 4.27

Problems 4.28–4.30
Sketch the top, front, and left side view of each problem.

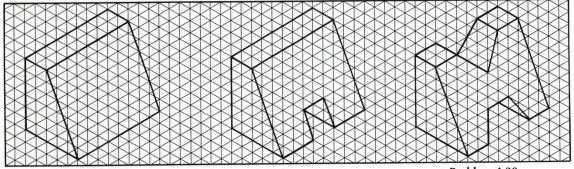

Problem 4.28   Problem 4.29   Problem 4.30

Problems 4.31–4.32
Sketch a top, front, and side view of the object shown. Identify with
color or code on each view the surfaces that are normal, inclined, and
oblique.

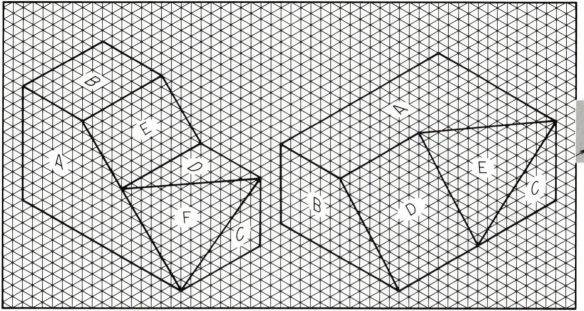

Problem 4.31   Problem 4.32

Problems 4.33–4.41

Draw with instruments the given views. Complete the missing view as it relates to the given views.

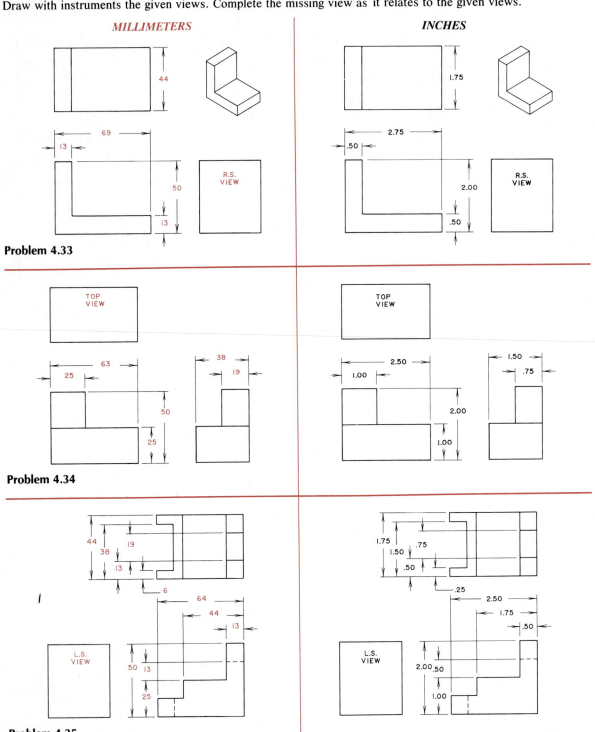

MILLIMETERS   *INCHES*

Problem 4.33

Problem 4.34

Problem 4.35

MILLIMETERS

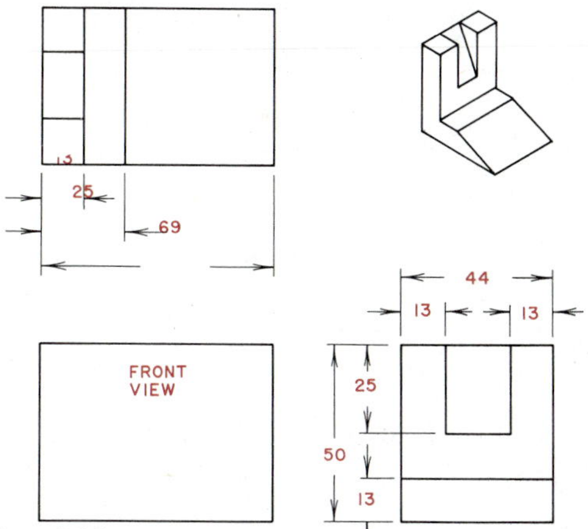

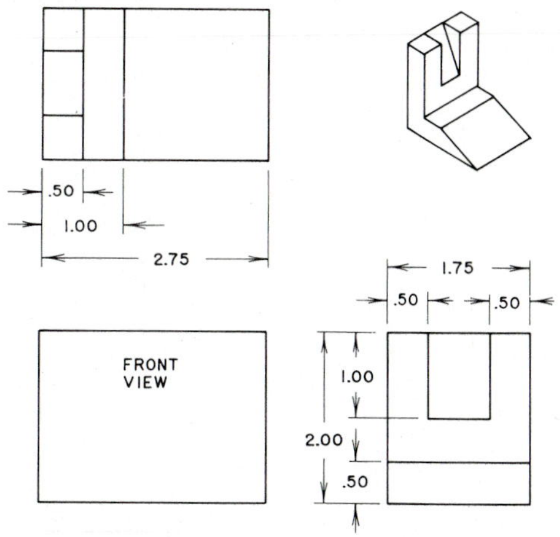

Problem 4.36

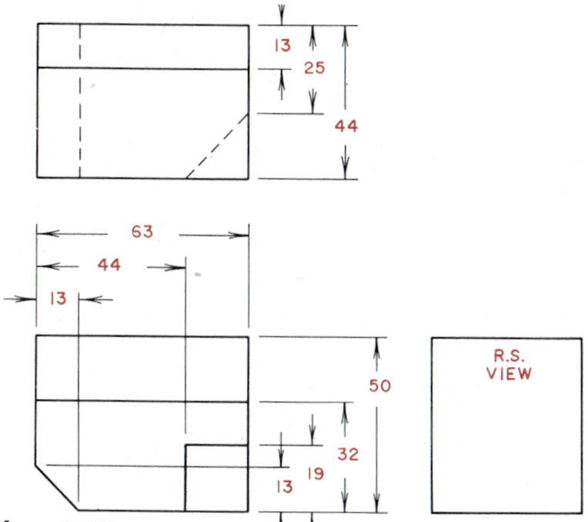

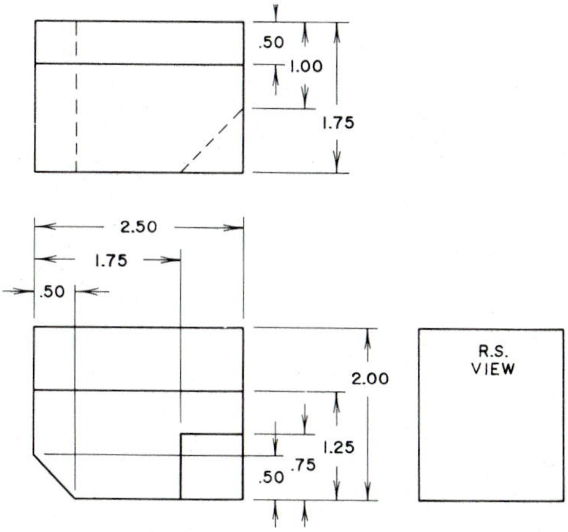

25
69
13

FRONT
VIEW

44
13 13
25
50
13

INCHES

.50
1.00
2.75

FRONT
VIEW

1.75
.50 .50
1.00
2.00
.50

Problem 4.37

13
25
44

63
44
13
50
32
19
13

R.S.
VIEW

.50
1.00
1.75

2.50
1.75
.50
2.00
1.25
.75
.50

R.S.
VIEW

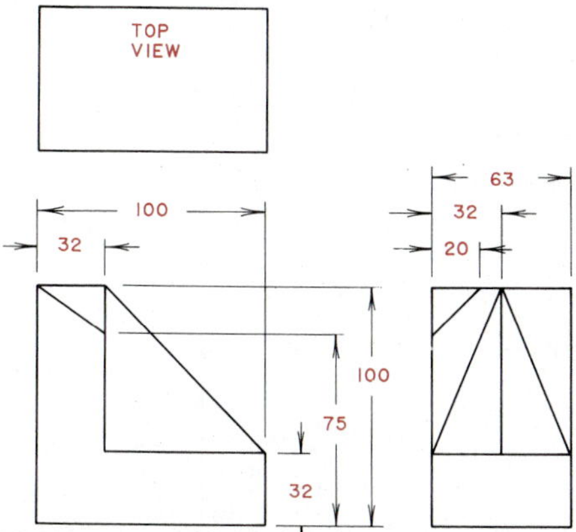

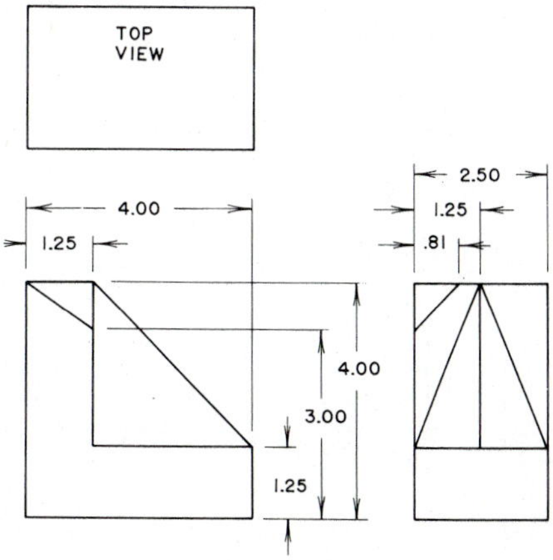

Problem 4.38

TOP
VIEW

100
32
100
75
32

63
32
20

TOP
VIEW

4.00
1.25
4.00
3.00
1.25

2.50
1.25
.81

Problems 4.33–4.41
Draw with instruments the given views. Complete the missing view as it relates to the given views.

MILLIMETERS   *INCHES*

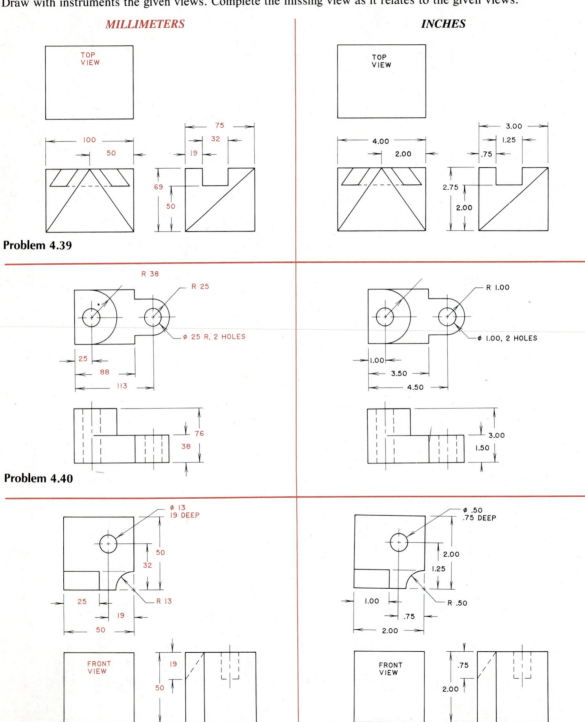

Problem 4.39

Problem 4.40

Problem 4.41

Problems 4.42–4.66

Draw three orthographic views of each of the pictorials shown. Specific problem sets are designed to give you practice with normal, inclined, and oblique surfaces as well as simple and complex curves.

MILLIMETERS   *INCHES*

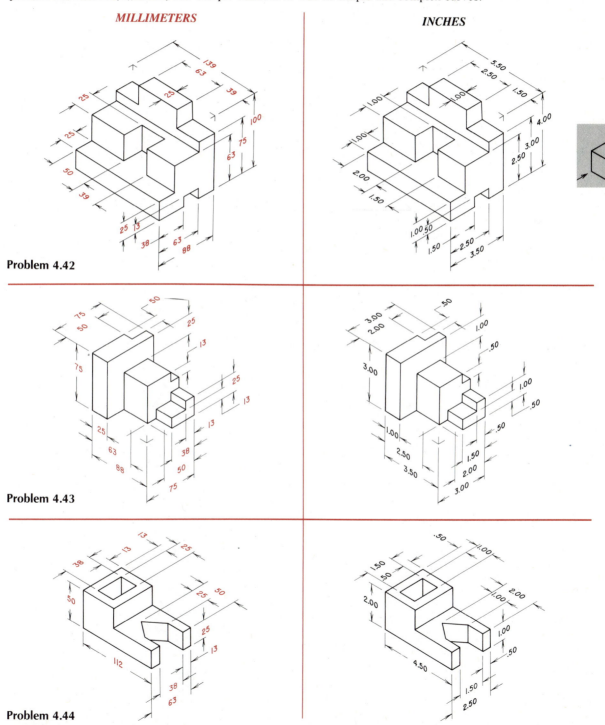

Problem 4.42

Problem 4.43

Problem 4.44

Problems 4.42–4.66

Draw three orthographic views of each of the pictorials shown. Specific problem sets are designed to give you practice with normal, inclined, and oblique surfaces as well as simple and complex curves.

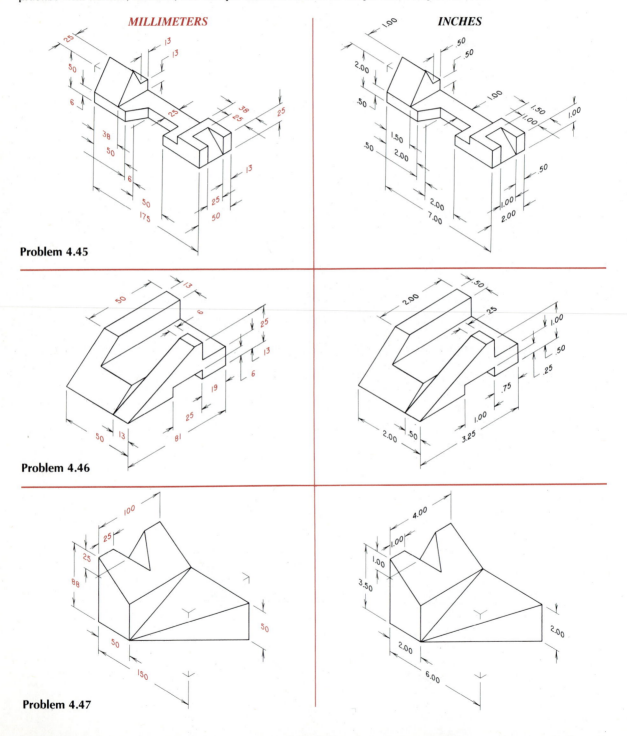

MILLIMETERS   *INCHES*

Problem 4.45

Problem 4.46

Problem 4.47

MILLIMETERS

INCHES

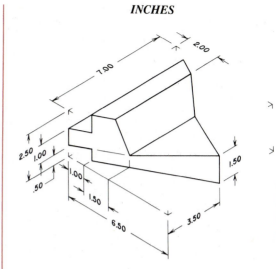

Problem 4.48

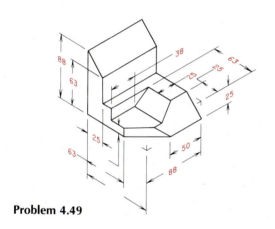

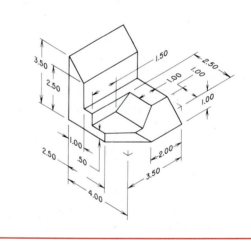

Problem 4.49

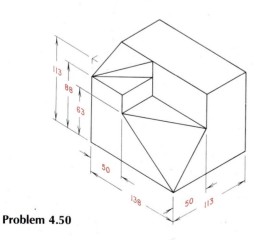

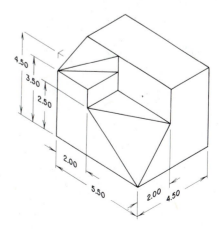

Problem 4.50

Problems 4.42–4.66

Draw three orthographic views of each of the pictorials shown. Specific problem sets are designed to give you practice with normal, inclined, and oblique surfaces as well as simple and complex curves.

MILLIMETERS   *INCHES*

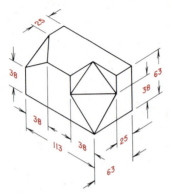

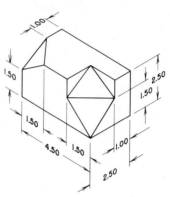

Problem 4.51

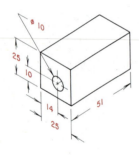

Problem 4.52

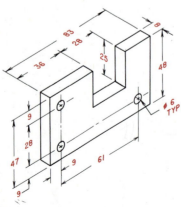

Problem 4.53

MILLIMETERS

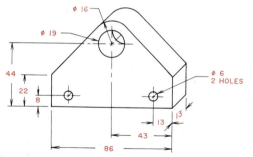

Problem 4.54

INCHES

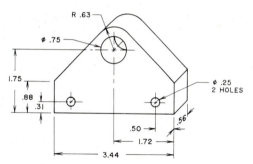

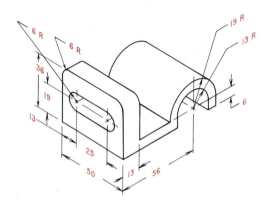

Problem 4.55

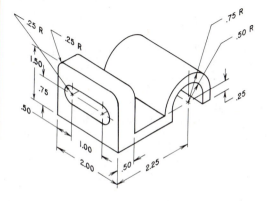

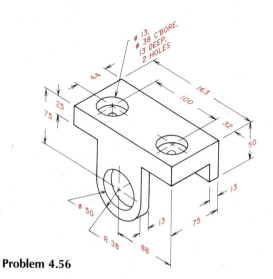

Problem 4.56

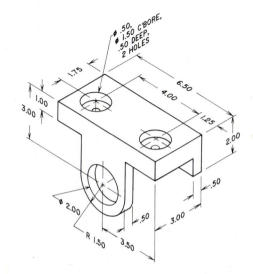

Problems 4.42–4.66

Draw three orthographic views of each of the pictorials shown. Specific problem sets are designed to give you practice with normal, inclined, and oblique surfaces as well as simple and complex curves.

MILLIMETERS   *INCHES*

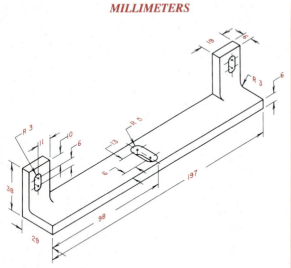

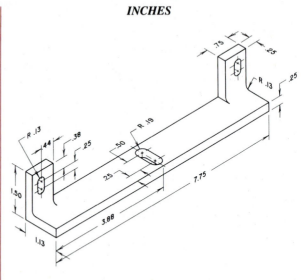

Problem 4.57

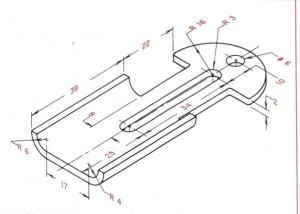

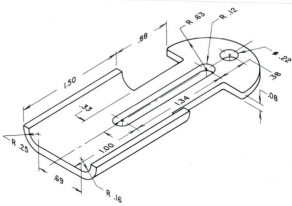

Problem 4.58

MILLIMETERS   *INCHES*

Problem 4.59

Problem 4.60

Problem 4.61

Problems 4.42–4.66
Draw three orthographic views of each of the pictorials shown. Specific problem sets are designed to give you practice with normal, inclined, and oblique surfaces as well as simple and complex curves.

MILLIMETERS   *INCHES*

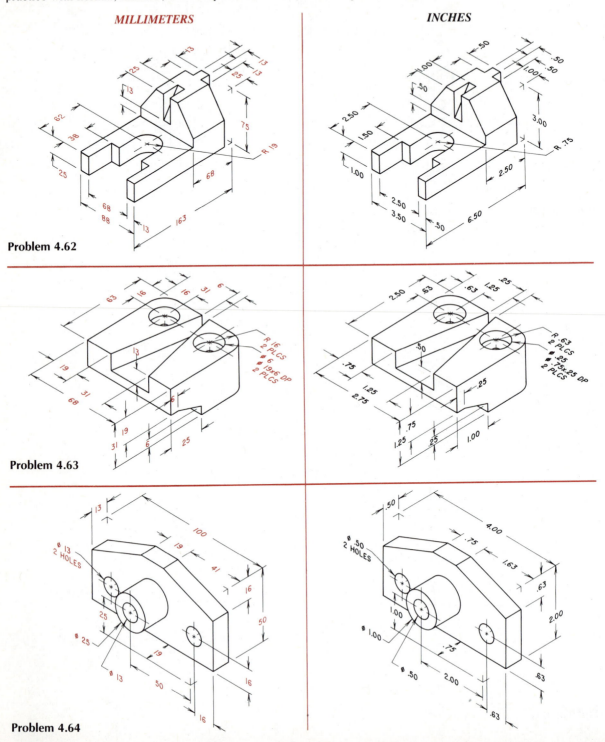

Problem 4.62

Problem 4.63

Problem 4.64

MILLIMETERS

INCHES

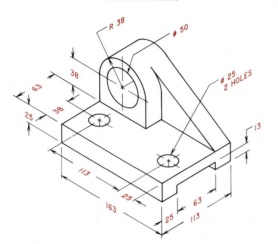

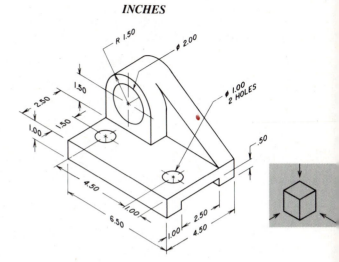

Problem 4.65

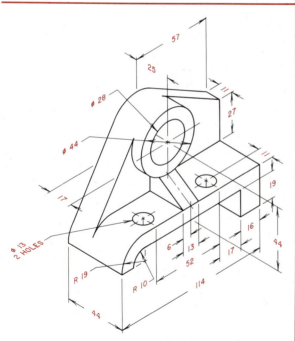

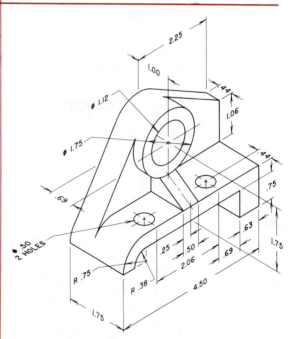

Problem 4.66

Problems 4.67–4.70
Draw with instruments all the views required to describe the objects shown. An auxiliary view is required to describe the inclined surface.

MILLIMETERS

INCHES

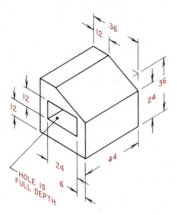

HOLE IS
FULL DEPTH

Problem 4.67

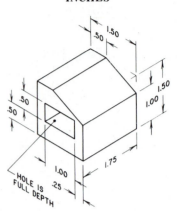

HOLE IS
FULL DEPTH

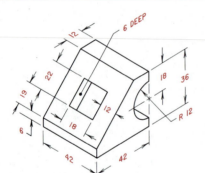

Problem 4.68

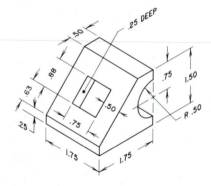

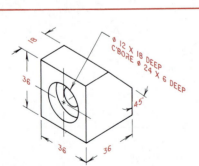

Problem 4.69

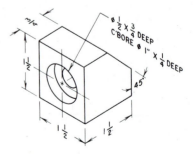

MILLIMETERS

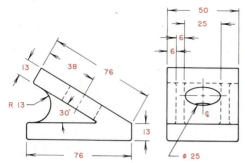

INCHES

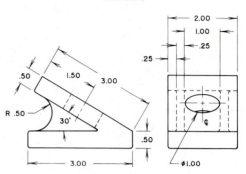

Problem 4.70

Problems 4.71–4.72
Draw with instruments all the views required to describe the objects shown.

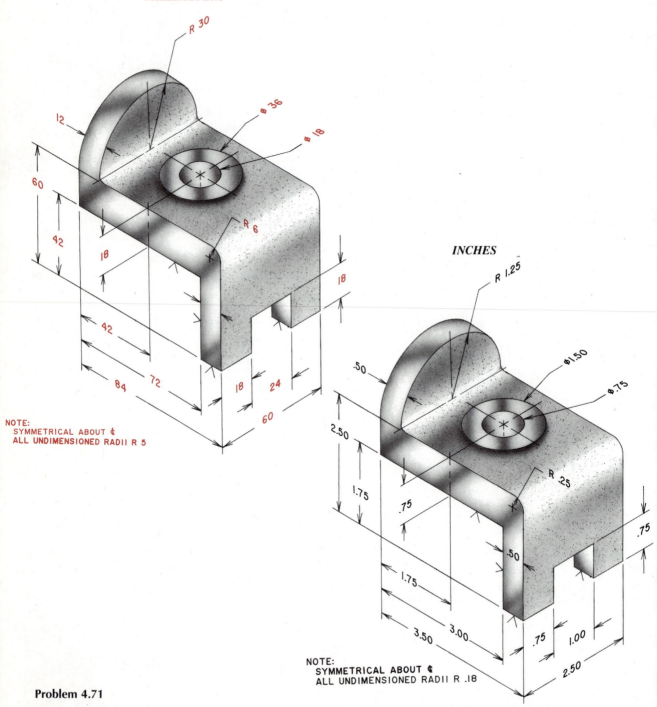

MILLIMETERS

R 30
12
Ø 36
Ø 18
60
42
18
R 6
18
42
18 24
84 72
60

NOTE:
SYMMETRICAL ABOUT ₵
ALL UNDIMENSIONED RADII R 5

INCHES

R 1.25
Ø 1.50
.50
Ø .75
2.50
R .25
1.75
.75
.50
.75
1.75
.75 1.00
3.50 3.00
2.50

NOTE:
SYMMETRICAL ABOUT ₵
ALL UNDIMENSIONED RADII R .18

Problem 4.71

MILLIMETERS

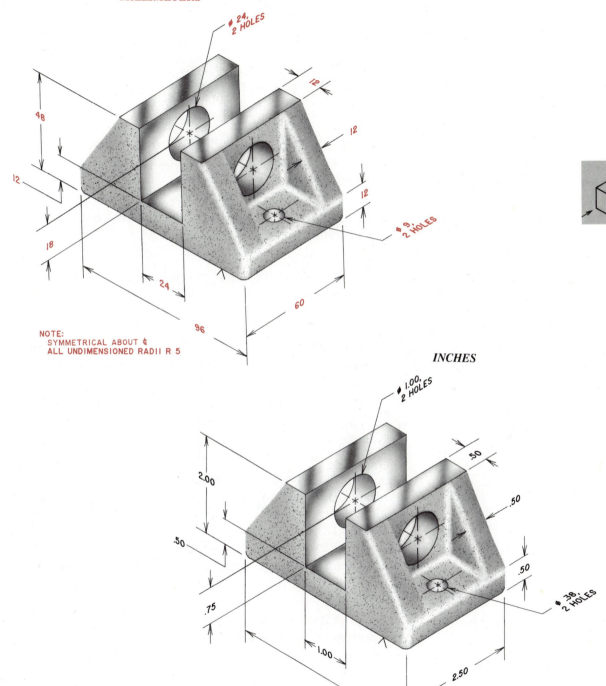

Ø 24,
2 HOLES

12

12

48

12

12

18

Ø 9,
2 HOLES

24

60

96

NOTE:
SYMMETRICAL ABOUT ₵
ALL UNDIMENSIONED RADII R 5

INCHES

Ø 1.00,
2 HOLES

.50

.50

2.00

.50

.50

.75

Ø .38,
2 HOLES

1.00

2.50

4.00

NOTE:
SYMMETRICAL ABOUT ₵
ALL UNDIMENSIONED RADII R .18

Problem 4.72

TERMS YOU WILL SEE IN THIS CHAPTER

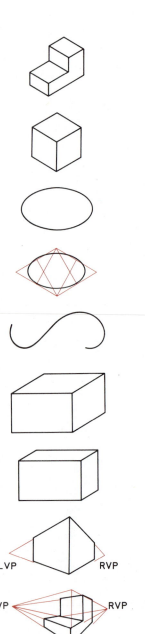

PICTORIAL DRAWING Graphic representation that shows features of width, height, and depth all in a single view.

ISOMETRIC DRAWING Pictorial drawing that shows the horizontal, frontal, and profile planes in the same view, with the three planes 120° apart.

ELLIPSE Closed symmetrical curve that does not have equal length radii.

FOUR-CENTER METHOD Approximate method of constructing isometric ellipses.

IRREGULAR CURVE Predetermined set of curves used to connect points on an irregular curve (curve that is not a function of a radius).

CAVALIER DRAWING Oblique drawing where the width, height, and depth are all measured using full-length dimensions.

CABINET Oblique drawing where the width and height are drawn at full length but the depth measurements are drawn at half their dimensioned size.

PERSPECTIVE DRAWING Pictorial view where width or depth lines are not parallel but converge at a vanishing point(s).

TWO-POINT EXTERIOR PERSPECTIVE Specific type of perspective having two vanishing points that show the exterior of an object.

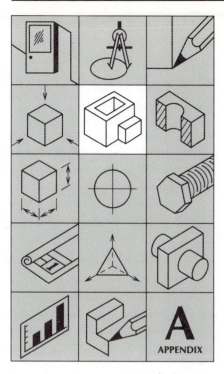

5

PICTORIAL DRAWINGS

5.1 INTRODUCTION

Pictorial drawings (Figure 5.1) make up a major portion of drawings used by the designer. There are many types and variations of pictorial drawings since they must serve different purposes. Some of the major types of pictorials are axonometric drawings, oblique drawings, and perspectives. After you have learned to draw these pictorials, variations may be applied such as sectioning a pictorial or exploding an assembly.

ISOMETRIC

OBLIQUE

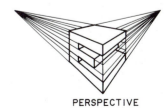

PERSPECTIVE

Figure 5.1 Samples of pictorial drawings.

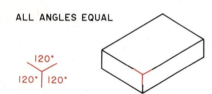

Figure 5.2 Three visible normal surfaces have equal angles between them (120°) in an isometric drawing.

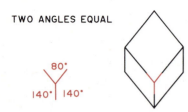

Figure 5.3 In dimetmric drawings, two of the three visible normal surfaces have equal angles not equal to 120°.

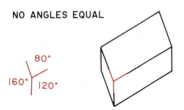

Figure 5.4 Trimetric drawing has the three visible normal surfaces in a position where none of the angles between the three surfaces are equal.

5.2 AXONOMETRIC DRAWINGS

5.2.1 Types of Axonometric Drawings

The general term *axonometric drawing* is used within the technical drawing field to describe three types of pictorials: isometric, dimetric, and trimetric drawings. **Isometric drawings** are the quickest and easiest of all the pictorials to draw and the most commonly used. In an isometric drawing the three normal surfaces of a rectangular solid will have equal angles between them (120°) (Figure 5.2). In dimetric drawings, two of the normal surfaces will be equally spaced, but the third surface will have an angle of a different number of degrees (Figure 5.3). A trimetric drawing will have the three normal surfaces of the rectangular solid positioned so none of the three angles have the same number of degrees (Figure 5.4).

The drawing of dimetric and trimetric drawings takes more time because uncommon angles are often used. This means the axes for this type of pictorial drawing are not the same as those for isometric, and generally the three axes are a different number of degrees from each other. Surfaces at uncommon angles require you to construct all the ellipses for circular features or to have an expensive set of ellipse templates at hand. Since dimetric and trimetric drawings are seldom used in engineering work, they will not be discussed in detail. Instead, the isometric drawing will be covered in depth.

5.2.2 The Isometric Axes

There are four primary positions for placing the axes when starting an isometric drawing or a rectangular solid. You may elect to look down on the object (the most common) or elect to look up to the object (seldom drawn). Further, you may elect whether the long dimension will extend to the left or to the right of the near corner of the object (Figure 5.5).

The center axis is vertical in all cases, and the left or right axes vary from the horizontal by 30° in the chosen direction. One of the nice aspects of drawing an isometric drawing is that true-length measurements may be made on or parallel to the three chosen isometric axes.

5.2.3 Normal Surfaces in Isometric

After the axes have been chosen and located, identify the overall width, depth, and height of the object to be drawn. Then measure these overall dimensions on the appropriate axes. From these

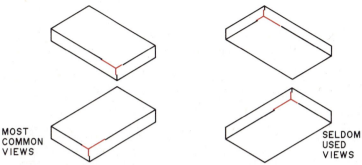

Figure 5.5 When drawing an isometric there are four primary positions for viewing the object. Looking down on the object with the long axis to the left or right are the most common views.

points you should be able to project the remaining corners of the rectangular solid that will contain the entire boundary surfaces of the object (Figure 5.6). It is important to first draw this box containing the extremes of the object and then remove desired portions in order to produce the pictorial representation of the object. You now have three normal surfaces on which measurements may be made, and you have the extremes within which the object must fit. Knowing this information, you should be able to draw objects having only normal surfaces by projecting and measuring on or parallel to the axes (Figure 5.7).

It should be remembered that with only rare exception, hidden lines are not used when drawing pictorials. You only draw the surfaces and edges of the object that your eye would see when viewing the object.

If normal surfaces are to be drawn within the box representing the extremes of the object, you will find it easy to measure on the box and project into the object. This is a good habit to get into because it will be an important key to drawing inclined and oblique surfaces later.

5.2.4 Inclined Surfaces in Isometric

If you need to draw inclined surfaces in isometric, you start by choosing and locating the axes. Then you measure the extreme width, height, and depth of the object and complete the rectangular solid representing the extremes of the object. Note that the entire box is drawn; then desired parts will be eliminated in order to produce the pictorial drawing of the object. It is best to measure the locations of the corners of the inclined surfaces on one of the outside three planes and then project and connect the points.

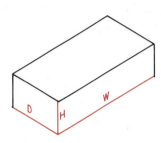

Figure 5.6 Locate and measure the extreme width, depth, and height of the object; then project to complete the rectangular solid in which the extreme sizes of the object must fit.

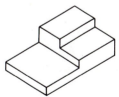

Figure 5.7 Measure true length on one of the three original isometric axes or on any line parallel to these three lines.

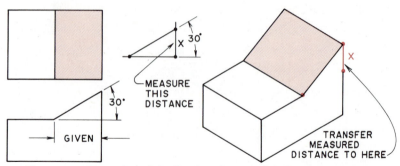

Figure 5.8 Even though inclined surfaces to be drawn in isometric may be given in degrees, it is mandatory to locate the inclined surface by locating the points, not transferring the angle in degrees.

It is very important that you remember that angles given in degrees on an orthographic view may *not* be transferred to the isometric drawing by using the same number of degrees. This is because the surface of the pictorial is at a different angle to the picture plane than in the orthographic view; therefore, you must locate the points representing the corners of the inclined surface and then connect the points (Figure 5.8). In order to know the isometric line length representing the sides of the angle given in degrees in the orthographic view, you must first construct the angle in orthographic, measure the length of the lines in this construction, and then measure and locate the lines on the pictorial view.

When two inclined surfaces intersect, first draw one of the inclined surfaces using light lines. Then draw the second inclined surface by identifying points on the normal surfaces. The points of the intersections can then be connected, and the correct lines may be darkened (Figure 5.9).

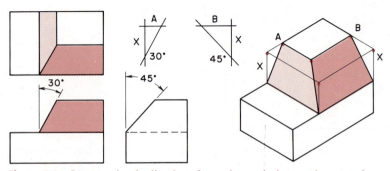

Figure 5.9 Intersecting inclined surfaces drawn in isometric must also be constructed by locating points from the orthographic views.

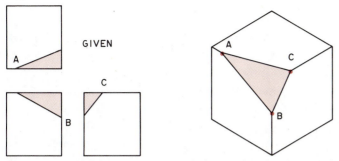

Figure 5.10 Oblique planes drawn in isometric must have their points located on normal surfaces or temporary construction on normal surfaces.

5.2.5 Oblique Surfaces in Isometric

Since an oblique surface is not parallel to any of the three normal surfaces drawn to outline the extremes of the object, all the points of the surface must be located on normal planes (or temporary construction of normal planes you must draw), and then the points must be connected (Figure 5.10).

To locate these points, you must make two measurements on each surface. On the frontal plane a height and width must be located. On the profile plane a height and a depth must be located, and on the horizontal plane the width and depth must be located. After the points have been located, the lines outlining the oblique surface may be drawn by connecting the points (Figure 5.11).

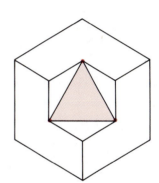

Figure 5.11 After points have been correctly located the oblique surface may be drawn.

5.2.6 Circles and Arcs Using an Isometric Template

When you draw a circle or arc on a normal surface of an orthographic drawing, you use a compass or a circle template because the normal surface is parallel to the plane of projection. In isometric drawing that plane of projection is not parallel to the three major planes of the rectangular solid; therefore, the circle will not appear round. The circles will appear as **ellipses** (Figure 5.12). Although we draw the planes of the rectangular solid at 30° from the horizontal, these planes are technically 35° 16′ back from the horizontal. This means that you need a special template to draw isometric ellipses. It is called an isometric ellipse template, and it is based on ellipses tilted back at an angle of 35° 16′ from the horizontal. The small isometric template will be helpful for drawing ellipses up to about 1.5 or 2 in. (38 or 50 mm) (Figure 5.13). Isometric ellipse templates are also available in metric sizes. If a

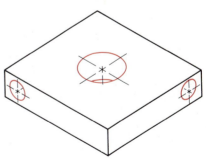

Figure 5.12 Circles on normal surfaces will appear as ellipses when normal planes are drawn in isometric.

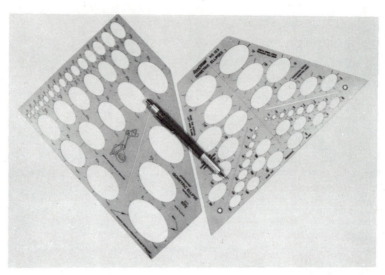

Figure 5.13 Small isometric templates are easy to use and save time.

Figure 5.14 Sets of large ellipse templates are more expensive and are used more as a specialty tool.

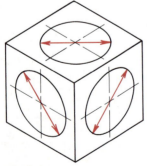

Figure 5.15 Note orientation of long axis for ellipses on each of the three normal surfaces of cube.

larger ellipse is required, then a large ellipse template may be used (Figure 5.14).

The isometric ellipse template is easy to use, but you must orient the template correctly or the drawn pictorial will look drastically wrong. First look at the orientation of the ellipse on each of the three surfaces of the pictured rectangular solid (Figure 5.15). Note the direction of the long axis of the ellipse to the

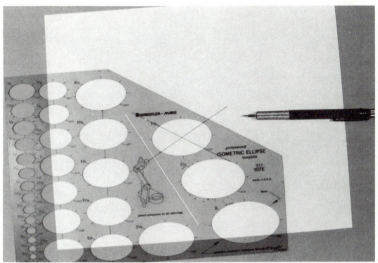

Figure 5.16 It is very important to learn to correctly orient the isometric template in order that the ellipse will appear correctly on drawing.

surface upon which it is pictured. You should be able to orient the template in the approximate position for drawing the ellipse on any of the three surfaces (Figure 5.16). Note also the position of the center lines for each of the three ellipses. The left and right planes have vertical center lines and one parallel to the 30° axis. The ellipse on the horizontal plane has two center lines parallel to the two 30° axes (Figure 5.17).

Once you have oriented the template to the approximate position, carefully line up the center line marks on the template with the center lines on the drawing. Finally draw the ellipse. Be careful to keep the pencil at a position perpendicular to the drawing surface or the ellipse will become distorted (Figure 5.18).

Arcs drawn on normal surfaces in isometric are simply a portion of the circumference of the ellipse. Therefore, you will first locate the center of the ellipse, draw the center lines, then align the template, and finally draw only the portion of the arc that is desired (Figure 5.19).

5.2.7 Circles and Arcs Using the Four-Center Method

To save time, use the ellipse template whenever possible to draw the isometric ellipses. However, sometimes a specific size ellipse (one that does not appear on a standard template) will be required or a very large ellipse will be required. In either of the these cases,

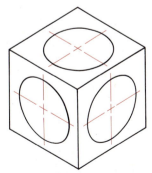

Figure 5.17 Note correct direction of center lines for ellipses that appear on the three visible normal surfaces.

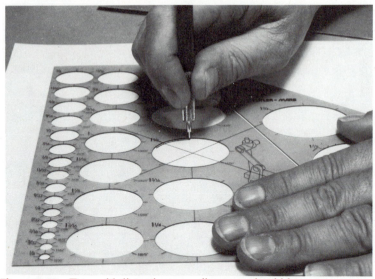

Figure 5.18 To avoid distortion, pencil or pen should be perpendicular to drawing surface when using the ellipse template.

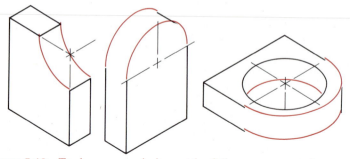

Figure 5.19 To draw an arc in isometric, follow same procedure as for drawing a circle but only darken the desired portion of the arc.

you will have to construct an ellipse using a compass. The most efficient way is to use the **four-center method.**

To construct an ellipse on any of the three visible normal surfaces of the rectangular solid, complete the following steps:

1. Draw the center lines on the selected plane.
2. Measure the distance of the radius away from the center on all four center lines.
3. Draw an isometric box as shown.
4. Draw the extra construction lines to locate the centers for the compass.

5. Set the compass for the distance R and strike the two large arcs using the compass.
6. Set the compass for distance r and strike the two small arcs (Figures 5.20–5.22).

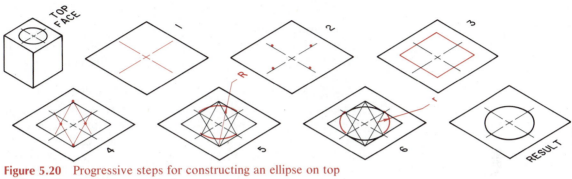

Figure 5.20 Progressive steps for constructing an ellipse on top surface using four-center method.

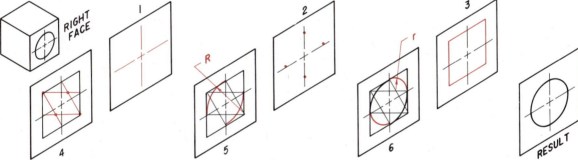

Figure 5.21 Progressive steps for constructing an ellipse on right surface using four-center method.

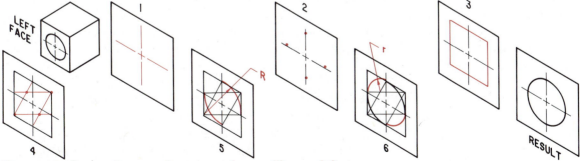

Figure 5.22 Progressive steps for constructing an ellipse on left surface using four-center method.

5.2.8   Drawing Ellipses on Nonisometric Surfaces

You will occasionally have to draw an ellipse on an inclined surface that appears in an isometric drawing. When this happens, you must first plot the circle (as an ellipse) on one of the two adjacent normal isometric surfaces. Then plot a grid starting at any point on the ellipse. Project this point to the adjacent isometric surface. Then project to the inclined edge and finally along the inclined surface. Repeat this procedure until enough points have been plotted to construct the ellipse on the inclined plane. You will have to use the **irregular curve** to draw the ellipse on the inclined surface (Figure 5.23).

A circle appearing on an oblique plane will appear as an ellipse. You must first locate and draw the isometric ellipse on a normal isometric plane. Then project selected points from the isometric ellipse to the oblique surface until they intersect each other creating a grid network. When enough points are located on the oblique surface, the irregular curve can be used to connect the points that produce the correctly shaped ellipse on the oblique surface (Figure 5.24).

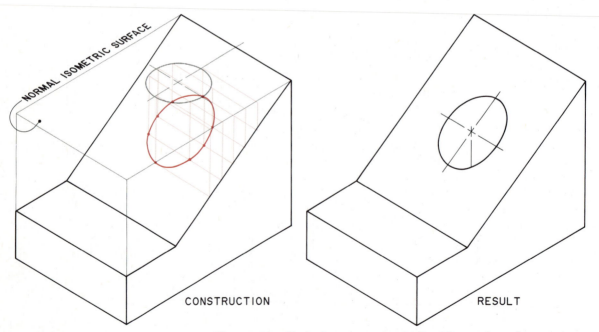

CONSTRUCTION   RESULT

Figure 5.23   Circle that appears on inclined isometric surface must first be constructed on normal plane and then projected to inclined isometric surface.

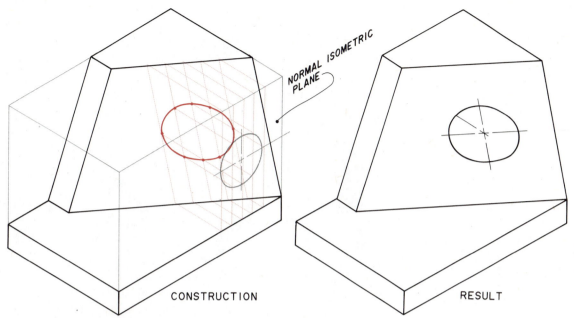

NORMAL ISOMETRIC PLANE

CONSTRUCTION

RESULT

Figure 5.24 Circle that appears on oblique isometric surface must first be constructed on normal plane and then projected to oblique isometric surface.

5.2.9 Using the Irregular Curve

The irregular curve is a difficult tool to master until you have had some practice using it. There are many sizes and shapes of irregular curves. You must select an appropriate size and shape to connect all the identified points (Figure 5.25).
Follow this procedure to use this tool:

1. Plot all the points you wish to connect.
2. Place the curve so that you align as many points as possible (at least four dots must be used except for the end spaces).
3. Connect all spaces between the points except for the space at each end.
4. Now reposition the irregular curve so the first space aligned overlaps the end space drawn last. Continue this procedure until the curve has been drawn as a continuous curved line.
5. Note that you cannot overlap the two end spaces; you will have to do the best you can with these two spaces.

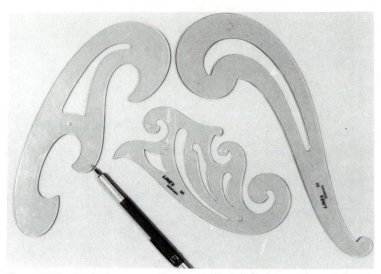

Figure 5.25 Irregular curves are available in a large number of sizes and shapes.

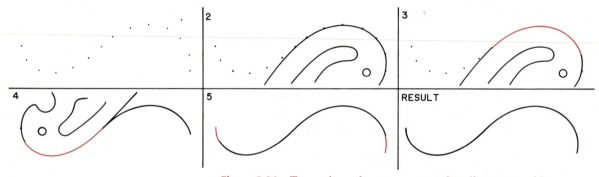

Figure 5.26 To use irregular curve correctly, align curve with as many locator dots as possible (minimum four dots except for end spaces). Then draw connecting line through all but the last space. When drawing next segment of curve, overlap last space.

6. If necessary, clean up the irregular curve so it looks like a continuous curve (Figure 5.26).

5.2.10 Exploded Views

Once you have mastered the drawing of objects in isometric, you can begin making exploded views. These views consist of isometric drawings of several parts that are displayed along their axes of assembly. This type of drawing allows the untrained eye

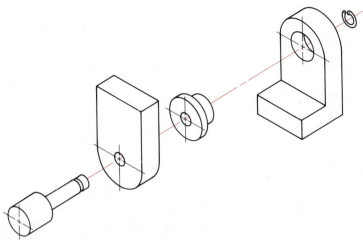

Figure 5.27 Exploded view shows parts of assembly separated along center line. This makes each part visible to viewer.

to immemdiately visualize how several parts of an assembly fit together. These drawings are particularly valuable when the viewer must either assemble or disassemble a number of parts that are manufactured to fit together. Exploded views are often used in maintenance manuals (Figure 5.27).

The easiest way to produce one of these drawings on the drawing board is to make rough drawings first. Then cut the pieces apart and reposition them with the spacing between the views being equal or pleasing to the eye. Then trace the original rough drawings to make the final presentation.

5.2.11   Isometric Assembly Drawings

There are times when a pictorial drawing is required to show the outside of an assembly made up of several parts. This type of drawing is known as an isometric assembly drawing. Depending on the complexity of the drawing, you will either make the drawing part by part in its assembled position or draw the objects individually and trace them, locating them correctly. These views tend to show how the entire assembly appears from the outside after the parts have been assembled (Figure 5.28).

5.2.12   Isometric Projections

Up to this point you have been making isometric drawings. Now you should be exposed to a little pictorial theory and try to

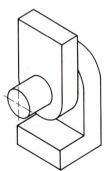

Figure 5.28   Assembly drawing is a pictorial view that shows visible features of a group of parts after they have been assembled.

understand the difference between an isometric drawing and an isometric projection. This is particularly important since most computer-generated pictorials are actually projections.

The biggest visual difference between isometric drawings and isometric projections is the size of the two images. The isometric drawing is drawn using 100% true-length measurements on the height, width, and depth axes (Figure 5.29). However, in isometric projection the height, width, and depth measurements are displayed at 82% of their true length (Figure 5.30). For the purpose of projection the X, Y, and Z axes remain fixed, and we think about the object rotating in space about the axes.

To understand why these are different, you must first visualize the axes as they relate to (1) the drawing board or the computer screen or (2) the axes of the object. Note that the X axis goes to the right, the Y axis is vertical, and the Z axis goes toward the observer (Figure 5.31).

In isometric projections the object is first rotated about the Y axis by −45° (Figure 5.32). Then the object is rotated about the

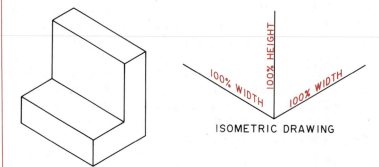

Figure 5.29 On isometric drawing, height, width, and depth measurements are made at 100%.

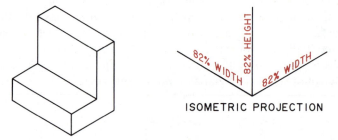

Figure 5.30 On isometric projection, height, width, and depth measurements are made at 82%.

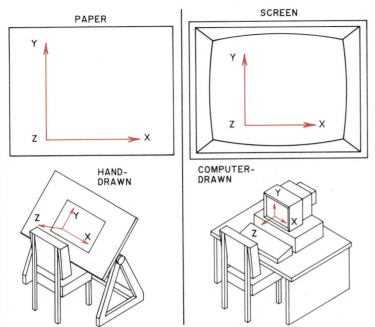

Figure 5.31 Note orientation of *X*, *Y*, and *Z* axes as they relate to observer at drawing board or at computer.

X axis by 35° 16' (Figure 5.33). Because the normal surfaces of the object are no longer parallel or perpendicular to the picture plane, the image edges will appear foreshortened on each axis by 18%. The foreshortened view is called an isometric projection.

When you draw a pictorial on a computer, the image that appears on the screen is a projection, and therefore it is foreshortened. The big advantage of drawing these pictorials on the screen is that the object may easily be rotated about the axes in a countless number of positions once the data for the object's features have been entered. This allows the observer to view the object from many positions while the computer redraws the object. Copies of the rotated object can be printed or plotted and used for a set of working drawings. The printout can also be the basis for a technical illustration of the object or assembly (Figure 5.34).

If an isometric drawing is made on the drawing board, you only get a view from one vantage point. To show the object rotated, it would have to be completely redrawn. In order to compensate for the difficulty of creating manually drawn isometric projections, drafters decided that if each of the isometric

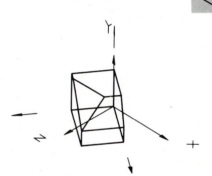

Figure 5.32 Object is first revolved around *Y* axis. (*Computer-generated image*)

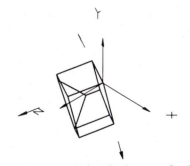

Figure 5.33 Object is then revolved about *X* axis. (*Computer-generated image*)

Figure 5.34   Computer-generated pictorials may easily be rotated to produce various revolved views. (*Computer-generated image*)

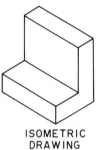

ISOMETRIC
DRAWING

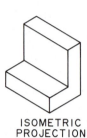

ISOMETRIC
PROJECTION

Figure 5.35   Resulting difference between isometric drawing and isometric projection.

axes was foreshortened the same amount, then a standard scale could be used for isometric drawings. For this procedure to work, it was assumed that any measurements made parallel to the object axes would be whatever was on the orthographic drawings of the principal views. This had the effect of creating a proportional but "larger-than-life" drawing. This results in the need for an 82% scale for making measurements on the X, Y, or Z axes if an isometric projection is created on the drawing board (Figure 5.35).

Some CAD software packages will allow you to rotate the object so you may view the object from several vantage points and allow you to get a hard copy of the rotated object.

5.3   *OBLIQUE DRAWINGS*

5.3.1   Oblique Views

Oblique drawings are very similar to isometric drawings, but there are some noticeable differences. The object has one face of the rectangular solid parallel to the picture plane. Two other faces do appear, but since the angle of the depth axis may vary, the angles of the other two surfaces may vary from drawing to drawing. The axes of an oblique drawing are horizontal, vertical, and a depth axis. You may select any angle for the depth axis, but a 45° angle is usually preferred. Notice how much change there is in the appearance of the object when the depth axis is altered (Figure 5.36). The main advantage of drawing an oblique drawing is that you may use the circle template on the plane that is parallel to the picture plane. This is possible because the circles will appear as true circles and not as ellipses on this one surface only. Remember, any plane parallel to the picture plane is true size and shape (Figure 5.37).

The computer is used quite often on a commercial basis for making oblique drawings of large exploded assemblies. The express purpose is for making parts catalogs for assemblies where repair on the assembly might be necessary. The parts are identified with a part number for easy, quick identification.

PREFERRED   SPECIAL USE ONLY

Figure 5.36   Oblique drawings generally have a front view that shows true width and height measured on horizontal and vertical axes. Depth measurements are generally on axis that recedes at 45° (although another angle may be selected).

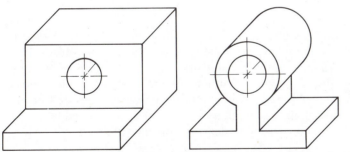

Figure 5.37   Circles appearing on front surface of an oblique drawing will be drawn as circles, not as ellipses.

5.3.2   Cabinet and Cavalier Oblique Drawings

Oblique drawings of objects drawn with a 45° depth axis that have a long depth measurement do not appear visually correct to the eye. These drawings are known as **cavalier drawings.** Notice that the cube, when drawn in cavalier, with true height, width, and depth, visually appears much too deep (Figure 5.38).

It is common for the person making an oblique drawing to take

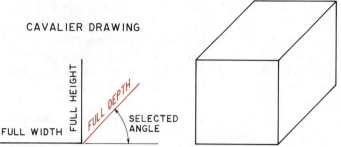

CAVALIER DRAWING

FULL HEIGHT

FULL WIDTH

FULL DEPTH

SELECTED ANGLE

Figure 5.38 Cavalier drawings are oblique drawings using full-size height, width, and depth measurements. With this type of drawing, depth appears visually exaggerated.

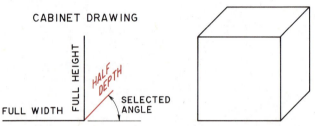

CABINET DRAWING

FULL HEIGHT

FULL WIDTH

HALF DEPTH

SELECTED ANGLE

Figure 5.39 Cabinet drawings are oblique drawings that use full width and height measurements but only half the size of the depth measurement.

some "artist's liberties." The furniture industry uses a type of oblique drawing that somewhat remedies this depth problem. They make a **cabinet** type of oblique drawing. For this type of drawing, the width and height measurements are made at full dimension. However, the depth axis is shortened and is measured at half dimension along the depth axis. This produces a completely different and more visually pleasing drawing (Figure 5.39).

5.3.3 Special Ellipse Templates

Those persons who frequently make oblique drawings will certainly have a complete set of ellipse templates that are commonly identified by 5° increments from 10° to 80°. Because these sets are expensive, every effort should be made to keep the sets intact. These ellipse templates are used to draw ellipses on the depth axis when the given depth axis angle equals one of the templates in the set. These templates are usually purchased in small, medium, and large size sets and range from $\frac{1}{2}$ to 8 in. (13 to 203 mm) or more in size (Figure 5.40).

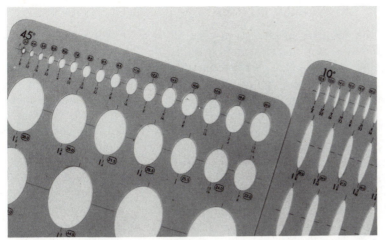

Figure 5.40 Ellipse templates for making oblique drawings range in increments from 10° to 80°. Note how the shape of ellipses changes as degree increment changes.

5.4 PERSPECTIVE DRAWINGS

5.4.1 Perspective Overview

The designer often makes a **perspective drawing** in order to best show how an object or building will look. Perspective drawings show the object in a way that is the easiest for the average viewer to visualize. The perpective drawing is as close as we come to showing an object graphically as it would be seen by the eye. All types of perspectives take a considerable amount of time to produce. Therefore, you must weigh the cost versus the need for perspective drawings. They are by far the most time-consuming (thus expensive) line drawings to produce.

In orthographic drawings the projectors from the object to the picture plane are parallel to each other because the eye of the viewer is assumed to be at an infinite distance away from the object. In perspective drawings, the eye is located at a predetermined distance, height, and orientation from the object. Therefore, a major difference with perspective drawings is the fact that projection lines called "projectors" are parallel (see Section 5.4.9), but projection lines that converge at a single point are called visual rays (see Section 5.4.10).

The three major classifications of perspective drawings are the one-, two-, and three-point perspective views. This means the views have either one, two, or three vanishing points. These classifications will be discussed in the following sections.

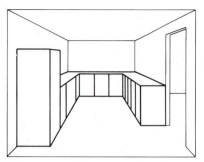

Figure 5.41 Linework of kitchen layout. (Courtesy of Utley Company.)

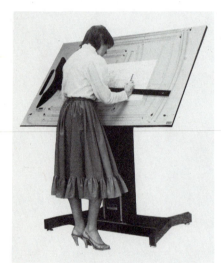

Figure 5.42 Using the Klok board to draw a perspective view. (Courtesy of Utley Company.)

5.4.2 One-Point Perspective

One-point perspectives are generally used to represent interior views of a room. Designers use this type of drawing to help the customer visualize how a room will look after construction. The kitchen planning contractor or cabinet maker uses this type of drawing to help the customer visualize how the finished kitchen will look after construction (Figure 5.41).

Time-saving commercial grids are available to the designer to help during the sketching, planning, or selling stages. A sheet of vellum or drawing film is placed over the preprinted grid, and the designer can accurately sketch a one-point perspective on the job site or while planning with the client.

The Klok Board is another time-saving device used to draw various types of perspective views (Figure 5.42). There are many sets of graduations printed on the surface of the Klok Board, but it is easy to use because you only have to use three sets of graduations (one height, one depth, and one width) when drawing any perspective view (Figure 5.43).

Drawing the one-point perspective is relatively easy. The height and width of the room are drawn with a vertical line and a horizontal line that represent the near edge of one wall and the floor. The near edge of the ceiling and the other wall are parallel to the

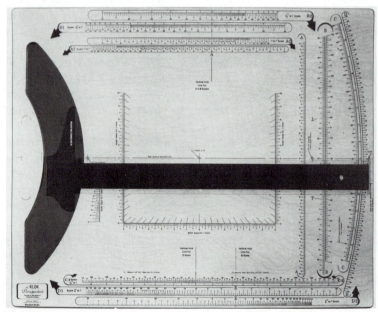

Figure 5.43 Klok board scales. (Courtesy of Utley Company.)

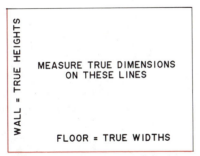

Figure 5.44 (*Step 1*) True dimension height and width dimensions.

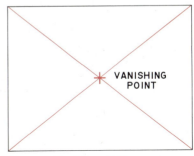

Figure 5.45 (*Step 2*) Locate vanishing point.

original wall line and the floor line. You may make true-length measurements of height or width on these lines only (Figure 5.44)!

In order to add depth to the view, you must establish a single vanishing point that is generally located in the center of the room. Simply construct two light construction lines connecting the opposite two corners of the rectangle to locate this vanishing point (Figure 5.45).

Generally, the plan view of a floor in a room is approximately square. To establish a proportional depth for the room, you next establish the back wall of the room. Locate a point half way between the vanishing point and one corner of the rectangle, and project this point through each of the diagonals within the rectangle. This smaller rectangle represents the back wall of the room and should be connected to the front edges of the room that already exist (Figure 5.46).

Remember, we said you could make true-length measurements on the near vertical and horizontal edges of the room only (the walls for vertical measurements and the floor or ceiling for the width measurements). In order to make true-length measurements of depth, you must establish a temporary measuring line. To do

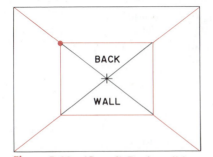

Figure 5.46 (*Step 3*) Back wall layout.

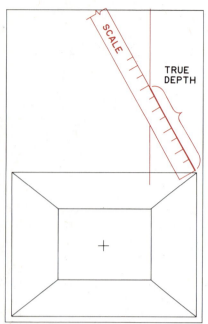

Figure 5.47 (*Step 4*) True dimension of depth measurement.

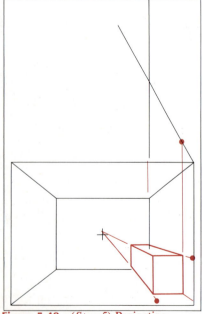

Figure 5.48 (*Step 5*) Projecting height, width, and depths.

so, project one of the rear wall vertical lines upward well beyond the outer rectangle. Position your scale so the zero graduation intersects the near corner of the outside rectangle. Now pivot the scale until the true-length depth of the room on your scale intersects the extended vertical line. After you draw this line, you may make true-length depth measurements on this line at any time and then project them directly to the adjacent wall (Figure 5.47).

You can locate any point in the room by measuring true height, width, and depth distances (*only* on the already identified lines upon which you can measure true-length measurements). Then project these points into the room until all three converge. As points are projected into the room, you can connect the points to indicate the outline of the interior features (Figure 5.48).

Because of the number of construction lines that will be needed to draw this type of view, it is best to do all your layout work on one sheet. Don't waste time erasing; just use different line weights to indicate the feature lines you want to remain. Then trace the final drawing of the perspective view on a clean sheet of vellum, film, or clay-coated paper using a tracing table. If it is necessary to present the final illustration on some sort of hard board that is not traceable, simply use spray adhesive to attach the vellum, film, or clay-coated paper to the hard board.

Rendering is the process of adding color, texture, shade, and/or shadow to a line drawing. Rendering makes drawings appear more realistic. The illustrator who needs a rendered version of the perspective view for presentation may elect to render the final line illustration or to skip the final line illustration and render directly from the rough layout.

5.4.3 Two-Point Perspective

The **two-point perspective** is the most commonly drawn perspective view. The purpose of this type of technical drawing is to show the exterior view of a building or object in a manner that closely parallels what the eye will actually see. Architects have two-point perspectives drawn to show a client what the finished commercial building will look like. These rendered versions of such perspectives are important tools when used for sales or promotional purposes (Figure 5.49).

There are two types of two-point perspectives: the ''footprint'' method and the ''standard'' method. The footprint method has the advantage of being able to make true-length measurements of height, width, and depth directly on the footprint. The footprint

Figure 5.49   Rendered perspective. (Courtesy of Jeffrey Wallace.)

method is a little more complicated to set up than the standard method, so we will not address the specifics of the footprint method in this edition other than to identify it.

You will need to understand several new terms in order to draw a two-point perspective: picture plane, plan view, measuring line, horizon, ground line, station point, vanishing points, projectors, and visual rays. These terms along with the standard method of drawing a two-point perspective are described in detail in the following sections.

There are four variables the designer can control when drawing a two-point perspective, each of which greatly affects the final line illustration. The relationship among these variables must be understood and applied as the perspective view is being generated. First, the *distance* the eye is positioned from the object can be established. Second, the designer can control the *size* of the perspective view by positioning the picture plane. Third, the *relative position* of the eye with respect to the surfaces of the object can be selected. Lastly, the *height* position of the eye with respect to the elevation of the object can be predetermined. To draw a standard two-point perpsective, you will need a copy of the plan view and a copy of the elevation view of the object. These four variables will be explained in detail throughout this chapter (see Figures 5.50, 5.58, 5.62, and 5.64).

For reasons given later, we like to use three colors of sharp pencil lines when drawing perspectives. You will need three pencils: black, red, and blue. Be sure to use the correct color of pencil as specified in the text. Sample illustrations in this book are printed in black and one color. The new lines in the step illustrations will be printed in color.

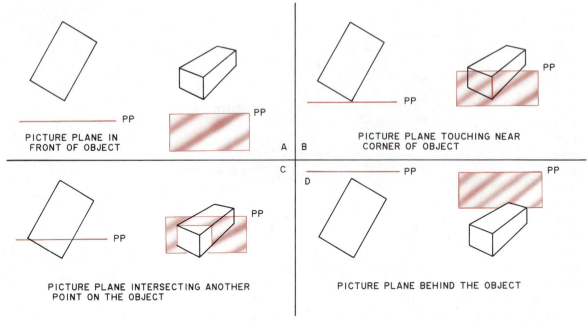

PICTURE PLANE IN
FRONT OF OBJECT

PP

A | B

PICTURE PLANE TOUCHING NEAR
CORNER OF OBJECT

PP

C

D

PP

PICTURE PLANE INTERSECTING ANOTHER
POINT ON THE OBJECT

PP

PICTURE PLANE BEHIND THE OBJECT

PP

Figure 5.50   Four positions of picture plane (PP) with respect to object.

5.4.4   Layout the Object and Picture Plane

The *picture plane* is an imaginary clear plane upon which the perspective view of the object will be projected. The picture plane can be drawn in one of four positions with respect to the object: (1) in front of the object (between the eye and the object) (Figure 5.50*a*); (2) with the picture plane touching the near corner of the object (Figure 5.50*b*); (3) with the picture plane penetrating the object (Figure 5.50*c*); or (4) with the picture plane behind the object (the entire object is between the eye and the picture plane) (Figure 5.50*d*).

Features of the object that are located behind the picture plane will be smaller and less distorted. Those objects or features located in front of the picture plane will appear larger, and the nearer features will be more pronounced.

By far the most common picture plane position to select is the one where the near corner of the object touches the picture plane. This position makes it much easier and less complicated at later stages of drawing the perspective view.

To help you understand the process of drawing a two-point perspective, an example will be illustrated step by step. You are encouraged to draw this example as it is described in the remainder of this chapter. The example will be presented within a thin border and identified by progressive step numbers.

Begin by laying out your drawing. Near the top of a large piece of paper draw a horizontal line the entire length of the paper with the black pencil to represent the edge view of the picture plane, and label this line PP (Figure 5.51).

Position a copy of the *plan view* (top view) of the object above the picture plane so that the near corner touches the picture plane. (If the object has a corner at the near corner, use it. If the near corner of the object is nonexistent, then draw a rectangle around the entire object to create the near corner). Now revolve the plan view so the long axis of the object is at a 30° angle to the picture plane. Slide the object toward the edge of the paper, which leaves a longer portion of the picture plane on the side of the long axis of the object. Tape the plan view to your large paper (Figure 5.52). At a later time (Section 5.4.6) you will need a copy of the elevation view at the same scale as the plan view in order to establish height measurements.

Figure 5.51 (*Step 1*) Establish picture plane line.

5.4.5 The Measuring Line

The *measuring line* is the *only* vertical line on the perspective view upon which you may make true-length height measurements. To help you remember this, the measuring line will be drawn with a thin, light red line. The measuring line will always be a vertical line extending down from the picture plane.

Earlier (Figure 5.50) we said the object can be placed in four positions with respect to the picture plane. The measuring line for these four positions is located as follows:

1. When the picture plane is in front of the object, use the red pencil to extend one of the visible sides of the object to the picture plane line; then extend it straight down to the bottom of the paper (Figure 5.53*a*).
2. When the picture plane touches the near corner of the object, use the red pencil to draw the measuring line straight down from this intersection (Figure 5.53*b*).
3. When the picture plane penetrates the object (hopefully through a visible corner), the measuring line is drawn straight down from the point of the object that intersects the picture plane (Figure 5.53*c*).
4. When the entire object is located in front of the picture plane, extend one of the visible sides of the object to the picture plane, and then draw it straight down (Figure 5.53*d*).

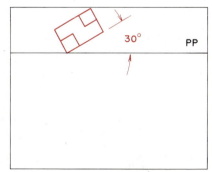

Figure 5.52 (*Step 2*) Establish plan view touching picture plane, rotated 30°.

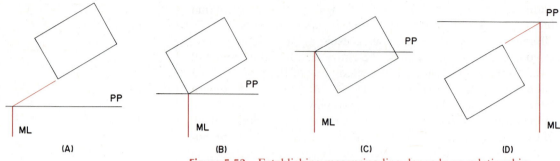

Figure 5.53 Establishing measuring line depends on relationship between location of plan view and picture plane.

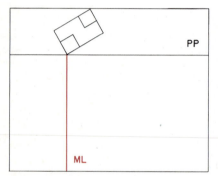

Figure 5.54 (*Step 3*) Draw and label measuring line.

Since our object is positioned with the near corner of the object touching the picture plane, use a thin red line and draw the measuring line straight down from this intersection. This will be the measuring line for your drawing. Remember, this is the only vertical line upon which true-length height measurements may be made (Figure 5.54).

5.4.6 The Horizon

The *horizon* represents the elevation level of the eye. You must relate this horizon with the elevation of the object when deciding how you want the final perspective view to look. There are four common horizon/elevation relationships. The bird's eye view (Figure 5.55*a*) has the horizon located above the object. In the person's eye view (Figure 5.55*b*), the horizon is generally placed 5 ft (scaled distance) above the *ground line*. With the ground level view (Figure 5.55*c*), the horizon is placed in line with the ground line. Finally, the worm's eye view (Figure 5.55*d*) has the horizon located well below the object. (You need to understand the re-

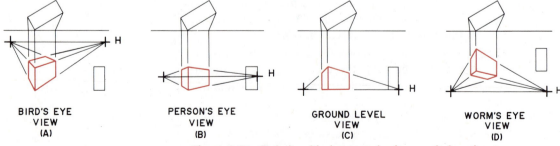

Figure 5.55 Relationship between horizon and elevation governs relationship between height of viewer's eye and object.

lationship between the horizon and the ground line of the elevation view.)

For this sample problem, we are selecting a bird's eye view. On your large sheet, draw a thin black horizontal line about 1–2 in. (25–50 mm) under the picture plane line across the entire width of the paper. Label this line H to identify the horizon (Figure 5.56).

For a bird's eye view, position the elevation view along the right edge of your paper in a position that is lower than the horizon, and tape it securely. Remember, the ground line (GL) is an extension of the bottom of the object (Figure 5.57). The elevation view must be drawn at the same scale as the plan view of the object.

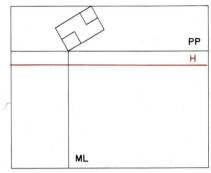

Figure 5.56 (*Step 4*) Draw the horizon.

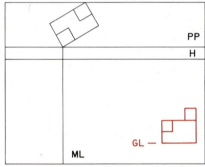

Figure 5.57 (*Step 5*) Place elevation view and label ground line.

5.4.7 The Station Point

The *station point* is the eye of the viewer; it involves two different relationships. First, the station point represents the distance the eye is located from the object. Second, the station point represents the relative position of the eye with respect to the orientation of the plan view. We shall illustrate these relationships by locating the station point on your sample drawing.

Since you are drawing a standard perspective, you may use a "rule of thumb" to locate the station point. Simply extend the widest extremes of the object down with thin, light black lines at an included angle of 30°. From this intersection, project horizontally to the measuring line. Place a small black cross at this intersection and label it SP to represent the station point. Actually, you have some degrees of freedom to move the station point along this horizontal line if you wish to emphasize a particular visible face of the object (Figure 5.58).

When using the footprint method for drawing building exteriors,

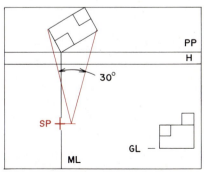

Figure 5.58 (*Step 6*) Locate the station point.

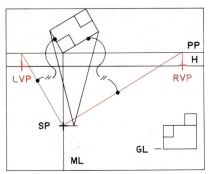

Figure 5.59 (*Step 7*) Identify location of left and right vanishing points.

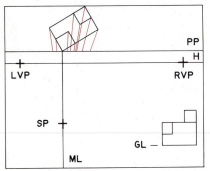

Figure 5.60 (*Step 8*) Project object intersections toward station point, but only draw as far as picture plane.

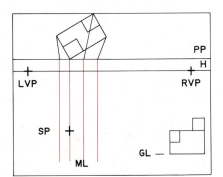

Figure 5.61 (*Step 9*) Drop projectors straight down from picture plane.

you would actually measure down from the picture plane line a specific distance you want the eye to be located from the object. For this method, the station point may be moved left or right a short distsance to position the eye relative to the plan view in order to show the perspective view from your selected point of reference to the object.

5.4.8   Vanishing Points

Next you must locate the *left vanishing point* and the *right vanishing point*. Using your red pencil, draw thin, light red lines from the station point to the picture plane line by drawing the lines parallel to the two visible surfaces on the plan view. With thin, light red lines, project these two intersections with the picture plane line down to the horizon line. The resultant two intersections become the left vanishing point and the right vanishing point. Construct a dark, small red cross at these intersections with the horizon, and label them LVP and RVP. You will not need the red pencil any more. The red pencil is only used to establish the measuring line and the vanishing points (Figure 5.59).

From this point on, only necessary projection lines will be shown in the step-by-step figures. This will make it easier for you to visualize the important features of that illustration.

5.4.9   Projectors

There are two steps for drawing the *projectors* from intersections in the plan view to the perspective view. First, use the blue pencil to draw light lines from the corners of the object in the plan view toward the station point, but *only draw them as far as the picture plane*. Note that these projectors will not be parallel (Figure 5.60).

Then using the black pencil, project with thin, light lines all of these intersections with the picture plane straight down below the level of the ground line. All visible vertical lines on the perspective view will fall somewhere on these vertical projectors that are now parallel to each other (Figure 5.61).

5.4.10   Visual Rays

Now that the setup is complete, start drawing the actual perspective view. This is accomplished totally by projecting points and connecting intersections from this point on. Use very light lines for your initial layout, and then begin to darken lines that you know will appear in the final view. We refer to these dark, final lines as "good" lines because they will appear on the final perspective view. Another helpful hint is to draw a light rectangular solid whose surfaces will include the extreme width, height, and depth of the object. By doing this, you have created a block within which all features of the object must fit. Also, this will help you establish the three visible planes that should help you to keep track of which lines should intersect with other lines on the same surface. This is described in the following paragraphs.

On your large paper, using the black pencil from now on, project horizontal lines from the top and bottom of the elevation view to the measuring line. Project these two points to the left and right vanishing points. These light lines are called *visual rays*. You should now be able to visualize the left and right principal planes of the perspective view by looking at the intersections of the existing visual rays and the appropriate vertical projectors. It is a good idea to make the outlines of these two surfaces a little darker so you can easily identify them (Figure 5.62).

To identify the top surface, project from the upper back corner of both the right and left visible planes to the opposite vanishing points. Slightly darken the portion of these new visual rays that outline the top surface of the perspective view. Note that the back corner in the perspective view intersects with the projector from the back corner of the plan view. You should now clearly see the three visible surfaces (top, left, and right). Later, reference will be made to this rectangular solid as the "big box." Some portions of these lines will become good lines shortly, but do not darken them yet (Figure 5.63).

Now that the exterior line figure exists, you should take a minute and study the object to be drawn. The object has three components: the base, the small block on top, and the notch. You want to draw each component as an individual entity. It is suggested you draw the base first, then the small block, and finally the notch. As portions of visual rays or projectors become good lines, darken them. This will help when visualizing subsequent parts of the object. One other helpful hint is to start at the correct height on the measuring line and follow the lines or points around the *exterior* surfaces of the *big box* and then project them into the interior of the big box.

To layout the base, project the height of the base horizontally from the elevation view to the measuring line. Project with visual rays to the vanishing points to establish the height of the right and left visible planes of the base. From the upper two corners of the left and right visible planes, project to the opposite vanishing points. These lines should be slightly darker than the visual rays (Figure 5.64).

Next, construct the small block that rests on the base. Since the top of the small block is on the same surface as the existing top of the original big box, construct the outline of the small block

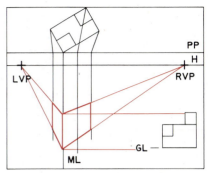

Figure 5.62 (*Step 10*) By projection, establish left and right visible planes.

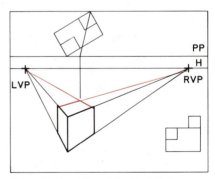

Figure 5.63 (*Step 11*) Establish top visible surface on big box.

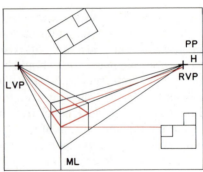

Figure 5.64 (*Step 12*) Project to establish top of base.

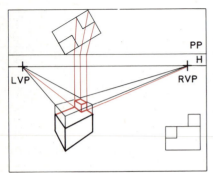

Figure 5.65 (*Step 13*) Project to outline small block.

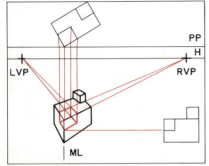

Figure 5.66 (*Step 14*) Projecting outline of notch.

on the top surface. Project the two outer edges down until they intersect the top edge of the base. Then project visual rays from the vanishing points through these intersections to form the bottom edge of the small block where it joins the top surface of the base. The near corner of the small block can now be added (Figure 5.65).

You should now be able to darken almost all the visible lines that comprise the perspective view. The exception is the top front corner where the notch will be located. Hidden lines are seldom, if ever, drawn. This type of view is designed to show only visible surfaces and features.

The notch should now be easy to lay out. Construct the top edge of the notch on the top surface of the base. Project the appropriate corners of the notch down to the bottom of the base. Then project the elevation of the notch to locate the bottom of the notch. From the location of the notch on the measuring line, draw visual rays to the vanishing points. Where these lines intersect the lines from the corners, provides enough information to complete the construction of the notch (Figure 5.66).

5.4.11 Presentation Variables

Earlier we recommended that you fasten the original layout drawing to the table (or better, a tracing table) and trace only the visible lines that make up the perspective view. If the finished drawing must be presented on hard board of some sort, the final tracing can be attached firmly to the hard board by one of several methods such as spray adhesive, dry mounting, and so on.

If the resultant perspective view is not the correct size for presentation when the original layout drawing is completed, simply enlarge or reduce this layout image using an office copy machine. Then trace the visible surfaces of the perspective at the correct size for presentation purposes (Figure 5.67).

By placing a piece of vellum over the line drawing, the illustrator can render the object with an airbrush (Figure 5.68).

Remember that perspectives take a lot of time to set up and draw but that they are pictorial drawings that closely represent what the eye sees. These drawings are generally used for presentations of architectural, construction, or product design projects. Mechanical parts and assemblies are generally drawn in pictorial views by using isometric or oblique drawings.

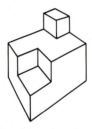

Figure 5.67 Line drawing of sample perspective that has been enlarged and traced.

5.4.12 Angular Surfaces

Once the setup has been completed and you know how to draw rectangular solids and visible surfaces, adding angular planes is not difficult. The only thing you must remember is to locate points on the angular plane (preferably on the outside visible planes) and then connect the points (Figure 5.69). Note that the points were first plotted onto the visible planes and then connected to other existing points drawn toward the appropriate vanishing points (Figure 5.70).

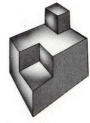

Figure 5.68 Sample perspective rendered with an airbrush.

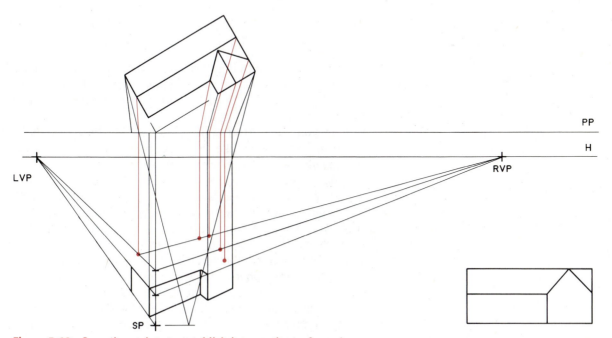

Figure 5.69 Locating points to establish intersections of angular surfaces.

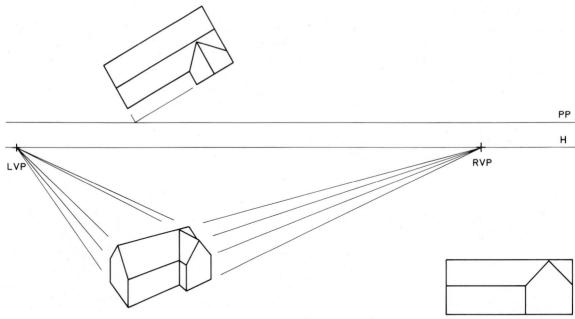

Figure 5.70 Line drawing of two-point perspective with intersecting angular planes.

5.4.13 Curved Surfaces

Drawing curved surfaces in perpsective is not too difficult either. However, this process involves creating a series of planes and points and the use of an irregular curve to connect the plotted points.

Assume you want to draw a curved surface in perspective. Draw all the setup lines like those described earlier. Construct the rectangular solid outlining the perspective view.

You must create the edge view of some planes that intersect the edge view of the curved surface in the plan view. You may select the number of planes that are necessary depending on how accurate you want the curve to be produced in the perspective view. Next, project these planes down to the perspective view. You are able to see that the end of each plane in the perspective view represents the upper and lower points (on that plane) where the plane intersects the curved surface (Figure 5.71).

After enough of these points have been located, use an irregular curve and connect the points to create the upper and lower edges of the curved surface (Figure 5.72).

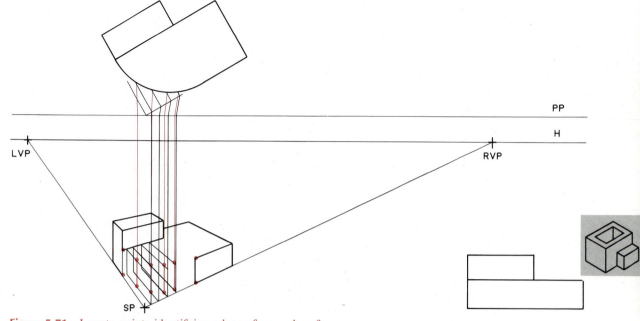

Figure 5.71 Locate points identifying edges of curved surfaces.

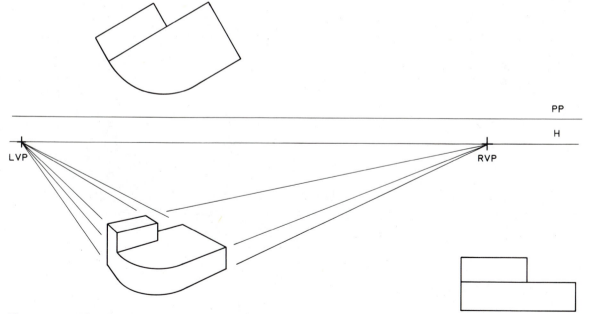

Figure 5.72 Line drawing of two-point perspective with curved surface.

5.4.14 Three-Point Perspective

Three-point perspective is seldom used commercially. It will not be emphasized or illustrated in this edition. In the three-point perspective, the vertical projectors become visual rays that converge either above the top or below the bottom of the perspective view. Architects, may sometimes select this type of drawing to emphasize a particular feature, but for all practical purposes, the three-point perspective is not used.

5.5 COMPUTER APPLICATIONS

There are professional firms that specialize in producing perspective line drawings for architects and engineers. When the need for such a service occurs, you send the elevations and plan views to the firm; they digitize the information and print out a line illustration in the type of perspective requested. Once the information has been digitized, it is easy to turn out different types of perspectives or different views based on different locations of the observer's eye (Figures 5.73 and 5.74).

Persons having access to computers that accommodate com-

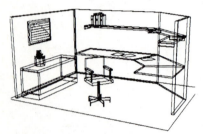

Figure 5.73 Computer-generated perspective view is generally used to save time on complicated perspectives. (*Computer-generated image*) (Courtesy of Ohio State University School of Architecture)

Figure 5.74 Computer-generated perspective view. (*Computer-generated image*) (Courtesy of Ohio State University School of Architecture)

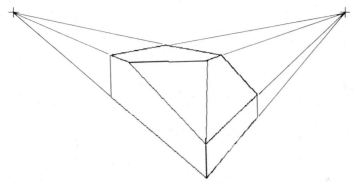

Figure 5.75 Computer software that allows for rubber banding can be used to draw perspective views on computer screen. (*Computer-generated image*)

puter graphics can make a perspective drawing of sorts. However, it is much easier if the rubber banding feature is available. This allows the person creating the drawing to start a line at the vanishing point and connect the other end of the line on the object where desired. This is a great time saver (Figure 5.75).

5.6 *LINE APPLICATIONS*

Very light construction lines are used when making a pictorial drawing. You will find that some drawings requiring many sets of center lines will result in a cluttered drawing during the layout process. If the drawing can easily be cleaned up, then use that piece of paper for the final drawing. If not, it is suggested you trace the final drawing on a piece of vellum or film using the original construction and center lines as guides for the tracing process (Figure 5.76).

Since hidden lines are hardly ever used on pictorial drawings, the only lines commonly drawn are object lines and some center lines. The object lines are wide lines, and the center lines are thin lines. The exception is when some dimensions are added to the pictorial view, in which case some additional thin lines are used.

5.7 *SAMPLE PROBLEM: TABLE SAW FIXTURE*

The data base developed in the orthographic views for Chapter 4 was used to make a pictorial view of the object shown in Figure 5.77. This was accomplished by combining the data points, re-

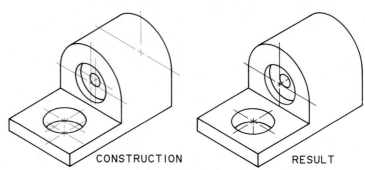

CONSTRUCTION   RESULT

Figure 5.76   If pictorial view construction requires many lines, it may be smarter to trace lines that make up final pictorial drawing on separate sheet.

volving the wire frame diagram, and eliminating nonvisible lines.

If a less powerful program is used, then the data points must be generated to make up the pictorial view. This also includes the manipulation of ellipses. In this case, the use of an isometric GRID feature is very helpful.

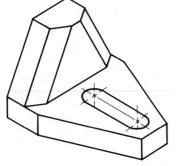

Figure 5.77   Isometric pictorial view of table saw fixture.

SUMMARY

Pictorial drawings are drawn to improve the communication process. They are likely to be less detailed than the orthographic or working drawings. Visualization for both the trained and untrained eye will be quicker when pictorial drawings are presented.

PROBLEMS

Problems 5.1–5.60

The following problems are given in one-, two-, or three-view orthographic drawings. Your instructor will assign these problems as pictorials. Where three orthographic views are given, these three views will become the three visible planes in the isometric drawing.

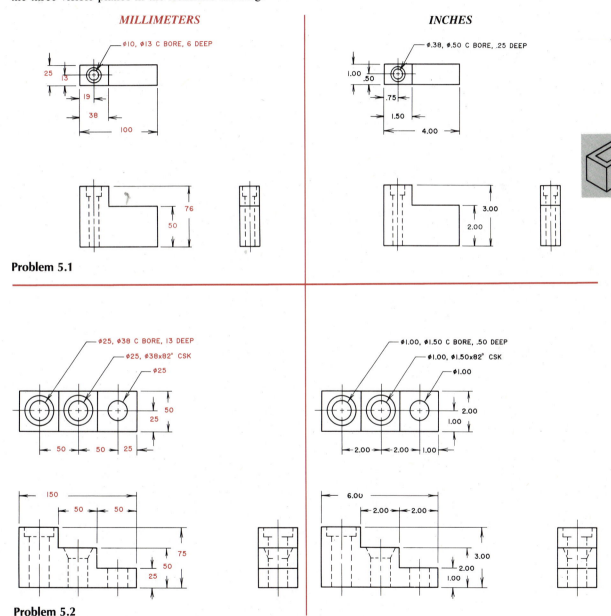

MILLIMETERS   *INCHES*

Problem 5.1

Problem 5.2

Problems 5.1–5.60

The following problems are given in one-, two-, or three-view orthographic drawings. Your instructor will assign these problems as pictorials. Where three orthographic views are given, these three views will become the three visible planes in the isometric drawing.

MILLIMETERS *INCHES*

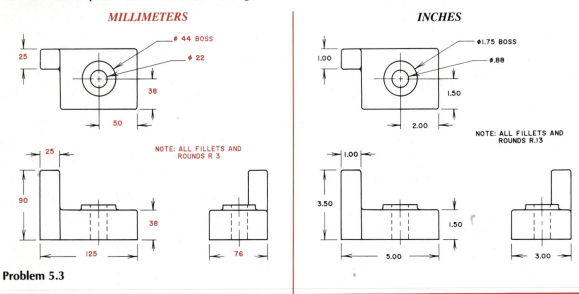

Problem 5.3

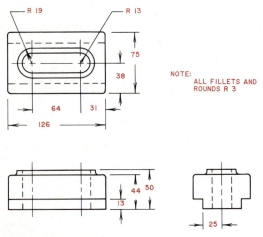

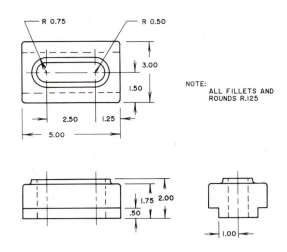

Problem 5.4

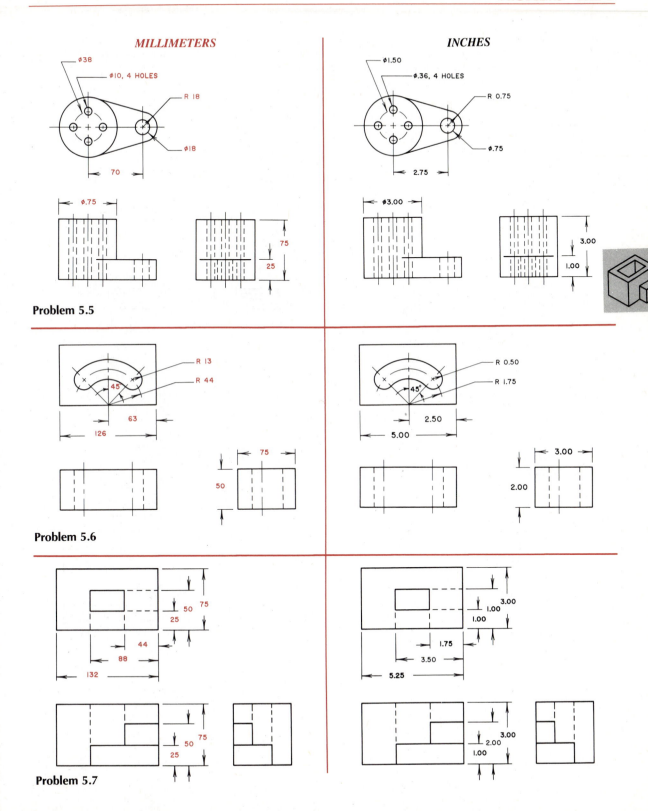

MILLIMETERS

ϕ38

ϕ10, 4 HOLES

R 18

ϕ18

70

ϕ.75

75

25

INCHES

ϕ1.50

ϕ.36, 4 HOLES

R 0.75

ϕ.75

2.75

ϕ3.00

3.00

1.00

Problem 5.5

R 13

R 44

45

63

126

75

50

R 0.50

R 1.75

45°

2.50

5.00

3.00

2.00

Problem 5.6

50 75

25

44

88

132

75

50

25

3.00

1.00

1.00

1.75

3.50

5.25

3.00

2.00

1.00

Problem 5.7

Problems 5.1–5.60

The following problems are given in one-, two-, or three-view orthographic drawings. Your instructor will assign these problems as pictorials. Where three orthographic views are given, these three views will become the three visible planes in the isometric drawing.

MILLIMETERS   *INCHES*

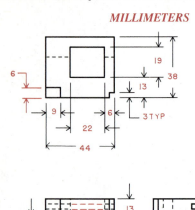

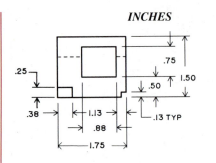

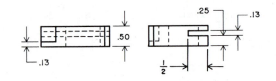

Problem 5.8 (*Computer-generated image*)

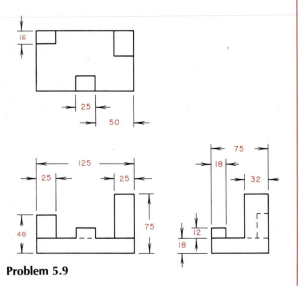

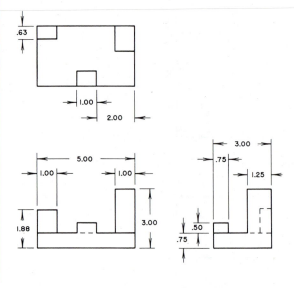

Problem 5.9

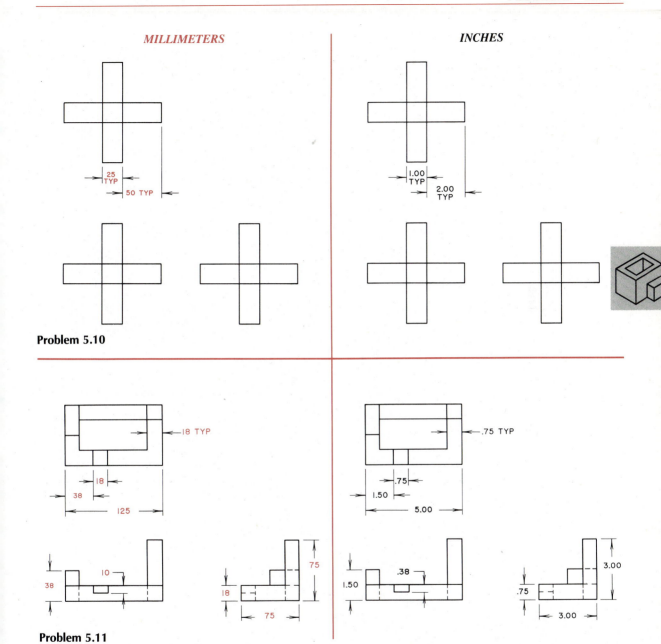

MILLIMETERS

INCHES

25 TYP
50 TYP

1.00 TYP
2.00 TYP

Problem 5.10

18 TYP
18
38
125

.75 TYP
.75
1.50
5.00

38
10
75
75

1.50
.38

18

3.00
75

.75
3.00

Problem 5.11

Problems 5.1–5.60

The following problems are given in one-, two-, or three-view orthographic drawings. Your instructor will assign these problems as pictorials. Where three orthographic views are given, these three views will become the three visible planes in the isometric drawing.

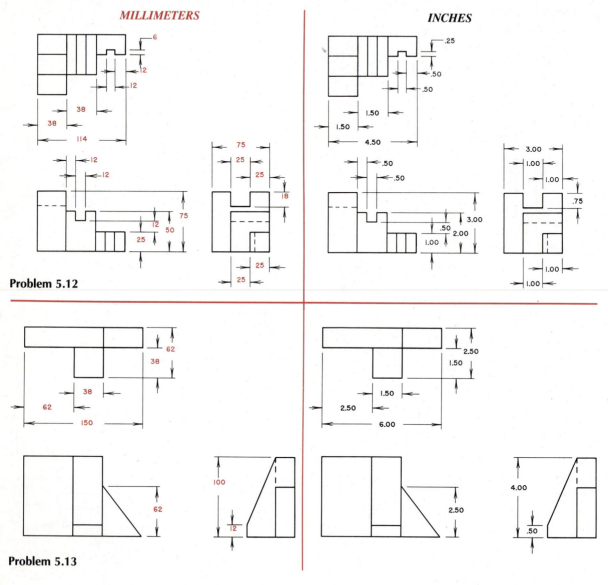

MILLIMETERS   *INCHES*

Problem 5.12

Problem 5.13

MILLIMETERS   *INCHES*

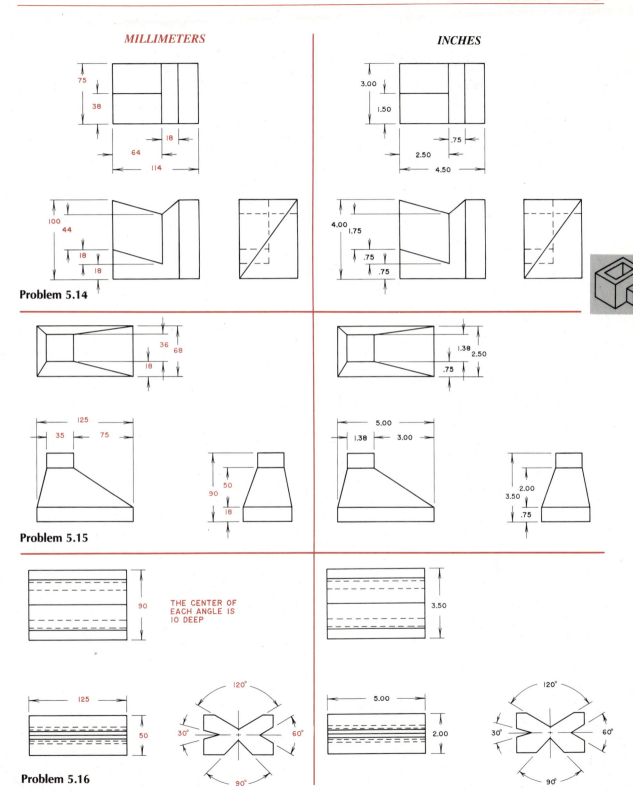

Problem 5.14

75
38
18
64
114
100
44
18
18

3.00
1.50
.75
2.50
4.50
4.00
1.75
.75
.75

Problem 5.15

36
68
18
125
35
75
90
50
18

1.38
2.50
.75
5.00
1.38
3.00
2.00
3.50
.75

Problem 5.16

90
THE CENTER OF
EACH ANGLE IS
10 DEEP
125
50
120°
30°
60°
90°

3.50
5.00
2.00
120°
30°
60°
90°

Problems 5.1–5.60

The following problems are given in one-, two-, or three-view orthographic drawings. Your instructor will assign these problems as pictorials. Where three orthographic views are given, these three views will become the three visible planes in the isometric drawing.

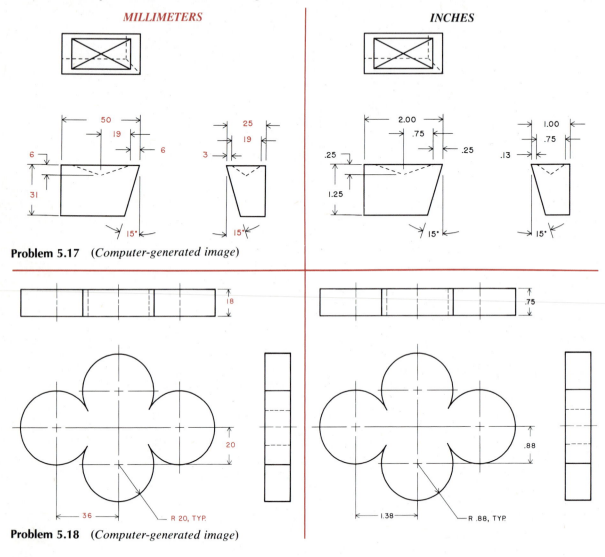

MILLIMETERS   *INCHES*

Problem 5.17   (*Computer-generated image*)

Problem 5.18   (*Computer-generated image*)

MILLIMETERS   *INCHES*

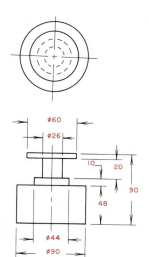

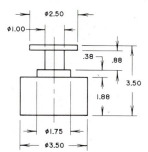

Problem 5.19

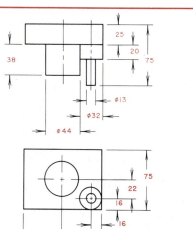

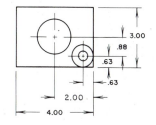

Problem 5.20

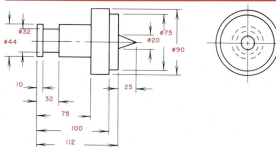

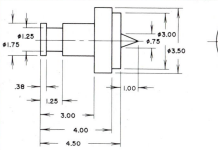

Problem 5.21

Problems 5.1–5.60
The following problems are given in one-, two-, or three-view orthographic drawings. Your instructor will assign these problems as pictorials. Where three orthographic views are given, these three views will become the three visible planes in the isometric drawing.

MILLIMETERS   *INCHES*

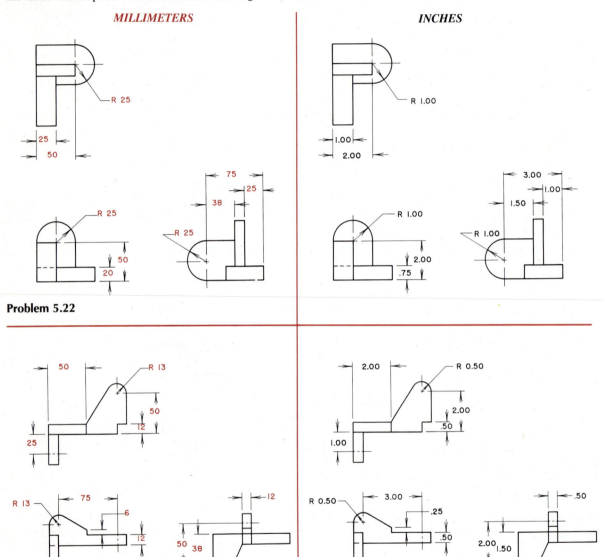

Problem 5.22

Problem 5.23

MILLIMETERS

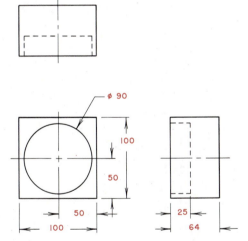

Problem 5.24

INCHES

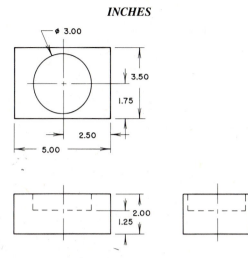

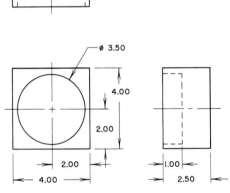

Problem 5.25

Problems 5.1–5.60
The following problems are given in one-, two-, or three-view orthographic drawings. Your instructor will assign these problems as pictorials. Where three orthographic views are given, these three views will become the three visible planes in the isometric drawing.

MILLIMETERS   *INCHES*

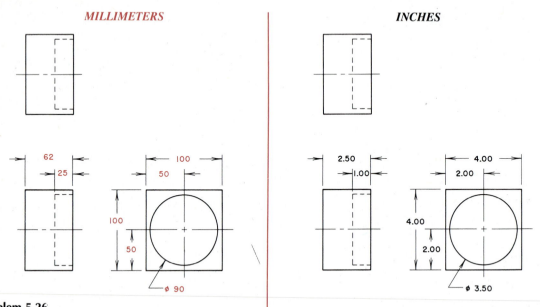

Problem 5.26

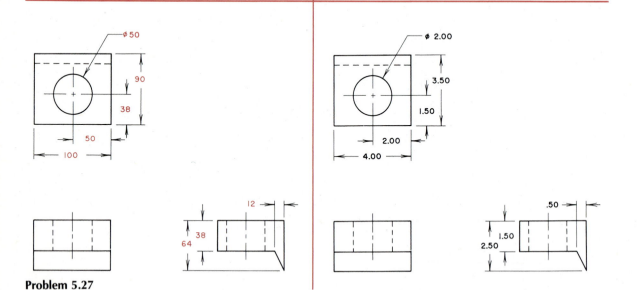

Problem 5.27

MILLIMETERS   *INCHES*

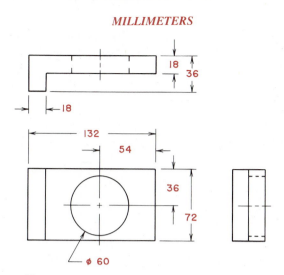

Problem 5.28

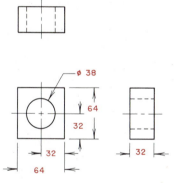

Problem 5.29

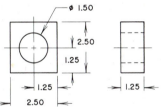

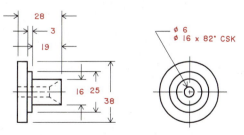

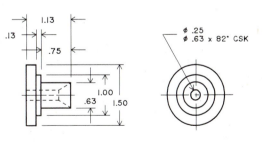

Problem 5.30   (*Computer-generated image*)

Problems 5.1–5.60

The following problems are given in one-, two-, or three-view orthographic drawings. Your instructor will assign these problems as pictorials. Where three orthographic views are given, these three views will become the three visible planes in the isometric drawing.

MILLIMETERS   *INCHES*

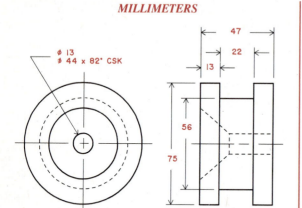

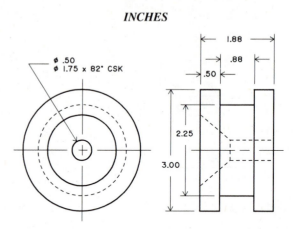

Problem 5.31 (*Computer-generated image*)

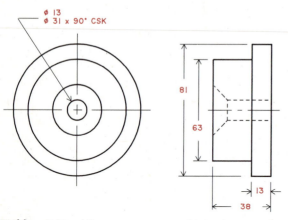

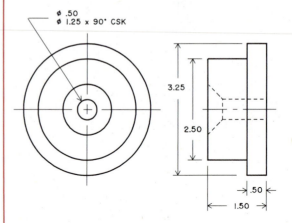

Problem 5.32 (*Computer-generated image*)

MILLIMETERS

INCHES

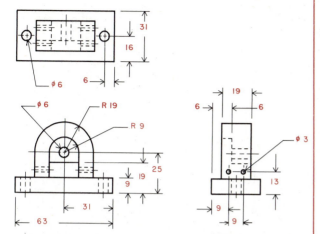

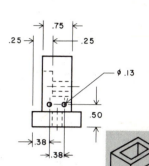

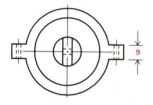

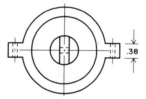

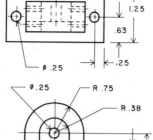

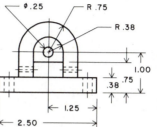

Problem 5.33 (*Computer-generated image*)

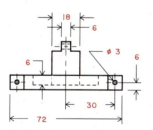

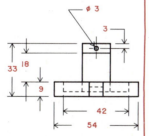

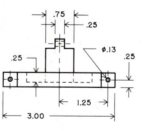

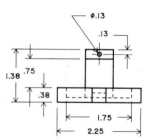

Problem 5.34 (*Computer-generated image*)

Problems 5.1–5.60

The following problems are given in one-, two-, or three-view orthographic drawings. Your instructor will assign these problems as pictorials. Where three orthographic views are given, these three views will become the three visible planes in the isometric drawing.

MILLIMETERS   *INCHES*

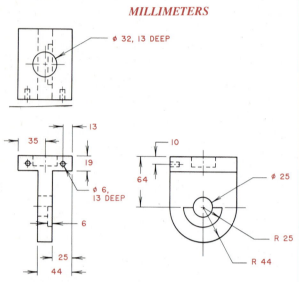

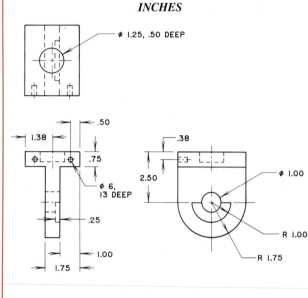

Problem 5.35   (*Computer-generated image*)

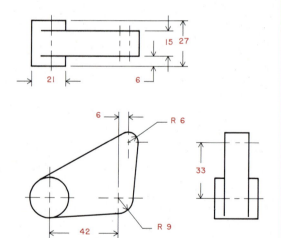

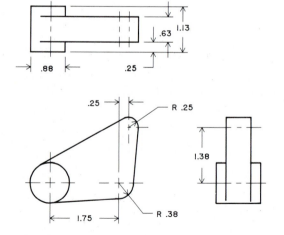

Problem 5.36   (*Computer-generated image*)

MILLIMETERS   *INCHES*

R 18   R 27   51   ⌀ 18   30   6   30   9

R .75   R 1.13   2.13   ⌀ .75   1.25   .25   1.25   .38

Problem 5.37   (*Computer-generated image*)

R 9   24   18   24   24   9   12   30

R .38   1.00   .75   1.00   1.00   .38   .50   1.25

Problem 5.38   (*Computer-generated image*)

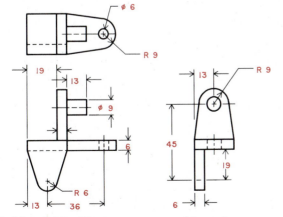

⌀ 6   R 9   19   13   ⌀ 9   6   13   R 9   45   19   R 6   13   36   6

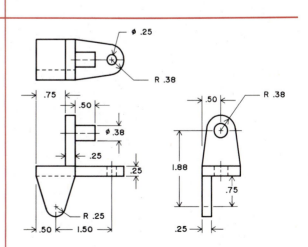

⌀ .25   R .38   .75   .50   ⌀ .38   .25   .25   .50   R .38   1.88   .75   R .25   .50   1.50   .25

Problem 5.39   (*Computer-generated image*)

Problems 5.1–5.60

The following problems are given in one-, two-, or three-view orthographic drawings. Your instructor will assign these problems as pictorials. Where three orthographic views are given, these three views will become the three visible planes in the isometric drawing.

MILLIMETERS   *INCHES*

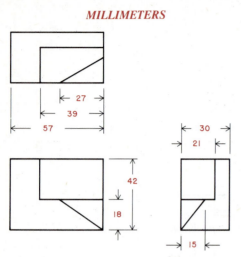
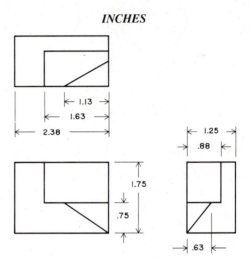

Problem 5.40   *(Computer-generated image)*

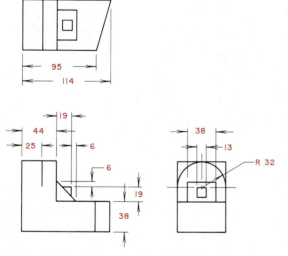
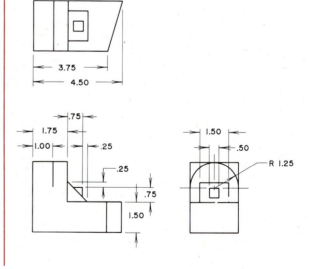

Problem 5.41   *(Computer-generated image)*

MILLIMETERS

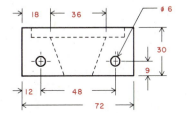
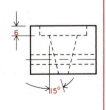
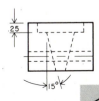

Problem 5.42 (*Computer-generated image*)

INCHES

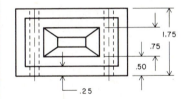
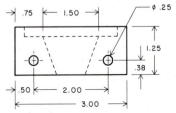

Problem 5.43 (*Computer-generated image*)

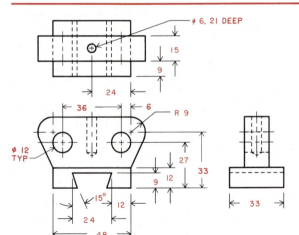
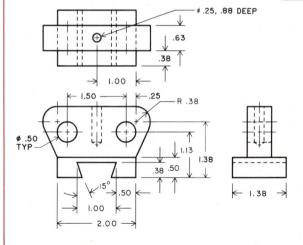

Problem 5.44 (*Computer-generated image*)

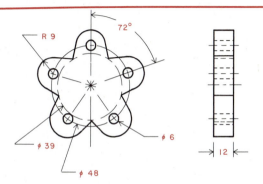
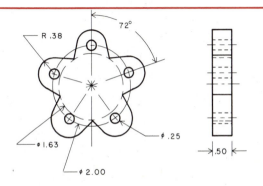

Problems 5.1–5.60

The following problems are given in one-, two-, or three-view orthographic drawings. Your instructor will assign these problems as pictorials. Where three orthographic views are given, these three views will become the three visible planes in the isometric drawing.

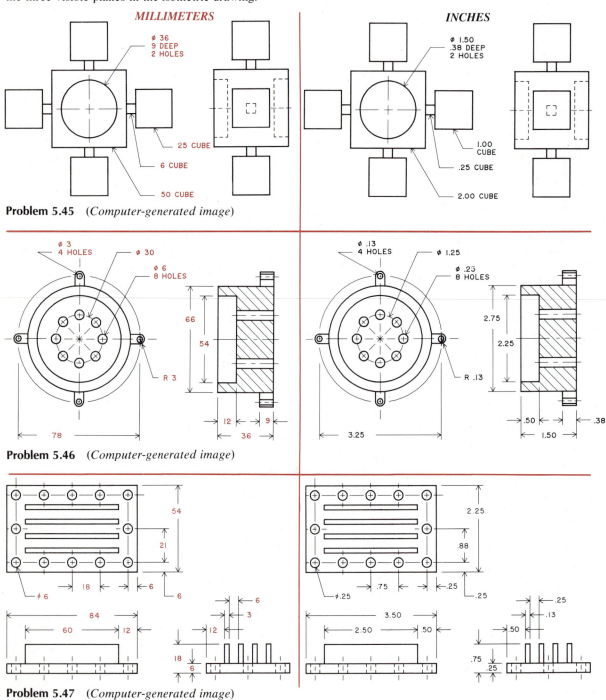

MILLIMETERS   *INCHES*

Problem 5.45   (*Computer-generated image*)

Problem 5.46   (*Computer-generated image*)

Problem 5.47   (*Computer-generated image*)

MILLIMETERS

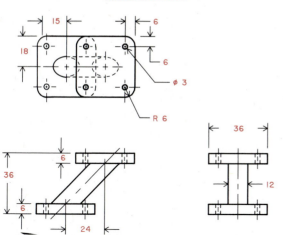

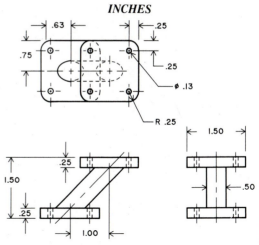

INCHES

Problem 5.48 (*Computer-generated image*)

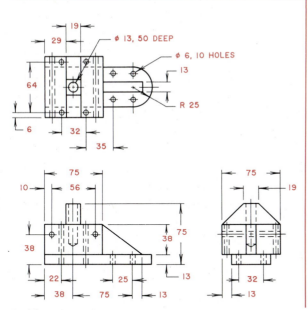

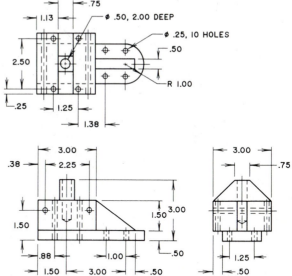

Problem 5.49

Problems 5.1–5.60

The following problems are given in one-, two-, or three-view orthographic drawings. Your instructor will assign these problems as pictorials. Where three orthographic views are given, these three views will become the three visible planes in the isometric drawing.

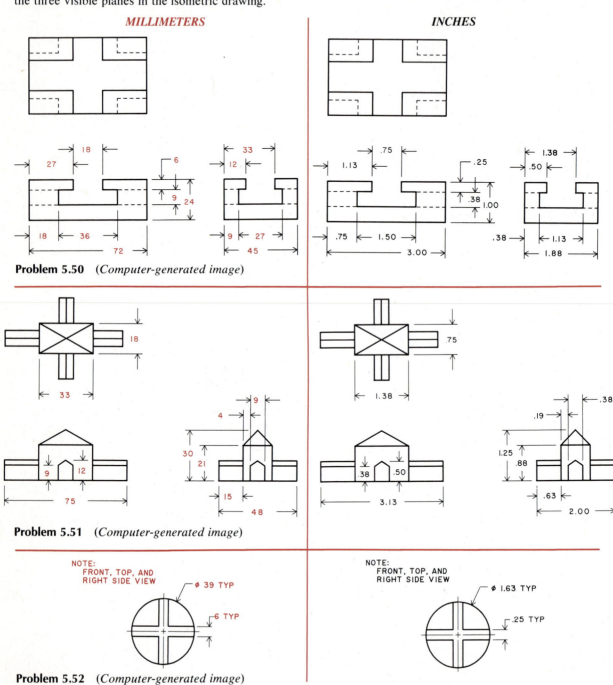

MILLIMETERS   *INCHES*

Problem 5.50 (*Computer-generated image*)

Problem 5.51 (*Computer-generated image*)

Problem 5.52 (*Computer-generated image*)

MILLIMETERS   INCHES

Problem 5.53 (*Computer-generated image*)

Problem 5.54 (*Computer-generated image*)

Problem 5.55 (*Computer-generated image*)

Problems 5.1–5.60
The following problems are given in one-, two-, or three-view orthographic drawings. Your instructor will
assign these problems as pictorials. Where three orthographic views are given, these three views will become
the three visible planes in the isometric drawing.

MILLIMETERS   *INCHES*

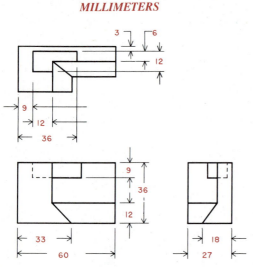

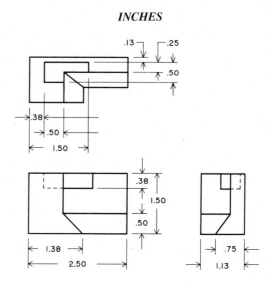

Problem 5.56 *(Computer-generated image)*

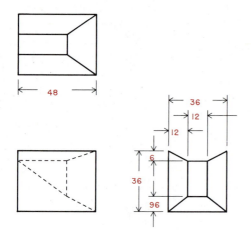

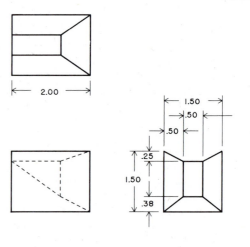

Problem 5.57 *(Computer-generated image)*

MILLIMETERS   *INCHES*

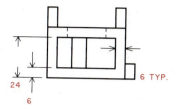

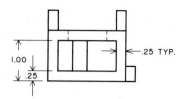

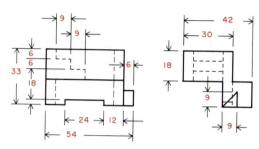

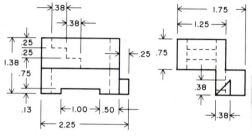

Problem 5.58

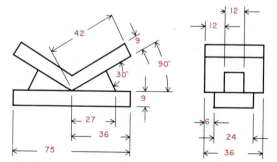

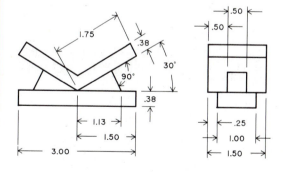

Problem 5.59   (*Computer-generated image*)

Problems 5.1–5.60

The following problems are given in one-, two-, or three-view orthographic drawings. Your instructor will assign these problems as pictorials. Where three orthographic views are given, these three views will become the three visible planes in the isometric drawing.

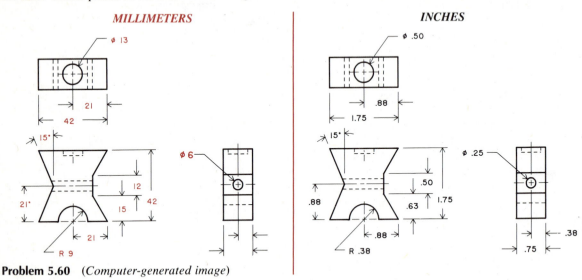

Problem 5.60   (*Computer-generated image*)

Problems 5.61–5.63

The following problems are to be drawn as one-point perspectives. The eye should be 5 feet above the floor in these three problems.

Problem 5.61

Draw an interior one-point perspective of a room 8 feet high that has a floor plan 20 feet square.

Problem 5.62

Draw an interior one-point perspective of a room that has a rectangular floor plan. On the back wall place a doorway, and on each of the side walls insert a window. When drawing the windows, show the thickness of the wall in which the window is located.

Problem 5.63

Draw an interior one-point perspective of a kitchen. Remember, your eye is 5 feet above the floor. In the kitchen, draw a refrigerator, floor-mounted cabinets, and counters that extend along two adjacent walls and have some overhead cabinets that touch the ceiling along one wall.

Problems 5.64–5.65

The following problems are to be drawn as two-point perspectives. You select whether you want a bird's eye view, person's eye view, ground line view, or a worm's eye view. Consider Problem 5.51 as a building. You establish the scaled heights, widths, and depths.

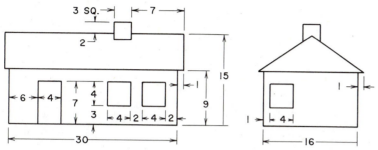

Problem 5.64

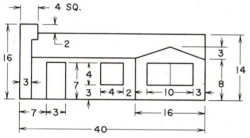

Problem 5.65

Problem 5.66   Design a kitchen using the information below and draw as a one-point perspective.

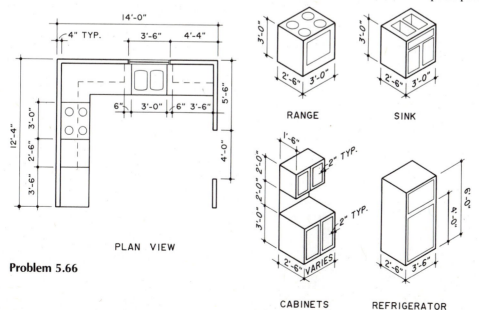

PLAN VIEW

RANGE   SINK

CABINETS   REFRIGERATOR

Problem 5.66

Problems 5.67 and 5.68

Draw these orthographic problems as isometric pictorials.

MILLIMETERS   *INCHES*

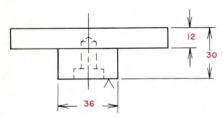

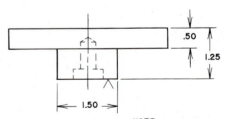

NOTE:
THE CENTER OF THE CYLINDRICAL
FEATURES IS EQUIDISTANT FROM
ALL SIDES OF THE TRIANGLE

NOTE:
THE CENTER OF THE CYLINDRICAL
FEATURES IS EQUIDISTANT FROM
ALL SIDES OF THE TRIANGLE

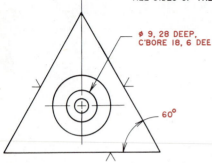

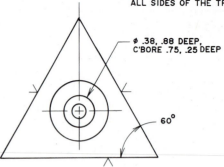

Problem 5.67

MILLIMETERS

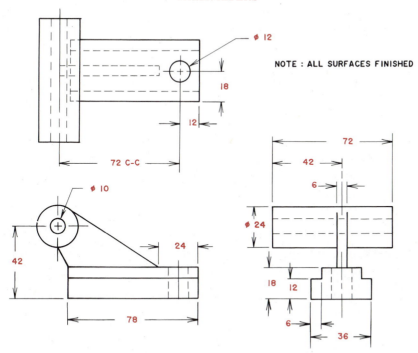

NOTE : ALL SURFACES FINISHED

Problem 5.68

INCHES

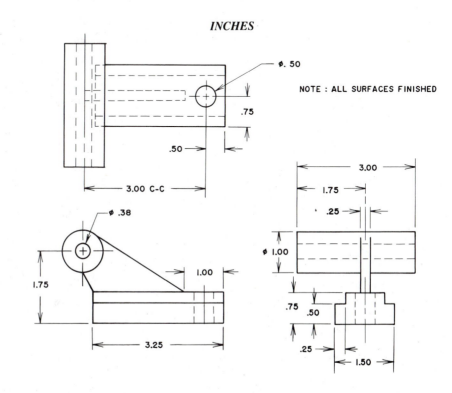

NOTE : ALL SURFACES FINISHED

TERMS YOU WILL SEE IN THIS CHAPTER

CUTTING PLANE Imaginary surface used to cut through an object to reveal its interior.

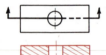

SECTION VIEW Orthographic view that shows the interior features of an object as visible lines.

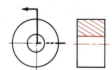

SECTION LINES Lines drawn on the cut surfaces of an interior portion of an object that has been cut by a cutting plane; also called crosshatch lines.

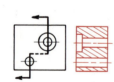

FULL SECTION Section view formed by passing a cutting plane entirely through an object to reveal its interior features.

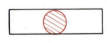

HALF SECTION Section view of symmetrical objects formed by passing the cutting plane in such a manner so as to show the interior of half the object.

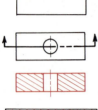

OFFSET SECTION Section view created with multiple cutting planes and used to show interior features that cannot be located with a single cutting plane.

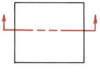

REVOLVED SECTION Section view of an elongated symmetrical feature where the cutting plane is passed perpendicular to the axis of symmetry and the interior feature is revolved 90° into the plane of the drawing.

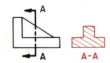

REMOVED SECTION Section view similar to a revolved section except that the section view is not drawn within the view containing the cutting plane.

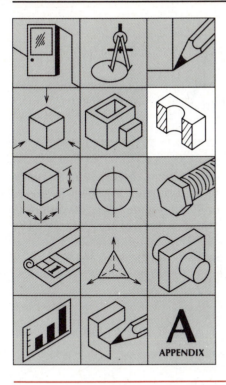

6

SECTIONS AND CONVENTIONS

6.1 INTRODUCTION

Standard orthographic views show all features of objects. Interior features may be very complex, and standard orthographic views may lead to confusion when several hidden lines are present. The confusion associated with these hidden lines may be reduced by cutting away part of the object and looking at the interior as a section view (see Figure 6.1 for samples).

In this chapter you will learn the concepts and techniques of sectioning. Also, there are conventional practices and standards used in making engineering drawings that you will learn and become familiar with so that you are able to read and produce technical drawings.

6.2 CUTTING PLANE

Figure 6.2*a* is a pictorial drawing with an imaginary **cutting plane** passing through an object. This cutting plane is similar to a knife

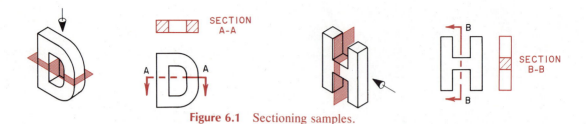

Figure 6.1 Sectioning samples.

or saw blade cutting the object. After the object is cut, a portion is removed to reveal the interior features, as shown in Figure 6.2b. The object shown pictorially in Figure 6.2 is shown orthographically in Figure 6.3. The interior features are represented by dashed lines in the front view. These interior features become visible and are shown more clearly in Figure 6.3b, which is a **section view** of the object. The section view is created by passing

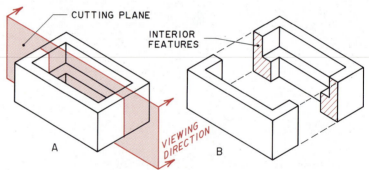

Figure 6.2 (a) Imaginary cutting plane passing through object. (b) Interior features of object are shown along line where cutting plane passed.

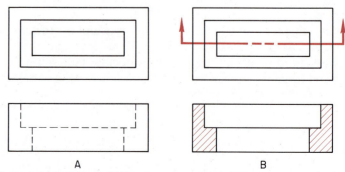

Figure 6.3 (a) Normal front and top orthographic views of object showing hidden features in front view. (b) Same object showing cutting plane in top view and resulting sectional view.

a cutting plane through the object and imagining that it is "cut" along this plane and a portion of the object is removed. The hidden features are now shown as visible lines, and the part of the object in contact with the cutting plane is indicated with crosshatching. When drawing section views, the edge view of the cutting plane is shown by a thick line, as shown in the top view of Figure 6.3*b*. Avoid hidden lines in section views. A second section view or additional orthographic views may be used to reveal additional interior or exterior details.

Figure 6.4 shows front, top, and right side section views. Each cutting plane line has arrows that are perpendicular to the cutting plane and point in the direction of the line of sight for the section view. Also, the cutting plane often is identified with alphabet characters such as *A–A* or *B–B*. When the cutting plane is identified in this manner, the view is called section *A–A* or section *B–B*. This identification technique is especially useful when showing removed sections (discussed in Section 6.8). Two types of cutting plane lines used in section views are shown in Figure 6.5. The line shown in Figure 6.5 (top marked *A–A*) is typically used

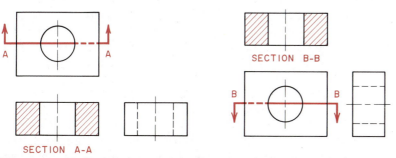

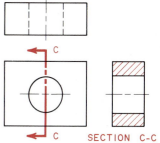

SECTION A-A

SECTION B-B

SECTION C-C

Figure 6.4 (*a*) Cutting plane and front section view. (*b*) Cutting plane and top or horizontal section view. (*c*) Cutting plane and profile or right side section view.

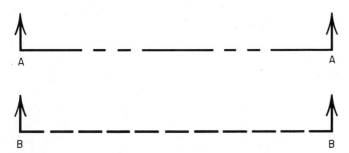

Figure 6.5 (*top*) Example of type of cutting plane line used when cutting plane is long. (*bottom*) Example of type of cutting plane line used when cutting plane is short.

when the cutting plane line is long. The cutting plane line shown in Figure 6.5 (bottom marked *B–B*) shows the line style that is used for short cutting plane lines.

6.3 SECTION LINES

Section lines, or crosshatch lines, are shown in the section view. Generally, the portion of the object that is in contact with the cutting plane is crosshatched to show the features that were once hidden but are now visible.

Historically, the section line pattern indicated a specific type of material to be used to make the object. For example, Figure 6.6 shows the patterns traditionally used for many fairly common materials. The pattern of uniformly spaced lines shown for cast iron in Figure 6.6 is now accepted as the standard for section lines regardless of the material from which the object is made; however, use of the other material patterns is optional. This standard is established by the American National Standards Institute (ANSI) in ANSI Y14.2M-1979.

Section lines are normally spaced approximately 0.06–0.12 in. (2–4 mm) apart by eye, although some triangles have a line etched on them to ensure uniform spacing of section lines. A line can be scribed on the triangle using the point of the dividers or compass (Figure 6.7). Simply place the triangle over a previously drawn section line, with the etched line covering the section line, to draw another section line at the correct spacing (Figure 6.8).

Section lines should be thin and dark, never varying in thickness. There should be a distinct contrast between the thickness of object (or visible) lines and the thickness of section lines. The 1982 ANSI standards call for object lines to be 0.7 mm thick and section lines 0.35 mm thick. Section lines must end exactly at the object outline, never extending beyond the object outline or stopping short of the object outline.

Section lines are normally slanted at a 45° angle, but any standard angle, such as 30° or 60°, can be used. Choose this angle so that the section lines will not be parallel to the object outlines of the part. Section lines that are perpendicular to the object outlines of the part are acceptable but not preferred. Horizontal and vertical section lines are not preferred as well (Figure 6.9).

Drawing section lines on a computer screen using a computer-aided design and drafting (CADD) program can be done with ease and precision. Most CADD programs allow the user to define a boundary area, and then by using a fill or hatch command, the area is filled with a pattern selected by the user. The user also selects the angle and spacing of the section lines. The

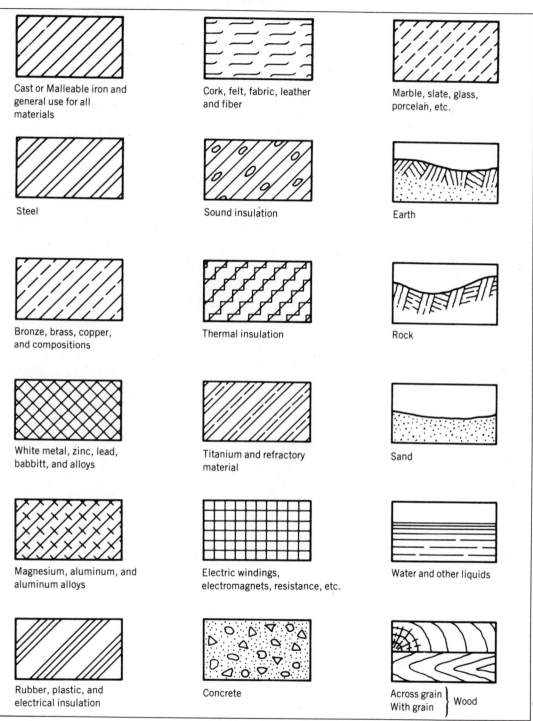

Figure 6.6 Section line patterns for various engineering materials. (Courtesy of ANSI)

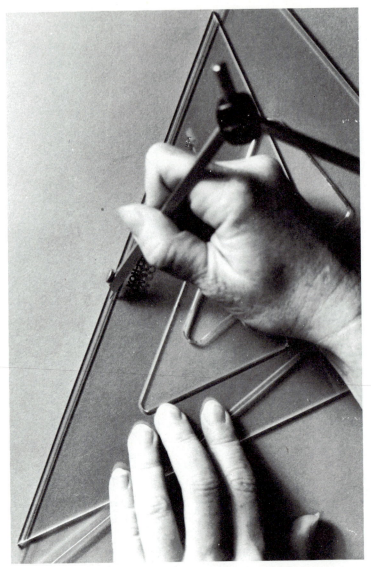

Figure 6.7 Etching triangle with line to aid in drawing uniform section lines.

more sophisticated CADD programs offer the user a wide range of hatch styles to accommodate engineering, architectural, and various other applications.

When using a CADD program that does not have an automatic fill or hatch command, a section view can still be drawn by carefully laying out a line at a 45° angle on the area to be sectioned. Using the repeat command, you can completely cross-hatch the section view by repeating the line at the desired spac-

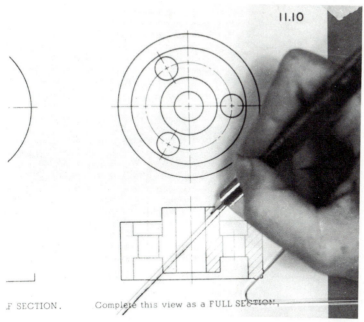

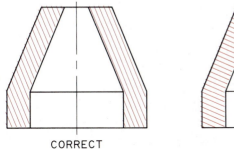

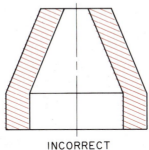

Figure 6.8 Using etched triangle to properly space section lines.

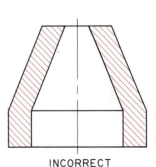

CORRECT INCORRECT INCORRECT

Figure 6.9 Choosing correct angle for section lines.

ing. The crosshatching will not only cover the desired area but also areas that you do not want to show as sectioned. By deleting the portion of the section lines that are not part of the section view using the line removal feature, a good section view can be made.

6.4 *FULL SECTION*

A **full section** is a section view formed by passing a cutting plane entirely through an object to reveal all of its interior features. An

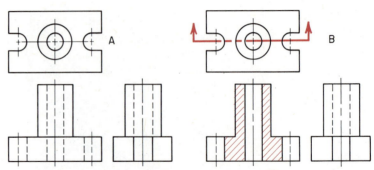

Figure 6.10 (*a*) Normal orthographic drawing of object. (*b*) Orthographic drawing of same object showing front view in full section (half the object is removed).

entire view (top, front, right side, etc.) is shown in section, and this view is said to be a full section. A full-section view generally shows half the object removed.

Figure 6.10 shows the top, front, and side views of an object. The front view can be drawn as a full section simply by passing a cutting plane fully through the top view and removing the front portion of the object. This procedure yields a full-section view in the front view, as shown in Figure 6.10*b*. The position of the cutting plane indicates the exact location of the section view, and the arrows on the cutting plane indicate the viewing direction. The area that is cut by the cutting plane is crosshatched with uniform section lines. For simple full sections, the cutting plane may be omitted when its position is obvious.

A full section through a cylindrical object introduces some common problems with sectioning. Figure 6.11 shows a cylindrical part in full section. A common error that should be avoided is to omit visible lines behind the cutting plane (Figure 6.11).

Hidden lines in full sections should be omitted unless they are necessary to provide a clear understanding of the sectioned view.

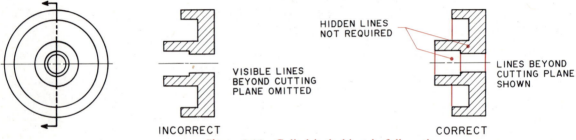

Figure 6.11 Cylindrical object in full section.

Including hidden lines in the section view is orthographically correct but may lead to confusion rather than clarity.

6.5 HALF SECTIONS

Symmetrical objects often can best be shown using **half-section** views. This method of sectioning simply requires that a quarter of the object be removed to expose the interior features of the part. Figure 6.12 shows an example of a half-section view. The cutting plane is really two cutting planes meeting perpendicular to one another at the axis of symmetry of the object. Removing one quarter of the object reveals the internal features of the object (Figure 6.12).

A half-section view has the advantage of exposing the interior features of half of the object and keeping the exterior features displayed on the other half. It is extremely important to keep in mind that one side of the half section shows the interior while the other side shows the exterior. The dividing line between the interior and exterior is not a solid line. It is a center line that denotes the axis of symmetry of the object. A common error made in drawing half sections is to include hidden lines on the half showing the exterior. This is wrong.

Since a half-section view shows both the interior and exterior in a single view, its usefulness is largely limited to symmetrical objects. Half-section views are very common in assembly drawings. Assembly drawings often require that both the interior and exterior features of the object be shown. Therefore, half-section views are very useful under these circumstances.

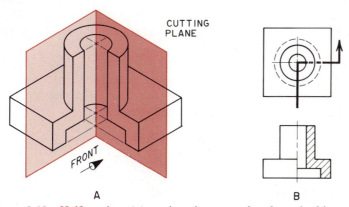

Figure 6.12 Half section: (*a*) cutting planes passing through object and (*b*) half-section view of object.

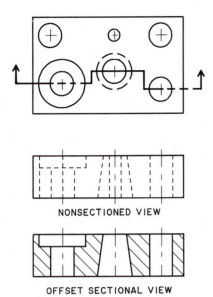

NONSECTIONED VIEW

OFFSET SECTIONAL VIEW

Figure 6.13 Offset section. (*Computer-generated image*)

6.6 OFFSET SECTIONS

Many times interior features are not located within a single plane. In order to include such features in a section view, the cutting plane may be stepped, or offset, to pass through all these features. The cutting plane is offset at right angles. The section view is then drawn as if the internal features were in a single plane or in a straight line. This type of section view is called an **offset section** and can be considered a special type of full section. Note that the offsets, or 90° turns, that the cutting plane makes to pass through each feature are not shown in the section view. Figure 6.13 shows an example of an offset section.

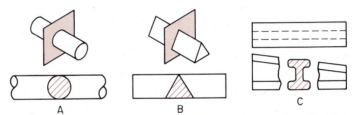

A   B   C

Figure 6.14 Revolved sections showing (*a*) solid cylinder, (*b*) solid triangular prism, and (*c*) conventional break to show section.

6.7 REVOLVED SECTIONS

The shape of a bar, arm, spoke, or other elongated symmetrical feature may be shown in the longitudinal view by means of a **revolved section** (Figure 6.14). The cutting plane in a revolved section is passed perpendicular to the axis of the elongated symmetrical feature and then revolved 90° into the plane of the drawing. A conventional break (discussed in Section 6.12) should be used to provide room for the revolved section if it conflicts with the original view (Figure 6.14*c*).

The revolved section is superimposed on the drawing. This means that lines that are significant in describing that view may interfere with the section view (Figure 6.15). The lines covered by the revolved section must be removed.

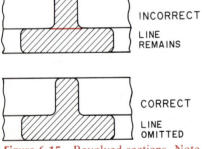

INCORRECT

LINE REMAINS

CORRECT

LINE OMITTED

Figure 6.15 Revolved sections. Note that line covered by the section view must be removed.

6.8 REMOVED SECTIONS

A **removed section** is similar to a revolved section but is not drawn within the view containing the cutting plane line. It is displaced from its normal projection position. The displacement from the normal projection position should be made without turning the

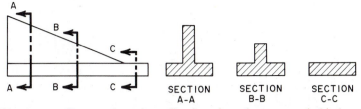

Figure 6.16 Removed sections. Section view always goes behind cutting plane or away from direction of view as defined by arrows on cutting plane.

section from its normal orientation. Figure 6.16 shows three examples of removed sections. In Figure 6.16, each removed section is labeled with letters and designated section *A–A*, section *B–B*, and section *C–C*. The section labels correspond to letters identifying the cutting plane that defines the exact location from which the removed section was taken.

6.9 BROKEN-OUT SECTIONS

Many times only a partial-section view is needed to expose the interior features of an object. This section, called a broken-out section, is limited by a break line. Figure 6.17 shows a broken-out section. A full or half section is not needed to explain the design. A small broken-out section is sufficient and takes less time to construct. Cutting planes are not shown for broken-out section views.

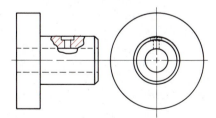

Figure 6.17 Broken-out section.

6.10 SECTIONED ASSEMBLIES

An assembly drawing showing several parts is sometimes sectioned to show the relationship of the parts to each other. When sectioning an assembly, it is important to separate the parts by varying the angle of the section lines. Figure 6.18 shows a typical

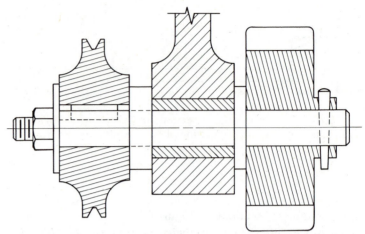

Figure 6.18   Sectioned assembly. Section lines vary in angle and spacing in adjacent parts. Standard parts such as nuts, bolts, keys, and washers (see Sections 9.4 and 9.6) do not get sectioned. Also, solid or hollow shafts do not get sectioned.

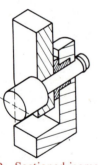

Figure 6.19   Sectioned isometric drawing shows how parts of assembly fit together by employing sectioning practices. Note that connective shafts, rods, bolts, and so on, are not sectioned.

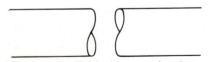

Figure 6.20   Typical conventional break.

sectioned assembly and the importance of distinguishing between the individual adjacent parts by varying the angle, direction, and spacing of the section lines. Small parts and fasteners that are not sectioned many times have their outline shown as hidden lines in the sectioned assembly.

6.11   SECTIONED ISOMETRIC DRAWINGS

Sometimes it is important to look inside a part drawn in pictorial. When a drawing is made of one part, it is a sectioned isometric drawing. If an assembly drawing is sectioned, it is called an isometric sectioned assembly. This might be used when the need arises to show how the internal parts in an assembly work together. Cylindrical and spherical parts such as shafts, bearings, bolts, and keys are not generally sectioned in this type of drawing (Figure 6.19).

6.12   CONVENTIONAL BREAKS

Conventional breaks are usually used to describe a long piece that has a uniform cross section. Figure 6.20 shows a typical conventional break. The shaft need not be drawn in true length; instead a conventional break can be used, and the dimension

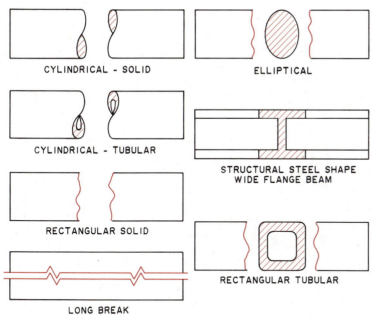

CYLINDRICAL - SOLID

ELLIPTICAL

CYLINDRICAL - TUBULAR

STRUCTURAL STEEL SHAPE
WIDE FLANGE BEAM

RECTANGULAR SOLID

RECTANGULAR TUBULAR

LONG BREAK

Figure 6.21   Conventional breaks of some cross-sectional shapes used in engineering.

showing the true length can be placed on the drawing. This procedure is a time and material saver.

A conventional break is shown in different ways depending on the geometry of the cross section. Figure 6.21 shows various cross-sectional geometries as conventional breaks.

Of the conventional breaks shown in Figure 6.21, three are most often used: the solid cylinder, the cylindrical tube, and the rectangular cross section. The conventional break for a solid cylinder is shown in more detail in Figure 6.22*a*. Generally the curved portion of the break is drawn carefully freehand; however, instruments can be used, particularly when the diameter is large.

The conventional break for a cylindrical tube is shown in more detail in Figure 6.22*b*. This conventional break resembles the conventional break used to show a cylindrical solid, except the interior space of the tube is shown without section lines. Again, the curved portions of the break are drawn freehand unless the diameter is especially large. In the case of both the conventional break for the cylindrical solid and the cylindrical tube, the area that represents the break is crosshatched with uniform section lines.

The conventional break for a rectangular uniform cross section is not as complex to draw as the cylindrical breaks. Two ''squig-

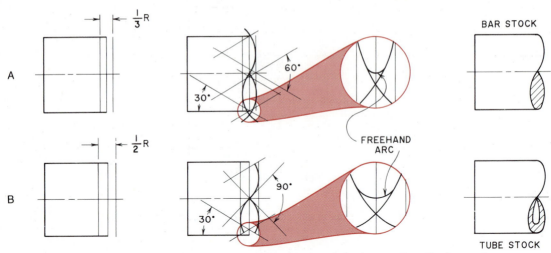

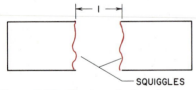

Figure 6.22 Cylindrical break for bar stock and tube stock.

Figure 6.23 Rectangular break.

gle'' lines approximately 1 in. apart are drawn transverse to the long dimension, and the area between these squiggles is erased (Figure 6.23).

6.12.1 Conventional Breaks by Computer

The computer can be used to do the construction of a conventional break. A CADD system may or may not have a primitive or a symbol to create the conventional break. A method that will work on most CADD systems involves the use of construction circles, the arc function, straight lines, and if available, the cross-hatching or hatching function. The method is discussed in what follows.

Begin by locating the position of the break on the object (Figure 6.24a). You may want to zoom in on this area until it fills the screen to make the construction easy to see. Add a grid to aid in the construction of the conventional break. Draw two dotted lines perpendicular to the object outlines to locate the center of the break symbols (Figure 6.24b). Then choose a circle with a diameter approximately one-half of the radius of the object. Draw the circles in the dotted line style so that they are just tangent to the upper and lower surfaces of the object (Figure 6.24b). It may help to do this construction on a separate drawing layer to aid in separating the construction lines and the finished symbol.

If the CADD system in use has a tangent function, draw two

Figure 6.24 (*a*) Construction of conventional cylindrical break using computer. Locate the position of the conventional break. (*b*) Draw construction lines perpendicular to object outlines to locate center of break symbol. Draw two construction circles with diameter of one-half the radius of the cylinder tangent to upper and lower surfaces of object. Draw two tangents to construction circles so they cross in center of object. (*c*) Darken in half of arcs and tangents on drawing layer to complete boundary of conventional break. (*d*) Locate evenly spaced construction lines on construction layer to crosshatch break if crosshatch function is not available on your system. (*e*) Zoom in on drawing layer and draw in crosshatch lines using arc and tangents as boundaries. (*f*) Zoom out and redisplay conventional break without construction layer to verify. (*Computer-generated image*)

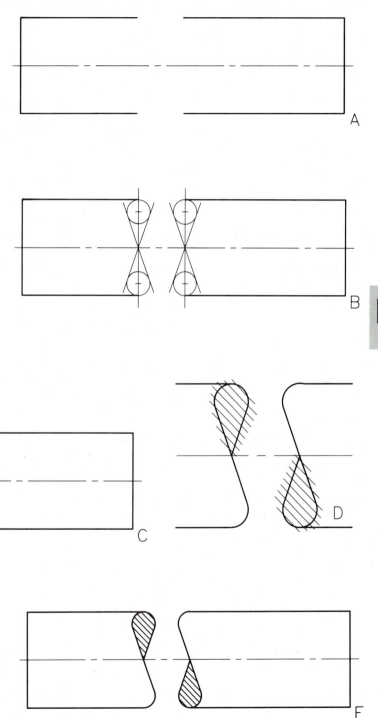

construction lines tangent to the two circles in dot style. They will cross in the middle of the object (Figure 6.24b). If a tangent function is not available, you can use the rubber band line mode to draw the tangents. Start lines where the vertical marking line crosses the center line of the object. Locate the estimated tangent point on the circle and draw in a construction line. Zoom in further and check the tangency (Figure 6.24b). If the line made in the rubber band mode does not touch the arc at one specific point, the tangency is not good; therefore, you should zoom out and try again. When your construction is complete for both tangents on one side of the break, repeat the procedure for the other two tangents.

Use the tangent points as the starting and ending points for the arc of the full symbol. Go back to the original drawing layer and draw in the solid arcs and half arcs first and then draw in the solid tangents (Figure 6.24c).

Select the crosshatch function if it is available and mark the arc and the tangent lines as the limits on the crosshatched area. Then using the crosshatch function, fill the area with the proper crosshatch symbol.

If a crosshatch function is not available, use the grid dots and the repeat line function to locate some evenly spaced construction lines that cross the boundaries of the break symbol (Figure 6.24d). Do this on the construction layer. Go back to the drawing layer and zoom in on the portion of the break to be crosshatched. Select the proper line thickness and use intersections of the construction lines as the beginning and ending points for the crosshatching lines (Figure 6.24e).

Zoom in on the other half of the symbol and crosshatch the bottom portion of that half. Zoom back out and redisplay without the construction layer to verify the construction (Figure 6.24f).

If the system being used allows the user to create, enlarge or shrink, and copy symbols, then the construction may be saved for future use on other breaks. It is probably not worth saving as a picture file only.

6.13 *CONVENTIONAL REVOLUTIONS*

Conventional revolutions are used to make section views clear and avoid confusion in interpretation. Conventional revolutions are associated with objects that have holes, ribs, spokes, or lugs in any combination equally spaced in circular fashion. The holes, ribs, spokes, or lugs may not project orthographically to the section view, but the convention allows for this practice and it is accepted as standard.

Figures 6.25–6.30 indicate full sections. Since the cutting plane line is horizontal and the arrows point to the rear, it is assumed the rear half of the object in the top view will be drawn in the section view, and the front half of the object in the top view will be removed. Therefore, only the features in the remaining rear portion of the top view will be discussed in the text for Figures 6.25–6.30. In the section view, you normally draw the features of the object that are visible on the cutting plane; however, sometimes it is necessary to revolve features into the cutting plane. You seldom, if ever, show hidden lines in section views. If a remaining feature on the given view does not fall on or is not visible on the cutting plane, then do not draw the feature in the section view; however, the exception to this practice concerns features that are not visible on the cutting plane, but fall on a circular center line or cylinder axis. These features are rotated about the cylinder axis to the cutting plane and are then drawn in the section view.

6.13.1 Conventional Hole Revolutions

Figure 6.25 shows three examples of holes in a circular plate. The first example shows four holes equally spaced around the circular plate. The cutting plane passes through two of the holes and a correct full-section view is shown. The holes that are not cut by

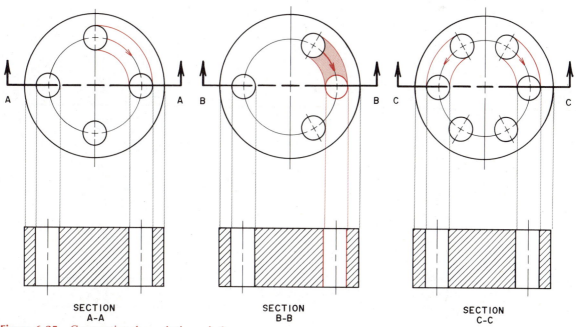

SECTION A-A	SECTION B-B	SECTION C-C

Figure 6.25 Conventional revolutions, holes.

the cutting plane are not shown. The remaining hole (or holes) that is not on the cutting plane line is revolved into the cutting plane and then projected to the sectioned view. The holes are not shown in the section view with hidden lines. The second example shows a circular plate with three holes equally spaced. The cutting plane is shown going through only one of these holes; however, the other hole that is not on the cutting plane line is revolved into the cutting plane before being projected to the section view. When comparing the section view of the first example with the section view of the second example, the views are exactly the same. The difference in the orthographic drawings is shown in the top views, which show the arrangement of the holes. The section view would look the same if there were six holes equally spaced in the circular plate, as shown in the third example.

6.13.2   Conventional Rib Revolutions

Ribs are simply stiffeners that anchor geometric objects to a supporting plate or surface. Ribs, like holes, may be equally spaced about a circular plate base. Figure 6.26 shows three examples of ribs equally spaced on a circular plate securing a cylinder on top of the plate. Like the nonvisible holes in the previous examples, the ribs are revolved into the cutting plane and projected to the section view. Since the ribs are not a continuous feature around

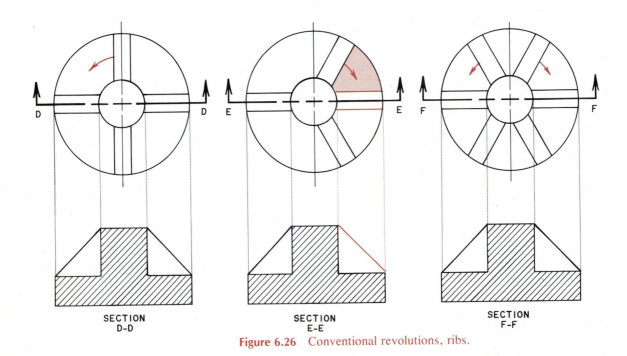

Figure 6.26 Conventional revolutions, ribs.

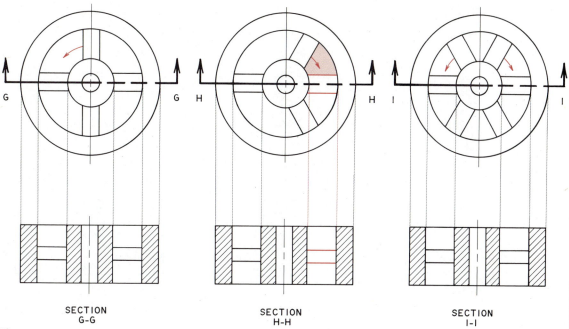

SECTION
G-G

SECTION
H-H

SECTION
I-I

Figure 6.27 Conventional revolutions, spokes.

the circular base, the section view shows them without section lines. This is an important convention used in all technical drawing.

6.13.3 Conventional Spoke Revolutions

Spokes are treated in exactly the same manner as ribs. Figure 6.27 shows three examples of a mechanical wheel with equally spaced spokes arranged in various configurations. The spokes are revolved into the cutting plane in each case, and the section views are shown to be identical.

6.13.4 Conventional Lug Revolutions

Lugs are similar to ribs except that they do not rest on a circular base. They are arranged about a cylinder. Figure 6.28 shows three examples of lugs and the section view that would be drawn for each case.

6.13.5 Hole, Rib, Spoke, and Lug Combination

When holes, ribs, spokes, and lugs are shown in combination, each is revolved into the cutting plane and projected to the section

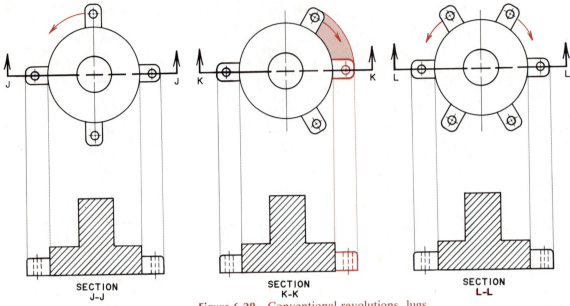

Figure 6.28 Conventional revolutions, lugs.

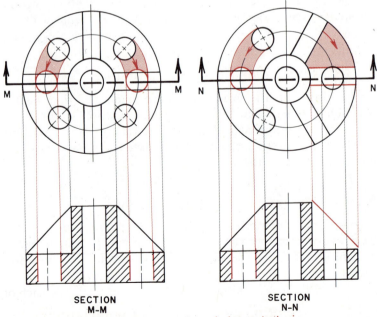

Figure 6.29 Conventional revolutions, holes and ribs in combination.

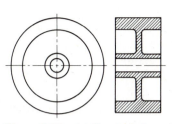
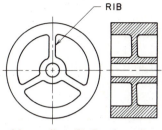
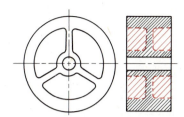

RIB

Figure 6.30 (*a*) Section view of part without ribs. (*b*) Section view of part with ribs that are flush with the top. (*c*) Section view with ribs identified with alternate crosshatching.

view. Figure 6.29 shows an example of holes and ribs occurring together and the resulting section view.

6.13.6 Alternate Crosshatching

Omitting crosshatching on ribs and similar features sometimes can lead to confusion and an unclear drawing. For example, Figures 6.30*a* and *b* show two full section views that look identical. Figure 6.30*a* is a part that does not have ribs, while Figure 6.30*b* shows the same part that includes ribs that are flush with the top portion. If you look at the full-section views, you can see that the ribs are not identified in Figure 6.30*b*. This is not clear. Greater clarity is achieved by the practice of alternate crosshatching (Figure 6.30*c*). Section lines (alternating) are drawn through the rib identifying it as a solid part of the object rather than air space. The lines separating the rib and the solid portions of the part are dashed.

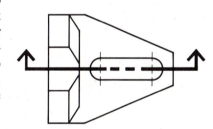

6.14 *SAMPLE PROBLEM: TABLE SAW FIXTURE*

The same data base developed in Chapter 4 when drawing in orthographic projection was used to make Figure 6.31. Notice that the right side view was eliminated. The top view was altered by adding the cutting plane line to indicate that a full-section view will be drawn in the front view. The front view was altered to reflect the fact that the near half has been removed. Also, thin section lines were added to indicate where the imaginary cutting plane actually pierced the object.

Entry level CADD programs may require you to draw each of the section lines. The more powerful CADD programs would fill any identified areas with a selected line pattern with the touch of a key.

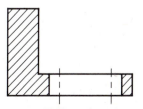

Figure 6.31 Full-section view with cutting plane line and section lines. (*Computer-generated image*)

6.15 *SUMMARY*

Standard orthographic views show all features of objects. Interior features shown as dashed lines may confuse the person interpreting the drawing. Sectioning cuts the object with an imaginary plane, making interior features, which were hidden to the viewer, visible. Sectioning makes the drawing clearer.

Many types of section views can be drawn depending on the complexity of the object. A full-section view shows the entire orthographic view as a section view, with half the object removed. A half-section view shows one-half of the orthographic view as a section view; however, a half-section view requires that the object have an axis of symmetry. An offset section view is a type of full section in which several cutting planes meet at 90° angles to show the interior features of the object. A revolved section view is a section shown within the longitudinal view of a long object. The cutting plane is located perpendicular to the long axis of the object, and the revolved section is then rotated 90° into the plane of the paper or computer screen to show the section view of the object.

Conventional breaks are usually used to describe a long piece with a uniform cross section. The long piece, such as a structural steel beam or a solid cylindrical shaft, does not need to be drawn in true length. A conventional break can be used to save time and drawing materials. The true-length dimension can be shown on the conventional break drawing.

Conventional revolutions are used to make section views clear and avoid confusion in interpretation. Conventional revolutions are associated with objects that have holes, ribs, spokes, or lugs in any combination symmetrically spaced in a circular fashion. When a section view of an object with these features is drawn, many times the cutting plane will not pass through one or more of the symmetrically spaced holes, ribs, spokes, or lugs. The features that are missed by the cutting plane are revolved into the cutting plane and projected to the section view from the revolved position. Conventional revolutions make the drawing clear and easily interpreted.

This chapter has presented techniques used in drawing section views as well as conventional practices and standards used in making engineering drawings. Mastery of these concepts will enable you to read and produce technical drawings and become literate in the language of the designer.

PROBLEMS

Problems 6.1–6.12

Sketch the given views and the indicated section for each problem. Note that the cutting plane in Problems 6.5 and 6.6 indicates a broken-out section.

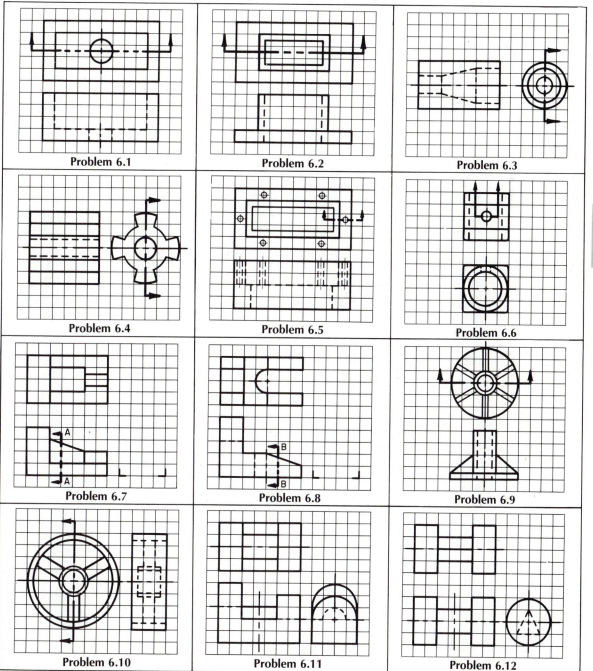

Problem 6.1

Problem 6.2

Problem 6.3

Problem 6.4

Problem 6.5

Problem 6.6

Problem 6.7

Problem 6.8

Problem 6.9

Problem 6.10

Problem 6.11

Problem 6.12

Problems 6.13 and 6.14
Given the top and front views of the object, draw the top view as shown and the front view in full section.

MILLIMETERS   *INCHES*

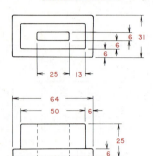
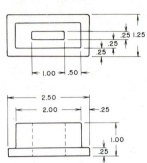

Problem 6.13

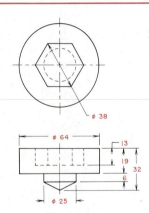
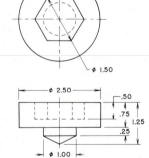

Problem 6.14

Problem 6.15
Given the top view and right side view (alternate position) of the object, draw the top view as shown and a front view in full section.

MILLIMETERS   *INCHES*

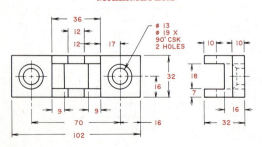
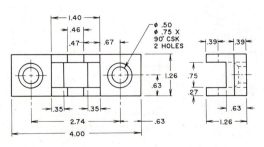

Problem 6.15

Problems 6.16–6.18
Draw the views that completely describe the object. One of the views must be a full section.

MILLIMETERS   *INCHES*

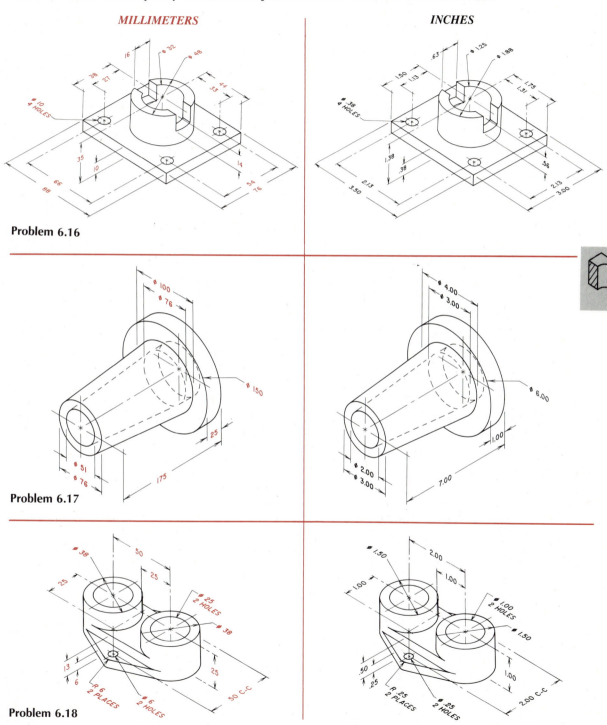

Problem 6.16

Problem 6.17

Problem 6.18

Problems 6.19–6.21
Given the two views of the object, draw the appropriate half section view.

MILLIMETERS   *INCHES*

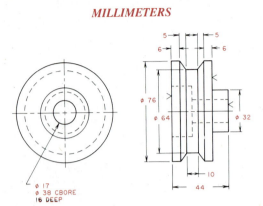

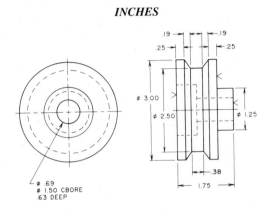

Problem 6.19

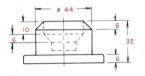

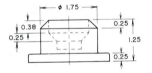

Problem 6.20

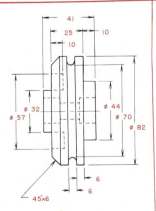

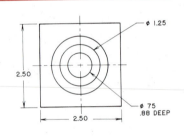

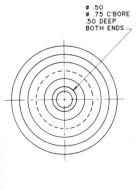

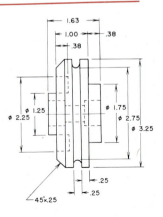

Problem 6.21

Problem 6.22
Draw the orthographic views that best describe the object. One of the views must be a half section.

MILLIMETERS   *INCHES*

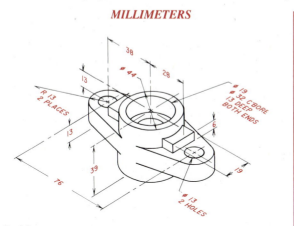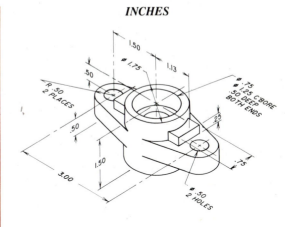

Problem 6.22

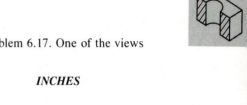

Problem 6.23
Draw the orthographic views that best describe the object shown for Problem 6.17. One of the views must be a half section.

MILLIMETERS   *INCHES*

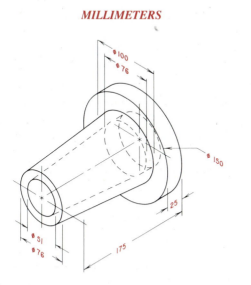

Problem 6.23

Problems 6.24 and 6.25
Draw the orthographic views that best describe the object shown. One of the views must be a full section.

MILLIMETERS

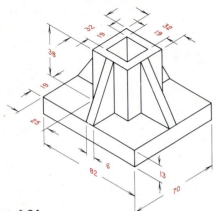

INCHES

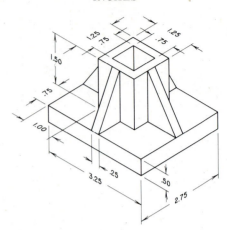

Problem 6.24

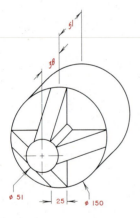

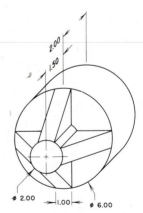

Problem 6.25

Problems 6.26–6.28

Given the front and right side views of the object, draw the front view as shown and a right side view as a full section.

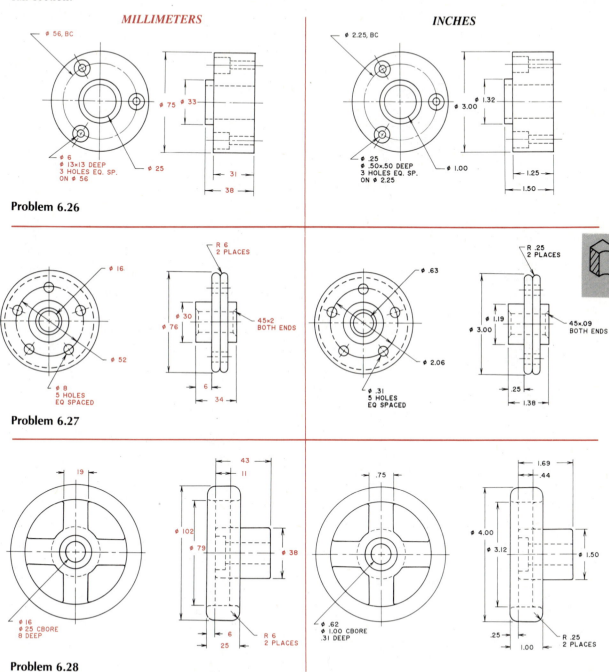

MILLIMETERS

Ø 56, BC

Ø 75 Ø 33

Ø 6
Ø 13×13 DEEP
3 HOLES EQ. SP.
ON Ø 56

Ø 25

31

38

Problem 6.26

INCHES

Ø 2.25, BC

Ø 3.00 1.32

Ø .25
Ø .50×.50 DEEP
3 HOLES EQ. SP.
ON Ø 2.25

Ø 1.00

1.25

1.50

R 6
2 PLACES

Ø 16.

Ø 30
Ø 76

Ø 52

45×2
BOTH ENDS

6

34

Ø 8
5 HOLES
EQ SPACED

Problem 6.27

Ø .63

Ø 3.00 1.19

Ø 2.06

Ø .31
5 HOLES
EQ SPACED

R .25
2 PLACES

45×.09
BOTH ENDS

.25

1.38

43
11

19

Ø 102
Ø 79
Ø 38

Ø 16
Ø 25 CBORE
8 DEEP

6
25

R 6
2 PLACES

Problem 6.28

.75

1.69
.44

Ø 4.00
Ø 3.12
Ø 1.50

Ø .62
Ø 1.00 CBORE
.31 DEEP

.25
1.00

R .25
2 PLACES

Problem 6.29
Draw the orthographic views required to best describe the object shown. One of the views must be a full section.

MILLIMETERS

INCHES

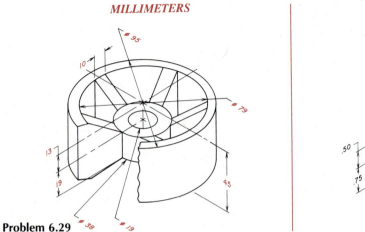
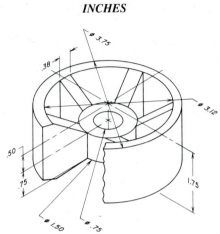

Problem 6.29

Problems 6.30 and 6.31
Draw the orthographic views required to best describe the object shown. One of the views must be a full section.

MILLIMETERS

INCHES

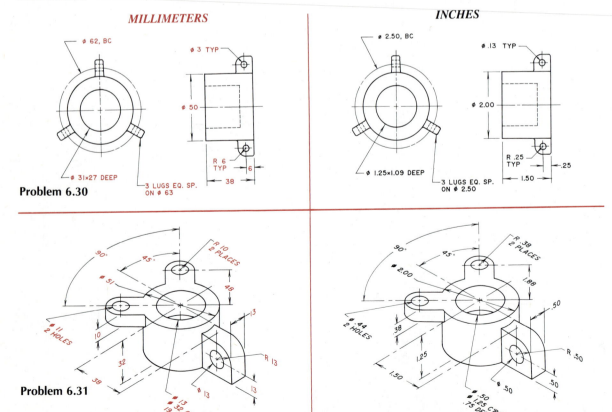

Problem 6.30

Problem 6.31

Problems 6.32–6.36
Given two views of the object, draw both views with the appropriate view showing the object in full section.

MILLIMETERS   *INCHES*

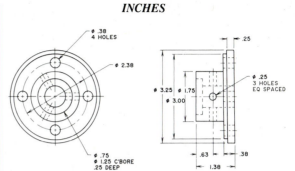

Problem 6.32

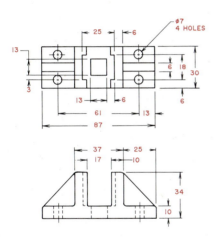

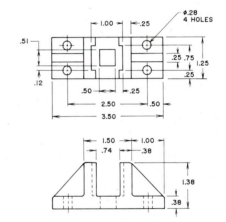

Problem 6.33

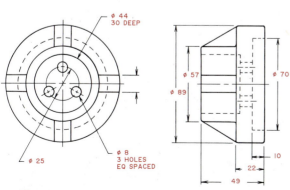

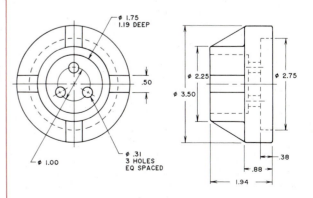

Problem 6.34

Problems 6.32–6.36
Given two views of the object, draw both views with the appropriate view showing the object in full section.

MILLIMETERS   *INCHES*

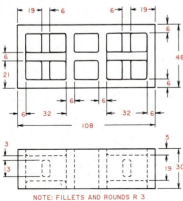

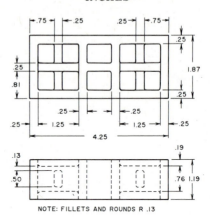

NOTE: FILLETS AND ROUNDS R 3   NOTE: FILLETS AND ROUNDS R .13

Problem 6.35

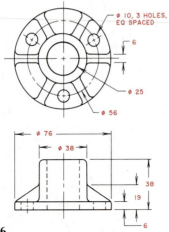

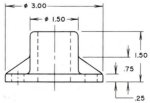

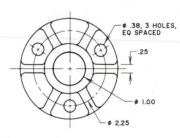

Problem 6.36

Problems 6.37 and 6.38
Select the most appropriate views to describe the object shown.

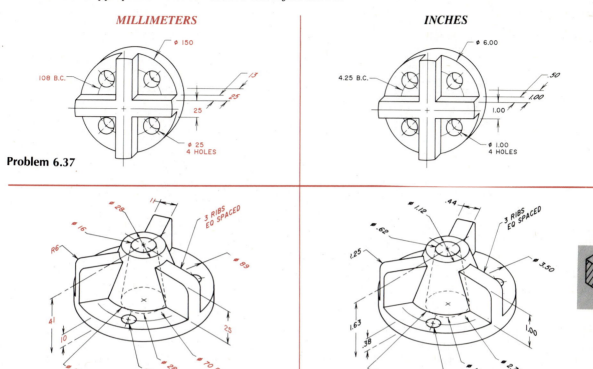

MILLIMETERS   *INCHES*

Problem 6.37

Problem 6.38

Problems 6.39 and 6.40
Draw the given views of the object shown. At the indicated location, draw a revolved section to indicate the interior shape.

MILLIMETERS

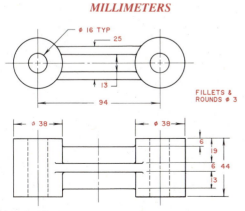

INCHES

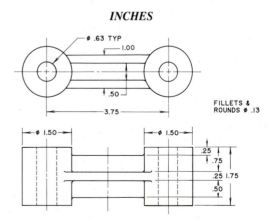

Problem 6.39

Problems 6.39 and 6.40
Draw the given views of the object shown. At the indicated location, draw a revolved section to indicate the interior shape.

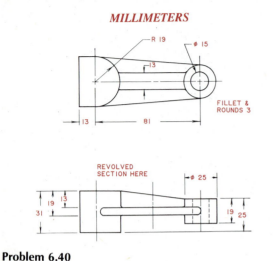

MILLIMETERS   *INCHES*

Problem 6.40

Problem 6.41
Complete the two views of the object shown. The top and bottom of the right arm are semicircular in shape while the top and bottom of the left arm are rectangular. The top view will show the conventional representations for runouts for these shapes.

MILLIMETERS   *INCHES*

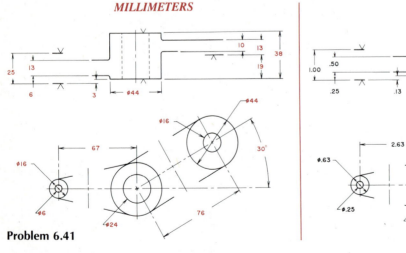

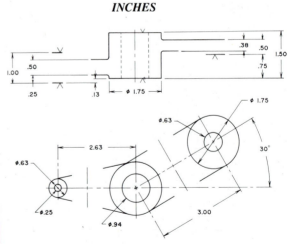

Problem 6.41

Problems 6.42–6.44
Draw the indicated views of the object shown.

MILLIMETERS   *INCHES*

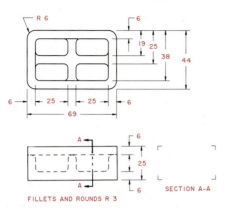

FILLETS AND ROUNDS R 3   SECTION A-A

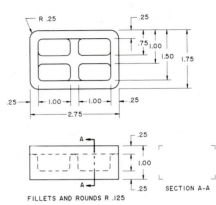

FILLETS AND ROUNDS R .125   SECTION A-A

Problem 6.42

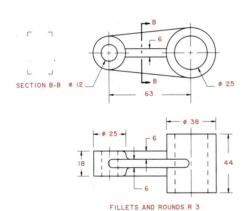

SECTION B-B ⌀ 12   FILLETS AND ROUNDS R 3

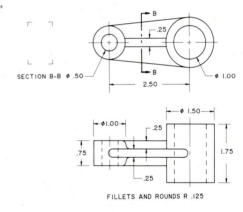

SECTION B-B ⌀ .50   FILLETS AND ROUNDS R .125

Problem 6.43

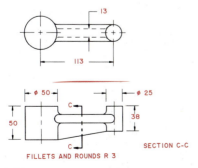

FILLETS AND ROUNDS R 3   SECTION C-C

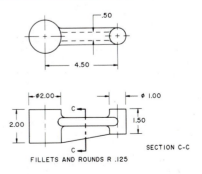

FILLETS AND ROUNDS R .125   SECTION C-C

Problem 6.44

Problems 6.45–6.47
Complete the views shown of the object. The cutting plane indicates that a broken-out section is required.

MILLIMETERS   *INCHES*

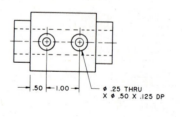

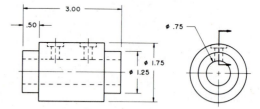

Problem 6.45

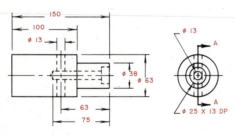

Problem 6.46

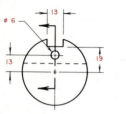

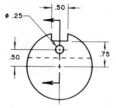

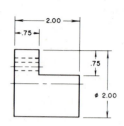

Problem 6.47

Problems 6.48–6.52
Draw the views that best describe the objects shown. One of the views must be an offset section.

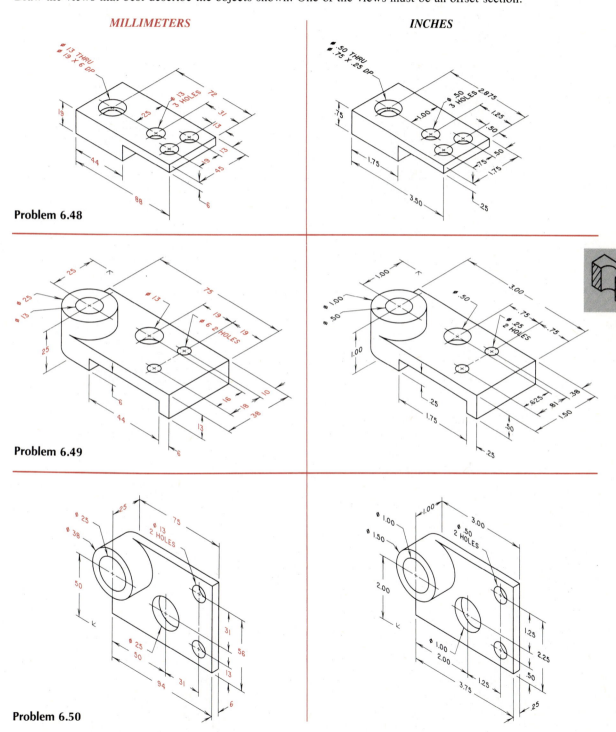

MILLIMETERS

INCHES

Problem 6.48

Problem 6.49

Problem 6.50

Problems 6.48–6.52

Draw the views that best describe the objects shown. One of the views must be an offset section.

MILLIMETERS   *INCHES*

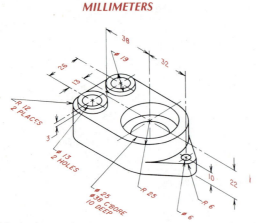

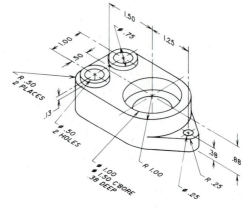

Problem 6.51

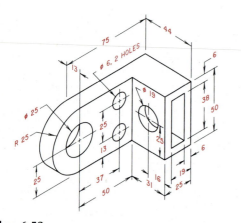

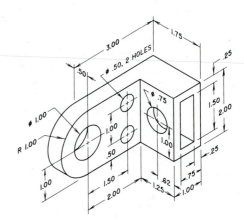

Problem 6.52

Problem 6.53
Draw the views that best describe the object shown.

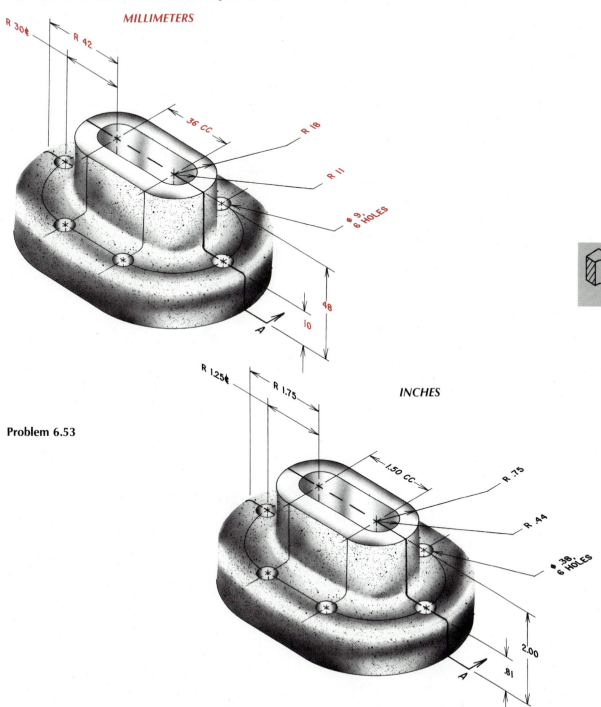

MILLIMETERS

R 30₡

R 42

36 CC

R 18

R 11

9
6 HOLES

48

10

A

R 1.25₡

R 1.75

INCHES

1.50 CC

R .75

R .44

.38
6 HOLES

2.00

.81

A

Problem 6.53

Problem 6.54

Draw the views that best describe the object shown.

MILLIMETERS

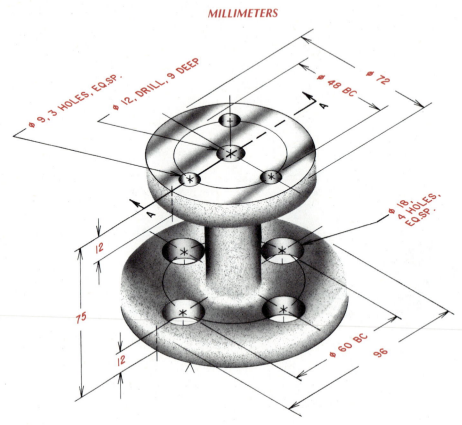

Problem 6.54

INCHES

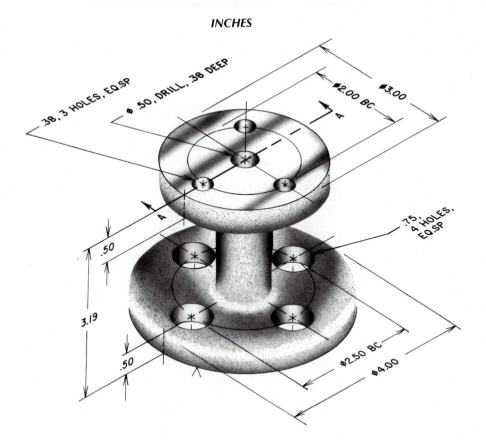

.38, 3 HOLES, EQ.SP

ø .50, DRILL, .38 DEEP

ø2.00 BC

ø3.00

A

A

.50

3.19

.50

.75, 4 HOLES, EQ.SP

ø2.50 BC

ø4.00

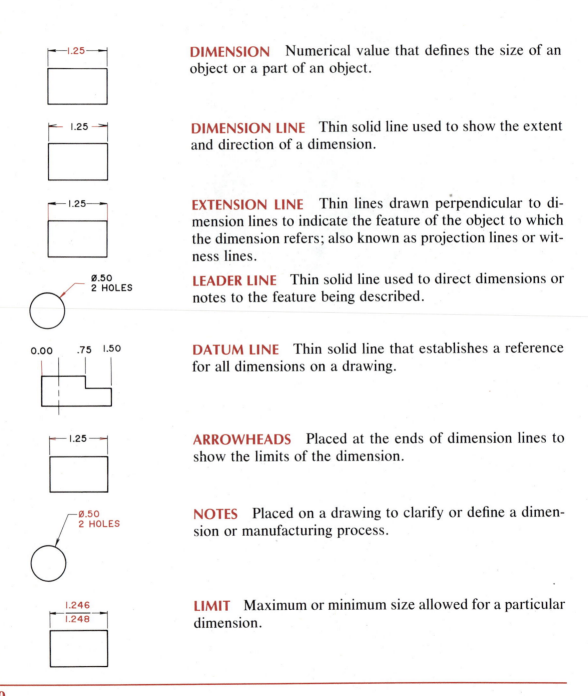

TERMS YOU WILL SEE IN THIS CHAPTER

DIMENSION   Numerical value that defines the size of an object or a part of an object.

DIMENSION LINE   Thin solid line used to show the extent and direction of a dimension.

EXTENSION LINE   Thin lines drawn perpendicular to dimension lines to indicate the feature of the object to which the dimension refers; also known as projection lines or witness lines.

LEADER LINE   Thin solid line used to direct dimensions or notes to the feature being described.

DATUM LINE   Thin solid line that establishes a reference for all dimensions on a drawing.

ARROWHEADS   Placed at the ends of dimension lines to show the limits of the dimension.

NOTES   Placed on a drawing to clarify or define a dimension or manufacturing process.

LIMIT   Maximum or minimum size allowed for a particular dimension.

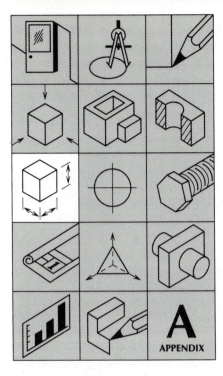

APPENDIX

7

DIMENSIONS AND TOLERANCES

7.1 INTRODUCTION

In earlier chapters you learned how to describe completely the shapes of objects on drawings following the principles of orthographic projection. Before the object can be built, however, the size of the object must also be completely described. Adding information to a drawing about the size of an object is known as *dimensioning* a drawing (Figure 7.1).

Dimensioning a drawing is not a simple task. In this chapter

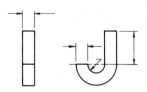 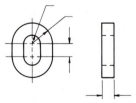 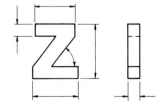

Figure 7.1 Adding size description to familiar shapes tells how big they are.

you will learn the standard vocabulary used, how drawing scale is specified, techniques for dimensioning basic geometric shapes, appropriate placement of dimensions on drawings, and standard dimensioning practices. This chapter also includes information regarding the proper specification of tolerances, the principles of good dimensioning, and techniques for checking a completed drawing for dimensioning accuracy.

7.1.1 Dimensioning Standards

It is important that all persons reading a drawing interpret it in exactly the same way. This is why the views in an orthographic drawing are always placed in the same relationship to each other. When reading size information (**dimensions**) on a drawing, it is also important that there be only one way to interpret the information. Misunderstanding will result in parts that may fail. This could cost loss of reputation, time and money, or even human life. To increase the clarity of dimensioning, standard practices have been developed.

The American National Standards Institute (known as ANSI) has established standards for dimensioning mechanical drawings. The material in this chapter (except where otherwise noted) conforms to the latest ANSI dimensioning standards known as ANSI Y14.5M.

7.1.2 Dimensioning Practices

You should note that there is more to describing the size (dimensioning) of an object than simply associating numbers with the geometry of the object. First, it should be remembered that there is no such thing as an "exact" measurement of a dimension. A carpenter installing the framework for the walls of a house may be satisfied if the wall is within a quarter of an inch of the specified measurement. A jet aircraft engine component, however, might have to be located to within less than .0001 in. of the specified measurement, and the location of conductor paths in a large-scale integrated circuit in a computer must be held to within fractions of a micron. Therefore, every dimension on a drawing will have a tolerance, or allowable variance, associated with it.

The selection and placement of dimensions will be influenced by the methods used to produce the object and the relationships between various surfaces of the object. The functional relationship between an object and other parts in an assembly (how parts fit together to do their job) will also influence selection and placement.

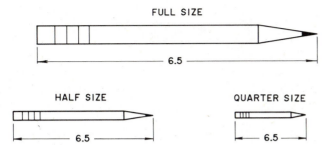

Figure 7.2 Pencil drawn and dimensioned to three different scales.

7.2 SCALING AND DIMENSIONS

More often than not it is impossible to make a drawing the same size as the finished product. For example, you would not expect the floor plan for your house to be drawn full scale (or life-size), but the dimensions (numerical values) on a drawing are always given as full size.

Let's suppose that you are making a half-scale drawing of an object that is 6.5 in. wide. When you lay out the width on your drawing, you will measure 3.25 in., or one-half of full size. Your half-size scale will enable you to do this with no calculations. When you place the dimensions on your drawing, however, you will dimension the width as 6.5 in., or actual size (Figure 7.2). The title block of the drawing will indicate the scale to which the drawing was made.

7.2.1 Types of Scales

There are many different types of scales used by specialists in different types of design. The *engineer's* scale, originally called the *civil engineer's* scale, is divided in decimal parts. On a triangular scale (Figure 7.3) the scales would be labeled 10, 20, 30, 40, 50, and 60. Ten would be full size with 10 parts per inch, or it could be used as $\frac{1}{10}$ size, or $\frac{1}{100}$, or $\frac{1}{1000}$. The 20 scale could mean decimal half size, or $\frac{1}{20}$, or any multiple of 10 times that ratio. As a civil engineer's scale it was used for drawings of large areas as $\frac{1}{1000}$, $\frac{1}{2000}$, and so on. As we use more decimal inch dimensioning, it is useful to many designers.

A special decimal scale is the 50-parts-per-inch scale (Figure 7.4). This scale allows you to interpolate .01 in.; it is valuable for decimal inch drawings.

The metric system is based on decimal progressions. A basic metric scale will be marked 1:1 or 1:100 (Figure 7.3). It would be used for full-size drawings or reduced or enlarged drawings

Figure 7.3 Various useful scales.

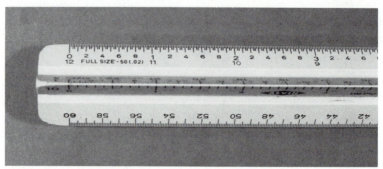

Figure 7.4 Decimal scale 50 parts/inch.

where the original size is shown as any multiplication of 10. The basic metric unit for machine parts is the *millimeter;* large structures would be described in meters; land measurement is in kilometers. Other common metric scales are 1:2, 1:3, 1:5, 1:25, 1:33.3, and 1:75. The scale 1:2 would give you half size and 1:33.3 could be used for one-third size.

Mechanical engineers' scales are labeled as $\frac{1}{2}$ size, $\frac{1}{4}$ size, or some related fraction (Figure 7.3). They are used to draw objects at reduced scales in a fractional inch system. Their use is declining as we go to decimal inch and metric dimensions on drawings. The full-size scale will be labeled 16 or 32, meaning 16 or 32 divisions per inch.

Architects' scales look much like *mechanical engineers'* scales but the divisions represent *feet,* not inches (Figures 7.3 and 7.5). The individual scales are labeled $\frac{1}{2}$, $\frac{3}{16}$, 1, or 3, meaning that this number of *inches* on the drawing represents that number of *feet* on the object. When you pick up a new scale, check to see if it is marked "$\frac{1}{2}$" or "$\frac{1}{2}$ size"; the former is an architect's scale, the latter a mechanical engineer's scale. Both are calibrated as "open" scales. This means that all of the integer units (feet or inches) are not subdivided into parts. Instead, one unit at the end of the scale is subdivided and all fractions of a unit are scaled here while integer units are measured from the zero at the end of the sub-divided unit (Figure 7.5).

The scales, or "measuring sticks," may be triangular in form or they may be flat with beveled edges (Figure 7.3). Some special scales made for students contain a fractional inch scale, a metric scale, decimal inches, half and quarter sizes, and 10 and 50 scales from the civil engineer's scale. This type of tool enables you to become familiar with many scales with a low investment.

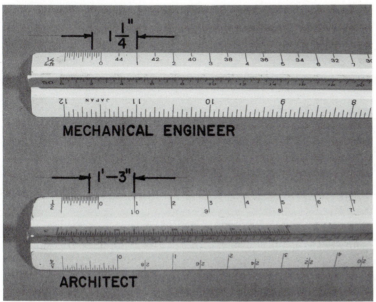

Figure 7.5   Open scales. (*Note:* Mechanical Engineer's scale = 16 divisions, Architect's scale = 24 divisions.)

7.2.2 Specifying Scales on a Drawing

A drawing that is made full size would be labeled FULL or FULL SIZE in the title block. Preprinted title blocks usually have an area reserved for stating the scale. A half-size drawing could be labeled HALF or 1:2 (metric) or 1 = 2 (inches) or 3″ = 1′–0″ (architectural). The size on the drawing is given *first* followed by the size of the real object. A map could be labeled 1:4000 or have a graphic scale.

7.3 DIMENSIONING BASIC SHAPES

There are a number of standard practices used in dimensioning basic shapes. The practices discussed in this section apply to mechanical drawings used in the manufacturing industries. Different practices are used in architectural drawings, structural drawings, or printed circuit board design.

7.3.1 Assumptions

A number of assumptions are made by those interpreting technical drawings. If there are exceptions to these assumptions, they must be clearly stated on the drawing. Two important ones concern perpendicularity and symmetry.

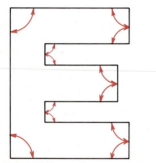

Figure 7.6 Right angles are assumed to be 90° and are not dimensioned.

Perpendicularity
All lines on a drawing that appear to be perpendicular (at a 90° angle) are assumed to be perpendicular to within the specified tolerance (Figure 7.6). All angles other than 90° must be dimensioned.

Symmetry
Objects, patterns of holes, or other features that appear to be symmetrical are assumed to be symmetrical unless otherwise specified. If there is any chance of confusion by the user of the drawing, symmetry (or the lack of symmetry) should be noted clearly on the drawing.

7.3.2 Simple Shapes

Even the most complex parts can be broken down into a collection of simple shapes. When you can dimension the simple shapes on a complex object, you will have dimensioned most of the object. Three shapes are most common: the rectangular prism, cylinder, and cone (or frustum of a cone).

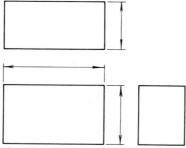

Figure 7.7 Three dimensions describe size of rectangular prism.

Rectangular Prisms

The height, width, and depth of a prism must be specified (Figure 7.7).

Cylinders

There are two types of cylinders to dimension: positive and negative. Positive cylinders are solid objects. Positive cylinders are dimensioned by giving the length and outside diameter of the cylinder. The numerical value of the diameter is always preceded by the diameter symbol (ϕ). It is usually placed on the longitudinal, or noncircular, view (Figure 7.8). A negative cylinder is a circular cavity (a hole). Negative cylinders are usually dimensioned on the circular view. When the hole goes all the way through an object, the depth is not given [the shape description shows this fact (Figure 7.9)]. In the case of a blind hole (a negative cylinder that does not pass completely through an object), the depth of the hole refers to the depth of the full diameter and may be given in a note (Figure 7.10).

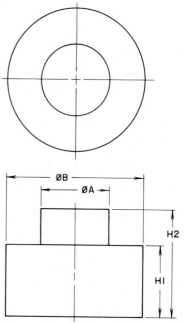

Figure 7.8 Positive cylinders with diameters and heights dimensioned.

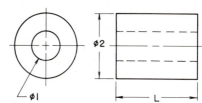

Figure 7.9 Positive cylinder with negative cylinder or hole through it.

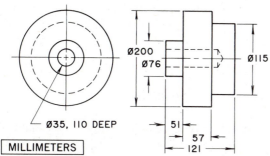

Figure 7.10 Cylindrical object with properly dimensioned blind hole.

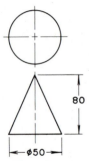

Figure 7.11   Size of cone totally described by its base diameter and height.

Cone and Frustum

A cone may be dimensioned by giving the diameter of the base and the height (Figure 7.11). Another method of describing a cone is to give the height and the included angle of the cone. Still another method is to specify the diameter of the base and the taper per foot of the sides. A frustum of a cone is a truncated cone, that is, a cone with its point cut off. It may be dimensioned by specifying the diameter of the base, the included angle, and the height; the diameters of both ends; and the height or the diameter of the base and height of the original cone and the height of the frustum (Figure 7.12).

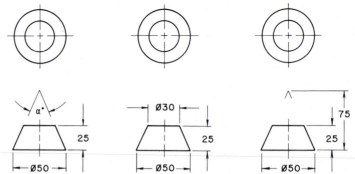

Figure 7.12   Frustum of cone may be described by three different sets of dimensions.

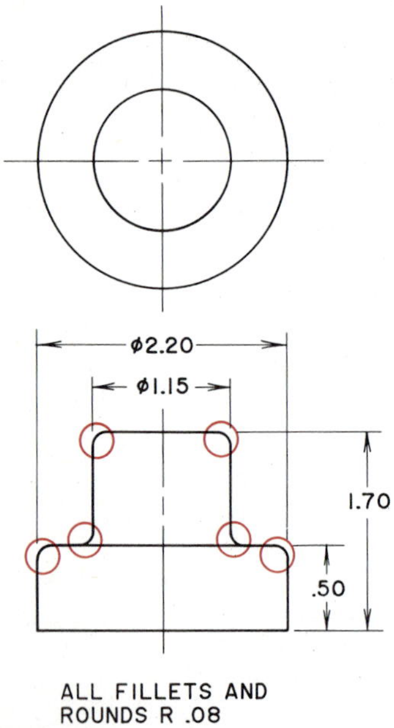

7.3.3   Simple Curved Shapes

In Chapter 4 you read about fillets and rounds, which are used at the intersection of surfaces to provide a curved contour and improve strength. You could dimension each of these arcs separately and would if they were all different sizes. Small arcs used for fillets and rounds are often the same size on a part. These are usually dimensioned with a single note (Figure 7.13). This note, similar to one of the following, would be placed in a prominent place on the drawing:

> All Fillets and Rounds R .08

> All Undimensioned Radii .08

We often find slots with rounded ends on mechanical parts. There are several accepted ways to describe the dimensions of these holes, as shown in Figure 7.14. Note that the curved ends have radii noted, but the radii are not dimensioned.

ALL FILLETS AND
ROUNDS R .08

Figure 7.13   All fillets and rounds of same size can be described in single note.

7.3.4 Circular Center Lines

Features on round objects are often located on circular center lines. Ribs for reinforcement and holes for connecting parts are two types of features that may be located on circular center lines. A group of holes arranged on a circular center line is often used for bolting two parts together; these center lines are called bolt circles. The diameter of a bolt circle or other circular center line is dimensioned on the circular view. Three acceptable ways to dimension them are shown in Figure 7.15. You may find the letters B.C.D. included with the size of the diameter on some drawings.

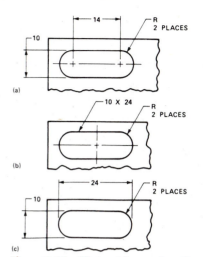

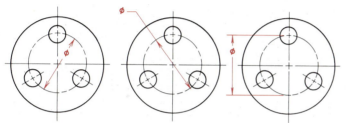

Figure 7.15 Diameter of circular center lines should be given by one of three methods shown.

Figure 7.14 Slots with round ends may be dimensioned in three different ways. (Courtesy of ANSI)

7.4 PLACEMENT OF DIMENSIONS

The proper placement of dimensions is important. If dimensions are placed in a haphazard manner, confusion and misunderstanding may result. Some general guidelines for proper placement of dimensions follow.

7.4.1 Characteristic View

Dimensions for a particular feature of an object should be placed in the most characteristic view of that feature. In the example shown in Figure 7.16, the notch shows most clearly in the front view. If you look at the top view, it is hard to tell if there is a notch or a series of steps. The front view, however, clearly shows the notch. Therefore, the width and height of the notch should be dimensioned in the front view rather than in the top or side view.

7.4.2 Dimensions Should Be Grouped

You should plan to group the dimensions on a drawing in such a way that the reader can locate them with a minimum of effort. It

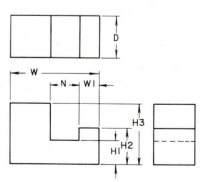

Figure 7.16 Dimensions placed on characteristic view and located from common corner.

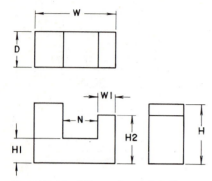

Figure 7.17 Dimensions placed haphazardly are hard to find and understand.

is also a good idea to start dimensions from a *common corner*. Study Figure 7.16 and you will note that all height dimensions are measured from the bottom and all width dimensions from the right side of the front view. Imagine trying to make these measurements on a block of metal—it is much easier if all measurements read from a common corner. Note how much harder it is to read the dimensions in Figure 7.17 compared to Figure 7.16. This problem gets worse as the complexity of the object increases.

7.4.3 Preferred Locations

The preferred locations for dimensions are, first, between the views; second, to the right of or below a view; and third, to the left of or above a view. The view should be one that describes the surface in true size and shape. Dimensions should not be located within the boundaries of an object unless other choices are less clear. Hidden lines should not be dimensioned since a hidden view of a feature is usually not the most characteristic view. It may be clearer if you construct a section view to expose hidden features for dimensioning.

7.5 DIMENSIONING SYSTEMS

A number of different dimensioning systems have been developed, and many are still in use. You should be aware of these systems in addition to the standards used in the classroom.

Figure 7.18 Aligned and unidirectional placement of dimensions; note height dimensions.

7.5.1 Alignment of Dimension Numbers

Older drawings were frequently dimensioned with the aligned dimensioning system. In the aligned system, the dimensions were placed in alignment with the dimension lines and were oriented to be read from either the bottom or right of the drawing. Current ANSI standards call for use of the unidirectional system where all dimensions are read from the bottom of the drawing (Figure 7.18).

7.5.2 Units of Measurement

The two most common systems standards are those published by the International Standards Organization (ISO) and the American National Standards Institute (ANSI Y14.5M). Drawings that follow ISO standards have the designation SI. The commonly used unit of measurement on SI drawings is the millimeter. The ANSI standards specify the decimal inch or the millimeter.

Dimensions on drawings made in the United States are usually in decimal inches unless otherwise specified on the drawing. The ANSI standards specify that a note of the following form should be used if there is any doubt about which units are used: UNLESS OTHERWISE SPECIFIED, ALL DIMENSIONS ARE IN MILLIMETERS (OR INCHES).

A convenient way to specify which system is used is to put the word *millimeters* or *inches* in a rectangle near the title block (Figure 7.10). Fractional dimensions are mostly found on older drawings, although fractional dimensions are sometimes used on new drawings. You should be aware that although the ANSI Y14.5M standards are the most commonly used, many companies maintain their own drawing and dimensioning standards. It is important, when starting a new job, to become familiar with the standards and practices in use in your new organization.

7.6 STANDARD PRACTICES

The purpose of technical drawings is to communicate information. As in any system of communication, consistency and clarity are important. There are several standard practices relative to lines, arrowheads, and notes with which you should be familiar.

7.6.1 Lines for Dimensioning

Relative line weights are important for improving the readability of drawings. The ANSI standards call for object lines to be 0.7 mm thick. **Dimension, extension, leader,** and center lines are to be 0.5 mm thick (your authors believe that the use of three line weights are preferable—the 0.5 mm thickness can be reserved for hidden lines and 0.35 mm used for dimension, extension, leader, and center lines). Computer-generated drawings may also be drawn with three weights of lines, as suggested in Chapter 4. All lines must be dark for proper reproduction. Remember, there are no light lines on finished copies of drawings—only thick, medium, or thin dark lines.

7.6.2 Arrowheads

Well-made **arrowheads** will give a drawing a professional appearance and make the drawing easier to read. Arrowheads should be $\sim\frac{1}{8}$ in. (3 mm) long, and the width should be about one-third of the length (Figure 7.19a). They may be a little longer on long dimension lines but should not be shorter. If too short, they become just tiny blobs on the drawing. Arrowheads may be open,

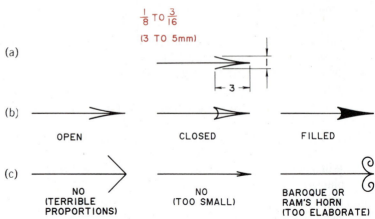

Figure 7.19 (*a*) Well-proportioned arrowheads improve readability of drawing. (*b*) Acceptable arrowheads may be open, closed, or filled. (*c*) Wide, tiny, or ornate arrowheads are unacceptable.

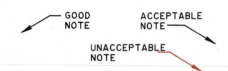

Figure 7.20 Leader lines for note are made up of straight lines; leader always goes to beginning or end of note, never in middle.

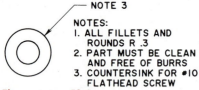

Figure 7.21 If there are many notes on a drawing, they may be lettered on a note list.

closed, or filled (Figure 7.19*b*). Do not attempt to "dress up" your drawing with fancy arrowheads—they reduce clarity rather than help it (Figure 7.19*c*).

7.6.3   Lettering and Notes

All of the numbers, letters, and words on your drawing must be readable. As you learned in Chapter 1, the minimum size for readable letters is $\frac{1}{8}$ in. (3 mm) high. None of your dimension numbers or letters should be smaller than this; guidelines should be used for all lettering.

    Notes are placed on drawings to indicate the size or depth of holes, the number of holes or spaces in a repeating pattern, or other information required in manufacturing an object. Notes may be placed at the end of a leader line, which has a short horizontal flag (Figure 7.20). If a note is too large to fit comfortably in the body of a drawing, it may be placed in a note list. In this case the leader line will have an indication as to which note applies (Figure 7.21). Leader lines should go to the beginning or the end of a note, never to a line in the middle of the note (Figure 7.20).

7.6.4   Readability

The principles and methods outlined in the preceding sections are not rules; they are the results of years of practical experience. Original drawings never leave the engineering office; only copies are sent to production or construction sites. These sites may be open to wind or rain; they may be using oily cutting fluids; and there may be no clean surface on which to lay the drawing for study. So drawings are folded to fit hip pockets, splashed with oil, and splattered with rain—after this treatment, *they must still be readable*. The principles outlined will help to make your drawings or those done under your supervision readable under shop and field conditions.

7.7 PLANNING AND LAYOUT

You will need to plan a dimensioned drawing more carefully than you planned only the views (Chapter 4). Keep in mind that most dimensions should be attached to the characteristic view and should be placed between the views. If you don't provide enough space for dimensions, the drawing will become crowded and hard to read. A good approach is to make a sketch of the object with all the dimensions before you begin the formal drawing. The views can then be placed to allow proper placement of dimensions. Review Chapter 4 for planning a drawing and now include space between the views and around the views for the dimensions and notes.

7.8 FUNCTIONAL RELATIONSHIPS

Selection of dimensions involves knowledge of the function as well as the shape of an object. Since no feature of an object can be made to an exact size, every dimension will have an associated tolerance. Sometimes a general tolerance note will be given that will apply to all dimensions on a drawing except those having a more precise tolerance. When an object is part of an assembly of parts, the tolerances must be such that the parts will fit together. For example, the largest permitted shaft must fit properly into the smallest permitted hole. Individual tolerances, more precise than general tolerances, would be specified for these parts.

7.8.1 Tolerance Accumulation

When a series of dimensions is involved, an accumulation of tolerances may cause problems. For example, if there were four uniform steps on an object and each step were dimensioned from the next step with a tolerance of .020, then there would be a tolerance accumulation of .080 in. (4 × .020) across the series of steps. This tolerance buildup could be avoided by dimensioning each step from one edge of the object. Look at Figure 7.22 to see how this is done. The tighter the tolerance, the more expensive

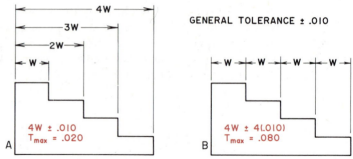

Figure 7.22 Total length of (*a*) has a tolerance of .020; total length of (*b*) has a tolerance of .080 (.020 × 4).

it will be to manufacture the part. Therefore, a good designer will select the method of dimensioning that will minimize tolerance accumulation.

7.8.2 Dimensioning to Finished Surfaces

You need to remember when dimensioning that finished surfaces are more important than unfinished surfaces. A finished surface has been machined (turned, milled, ground, or polished). Finished surfaces are shown with a small V, which resembles a cutting tool touching the surface. Because they have been cut, finished corners are shown as sharp corners where they intersect unfinished surfaces or other finished surfaces (Figure 7.23). Parts for machinery or buildings are made by forging, casting, rolling, welding, and other methods that result in a rough exterior surface. When parts must fit together or be precise, we invest extra time, machinery, and people to make the rough surfaces smoother and more exact.

Figure 7.24 shows two methods for dimensioning the height of the same object. Figure 7.24*a* shows dimensioning the height as a chain of dimensions, while Figure 7.24*b* shows an overall height dimension with the upper and lower protrusions dimensioned separately. At first glance the two methods might appear equal since a simple calculation indicates the same overall height. When tolerances are considered, however, there is a significant difference.

If the designer has specified a general tolerance for dimensions of .01 in., then the maximum and minimum permissible heights for the object in Figure 7.24*a* would be 2.03 and 1.97 in. The maximum and minimum for Figure 7.24*b* would be 2.01 and 1.99 in. Since there are finish marks on the upper and lower surfaces of the object, we must assume the distance between these surfaces is more important than the size of the cavity. You should select the method shown in Figure 7.24*b*.

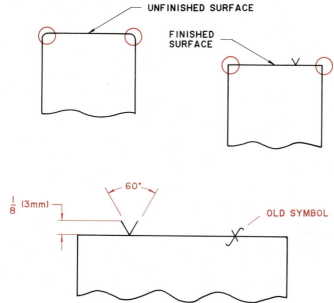

Figure 7.23   Finished surfaces (note square corners) and finish marks—new and old.

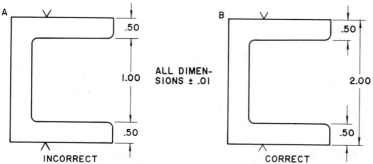

Figure 7.24   Finished surfaces should always be used to locate features on drawing.

7.8.3   Dimensions for Locating Features on an Object

Finished surfaces have some important function (otherwise why go to the additional expense of finishing the surface?). When you locate the centers of holes such as those found in the locator block in Figure 7.25, the center should be dimensioned to the finished rather than the unfinished surfaces.

A pattern of holes such as the counterbored holes in Figure 7.25 usually will be matched to similar holes in an assembly.

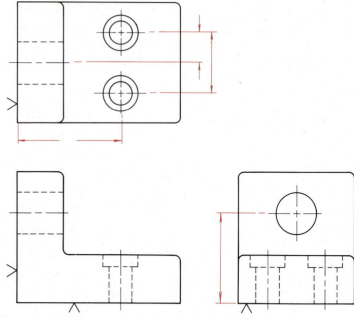

Figure 7.25 Holes need to be located relative to each other or to finished surfaces.

Therefore, the distance between holes should be specified. The reason for this is the same as that discussed in Section 7.8.2. Although simple arithmetic might lead you to believe that the results would be the same if we dimensioned from the outside edges to the centers of the holes, the tolerance accumulation will either cause the parts to not fit together or require specification of tighter tolerances, which will add to costs.

7.8.4 Procedure for Functional Dimensioning

A part that has finished surfaces or holes should be dimensioned so that these features are located with respect to each other. Unfinished surfaces are less important and may be located relative to finished surfaces or other unfinished surfaces. A procedure for dimensioning that has worked for many designers is outlined:

1. Dimension the basic shapes
2. Dimension the cavities
3. Locate the basic shapes and cavities
   a. Finished surface to finished surface
   b. Finished surface to center line
   c. Center line to center line
   d. Other locations

4. Add the notes
5. Check your work
    a. Are all dimensions given that are needed?
    b. Are the numbers correct?
    c. Do the notes describe all special needs?
    d. Will this part fit with matching or mating parts?

7.9   DIMENSIONING ON THE COMPUTER SCREEN

You can use the principles outlined for dimensioning drawings made on a computer. When you plan the drawing, you will need to leave room between the views for the dimensions. If you don't leave enough, it is easier to redraw or move a view on the computer screen and make more space for the dimensions than it is on paper.

7.9.1   A Layer for Dimensions

It is a good idea to put the dimensions and notes on a different layer or level than the views or construction lines. This was mentioned in Chapter 4. If you need to make several changes, you can change dimensions without the danger of changing your views or you can modify the views and then overlay the dimensions and change only the affected values.

7.9.2   Scaling a CADD Drawing

The concept of a drawing scale is somewhat different with CADD systems than when creating a drawing manually. When you create a drawing on a computer, you draw it at full scale. If an object is 10 ft long, you draw it as 10 long. You select an X–Y origin on the computer's electronic drawing board from which all dimensions will be measured. Then the coordinates you specify for line ends, circle centers and diameters, and other features are full-scale measurements from that origin. Of course, a 10-ft-long object cannot be displayed on a computer screen without scaling it down. Fortunately, CADD programs provide ZOOM capability. ZOOM allows you to change the scale at which your drawing is displayed on the screen. Remember that ZOOM controls only the size at which the drawing is displayed on the computer screen, not the dimensions that are stored in the drawing data file for the object. The file always contains the full-scale real-world dimensions of the object. As a result, you can zoom in or out of your drawing as often as you wish to view all or a part of it at different display scales. The data file does not change.

If you wish to plot your drawing, you can specify a plotting scale that will allow you to plot it on any size paper the plotter will accept without changing the drawing at all.

Procedures for establishing a drawing size or scale vary somewhat from one CADD program to another. The first step is to select the system of units in which you wish to work. The program may offer some or all of the following choices: decimal inches, fractional inches, feet, miles, millimeters, meters, kilometers, or user-defined units. Some programs offer only decimal units of unspecified type. It is then up to you, the user, to decide whether the units are inches, millimeters, miles, or some other unit of measure.

Let us assume that you have selected decimal inches as the desired units and that your drawing will fit in an 80 × 100-in. area. You may now select a display scale (zoom in or out) to display an 80 × 100-in. area on the screen. If you now draw a rectangle that is 80 × 100 in., it will outline your entire drawing area. Of course, features of your drawing are going to appear small on the screen. If you wish to work on only a portion of the drawing, you can select the area you are working on and zoom to fill the screen with that portion of your drawing.

When it comes time to plot your drawing, you can select a plotting scale that will fit the drawing or a selected part of it on the size plotting paper you have. For instance, the entire drawing could be plotted on a C size (24 × 36) sheet at quarter scale. If you are planning to plot your drawing, it is wise to check on the paper sizes your plotter will accept when planning the layout of your drawing. Remember that the ratio of paper width to height varies for different sizes. Plan accordingly.

7.9.3 Automatic Dimensioning

Some graphics programs have "automatic" dimensioning. This means that the length of the lines you specified on your orthographic views will be lettered on the drawing if you use this feature. This is one of the reasons the drawing is created full scale! If you were a little sloppy when making the views, the dimensions could be in error. You may get the dimensions you drew rather than the ones you wanted.

7.9.4 Commercial Software and Educational Software

Graphics packages designed for professional drafters have many useful features—these are often specified by picking options from a series of menus. This can mean finding your way through a series of four or five menus. Professionals will soon become used to the options needed for their type of work. Some programs will allow them to specify a path that automatically selects the needed option from consecutive menus.

Some commercial CADD software supports use of digitizing tablets as menu boards. The puck or stylus on the tablet is used to select items from menu overlays. The overlays can display more selections than can be displayed at one time on the screen. They can even offer macros (menu items that execute a sequence of commands) to speed up the selection process.

There are other CADD programs designed for classroom use that do not have so many automatic features and options. They require that you draw each line and select each number and letter as you would when making a drawing by hand. This might prove faster for a student who does not spend every day with a graphics package. Try the packages you have available and become as familiar with the computer screen as you have with a drawing board or sketch pad. The problems at the end of this chapter are designed to be done on the computer screen or manually.

7.10 *OTHER STANDARD DIMENSIONING SYSTEMS*

New manufacturing methods, including computer-controlled machine tools, have created the need for additional dimensioning systems. Computer-controlled machines sometimes allow us to transfer information directly from the data base for computer-aided design and drafting systems (CADD) to computer-aided manufacturing systems (CAM). Engineering disciplines other than mechanical design have evolved their own particular dimensioning systems. You need to be aware of the existence of these different systems.

7.10.1 Datum Lines and Surfaces

Numerically controlled machine tools and computer-aided machining frequently employ a system of dimensioning using rectangular coordinates. Figure 7.26 shows an example of a drawing using rectangular coordinates. In this drawing horizontal and vertical **datum lines** are established. Sometimes the datum lines will correspond to the edge of the object, as in Figure 7.26, and sometimes the center lines of the first hole are used as datum lines. The datum lines are labeled 0, and all other numbers represent the distance in the horizontal or vertical direction from the datum lines. Datum lines or surfaces may be established for all the prin-

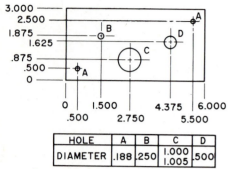

HOLE	A	B	C	D
DIAMETER	.188	.250	1.000 1.005	.500

Figure 7.26 Rectangular coordinates measured from datum lines are used to dimension parts for computer-aided manufacturing.

cipal directions—width, height, and depth. In the example shown in Figure 7.26, the sizes of the holes are shown in a hole table. In some drawings, the hole table may also show the required number of each size hole.

7.10.2 Tabular Dimensioning

Tabular dimensioning is used when several parts have the same general shape but a dimension, or series of dimensions, changes. Figure 7.27 is an example of a drawing with tabular dimensions. In this case there are three different washers required. Rather than make three different drawings, one drawing is made and a table is used to indicate the numerical values for each dimension. Tabular dimensioning is used to reduce the number of drawings required without reducing the amount or clarity of information.

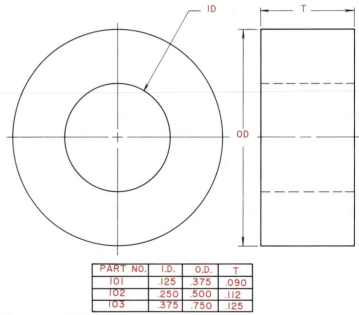

PART NO.	I.D.	O.D.	T
101	.125	.375	.090
102	.250	.500	.112
103	.375	.750	.125

Figure 7.27 Tabular drawings allow description of several sizes of same shape on one drawing.

7.10.3 Other Design Disciplines

The dimensioning systems and examples given have all been taken from mechanical engineering practices used for producing machine parts. Different dimensioning systems and practices are used in other types of drawings such as architectural, structural, or electrical. Figures 7.28–7.30 show examples of dimensioning practices for these types of drawings.

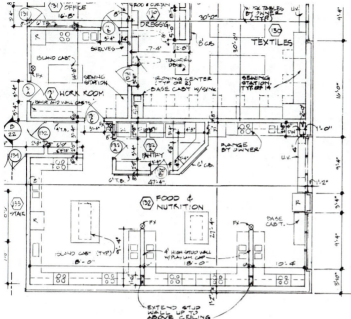

Figure 7.28   Some architects use slashes or large dots in place of arrowheads. They don't break dimension lines and specify feet and inches rather than just inches. (Courtesy of McDonald-Cassel & Bassett, Inc.)

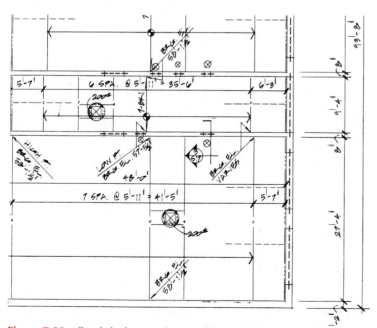

Figure 7.29   Steel designers also use feet and inches and don't break dimension lines. Slashes or double slashes may be used rather than arrowheads. (Courtesy of McDonald-Cassel & Bassett, Inc.)

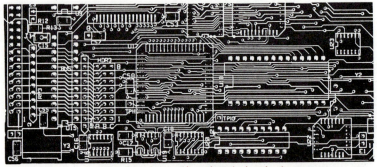

Figure 7.30 Drawings for electronic printed circuits have no dimensions but are drawn many times full size with great precision and copied photographically.

7.11 DIMENSIONING FOR INTERCHANGEABLE PARTS

Many industries rely on manufacturing large numbers of items that consist of assemblies of parts. (Automobiles, home appliances, and weapons are examples.) To be cost effective, mass production requires the use of interchangeable parts. No object can be made exactly to a given dimension. Therefore, tolerances are specified to indicate the permissible variation from the given dimension. The smaller the tolerance, the more expensive the object is to manufacture. Dimensions with greater tolerances are less expensive to produce but may result in parts that cannot be assembled or will not perform the function for which they were intended. There is a need, therefore, for dimensioning and tolerancing systems for interchangeable assembly. Eli Whitney, the inventor of the cotton gin, is given credit for the first application of interchangeable parts when his shop produced weapons for the U.S. government before the Civil War.

7.11.1 Types of Fits

There are three types of fits of mating parts. They are clearance, interference, and transition fits. Each of these fits will be illustrated using the groove block shown in Figure 7.31. The slot in the block has a width with a tolerance of .020 in. Remember, the tolerance is the difference between the minimum and maximum permitted sizes (**limits**). Three different sliders are shown in Figure 7.31: A, B, and C. Figure 7.32 shows the slider in place in the groove block. It is possible to write an expression to calculate

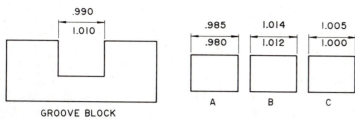

GROOVE BLOCK

Figure 7.31 Grooved block with three different sliders that may fit in it.

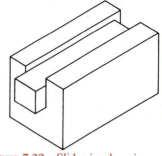

Figure 7.32 Slider in place in groove.

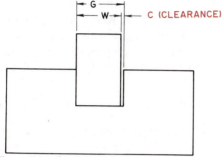

Figure 7.33 When slider fits in groove, space left is called clearance.

the maximum and minimum clearance between the slider and the slot (Figure 7.33). Maximum clearance will occur when the slider is small and the slot is large. Minimum clearance will occur when the slider is large and the slot is small. The relationships can be expressed as follows:

$$C_{max} = G_{max} - W_{min}$$

Where C is the clearance, G is the groove size, and W is the width of the slider.

Clearance Fits

If we calculate the C_{max} and C_{min} for slider A and the block (Figure 7.31), the results are as follows:

$$C_{max} = 1.010 - .980$$
$$= .030$$
$$C_{min} = .990 - .985$$
$$= .005$$

Therefore, the clearance will vary between .005 and .030 in.

When slider A is assembled with the groove block, we have determined that there will be clearance no matter what combination of permissible sizes of slider and block we choose. This type of fit is known as a clearance fit. Some clearance fits are called running or sliding fits.

Checking Fit Calculations

Slider A has a tolerance of .005 in., while the tolerance on the slot is .020. Note that the difference between the maximum and minimum clearances in this case is .025 in. (.030 − .005), which is equal to the sum of the tolerances of the two parts generating the clearance. It will always be true that the difference between the maximum and minimum clearances will equal the sum of the tolerances of the parts generating the clearance. This provides a quick way to check your work when calculating fits. Simply compare the sum of the tolerances with the difference between the maximum and minimum clearances. If they are not equal, there is an error. The tolerance of the system may be expressed mathematically as

$$T_s = \Sigma \, (T_1 + T_2 + \cdots + T_n) = C_{max} - C_{min}$$

We calculate or examine the maximum and minimum conditions of a fit (the extremes). If these values are acceptable, we know that every value between the extremes will be acceptable.

Interference Fits

Now consider the case of assembling slider B with the block (Figure 7.31). You can use the same expression to calculate the maximum and mininum clearances:

$$C_{max} = 1.010 - 1.012$$
$$= -.002$$
$$C_{min} = .990 - 1.014$$
$$= -.024$$

In both extremes we get negative clearance. Negative clearance is another way of saying interference. This type of fit is called an interference fit.

Check the calculations as follows:

$$\text{Tolerance on slot} = .20$$
$$\text{Tolerance on slider} = .002$$
$$T_s = \Sigma \, (T_1 + T_2) = C_{max} - C_{min}$$
$$\text{Sum of tolerances } T_1 + T_2 = .020 + .002 = .022$$
$$\text{Difference between } C_{max} \text{ and } C_{min} = (-.002) - (-.024)$$
$$= .022$$

Interference fits are used to permanently (or semipermanently) assemble parts. Some interference fits are called force or shrink fits.

Transition Fits

Now assemble slider C with the block (Figure 7.31). The maximum and minimum clearances are

$$C_{max} = 1.010 - 1.000$$
$$= .010$$
$$C_{min} = .990 - 1.005$$
$$= -.015$$

In this case the clearance varies from .010 to $-.015$. At one extreme we have clearance and at the other we have interference. This type of fit is known as a transition fit. Transition fits are sometimes called locational fits.

Checking calculations you see the following:

$$\text{Tolerance on slot} = .020$$
$$\text{Tolerance on slider} = .005$$
$$\text{Sum of the tolerances } T_1 + T_2 = .025$$
$$\text{Difference between } C_{max} \text{ and } C_{min} = .010 - (-.015)$$
$$= .025$$

7.11.2 Standard Fits

In the examples presented, the objects were dimensioned and the clearances were calculated. In most cases, however, the clearances will be specified and the dimensions must be calculated. The important thing to recognize is that the mathematical relationship between the clearance and the dimensions remains constant. By first writing a mathematical expression and then solving that expression for the required variables, clearance calculation problems can be solved with a minimum of confusion. You should be able to analyze an assembly, write an expression for the relationship, determine maximum and minimum clearances, and determine the type of fit.

Since many people are concerned with specifying types of fits, standard fits have been developed by various organizations such as ISO and ANSI. Tables for standard clearance, interference, and transition fits are available from the standards organizations and included in many engineering handbooks. Samples of the ANSI tables for fits are included in the Appendix.

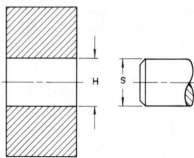

Figure 7.34 Cylindrical fit system with a round shaft fitting round hole, shown in section.

7.11.3 Cylindrical Fits

There are two systems for determining the dimensions for cylindrical fits—the fit between round holes and shafts (Figure 7.34). These systems are known as the basic hole system and the basic shaft system. To understand both systems, you must be familiar with the terms *allowance* and *nominal*. Allowance is defined as the difference in size between the hole and shaft for the tightest fit. The allowance always equals the minimum hole minus the maximum shaft. Nominal refers to the general size of a dimension independent of the tolerance. For example, in Figure 7.31 the slot in the groove block and the width of the sliders all would be considered to have a nominal size of 1 in.

Basic Hole System

The basic hole system is used when the size of the hole is considered more important or more expensive to produce than the size of the shaft. If a machine shop only has certain sizes of reamers (tools for finishing round holes very accurately), it would be desirable to adjust the shaft size to fit the hole. In the basic hole system, the minimum hole diameter is taken as the basic hole size. For example, for a nominal $\frac{1}{2}$-in. hole and shaft, a $\frac{1}{2}$-in. reamer might be used. Because reamers and other tools cannot be operated perfectly, the minimum hole size would be .500 in., and the largest hole likely to be produced would be a little larger. Let's calculate the limit dimensions for the hole and shaft with a nominal size of $\frac{1}{2}$ in.

Assume that the following design specifications are to be met:

Nominal size	$\frac{1}{2}$ in.	Hole tolerance	.003
Allowance	.005	Shaft tolerance	.001

To find the limit dimensions for the hole:

Basic size (minimum hole diameter)	.500
Assigned hole tolerance	.003
Maximum hole (minimum hole plus tolerance)	.503

To find limit dimensions for the shaft:

Basic size	.500
Allowance	.005
Maximum shaft (basic size minus allowance)	.495
Assigned shaft tolerance	.001
Minimum shaft (maximum shaft minus tolerance)	.494

Basic Shaft System

Sometimes the diameter of the shaft may be more important than the diameter of the hole. For example, it may be desirable to use standard size stock shafts, which can be purchased with carefully controlled, precise diameters. In these situations, the basic shaft system would be used. The basic shaft system is the least common of the two systems. The basic shaft system is similar to the basic hole system except that the maximum shaft diameter is used as the basic size. Here is an example calculation of maximum hole and shaft diameters using the basic shaft system.

Assume the following design specifications:

Nominal size	$\frac{1}{2}$ in.	Shaft tolerance	.002
Allowance	.010	Hole tolerance	.006

To find the limit dimensions for the shaft:

Basic size (maximum shaft diameter)	.500
Assigned shaft tolerance	.002
Minimum shaft diameter (maximum minus tolerance)	.498

To find the limit dimensions for the hole:

Basic size	.500
Allowance	.010
Minimum hole diameter (minimum hole plus allowance)	.510
Assigned hole tolerance	.006
Maximum hole diameter (minimum hole plus tolerance)	.516

7.12 *ADVANCED DIMENSIONING TOPICS*

The information you have studied about dimensioning is only a small part of the total you would need to do a professional job of specifying parts for manufacture. What you have studied is basic to all size descriptions. However, there are several other topics with which you need to become familiar to be a professional designer.

7.12.1 Metric Dimensioning with International Standards

All of the limit dimension examples given can be done using metric values. However, a standard for metric tolerances exists that considers tolerances in a different way: a system of tolerances has been established that uses code letters (Hole: 20H8, Shaft: 20f7). This system is described in ANSI standard B4.2. The standard and its application are discussed in Chapter 8.

7.12.2 Surface Control

Another area of interest is surface control. When we design parts for interchangeable assembly, we need to be concerned with not only the size but also the surface roughness and quality. Imagine trying to measure the height of coarse sandpaper to a thousandth of an inch! We cannot achieve precise tolerances unless the surfaces are smooth, and sometimes the nature of the tool marks or roughness affects the function of the part. ANSI standards B46.1 and Y14.36 provide information about surface texture and specification for controlling it. Some of this information is presented in Chapter 8.

7.12.3 Geometric Tolerancing

Geometric tolerancing is another advanced topic. It concerns the control of factors affecting parts to be assembled such as straightness, flatness, concentricity, parallelism, and others. There is a complete set of symbols available to describe on the drawing what is required for a part to function as needed. The use and application of these symbols is a complete course in itself. Some of this information is given in Chapter 8 so that you may recognize the symbols when you find them on a drawing.

7.13 *PRINCIPLES OF GOOD DIMENSIONING*

A number of fundamental principles have been developed that will help you in selection and placement of dimensions. These principles, understood by designers and craftsmen, will avoid possible misinterpretations. They have been summarized below for easy reference. Remember, however, that the overriding principle is that of *clarity*. Whenever it seems that the fundamental principles are in conflict, the conflict should always be resolved in a way that promotes maximum clarity.

1. *Each feature of an object is dimensioned once and only once.* Remember that every dimension has a tolerance associated with it. In Figure 7.35a, the overall height, the height of the step, and the depth of the notch are given. Assume for the moment that the tolerance on all three dimensions is ±.010 in. This leads to a dilemma relative to the permissible variation in the height of the object. If the height is determined to be the sum of the notch and the step, the permissible variation would be ±.020 in. If the height dimension is used, however, the variation could only be ±.010. To resolve this problem, one of the dimensions must be left out. In practice, the least important dimension is the one left out, and error is allowed to accumulate in a part of the object that is not critical. Unless there is some functional requirement, it is usually better to dimension material rather than space. Figure 7.35b shows the preferred way to dimension the object.

2. *Dimensions should be selected to suit the function of the object.* If, as in the case of the object in Figure 7.35b, the object were used in an assembly where the depth of the slot were more important to the assembly than the height of the step, then the step height dimension would be replaced by the slot depth dimension.

3. *Dimensions should be attached to the most descriptive view of the feature being dimensioned.* In Figure 7.36 the most descriptive view of the notch is the front view. It is incorrect, therefore, to dimension this notch in the less descriptive top or side views.

4. *Dimensions should not specify the manufacturing method.* The manufacturing method should only be specified if it is a mandatory design requirement. A common practice found in older drawings is to see hole diameters specified by the drill size, as in Figure 7.37a. The preferred method is shown in Figure 7.37b. The only time a manufacturing method should be specified is when it is the only acceptable way to produce the object. If, for example, the tolerance could be met by either punching, drilling, laser machining, chemical milling, or electrical discharge machining, the production department should have the flexibility to select the most economical manufacturing process. On the other hand, if a particular process is important to the function of the object, for example, the surface obtained by electric discharge machining rather than drilling, the process should be specified on the drawing.

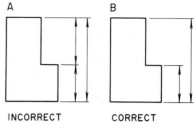

INCORRECT   CORRECT

Figure 7.35 Each feature of object is dimensioned only once, with the dimensions selected to suit function.

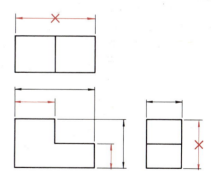

Figure 7.36 Dimensions should be placed on most descriptive views.

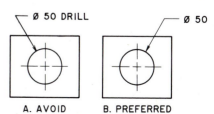

A. AVOID   B. PREFERRED

Figure 7.37 Dimensions should specify only size, not process.

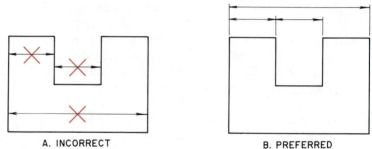

A. INCORRECT   B. PREFERRED

Figure 7.38 Dimensions and dimension lines should not be placed on views.

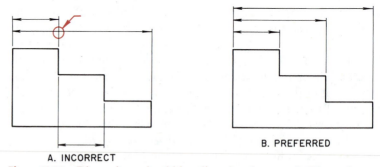

A. INCORRECT   B. PREFERRED

Figure 7.39 Dimensions should be aligned and grouped. Dimension lines should never be crossed.

5. *Angles shown on drawings as right angles are assumed to be 90° unless otherwise specified.*

6. *Dimensions should be located outside the boundaries of the object whenever possible.* When dimensions are located within the boundary of an object, such as in Figure 7.38a, they are difficult to find and read. Figure 7.38b shows the preferred way to locate them.

7. *Dimension lines should be aligned and grouped to promote uniform appearance.* Unnecessary crossing of object and extension lines as shown in Figure 7.39a should be avoided. By aligning and grouping the dimensions, as in Figure 7.39b, the drawing not only looks neater but is also much easier to read.

8. *Dimension lines should be unbroken except for the number between the arrowheads.* It is very easy to misread a dimension if the dimension line is crossed by any line other than the extension lines at the ends (Figure 7.39a).

9. *The space between the first dimension line and the object should be at least $\frac{3}{8}$ in. (10 mm). The space between dimension lines should be at least $\frac{3}{8}$ in. (10 mm).* Proper spacing is shown in

Figure 7.40. Remember, the spacing shown is minimum. The spacing can be increased to improve readability.

10. *There should be a visible gap between the object and the origin of an extension line.* This gap of about $\frac{1}{16}$ in. (2 mm) is shown in Figure 7.40.

11. *Extension lines should extend $\frac{1}{8}$ in. (3 mm) beyond the last dimension line* (Figure 7.40).

12. *Extension lines should be broken if they cross or are close to arrowheads* (Figure 7.41).

13. *Leader lines used to dimension circles or arcs should be radial.* Radial lines are those that, if extended, pass through the center of a circle or arc. Figure 7.42*a* shows a radial line; Figure 7.42*b* shows a nonradial line.

14. *Dimensions should be oriented to be read from the bottom of the drawing.* This orientation is known as the unidirectional dimensioning system. Older drawings frequently used the aligned system. These systems were shown earlier in this chapter in Figure 7.18.

15. *Diameters are dimensioned with a numerical value preceded by the diameter symbol (ϕ).* You may find many drawing examples that were produced before publication of the 1982 ANSI standards that do not conform to this practice. All new drawings should use the diameter symbol as shown in Figures 7.37*b*, 7.42*b*, and 7.43.

16. *Positive cylinders should be dimensioned in the longitudinal view.* Figure 7.43 shows the proper dimensioning of concentric cylinders. The hole was not dimensioned in the longitudinal view because this would have resulted in dimensioning to a hidden line. Hidden lines are not considered to be the most descriptive view of features. If a section view of the hole in Figure 7.43 were shown instead of a hidden view, the hole could be dimensioned in the longitudinal view.

Even more important is the manner in which a machinist usually makes a circular hole. It is most often made with a tool, like a drill, that requires the machinist to locate the center as a starting point. Therefore, the machinist needs to know the location of the

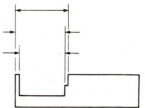

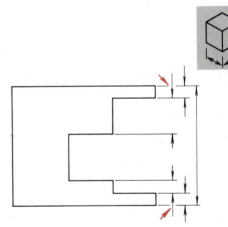

Figure 7.40   Dimension lines should be spaced for readability; extension lines should have visible gap at origin and extend beyond last dimension line.

Figure 7.41   Extension lines should be broken if they cross arrowheads.

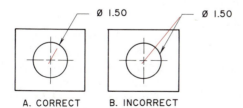

Figure 7.42   Leaders for notes on cylindrical parts should be radial; diameters should always be preceded by symbol (ϕ).

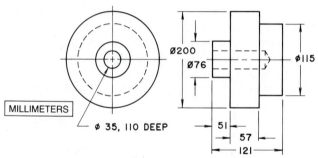

Figure 7.43 Positive cylinders should be dimensioned in longitudinal view; depth of blind hole is given in note with diameter.

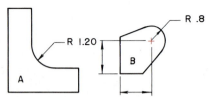

Figure 7.44 Internal and external radii should be dimensioned with numerical value preceded by symbol R.

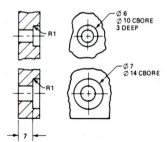

Figure 7.45 Counterbored holes: depth may be given with diameter or to visible line on longitudinal view, when sectioned. (Courtesy of ANSI)

center point, which is shown on the characteristic, circular view of the hole. It is here that the size should be given, so that all information about the hole is grouped in one place. Positive cylinders are most frequently produced by turning on a lathe. The machinist is looking at the work turning between centers in a longitudinal position. It is in this position that measurements of diameters are taken, so it is in this position on the drawing that these diameters should be given. See Chapter 8 for more information on manufacturing processes.

17. *Radii are dimensioned with a numerical value preceded by the radius symbol (R).* Older drawings have the R following rather than preceding the numerical value. All new drawings should conform to the current standards. Figure 7.44 shows how to dimension a radius.

18. *When a dimension is given to the center of an arc or radius, a small cross is shown at the center.* Figure 7.44*B* shows a radius with a located center. In Figure 7.44*A* it is not necessary to locate the center of the radius since it can be located from the horizontal and vertical lines to which it is tangent. It is not necessary to locate the centers for small fillets and rounds.

19. *The depth of a blind hole may be specified in a note and is the depth of the full diameter from the surface of the object* (Figure 7.43).

20. *Counterbored, spotfaced, or countersunk holes should be specified in a note.* Figures 7.45–7.47 show dimensioning of counterbored, countersunk, and spotfaced holes. All are concentric with a starting hole. A counterbored hole is usually intended to provide space for the head of a socket head or fillister head screw; a countersunk hole is the negative frustum of a cone and provides space for the head of a flat head screw. A spotface is a very shallow cavity intended only to provide a smooth bearing area for a fastener. You will learn more about fasteners in Chapter 9.

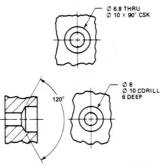

Figure 7.46 Countersunk and counterdrilled holes. (Courtesy of ANSI)

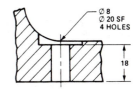

Figure 7.47 Spotfaced holes: either depth or remaining thickness of material may be specified. (Courtesy of ANSI)

7.14 CHECKING A DIMENSIONED DRAWING

Dimensions on drawings are very important for communicating technical information. Incorrect or missing dimensions could result in faulty parts. This can be costly in terms of time, money, or professional reputation. Therefore, you should check all dimensions before completing a drawing.

You should first check to see that every feature has been dimensioned once and only once. This can be done by moving a straightedge from left to right and from top to bottom across the drawing, stopping at each feature and making sure it has been properly dimensioned.

Next make sure that all numerical values are correct. Take a scale and lay it on every dimension, checking for accuracy. These two steps will only take a couple of extra minutes in preparing a drawing, and by getting into the habit of checking your dimensions, you will avoid many embarassing and costly mistakes.

7.15 SAMPLE PROBLEM: TABLE SAW FIXTURE

The orthographic views of the fixture developed in Chapter 4 give us the complete shape description for the part. We can complete the drawing with size description using a CADD package. It may be necessary to change the space between the views in order to provide enough space for the dimensions.

Advanced CADD packages provide automatic dimensioning; you call for dimensions and the screen shows the dimension, extension and leader lines, and numerical values. You will need to be careful when making the drawing for the automatic dimensions will be those you used to draw the views; if those

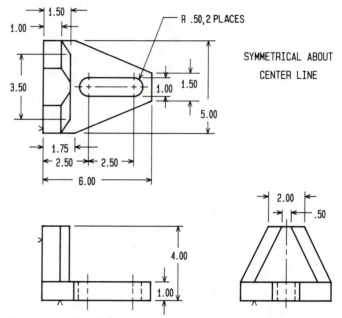

Figure 7.48 Multiview drawing produced and dimensioned with enhanced GraphiCad program, including automatic dimensioning.

values were not accurate, your dimensions will be incorrect. Automatic dimensioning systems will not do tolerancing well. The designer still needs to use judgment to select the appropriate tolerances; then they are installed as dimension lines and text.

Simpler CADD packages do not provide automatic dimensioning. You can place the lines and use text commands to enter the numerical values and notes. Arrowheads are often provided, even in simple CADD packages.

Figure 7.48 was prepared with an advanced version of GraphiCad, which provides automatic dimensioning. An up-to-date program with automatic dimensioning such as this one will give you dimensions that meet the ANSI standards for form and placement.

7.16 SUMMARY

Describing the size of objects is important. Without size information, nothing could be built. Without the concept of tolerances, there could be no interchangeable parts, and manufactured products would be prohibitively expensive.

Because of the importance of dimensioning drawings, standard practices have been developed. Since the primary reason for producing drawings is that of communicating information, these standard practices must be learned and followed. Proper attention to line weights and selection and placement of dimensions will result in drawings that communicate clearly and accurately the requirements of a design. If you take a few minutes to review the principles of good dimensioning before making a drawing and then check each completed drawing, you will soon be making clear, professional-quality, engineering drawings.

PROBLEMS

Problems 7.1–7.8
Prepare the needed orthographic views of each assigned problem on $8\frac{1}{2} \times 11$ paper or on your computer. Let each unit on the sketch equal $\frac{1}{4}$ in., or 6 mm, on your drawing, as designated by your instructor. Completely dimension the object so that it could be built from your drawing. Be sure to allow enough space around the views for good size description.

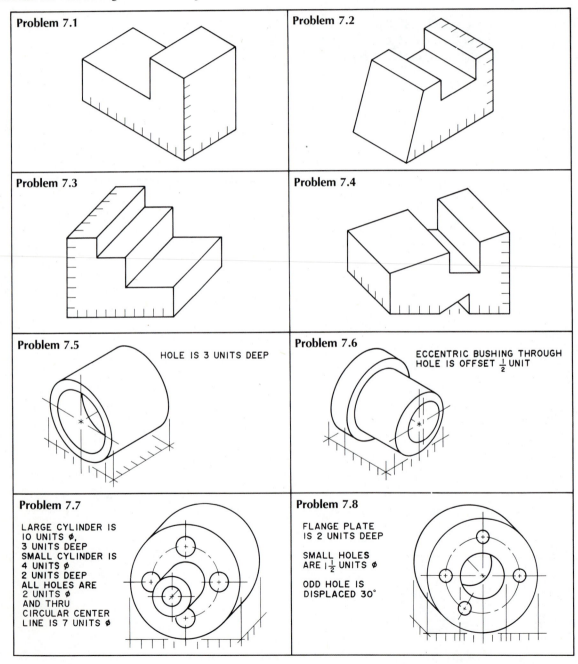

Problem 7.1

Problem 7.2

Problem 7.3

Problem 7.4

Problem 7.5 HOLE IS 3 UNITS DEEP

Problem 7.6 ECCENTRIC BUSHING THROUGH HOLE IS OFFSET $\frac{1}{2}$ UNIT

Problem 7.7
LARGE CYLINDER IS
IO UNITS ø,
3 UNITS DEEP
SMALL CYLINDER IS
4 UNITS ø
2 UNITS DEEP
ALL HOLES ARE
2 UNITS ø
AND THRU
CIRCULAR CENTER
LINE IS 7 UNITS ø

Problem 7.8
FLANGE PLATE
IS 2 UNITS DEEP

SMALL HOLES
ARE $1\frac{1}{2}$ UNITS ø

ODD HOLE IS
DISPLACED 30°

Problems 7.9–7.14

The views of each of these objects should be drawn on 8½ × 11 paper or on your computer. Determine dimensions with your dividers and the scale printed in color on this page. Dimension your drawing in one-place millimeters or two-place inches as specified by your instructor. Include a note on each drawing that states which system is used.

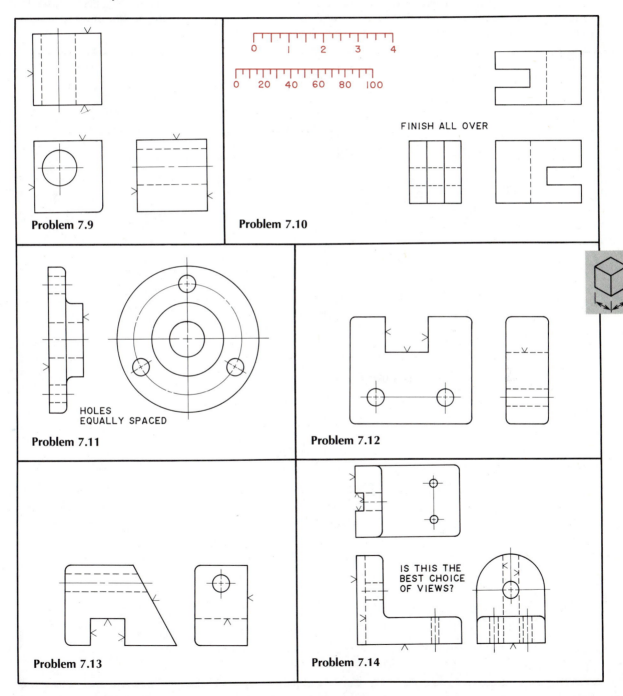

Problem 7.9

FINISH ALL OVER

Problem 7.10

HOLES
EQUALLY SPACED

Problem 7.11

Problem 7.12

Problem 7.13

IS THIS THE
BEST CHOICE
OF VIEWS?

Problem 7.14

Problems 7.15–7.21

Scale the drawings in problems 7.9–7.14 and show the principal dimensions on a sketch. Use the scales assigned by your instructor.

Problems 7.22–7.25

Complete the required information for each tolerance problem assigned by your instructor. You may be asked to letter the information neatly on a drawing sheet or a sheet of note paper or to reproduce the drawing manually or on the computer and complete the information on your drawing.

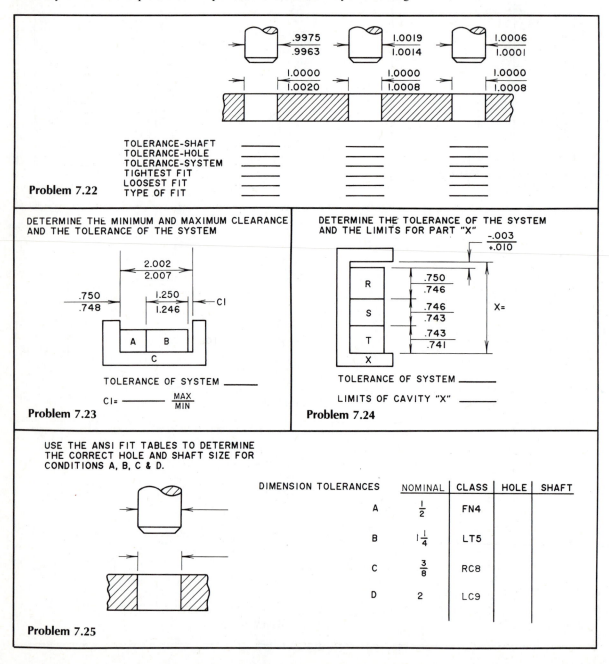

Problem 7.22

TOLERANCE-SHAFT _____
TOLERANCE-HOLE _____
TOLERANCE-SYSTEM _____
TIGHTEST FIT _____
LOOSEST FIT _____
TYPE OF FIT _____

DETERMINE THE MINIMUM AND MAXIMUM CLEARANCE AND THE TOLERANCE OF THE SYSTEM

TOLERANCE OF SYSTEM _____

$$CI = \frac{MAX}{MIN}$$

Problem 7.23

DETERMINE THE TOLERANCE OF THE SYSTEM AND THE LIMITS FOR PART "X"

TOLERANCE OF SYSTEM _____

LIMITS OF CAVITY "X" _____

Problem 7.24

USE THE ANSI FIT TABLES TO DETERMINE THE CORRECT HOLE AND SHAFT SIZE FOR CONDITIONS A, B, C & D.

DIMENSION TOLERANCES	NOMINAL	CLASS	HOLE	SHAFT
A	$\frac{1}{2}$	FN4		
B	$1\frac{1}{4}$	LT5		
C	$\frac{3}{8}$	RC8		
D	2	LC9		

Problem 7.25

Problems 7.26–7.28
Complete the required information for each tolerance problem assigned by your instructor. You may be asked to letter the information neatly on a drawing sheet or a sheet of note paper or to reproduce the drawing manually or on the computer and complete the information on your drawing.

DETERMINE LIMITS OF INSIDE DIAMETER AND
LENGTH OF WHEEL.

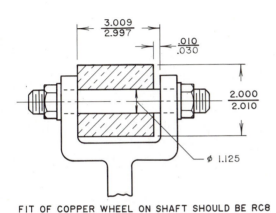

FIT OF COPPER WHEEL ON SHAFT SHOULD BE RC8
Problem 7.26

A STEEL TIRE IS TO BE FORCE FIT ON AN IRON
WHEEL. THE WHEEL HAS A BRONZE BEARING IN
IT TO RECEIVE THE SHAFT. COMPLETE THE
FOLLOWING INFORMATION:

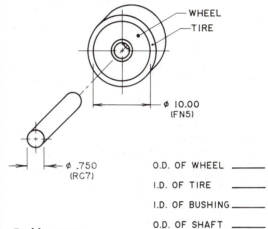

O.D. OF WHEEL _____

I.D. OF TIRE _____

I.D. OF BUSHING _____

O.D. OF SHAFT _____

Problem 7.27

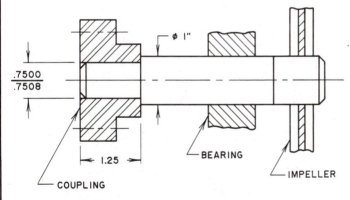

GENERAL TOLERANCE ± .03
FITS REQUIRED:

    SHAFT W/COUPLING - FN2
    SHAFT W/BEARING - RC3
    SHAFT W/IMPELLER - FN4

DIMENSIONS OF COUPLING ARE
SHOWN. SHAFT IS NEVER
TO EXTEND BEYOND END
OF COUPLING

Problem 7.28

LENGTH OF $\phi\frac{3}{4}$ SHAFT _____

LIMITS OF $\phi\frac{3}{4}$ SHAFT _____

TOLERANCE OF SYSTEM _____

BEARING:
LIMITS OF SHAFT _____

LIMITS OF HOLE _____

TOLERANCE OF SYSTEM _____

IMPELLER:
LIMITS OF SHAFT _____

LIMITS OF BEARING _____

TOLERANCE OF SYSTEM _____

Problems 7.29–7.33
Draw the views necessary to describe the object and dimension them for production. Use decimal inches or millimeters as specified by your instructor; note on your drawing which system is used. Remember that the dimensions used to describe an object in a pictorial drawing may not be the dimensions most appropriate for an orthographic multiview drawing.

Problem	Reference
7.29	4.43
7.30	4.53
7.31	4.56
7.32	4.64
7.33	4.66

Problems 7.34–7.43
Draw the views necessary to describe the object and dimension them for production. Use decimal inches or millimeters as specified by your instructor; note on your drawing which system is used. Remember that the dimensions used to describe the object in a pictorial drawing may not be the dimensions most appropriate for an orthographic multiview drawing.

MILLIMETERS | *INCHES*

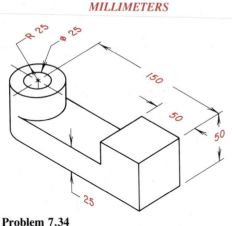

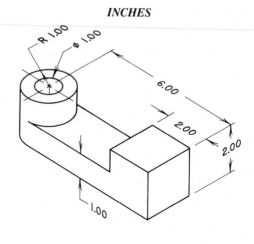

Problem 7.34

Problems 7.34–7.43
Draw the views necessary to describe the object and dimension them for production. Use decimal inches or millimeters as specified by your instructor; note on your drawing which system is used. Remember that the dimensions used to describe the object in a pictorial drawing may not be the dimensions most appropriate for an orthographic multiview drawing.

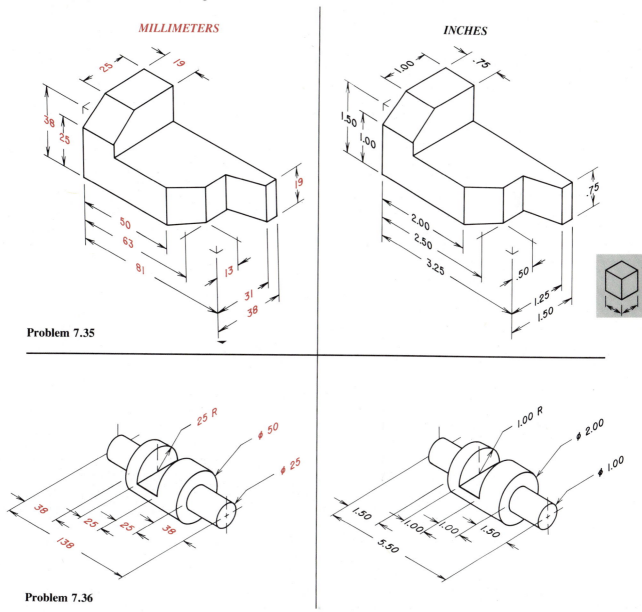

MILLIMETERS

INCHES

Problem 7.35

Problem 7.36

Problems 7.34–7.43
Draw the views necessary to describe the object and dimension them for production. Use decimal inches or millimeters as specified by your instructor; note on your drawing which system is used. Remember that the dimensions used to describe the object in a pictorial drawing may not be the dimensions most appropriate for an orthographic multiview drawing.

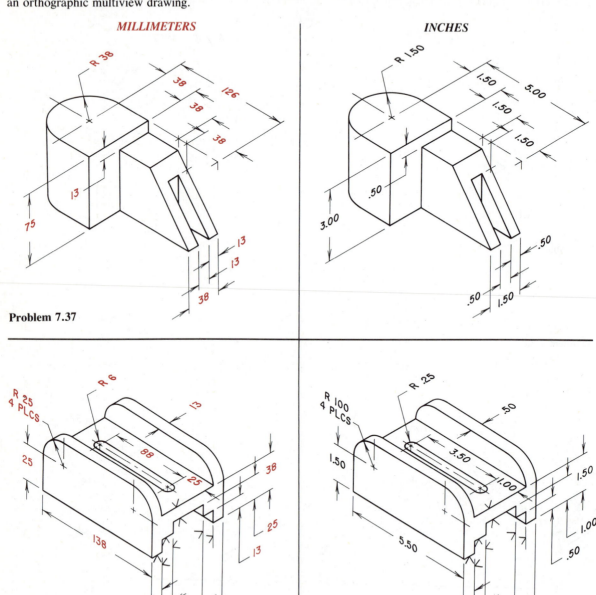

MILLIMETERS

INCHES

Problem 7.37

Problem 7.38

MILLIMETERS

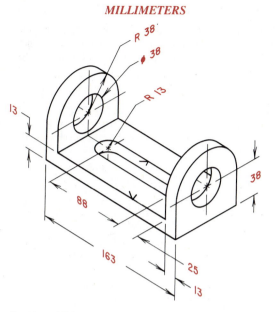

R 38

⌀ 38

R 13

13

38

88

163

25

13

INCHES

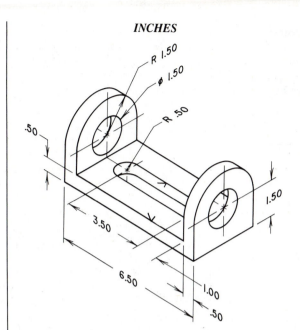

R 1.50

⌀ 1.50

R .50

.50

1.50

3.50

6.50

1.00

.50

Problem 7.39

Problems 7.34–7.43

Draw the views necessary to describe the object and dimension them for production. Use decimal inches or millimeters as specified by your instructor; note on your drawing which system is used. Remember that the dimensions used to describe the object in a pictorial drawing may not be the dimensions most appropriate for an orthographic multiview drawing.

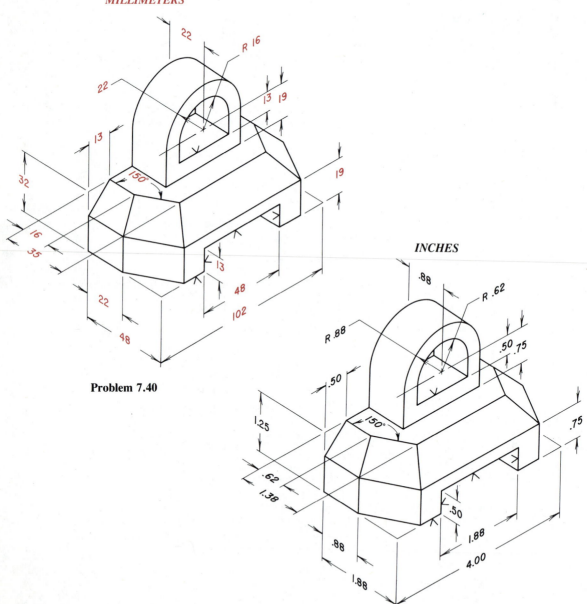

MILLIMETERS

INCHES

Problem 7.40

MILLIMETERS

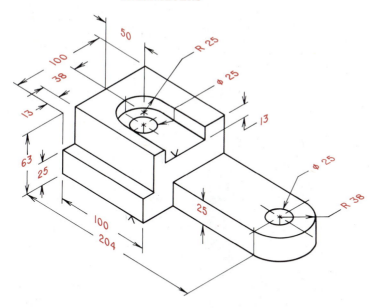

Problem 7.41

INCHES

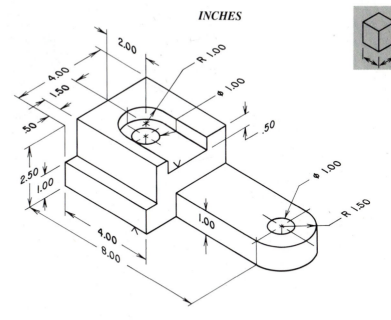

Problems 7.34–7.43
Draw the views necessary to describe the object and dimension them for production. Use decimal inches or millimeters as specified by your instructor; note on your drawing which system is used. Remember that the dimensions used to describe the object in a pictorial drawing may not be the dimensions most appropriate for an orthographic multiview drawing.

MILLIMETERS

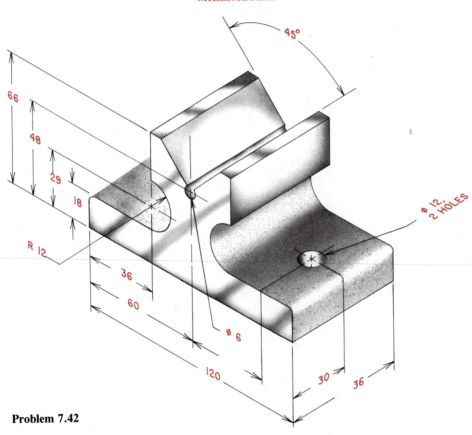

Problem 7.42

INCHES

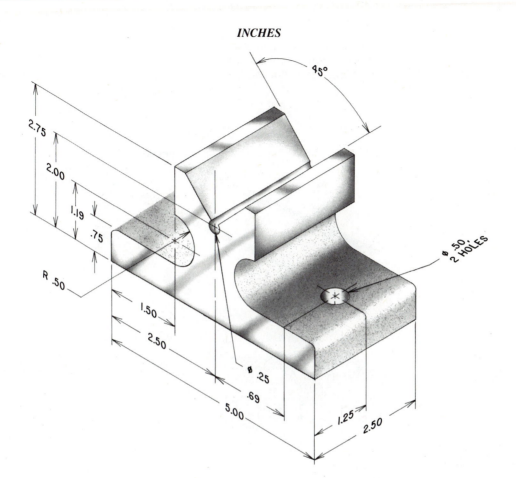

45°

2.75

2.00

1.19

.75

R .50

1.50

2.50

Ø .25

.69

5.00

Ø .50
2 HOLES

1.25

2.50

Problems 7.34–7.43

Draw the views necessary to describe the object and dimension them for production. Use decimal inches or millimeters as specified by your instructor; note on your drawing which system is used. Remember that the dimensions used to describe the object in a pictorial drawing may not be the dimensions most appropriate for an orthographic multiview drawing.

MILLIMETERS

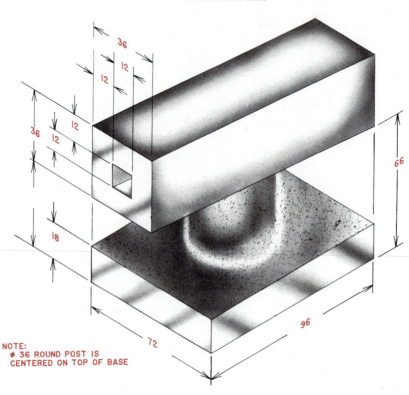

NOTE:
Ø 36 ROUND POST IS
CENTERED ON TOP OF BASE

Problem 7.43

INCHES

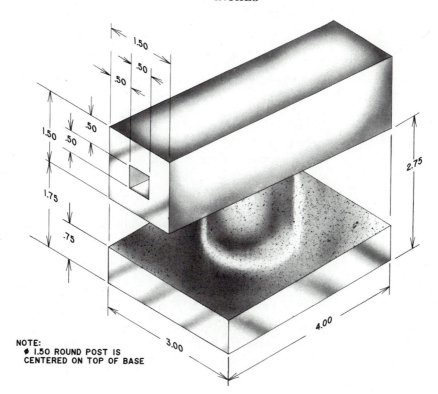

1.50

.50

.50

.50

1.50   .50

1.75

.75

2.75

4.00

3.00

NOTE:
Ø 1.50 ROUND POST IS
CENTERED ON TOP OF BASE

TERMS YOU WILL SEE IN THIS CHAPTER

SYMBOLS Graphical shorthand used to provide specifications on a drawing.

DATUM REFERENCE Theoretically perfect surface or line on an object used as a starting point to relate important properties of the object.

GEOMETRIC TOLERANCING Specification of properties other than length; may include form, profile, orientation, or location.

FORMING To give a particular shape to materials.

ASSEMBLY Fitting parts together to produce a complete machine.

DIE Tool used to give form to an object by pressing or squeezing.

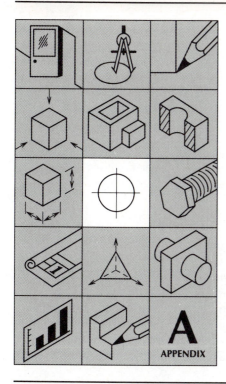

8

DIMENSIONING FOR PRODUCTION

8.1 INTRODUCTION

You learned about dimensioning and tolerancing in Chapter 7. The material you studied there is an introduction to the fascinating world of production. It is only an introduction, for in actual production environments designers must consider many other variables besides those described in Chapter 7. The standard for dimensioning practice, ANSI Y14.5M, includes information on topics that are beyond the scope of that chapter. It is one of the many references designers and production engineers use to develop a concept into useful products or processes.

The standard provides **symbols** for describing types of tolerances; this goes beyond just limits on numerical values. It also presents the concept of **datum references;** planes to which we can tie all the critical dimensions of a part. **Geometric tolerancing** includes tolerances for location, form, profile, orientation, and runout. This system for describing the functional relationships among features of a part provides

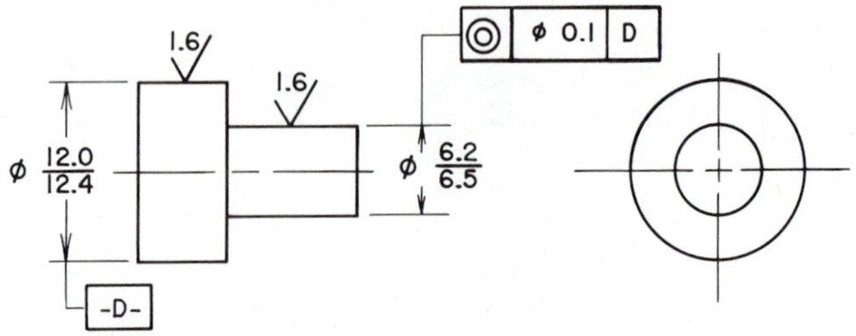

Figure 8.1 Geometric dimensioning, tolerancing, and surface control provide additional information to decrease manufacturing costs and improve productivity.

additional information to decrease manufacturing costs and improve productivity (Figure 8.1).

Tolerancing for metric dimensions will also be reviewed. This information is based on ANSI B4.2 and B14.5. The methods for displaying metric tolerances are much different from those you have learned so far. Fortunately, the standards provide for transition steps in dimensioning so that designers, tool makers, and manufacturing personnel can learn the system gradually.

You learned in earlier dimensioning studies that the accuracy of measurements is related to the accuracy of a surface. (You cannot measure to within .0001 between two surfaces rough as sandpaper.) Important dimensions are always referenced to datums; key finished surfaces are used as datum features. However, the quality and character of finished surfaces varies. This variability is described in ANSI Y14.36 and B46.1. Material from the standards is included in this chapter.

If you have not worked in a manufacturing environment or had laboratory courses in manufacturing practices, the terms and concepts to be discussed will be unfamiliar to you. For this reason, this chapter includes a very short section on basic production processes before you get into the advanced topics.

8.2 MANUFACTURING PROCESSES

Manufacturing processes include methods for **forming** materials and methods for assembling parts made of various materials. Processes for forming can be classified as reformation, removal, and addition. **Assembly** processes include manual, mechanized, and automated.

8.2.1 Material Reformation

Historic methods for forming metals are casting and forging. In the casting process metal is heated until it is a liquid and the molten metal is poured into a mold. After it cools, the solid metal is taken from the mold. It will have the shape of the cavity in which it is cooled (Figure 8.2). Casting involves making patterns of the object to be produced; enclosing the pattern in a mold of treated sand, ceramics, or metal of a higher melting point than the one to be cast; melting the metal; pouring it into the mold; cooling the cast part; and removing it from the mold. The part, as cast, may be very rough and have attached to it material needed for the casting process that is not a part of the finished object; this needs to be removed. All of these steps have been studied in detail, and there are texts, magazines, and technical papers providing information in depth. The casting process may also be used for nonmetallic materials such as ceramics and polymers. These plastic materials require less heat or no heat at all for casting. Some will melt at low temperatures while others will solidify through chemical action or without the addition of heat. They may be cast in precise metal molds and require little finishing after removal from the molds.

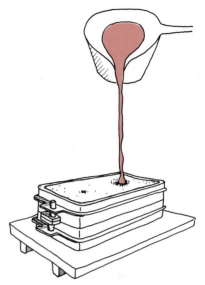

Figure 8.2 Casting.

Forging is the method of the blacksmith (Figure 8.3). Metal is heated until its structure changes and it becomes malleable (steel is usually red hot). Then it is hammered or pressed into the desired shape. The hammer may be wielded by a human arm or it may be powered by air or hydraulic fluid. The shape may be obtained by moving the forged part under the hammer to obtain the desired shape or it may be pressed into a mold known as a forging **die.** If the material is a thin sheet, it may be formed cold using only pressure. This related process is called pressing. Auto body parts such as roofs, hoods, and doors are made by pressing. Forging and pressing are also complete engineering disciplines under constant study.

Extrusion is related to forging in that the material is softened but not usually liquid. The softened material is squeezed through a shaped hole (a die) and comes out as a continuous bar having the shape of the orifice through which it passed (Figure 8.4). Extrusion may be used for metals, usually with heat to soften them. It may be used with plastic materials with heat or often without heat. Good extrusion processes can produce a very accurate shape with a smooth surface. The extruded bars are cut into pieces with each piece becoming an individual part.

Processes developed in more recent years include powder metallurgy and hot isostatic pressing. In the former the metal is ground to a powder and the powder pressed into a mold. Different metals

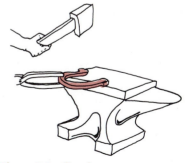

Figure 8.3 Forging.

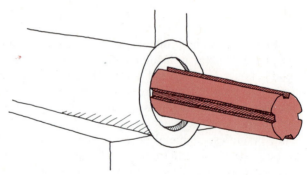

Figure 8.4 Extrusion.

or a mixture of metallic and nonmetallic materials may be used. The materials may be bonded by pressure only or they may be heated until they soften and fuse together. Powder metallurgy can produce a porous structure ideal for bearings: The cavities between the metal particles can be impregnated with oil or other lubricants that will be squeezed to the surface when the part is under load.

Hot isostatic pressing is a sophisticated process that immerses the heated raw material in a bath of high-pressure fluid that squeezes it uniformly from all sides into the desired shape. There are other reformation processes available. You will need to learn about new developments and the processes applicable to the products and process with which you work.

8.2.2   Material Removal

Material removal always produces scrap or waste material in addition to the desired part. These processes are still among the most used and useful of manufacturing methods. They include a tool for removing material and a means of moving the tool against the part or the part against the tool or both. Material removal can produce a new geometric shape, for example, a round hole—a negative cylinder. It can also refine an existing shape—make the surface smoother or change a dimension. Classic processes include drilling, reaming, boring, turning, milling, broaching, shaping, and grinding. More recent processes include superfinishing, chemical milling, and electron discharge machining. Many other processes exist that are specialized for a single industry or are part of a production line making many units of a product; here they may be developed to perform a single operation very quickly with great precision.

Drilling uses a cylindrical tool with a pointed end and spiral cutting flutes (grooves) cut into the cylindrical surface. The tool

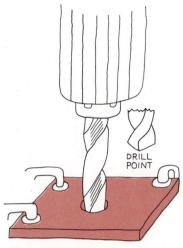

Figure 8.5 Drilling.

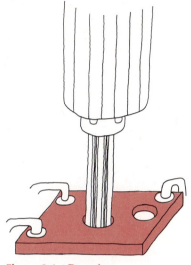

Figure 8.6 Reaming.

is made of hardened steel. It is pressed against the surface to be formed and rotated (Figure 8.5). The tool is usually rotated, but in some cases the part may be rotated. The drill enters the part and generates a round hole as it turns and advances.

Reaming is a finishing process: You start with a drilled hole and make it smoother and more accurately sized. The tool is a cylinder with a flat end slightly tapered or chamfered (you do not need a conical point to start the hole) and straight cutting flutes parallel to the axis of the reamer. It is rotated as it advances into the work. It removes only a small quantity of material as it smooths and sizes the hole (Figure 8.6).

The process of boring is related to drilling and reaming. It uses a single-point tool to generate a negative cylinder. The tool is usually fixed and the work revolves (Figure 8.7). The tool holder is fed axially into the work to the depth needed.

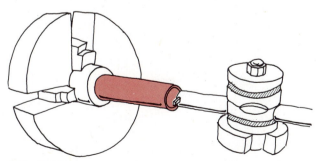

Figure 8.7 Boring.

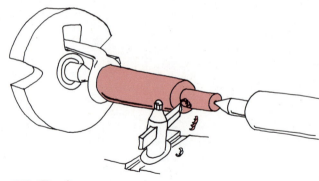

Figure 8.8 Turning.

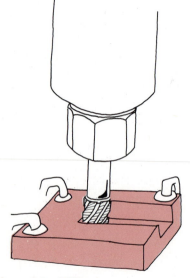

Figure 8.9 Milling.

Turning is the classic method for producing positive cylinders. The work is held between centers, or by one end, and rotated. As it turns, a single-point tool is pressed into the work and advanced axially to generate the cylindrical shape (Figure 8.8). Tapered or conical shapes can be generated by offsetting the axis of the work from the path of travel of the tool. Multipoint tools may be used to generate special shapes.

Milling is one of the most versatile of machining processes. Shaped cavities, external shapes, or surfaces can be produced by milling. The work is fastened to a table that can be moved sideways, forward, backward, vertically, or tipped at an angle. The rotating tool generally moves vertically, but some move on a horizontal axis. The concept of milling is shown in Figure 8.9.

The work is held still in the broaching process. A special tool, often resembling a large file, is pulled past the work. Each cutting edge on the broach removes a little material; with many teeth the broach can remove a lot of material in one pass (Figure 8.10).

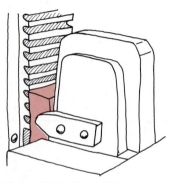

Figure 8.10 Broaching.

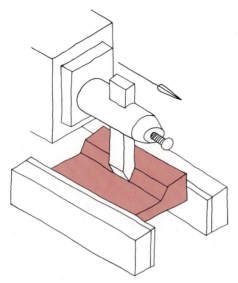

Figure 8.11 Shaping.

Shaping can produce flat parts or parts of a constant contour. The work is clamped to a fixed table that can be adjusted vertically while the tool is pushed across it. The tool cuts on the forward stroke and is hinged to slide back over the work on the back stroke (Figure 8.11).

Grinding uses an abrasive wheel or disk to remove material (Figure 8.12). The abrasive tool may also have a special shape. The work is fed into the tool, and the abrasive removes small amounts of material. A course abrasive coupled with a rapidly moving work piece can remove a lot of material in a short time. A fine abrasive coupled with a slow movement of the work can result in a very fine and accurate finish.

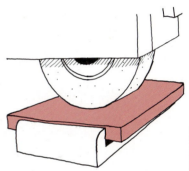

Figure 8.12 Grinding.

8.2.3 Material Addition

Welding is discussed in Chapter 9 as a method for joining parts. Welding can also be used to add material to a single part. It is often used to repair or salvage a part that is broken or undersize in one area. A small amount of material may be deposited from the weld rod in one pass; multiple passes can be used to build up greater thicknesses.

Plating is another method for adding material to a part. This method adds a very thin layer of material to a part, often to provide new properties to the surface. Common plating materials are nickel, chromium, cadmium, silver, rhodium, and gold. Some plating is accomplished by dipping; many materials are applied by electroplating. The objective of plating may be to bring an undersize part back to specification, but more commonly it is to provide protection against corrosion. Plating with silver or gold is used to improve the electrical conductivity of electrical parts. These precious metals are also used to improve appearance. If a part is to be plated, the thickness of the plating must be taken into account in the design. The basic part must be finished undersize, so that when the plating thickness is added, the part is within needed tolerances.

Coating is another way to add material to a part. Some coating materials are nonmetallic and cannot be applied by electroplating. Coatings may be used to protect against corrosion or to improve appearance or both. A common method of coating is the bluing applied to the steel parts of firearms. It is wiped on the steel surface and bonds to the metal to provide protection against corrosion. Aluminum is anodized. This is a method of electrical deposition that provides a thin, protective layer over the aluminum. It also can provide color. Coatings are relatively thin and most often applied to exterior surfaces rather than mating surfaces, so tolerances are not so critical. The important consideration of thickness is getting a coating thick enough to protect or color the part while minimizing cost.

Material addition by welding will usually produce a surface that needs to be machined: The surface will be rough and irregular. Chromium plating is often a two- or three-step process. A coating of nickel is usually deposited on the steel to seal it against corrosion and the chromium deposited over the nickel to provide the shiny, hard surface. Chromium plating, by itself, is porous and would allow pinpoint corrosion. Coatings are thin and usually duplicate the surface quality of the underlying part.

8.2.4 Assembly Processes

Over half of the work required to produce useful products may be in the assembly of the component parts. This is true for electronic equipment; farm machinery typically has less than a quarter of the work in assembly. It is an area that production engineers constantly study.

Originally all assembly work was done by hand. One craftsman conceived the device, made the parts, and put them together. If the parts did not fit, he would adjust or remake parts until the device functioned properly. Beginning in the nineteenth century, the concept of interchangeable parts was introduced (thus the need for limit dimensions). A worker or team of workers received a collection of parts and an *assembly drawing*. Their job was to put the parts together according to the drawing. This method is still used for one-of-a-kind machines or in small shops building special equipment.

At the beginning of the twentieth century the concept of assembly lines was introduced. Frederick Winslow Taylor, the father of scientific management, developed time and motion study. Henry Ford conceived the assembly line and applied it to the building of the Model T Ford car. The work was subdivided into individual tasks, and each worker did their task as the cars moved by on a conveyor. Parts were delivered to the worker and tools were provided for the needed tasks.

The next step was to mechanize the operations so that they were independent of the strength and skill of the worker. Powered wrenches tightened fasteners to exactly the same torque every time. Parts feeders supplied the fasteners and other parts to the point of use without need for the worker to reach for them. The worker was still there, but the need for high skills was reduced for many operations. Imagine the consternation if the parts supplied were out of tolerance or the power tools were not adjusted properly! The designers and production engineers had to transfer their knowledge directly to the production process; the craftsman was no longer using his skills to correct errors.

We are now in the era of automated assembly. Parts are delivered to an assembly line by automatic feeders and put in place by robotic-powered tools. Positioning, fastening, and welding are all performed by robotic elements that sense position and the presence of the part and perform the same task in the same way every time without tiring. Now the designers have no one to correct their errors; they must design the parts for feeding and orienting and design the end product so that it can be assembled easily without human intervention. The workers on the line now become supervisors of the machines. Their responsibility in the best manufacturing operations is to monitor quality and note ways to improve methods and product quality.

Design for automatic assembly places greater demands on the designers. The end product must not only perform but also be capable of being assembled by machines. This requires a different perspective on component design and assembly methods. The good designer will need to study research and current literature in both areas.

8.3 STANDARD TABLES FOR FITS

The application of tolerances to metric dimensions is prescribed by the International Standards Organization and by ANSI standards Y14.5 and B4.2. Tolerances for cylindrical fits in inches are

Preferred Basic Sizes
Decimal

0.010	2.00	8.50
0.012	2.20	9.00
0.016	2.40	9.50
0.020	2.60	10.00
0.025	2.80	10.50
0.032	3.00	11.00
0.040	3.20	11.50
0.05	3.40	12.00
0.06	3.60	12.50
0.08	3.80	13.00
0.10	4.00	13.50
0.12	4.20	14.00
0.16	4.40	14.50
0.20	4.60	15.00
0.24	4.80	15.50
0.30	5.00	16.00
0.40	5.20	16.50
0.50	5.40	17.00
0.60	5.60	17.50
0.80	5.80	18.00
1.00	6.00	18.50
1.20	6.50	19.00
1.40	7.00	19.50
1.60	7.50	20.00
1.80	8.00	

Figure 8.13 Preferred basic sizes (in.). (Courtesy of USAS, B4.1)

prescribed in USA Standard B4.1. These standards are based upon years of experience and should be used for selecting tolerances unless special conditions require other choices.

8.3.1 Preferred Limits and Fits in Inches

Standard B4.1 includes all of the terms and concepts you learned in Chapter 7. It includes some new terms and concepts you will need to understand for the development of the tables in Appendix B. However, you can apply the tables starting with the concept of *basic size*. Although the tables may be applied to any basic size you choose, the standards also list preferred basic sizes for fractional and decimal dimensions (Figure 8.13).

As the designer develops an assembly, he or she knows what kind of fit must exist between adjacent parts. There are three general types of fits: running fits, locational fits, and force fits. Each of these is further subdivided and identified with a letter code:

RC: running or sliding clearance fit.

LC: locational clearance fit.

LT: transition clearance or interference fit.

LN: locational interference fit.

FN: force or shrink fit.

There are nine classes of RC fits ranging from the closest, RC1, to the loosest, RC9. Each letter group is similarly subdivided. You will need experience to do a good job of selecting which class is needed for a particular pair of mating parts. The classes are described in the standard; an example follows: RC 7 *free-running fits* are intended for use where accuracy is not essential or where large temperature variations are likely to be encountered or under both of these conditions.

A portion of one of the tables is shown in Figure 8.14; more complete tables are included in Appendix B. Note that all limits are given in thousandths of an inch. You start with the basic size to apply the tables. Note the ABC agreements are mentioned; they are also referred to in Chapter 9 as an agreement among the United States, Great Britain, and Canada. The symbols below the words *hole* and *shaft* are taken from the ABC System and are related to metric tolerancing, which is reviewed in the next section.

Nominal Size Range Inches		Class RC 1			Class RC 2			Class RC 3			Class RC 4		
		Limits of Clearance	Standard Limits		Limits of Clearance	Standard Limits		Limits of Clearance	Standard Limits		Limits of Clearance	Standard Limits	
Over	To		Hole H5	Shaft g4		Hole H6	Shaft g5		Hole H7	Shaft f6		Hole H8	Shaft f7
0	– 0.12	0.1 0.45	+ 0.2 0	– 0.1 – 0.25	0.1 0.55	+ 0.25 0	– 0.1 – 0.3	0.3 0.95	+ 0.4 0	– 0.3 – 0.55	0.3 1.3	+ 0.6 0	– 0.3 – 0.7
0.12	– 0.24	0.15 0.5	+ 0.2 0	– 0.15 – 0.3	0.15 0.65	+ 0.3 0	– 0.15 – 0.35	0.4 1.2	+ 0.5 0	– 0.4 – 0.7	0.4 1.6	+ 0.7 0	– 0.4 – 0.9
0.24	– 0.40	0.2 0.6	+ 0.25 0	– 0.2 – 0.35	0.2 0.85	+ 0.4 0	– 0.2 – 0.45	0.5 1.5	+ 0.6 0	– 0.5 – 0.9	0.5 2.0	+ 0.9 0	– 0.5 – 1.1
0.40	– 0.71	0.25 0.75	+ 0.3 0	– 0.25 – 0.45	0.25 0.95	+ 0.4 0	– 0.25 – 0.55	0.6 1.7	+ 0.7 0	– 0.6 – 1.0	0.6 2.3	+ 1.0 0	– 0.6 – 1.3
0.71	– 1.19	0.3 0.95	+ 0.4 0	– 0.3 – 0.55	0.3 1.2	+ 0.5 0	– 0.3 – 0.7	0.8 2.1	+ 0.8 0	– 0.8 – 1.3	0.8 2.8	+ 1.2 0	– 0.8 – 1.6
1.19	– 1.97	0.4 1.1	+ 0.4 0	– 0.4 – 0.7	0.4 1.4	+ 0.6 0	– 0.4 – 0.8	1.0 2.6	+ 1.0 0	– 1.0 – 1.6	1.0 3.6	+ 1.6 0	– 1.0 – 2.0

Figure 8.14   Portion of standard fit table (USA Standard). (Courtesy of USAS, B4.1)

Assume you have two mating parts that need to move freely and to fit precisely with little temperature difference. This is the description for an RC 3 fit. If the nominal size is 1.40, then the basic size is 1.4000. This is true because the table provides values to 0.0001 in. in the range we are using. Reading down the table we are in the RC 3 column. Reading across the table we are in the nominal-size row "over 1.19 to 1.97." This zeroes us in on one cell of the table, which looks like this:

$$1.0 \quad +1.0 \quad -1.0$$
$$2.6 \quad 0 \quad -1.6$$

The first column gives us the limits for the fit, the second gives us the limits for the hole, and the third gives us the limits for the shaft. Applying these limits to the basic size, we get:

Limits for hole: $\underline{\dfrac{1.4000}{1.4010}}$ Difference = .0010

Limits for shaft: $\underline{\dfrac{1.3990}{1.3984}}$ Difference = .0006

Tolerance of system = clearance$_{max}$ – clearance$_{min}$
= .0026 – .0010 = .0016

Tolerance of system = sum of individual tolerances
= .0010 + .0006 = .0016   *Check!*

You can apply the tables to solving any tolerance problem for a pair of parts in the same manner.

8.3.2 Preferred Metric Limits and Fits

Standard B4.2 describes the system for tolerancing metric dimensions. It includes an abbreviated way for designating tolerances that assumes that those using the information have access to the tables in the standard. A 40-mm hole can be completely toleranced by the description 40H8; a mating shaft could be described as 40f7.

Some terms you will need to know to understand the system follow:

Deviation The algebraic difference between a size and the corresponding basic size.

Upper Deviation The algebraic difference between the maximum limit of size and the corresponding basic size.

Lower Deviation The algebraic difference between the minimum limit of size and the corresponding basic size.

Fundamental Deviation That one of the two deviations closest to the basic size. The letter H in the hole specification 40H8 designates the fundamental deviation.

International Tolerance Grade (IT) A group of tolerances that vary depending on the basic size but provide the same relative level of accuracy within a given grade. It is designated by the number 8 in 40H8; it would be described as IT8.

Note again that the basic size is the starting point for all these concepts. Preferred sizes also exist in the metric system (Figure 8.15). The combination of a fundamental deviation and an international tolerance grade describes a *tolerance zone:* H8 or f7. Note that the fundamental deviation is a capital letter for cavities and a lowercase letter for external shapes. There are also preferred tolerance zones for both internal dimensions and external dimensions. Although there are over 100 tolerance zones for both holes and shafts, only 13 zones are preferred zones for each. Combinations of preferred holes with preferred shafts reduce the usual choices to 10 preferred fits (Figure 8.16). Now the system begins to look very manageable and easier to specify than the inch system.

Assume that your design calls for a nominal size of 40 mm for a shaft and hole. If it is to be a free-running fit, you can see from Figure 8.16 that the fit would be specified as 40H9/d9. The hole would be dimensioned 40H9 and the shaft 40d9. When we read the numerical tolerances from the tables, it looks like this:

Basic Size, mm		Basic Size, mm		Basic Size, mm	
First Choice	Second Choice	First Choice	Second Choice	First Choice	Second Choice
1		10		100	
	1.1		11		110
1.2		12		120	
	1.4		14		140
1.6		16		160	
	1.8		18		180
2		20		200	
	2.2		22		220
2.5		25		250	
	2.8		28		280
3		30		300	
	3.5		35		350
4		40		400	
	4.5		45		450
5		50		500	
	5.5		55		550
6		60		600	
	7		70		700
8		80		800	
	9		90		900
				1000	

Figure 8.15 Preferred basic sizes (mm). (Courtesy ANSI, B4.2)

	Hole H9	Free-running Shaft d9	Fit
Maximum	40.062	39.920	0.204
Minimum	40.000	39.858	0.080

No addition or subtraction is required. The tables are longer, but no arithmetic is needed. Now, what do you put on your drawing? The following notes are recommended in the standards.

(a) $\dfrac{40.000}{40.062}$ (40H9)   (b) 40H9 $\dfrac{40.000}{40.062}$ (c) 40H9

Note (a) is recommended when the system is new to an organization; as experience is gained, note (b) is suggested; and when everyone is used to metric notation, note (c) may be used.

ISO SYMBOL		DESCRIPTION
Hole Basis	**Shaft[1] Basis**	
H11/c11	C11/h11	*Loose running* fit for wide commercial tolerances or allowances on external members.
H9/d9	D9/h9	*Free running* fit not for use where accuracy is essential, but good for large temperature variations, high running speeds, or heavy journal pressures.
H8/f7	F8/h7	*Close running* fit for running on accurate machines and for accurate location at moderate speeds and journal pressures.
H7/g6	G7/h6	*Sliding* fit not intended to run freely, but to move and turn freely and locate accurately.
H7/h6	H7/h6	*Locational clearance* fit provides snug fit for locating stationary parts; but can be freely assembled and disassembled.
H7/k6	K7/h6	*Locational transition* fit for accurate location, a compromise between clearance and interference.
H7/n6	N7/h6	*Locational transition* fit for more accurate location where greater interference is permissible.
H7/p6	P7/h6	*Locational interference* fit for parts requiring rigidity and alignment with prime accuracy of location but without special bore pressure requirements.
H7/s6	S7/h6	*Medium drive* fit for ordinary steel parts or shrink fits on light sections, the tightest fit usable with cast iron.
H7/u6	U7/h6	*Force* fit suitable for parts which can be highly stressed or for shrink fits where the heavy pressing forces required are impractical.

The vertical labels on the left indicate: Clearance Fits, Transition Fits, Interference Fits. The vertical labels on the right indicate: More Clearance, More Interference.

[1] The transition and interference shaft basis fits shown do not convert to exactly the same hole basis fit conditions for basic sizes in range from 0 through 3 mm. Interference fit P7/h6 converts to a transition fit H7/p6 in the above size range.

Figure 8.16 Description of preferred metric fits. (Courtesy of ANSI, B4.2)

Metric dimensioning may be applied in a basic shaft system as well as in the basic hole system. The basic hole system is preferred unless you are dealing with standard size shafts and must dimension the holes to fit the shafts.

8.4 SURFACE CONTROL

We need to control surface characteristics for two reasons: Rough surfaces cause friction and wear with moving parts and it is hard to get precise dimensions when the surfaces are irregular. Surface control is addressed in ANSI Standard B46.1; the symbols for surface texture are defined in ANSI Y14.36.

There are three surface characteristics defined in the standards. *Roughness* refers to the small hills and valleys that can be found on a surface. They may be obvious on a surface such as sandpaper, but they still exist in smaller form on smooth-looking chromium-plated hardware. *Lay* refers to the direction of the tool marks on a machined surface. This factor can be important on mating surfaces that move relative to each other. *Waviness* refers to surface irregularities greater than roughness; they can usually be seen as a pattern on a machined surface.

Roughness is defined on a drawing as the arithmetic average of the deviations above and below a mean height of a surface. It is expressed in microinches or micrometers. Preferred values for surface roughness are shown in Figure 8.17. Surface roughness

Micrometers—μm		Microinches—μin	
μm	μin	μm	μin
0.012	0.5	1.25	50
0.025*	1*	1.60*	63*
0.050*	2*	2.0	80
0.075	3	2.5	100
0.10*	4*	3.2*	125*
0.125	5	4.0	160
0.15	6	5.0	200
0.20*	8*	6.3*	250*
0.25	10	8.0	320
0.32	13	10.0	400
0.40*	16*	12.5*	500*
0.50	20	15	600
0.63	25	20	800
0.80*	32*	25*	1000*
1.00	40		

*Recommended

Figure 8.17 Preferred numbers for expressing roughness. (Courtesy of ANSI, Y14.36)

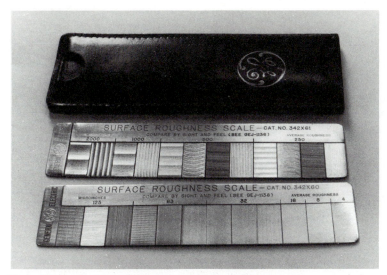

Figure 8.18 Surface roughness scales for comparison.

is measured with electrical instruments or it may be estimated by comparison with standard samples (Figure 8.18). The manufacturing processes referred to earlier in this chapter produce surfaces of different roughness; Figure 8.19 provides some comparative ranges. If we want a shaft to run freely in a bearing hole, we would not put a cast shaft in a plain drilled hole. The surfaces would be rough—friction would be high, wear would be fast, and the combination would generate a lot of heat. If you were to specify appropriate tolerances for a precision running fit, you could not even measure them with these kinds of surfaces.

Lay is defined by the direction of the tool marks on a surface. Standard symbols and their meanings are shown in Figure 8.20. Parallel and perpendicular lay would be the result of milling or shaping operations. Facing the end of an object held in a lathe would produce a circular pattern. Some methods of grinding would produce a multidirectional lay. Sand casting would reproduce the texture of the sand in the surface of the part and the lay would be P. Lay is independent of roughness but can also be a criterion for surface quality.

Waviness is measured in inches or millimeters. It is an order of magnitude greater than roughness. Roughness is imposed on top of waviness. Figure 8.21 shows the preferred values for expressing waviness.

Surface control is noted on drawings with standard symbols. The basic symbol is a 60° V with the right leg extended up. For a simple designation of surface roughness the maximum allowable average value is entered in the open area of the extended V. For

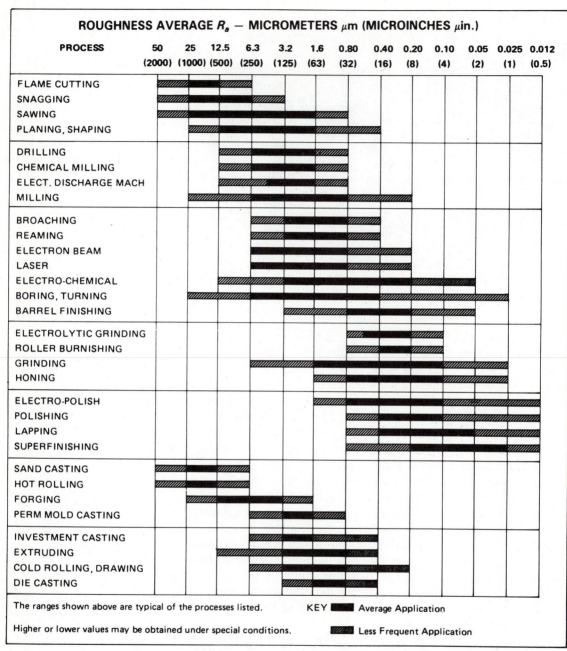

Figure 8.19   Surface roughness produced by common production methods. (Courtesy of ANSI, B46.1)

Lay Symbol	Meaning	Example Showing Direction of Tool Marks
—	Lay approximately parallel to the line representing the surface to which the symbol is applied.	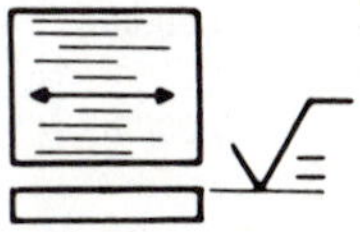
⊥	Lay approximately perpendicular to the line representing the surface to which the symbol is applied.	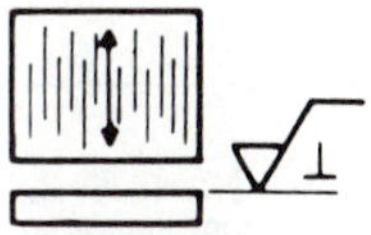
X	Lay angular in both directions to line representing the surface to which the symbol is applied.	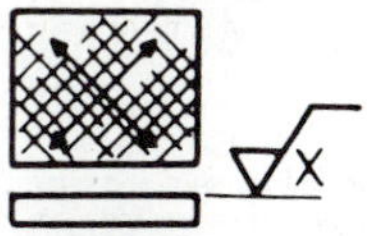
M	Lay multidirectional.	
C	Lay approximately circular relative to the center of the surface to which the symbol is applied.	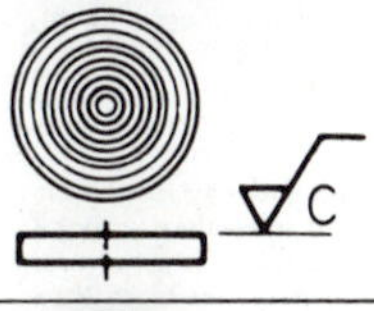
R	Lay approximately radial relative to the center of the surface to which the symbol is applied.	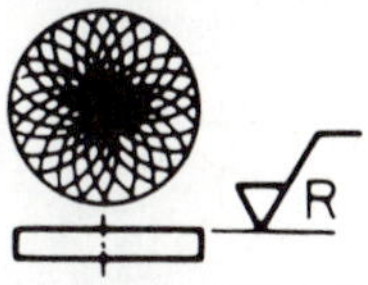
P	Lay particulate, non-directional, or protuberant.	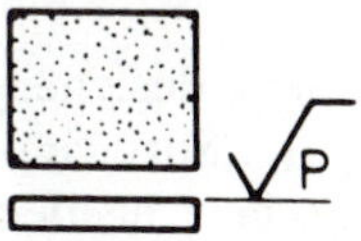

Figure 8.20 Standard symbols and examples for lay. (Courtesy of ANSI, Y14.36)

more complete specifications a horizontal bar is added to the extended leg and additional information given above and below the bar. A small circle in the V means that material removal is prohibited; closing the V to form a triangle means that material removal by machining is required. Several applications of the symbol are shown in Figure 8.22.

mm	in.	mm	in.	mm	in.
0.0005	0.00002	0.008	0.0003	0.12	0.005
0.0008	0.00003	0.012	0.0005	0.20	0.008
0.0012	0.00005	0.020	0.0008	0.25	0.010
0.0020	0.00008	0.025	0.001	0.38	0.015
0.0025	0.0001	0.05	0.002	0.50	0.020
0.005	0.0002	0.08	0.003	0.80	0.030

Figure 8.21 Preferred numbers for expressing maximum waviness. (Courtesy of ANSI, Y14.36)

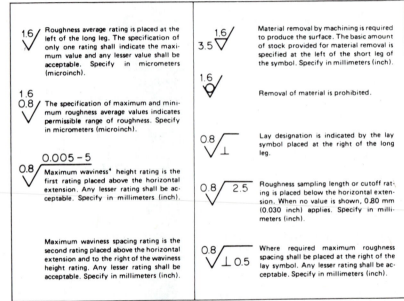

Figure 8.22 Applications of surface texture symbol. (Courtesy of ANSI, Y14.36)

The beginning designer should not apply surface control symbols frequently. Surface control is expensive, and it should only be required where needed. It is best left to the judgment of experienced engineers and designers. In industries producing equipment such as rough machinery and conveyor frames, you can inspect thousands of drawings and never find a symbol for surface texture. Use only where needed!

8.5 GEOMETRIC TOLERANCING

Geometric tolerancing includes specification of form, profile, orientation, location, and runout. These are factors we have not considered yet; Chapter 7 considered lengths, diameters, and ra-

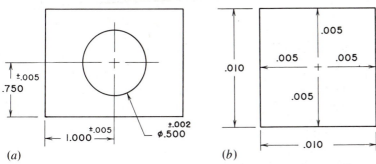

(a)

(b)

Figure 8.23 (a) Toleranced hole diameter with rectangular coordinate location tolerances. (b) Tolerance zone for the hole in (a).

dii. We also applied tolerances to obtain maximum and minimum values for critical dimensions. Now we can look at the geometric relationships among the elements of a part.

This subject includes a philosophy of design that probes the function of the part and the interrelationships of its features. It provides a more accurate appraisal of the tolerance zone allowed in a given machining operation.

It can result in a "bonus" tolerance when properly applied and understood. Consider a hole located from two surfaces in a part, as shown in Figure 8.23a. The location dimensions indicate a tolerance zone as shown in Figure 8.23b, a square whose dimensions are .010 × .010 in. The hole center can vary in a vertical or horizontal direction by .005 and still be within the tolerance zone. Consider what happens when the hole moves the full amount of .005 in both a horizontal and vertical direction: It is now not .005 away from the center but .007 from the true center. We would have to inscribe a circle of radius .005 within the square to keep the center within .005 of the true center. Or if the allowable variation is every point within the square, we could circumscribe a circle around the square and have every point within the circle no farther away from true center than the corners of the square. This circle would have a diameter of .014 (Figure 8.24). The area of the circle outside the original square represents a bonus tolerance we have found by a logical interpretation of the original tolerance specification. Geometric dimensioning and tolerancing recognizes this bonus and allows the specified hole center to vary from its true position *in any direction* by the maximum allowed.

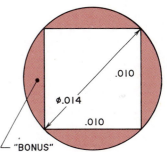

Figure 8.24 Bonus tolerance provided by circular tolerance zone.

8.5.1 Definitions and Symbols

Some of the terms you will need you already know. Some of the new terms follow here:

Basic Dimension: A numerical value used to describe the theoretically exact size or location. Permissible variations from this theoretical value are a part of geometric tolerancing.

True Position: The theoretically exact location of a feature established by basic dimensions.

Datum: A theoretically exact point, axis, or plane used as the origin from which location or geometric characteristics of features are located.

Datum Target: A specified point, line, or area on a part used to establish a datum.

Datum Feature: An actual feature of a part used to establish a datum.

Maximum Material Condition (MMC):   The condition in which a feature of size contains the maximum amount of material within the stated limits of size, for example, minimum hole diameter and maximum shaft diameter.

Least Material Condition (LMC):   Opposite of MMC, the feature contains the least material, for example, maximum hole diameter and minimum shaft diameter.

Regardless of Feature Size (RFS):   Indicates that a geometric tolerance or datum reference applies at any increment of size of the feature within its size tolerance.

Virtual Condition: The envelope or boundary that describes the collective effects of all tolerance requirements on a feature (Figure 8.25).

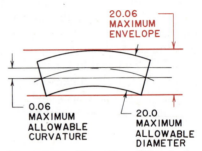

Figure 8.25   Virtual condition is the envelope defined by the application of all required tolerances to a feature.

	TYPE OF TOLERANCE	CHARACTERISTIC	SYMBOL
FOR INDIVIDUAL FEATURES	FORM	STRAIGHTNESS	—
		FLATNESS	▱
		CIRCULARITY (ROUNDNESS)	○
		CYLINDRICITY	⌀
FOR INDIVIDUAL OR RELATED FEATURES	PROFILE	PROFILE OF A LINE	⌒
		PROFILE OF A SURFACE	⌓
FOR RELATED FEATURES	ORIENTATION	ANGULARITY	∠
		PERPENDICULARITY	⊥
		PARALLELISM	//
	LOCATION	POSITION	⊕
		CONCENTRICITY	◎
	RUNOUT	CIRCULAR RUNOUT	↗
		TOTAL RUNOUT	↗↗

Figure 8.26   Basic symbols for geometric characteristics. (Courtesy of ANSI, Y14.5M)

TERM	SYMBOL
AT MAXIMUM MATERIAL CONDITION	Ⓜ
REGARDLESS OF FEATURE SIZE	Ⓢ
AT LEAST MATERIAL CONDITION	Ⓛ
PROJECTED TOLERANCE ZONE	Ⓟ
DIAMETER	⌀
SPHERICAL DIAMETER	S⌀
RADIUS	R
SPHERICAL RADIUS	SR
REFERENCE	( )
ARC LENGTH	⌒

Figure 8.27   Modifying symbols for geometric characteristics. (Courtesy of ANSI, Y14.5M)

Geometric tolerances are specified with symbols (Figures 8.26 and 8.27). Several symbols and numbers may be combined in a sort of shorthand to describe what is required. The collection of symbols is presented in a rectangular shape called a feature control frame (Figure 8.28). The frame shown is to control perpendicularity of a hole relative to surface *C* and at maximum material condition. The axis of the hole must lie within a cylindrical zone with a diameter of 0.05 mm that is perpendicular to and projects from surface *C*. A lot of information in a little rectangle!

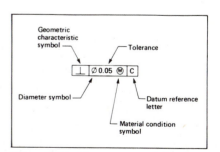

Figure 8.28   Feature control frame used to combine geometric tolerance information. (Courtesy of ANSI, Y14.5M)

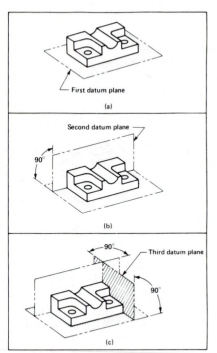

Figure 8.29 Three datum reference planes at right angles. (Courtesy of ANSI, Y14.5M)

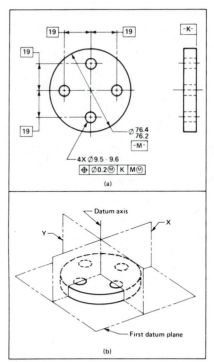

Figure 8.30 Cylindrical part with datum plane and datum axis. (Courtesy of ANSI, Y14.5M)

8.5.2 Datum Referencing

A datum indicates the origin of a dimensional relationship between a toleranced feature and a designated feature or features on a part. A theoretically exact datum would be difficult to establish, so very flat surfaces such as machine tables or stone surface plates are used. Enough datum references should be selected to control all needed features. An example is shown in Figure 8.29. Cylindrical parts should have datums passing through the center (Figure 8.30). Datums may be referenced to a surface on a part, a line, a point, a single feature, or a collection of features (a group of holes).

8.5.3 Tolerances of Location

Location includes position, concentricity, and symmetry. Basic dimensions are used to establish true positions from datums. The basic dimensions are put in a box on the drawing. Datum references are usually designated by a letter preceded and followed by a dash. Examine Figure 8.31, and you will find several basic

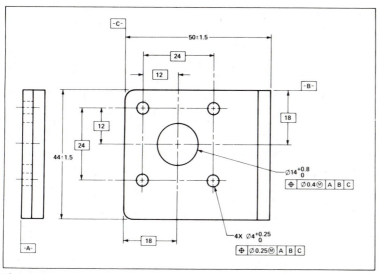

Figure 8.31 Example of positional tolerancing. (Courtesy of ANSI, Y14.5M)

dimensions, three datum references, and two feature control frames. The true position of the large hole is fixed relative to datums *A*, *B*, and *C*; the true positions of the small holes are fixed relative to the center lines of the large hole and datums *A*, *B*, and *C*.

If we go back and look at our early illustration that demonstrated bonus tolerance (Figure 8.23), we can now express the location with geometric dimensions and tolerances that truly describe the allowable conditions. Examine Figure 8.32, and you will note that the datums are now specified, the location dimensions are flagged as basic dimensions, and the positional tolerance at maximum material condition is shown as a circle with a di-

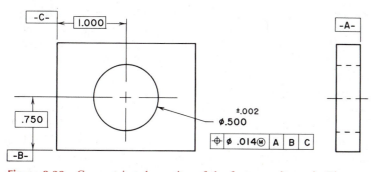

Figure 8.32 Geometric tolerancing of the features shown in Figure 8.23. Datums are established for control in three planes and the circular tolerance zone with a specified diameter of .014.

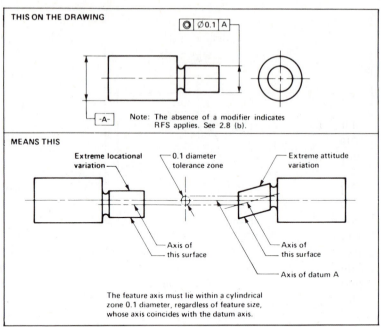

Figure 8.33 Concentricity tolerance note and its meaning. (Courtesy of ANSI, Y14.5M)

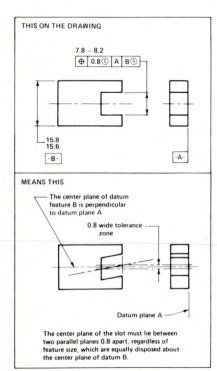

Figure 8.34 Symmetry tolerance note and its meaning. (Courtesy of ANSI, Y14.5M)

ameter of .014. The tolerance on the hole has not changed, but the way we describe its position is more specific and provides *more* tolerance than we had before.

Concentricity is specified by the small concentric circle symbol. An example is shown in Figure 8.33. Symmetry is shown in Figure 8.34.

8.5.4 Tolerances of Form

Tolerances of form include straightness, flatness, circularity, and cylindricity. Orientation tolerances control angularity, parallelism, and perpendicularity. Form and orientation are used together where tolerances for size and location do not provide enough control.

Straightness means being in a straight line. Figure 8.35 shows a straightness requirement for the elements of a cylinder. The tolerance on the diameter is 0.11 mm, but the elements of the cylinder must all be straight within 0.02 mm. Flatness means all points on a surface are in the same plane. Flatness control is illustrated in Figure 8.36. Note that the plane may not be truly horizontal, but all points on the top surface must be within 0.25 mm of the plane. Other tolerances for size are not shown; they must be met also. Figure 8.37 shows control of circularity. A right section cut anywhere must contain all elements with the tolerance zone defined by the two concentric circles. Figure 8.38 shows control of cylindricity.

Orientation control includes angularity, parallelism, and perpendicularity. Specification of these features is usually tied to a datum. Note that datum plane *A* is the reference for angular control in Figure 8.39. Parallelism is demonstrated in Figure 8.40 with datum *A* as the control reference. Perpendicularity is shown in Figure 8.41; the bottom surface of the angle is specified as datum *A*.

A profile is the outline of an object in two dimensions: a projection or a section. The boundaries of the profile need to be shown on the drawing along with a datum and the control tolerance. Figure 8.42 shows the application of profile tolerance to a curved shape. If the allowable deviation can occur on both sides of the true profile line, it is a bilateral tolerance and no special symbol is added. If the deviation can only be allowed on one side of the true profile line, a phantom line is added to show in which direction the deviation is allowed.

Runout is the deviation from an axis or datum surface along some given length. Runout for both a conical surface and a cylindrical surface is shown in Figure 8.43. Note that runout control applies to only a 17-mm-long portion of the large cylinder (it probably has to mate with another part in this zone). A measuring

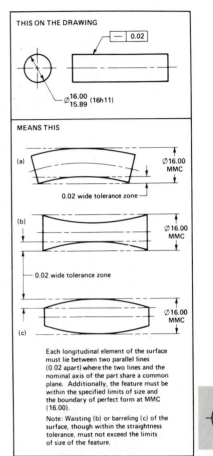

Figure 8.35 Straightness note and its meaning. (Courtesy of ANSI, Y14.5M)

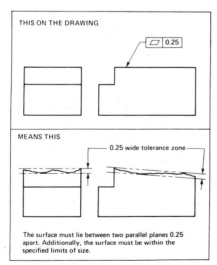

Figure 8.36 Flatness note and its meaning. (Courtesy of ANSI, Y14.5M)

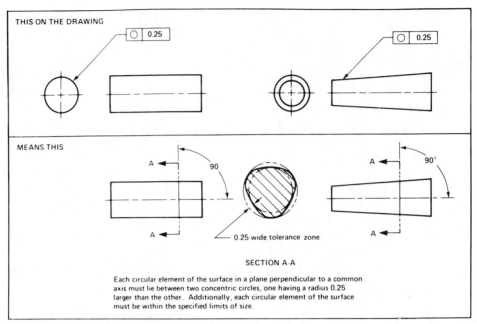

Figure 8.37 Circularity note for cylinder or cone and its meaning. (Courtesy of ANSI, Y14.5M)

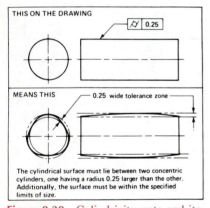

Figure 8.38 Cylindricity note and its meaning. (Courtesy of ANSI, Y14.5M)

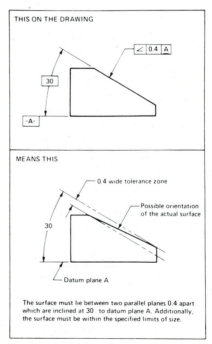

Figure 8.39 Angularity note and its meaning. (Courtesy of ANSI, Y14.5M)

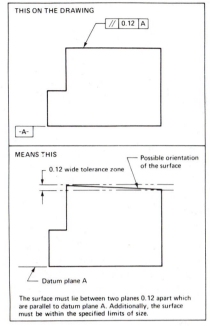

Figure 8.40 Parallelism note and its meaning. (Courtesy of ANSI, Y14.5M)

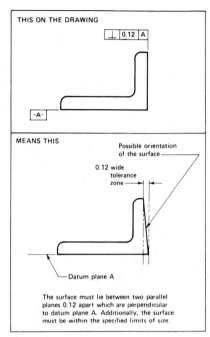

Figure 8.41 Perpendicularity note and its meaning. (Courtesy of ANSI, Y14.5M)

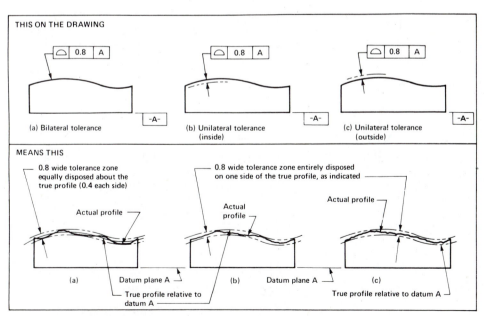

Figure 8.42 Profile note given three ways and their meanings. (Courtesy of ANSI, Y14.5M)

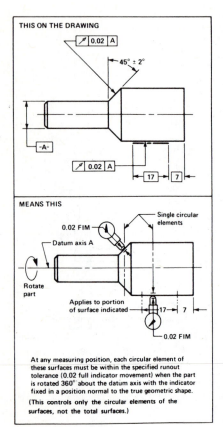

Figure 8.43 Circular runout note and its meaning. (Courtesy of ANSI, Y14.5M)

tool with a dial reading parts of a millimeter or inch is used to take several readings in the control area as the part is rotated about axis *A*. In this example all the readings at each section must be within 0.02 mm.

8.6 *SAMPLE PROBLEM: TABLE SAW FIXTURE*

The illustration at the end of Chapter 7 left us with a drawing complete with shape and size description. This drawing may not be ready for production: If some relationships of parallelism, perpendicularity, and concentricity are important, they will need to be defined on production drawings. Surface control may also need to be defined.

Some CADD packages contain symbols you can use to build surface control and geometric tolerance notes. If you do not have them in your package, you can create the basic symbols such as the 60° ''V'' with an extended leg, geometric tolerance symbols from Figures 8.23, and a feature control frame. These symbols can then be combined to build complete specifications for production drawings.

Figure 8.44 shows our basic fixture with only a few notes. The dimensions developed in Chapter 7 have been removed for the sake of clarity, and only those notes needed for geometric tolerancing and surface control are shown. These notes were placed using symbols developed for the author's use. The symbols were drawn at a large scale for accuracy and filed at a size that would be appropriate for most CADD drawings. The feature control frame symbol was recalled and symbols for geometric characteristics added to it.

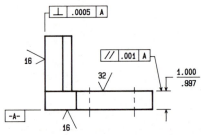

Figure 8.44 Table saw fixture orientation control and surface control added (all other dimensions have been removed to highlight these notes). Notes were drawn with CADD system by assembling symbols or patterns stored in data files. (*Computer generated image*)

8.7 *SUMMARY*

You have been introduced to a few of the basic manufacturing processes. Drilling and turning have been practiced for centuries. Swords and other weapons were made by forging since metals were discovered. New means for forming materials are being developed every year. If you become involved in manufacturing, you will need to be aware of them.

Specifying tolerances is not difficult when you use the standard tables. Tolerances should be selected from the tables based on the function of the parts. Checking tolerances can still be done by calculating the tolerance of the system. Metric tolerances may be specified in several ways depending on the familiarity of the people using them.

In addition to specifying tolerances, the designer may need to specify the quality of the surfaces of a part. Standard measures and symbols exist for controlling surface roughness, lay, and waviness.

Geometric tolerancing is a sophisticated method for specifying form and position. You have a detailed method and symbols for controlling all the geometric elements of a part. Datum references are an important part of the methods.

Tolerances of size, geometric tolerances, and surface control can improve the quality and performance of machinery. These controls, properly used, can save time and money while improving the quality of the production process. Improperly used, they add confusion and cost; learn to use them properly when you are in a position to apply them.

The information in this chapter is a brief summary of the information necessary to do a complete job of dimensioning. Most readers will not need to acquire total information in all of the areas mentioned; those who become practicing designers will need to learn more about some of the areas. For more information start with the standards noted with each topic; books and professional magazines and journals are available that cover many areas of manufacturing and design. In an advanced course in design, you may study some of these subjects in depth; as a practicing designer you may need to take short courses or graduate work to gain the knowledge necessary to fulfill your professional obligations.

PROBLEMS

Problem 8.1
Show three ways to specify a metric fit that is a nominal 10 mm close running fit. Assume a hole basis for your fit; sketch a shaft and hole and attach three notes to each part.

Problem 8.2
Sketch the parts for a 40-mm medium-drive fit. Show tolerance notes on the drawing as if metric tolerancing were new to your organization. Calculate the clearances and part tolerances. Check your calculations with a tolerance of the system equation. Show all work on your drawing.

Problem 8.3
Copy the following sketch and show that the upper surface is to have an average roughness of 16 microinches with parallel lay. The bottom surface should have an 8-microinch average roughness with perpendicular lay.

Problem 8.4
Do problem 8.3 with equivalent metric specifications.

Problem 8.5
Prepare a sketch, including a feature control frame, for the cylinder. It has a nominal diameter of 20 + .8 − 0 mm. The straightness of the cylinder must be within .2 mm at the maximum material condition.

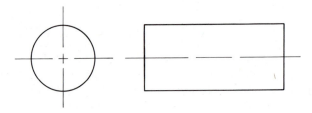

Problem 8.6

Copy the sketch and complete the feature control frame to specify a
cylindricity tolerance of .0002. Add limit dimensions to the length of
the pin to make it 4 long with a tolerance of plus .005 and plus .010.

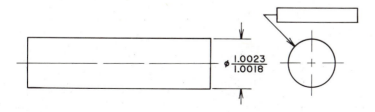

Problem 8.7

Make the lower surface of the part a datum feature. The height should
be 20 + .2 − .1 mm. Show a feature control frame that specifies
parallelism of the upper to the lower surface within .1 mm. The right
side must be perpendicular to the lower surface within .2 mm.

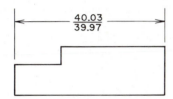

Problem 8.8

Letter a paragraph explaining the meaning of the given geometric
tolerances.

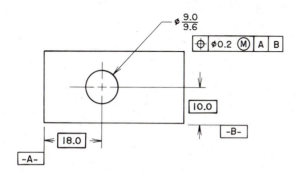

Problem 8.9

Prepare a drawing that specifies that surface X must be flat within .0005, with an average surface roughness of 8 μm, and parallel to surface Y within .001. The distance between is .75 $\pm$ .002. Surface Y should have an average roughness of 16 μm with multidirectional lay.

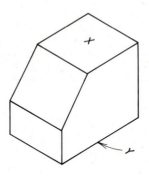

Problem 8.10

Letter a paragraph in which you interpret the specifications on this sketch. Which diameter is a clearance fit? Which is an interference fit? Give the references for your choices.

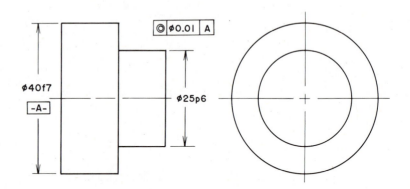

Problems 8.11

Prepare three-view orthographic drawings of these parts. You may dimension them one of two ways; your instructor will specify which option to use.

    (a) Completely dimension the objects, including geometric dimensions and tolerances needed to meet the noted specifications.

    (b) Partially dimension the objects with the information needed to meet the noted specifications.

MILLIMETERS

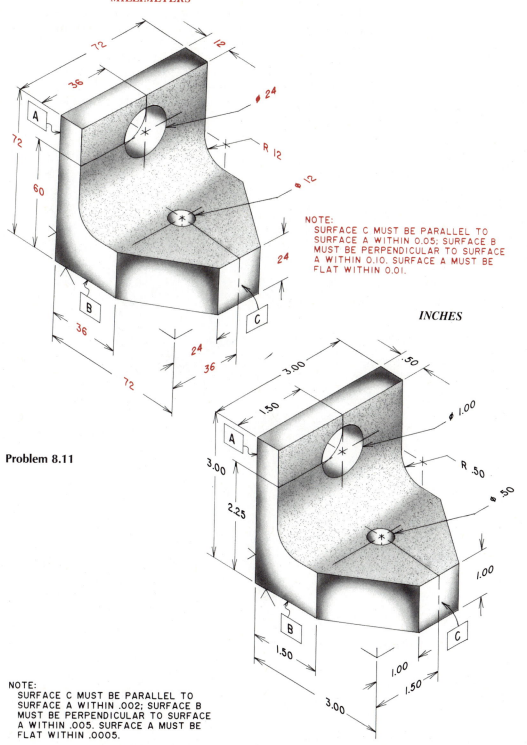

Problem 8.11

INCHES

NOTE:
SURFACE C MUST BE PARALLEL TO
SURFACE A WITHIN 0.05; SURFACE B
MUST BE PERPENDICULAR TO SURFACE
A WITHIN 0.10. SURFACE A MUST BE
FLAT WITHIN 0.01.

NOTE:
SURFACE C MUST BE PARALLEL TO
SURFACE A WITHIN .002; SURFACE B
MUST BE PERPENDICULAR TO SURFACE
A WITHIN .005. SURFACE A MUST BE
FLAT WITHIN .0005.

Problems 8.12

Prepare three-view orthographic drawings of these parts. You may dimension them one of two ways; your instructor will specify which option to use.

(a) Completely dimension the objects, including geometric dimensions and tolerances needed to meet the noted specifications.

(b) Partially dimension the objects with the information needed to meet the noted specifications.

MILLIMETERS

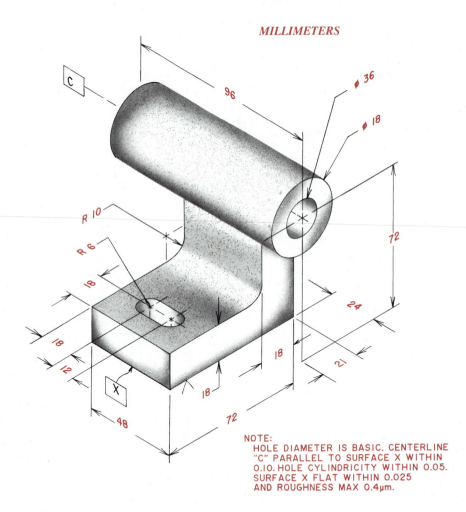

NOTE:
HOLE DIAMETER IS BASIC. CENTERLINE
"C" PARALLEL TO SURFACE X WITHIN
0.10. HOLE CYLINDRICITY WITHIN 0.05.
SURFACE X FLAT WITHIN 0.025
AND ROUGHNESS MAX 0.4μm.

Problem 8.12

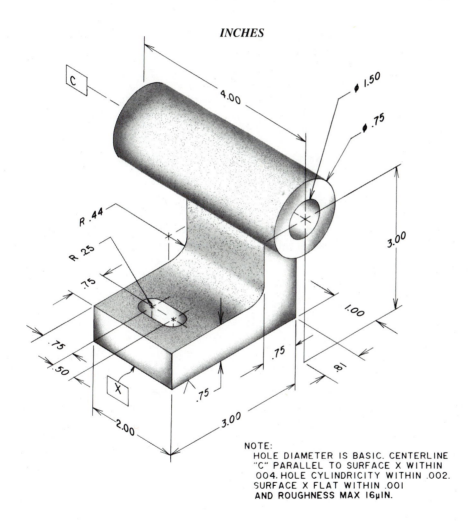

INCHES

C

4.00

⌀ 1.50

⌀ .75

3.00

R .44

R .25

.75

1.00

.75

.75

.81

X

.75

.75

.50

2.00

3.00

NOTE:
HOLE DIAMETER IS BASIC. CENTERLINE
"C" PARALLEL TO SURFACE X WITHIN
.004. HOLE CYLINDRICITY WITHIN .002.
SURFACE X FLAT WITHIN .001
AND ROUGHNESS MAX 16μIN.

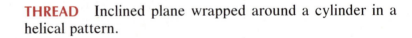

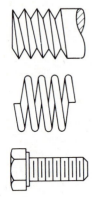

THREAD Inclined plane wrapped around a cylinder in a helical pattern.

HELIX Curve traced on a cylinder by the rotation of a point crossing its right section at a constant oblique angle.

SCREW Fastener having an external thread.

NUT Fastener having an internal thread.

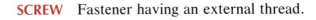

WELD Joint produced in materials by heat or pressure, often with the addition of a filler material.

ADHESIVE Material applied in thin films that bonds parts together by sticking to both.

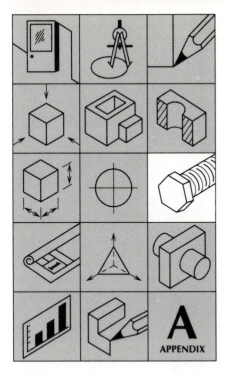

9

FASTENING, JOINING, AND STANDARD PARTS

9.1 INTRODUCTION

Imagine a world without fasteners! Automobiles, airplanes, and appliances would be only piles of loose parts; your clothes would be "swaddling clothes"—just wrapped around you; and your typewriter, telephone, and television could no longer communicate. Fasteners of many types hold the tools of our civilization together (Figure 9.1). The most common fasteners are threaded fasteners such as bolts, nuts, and screws. Other

Figure 9.1 Fasteners are all around us.

mechanical fasteners include pins, rivets, and keys. Joining processes you will learn about in this text are welding and adhesion. New designs of fasteners and new methods for joining parts together are constantly being developed. This chapter will introduce you to the basics of fastening and joining; as you pursue your professional career, you will need to learn more about existing fasteners and new developments that may solve the fastening and joining problems you encounter.

9.2 SCREW-THREAD TERMS

Screw **threads** look like the pictures in Figure 9.2. In order for them to function as fasteners, or adjustment devices, you need two parts: an external thread and an internal thread. The external thread is a V-shaped strip around a shaft in the form of a **helix**. (The **screw** is one of the basic machines described in the history of technology and was conceived by the Greek mathematician Archimedes as an inclined plane wrapped around a cylinder.) The internal thread is the complement of the external thread—a cavity with the screw thread removed from the material.

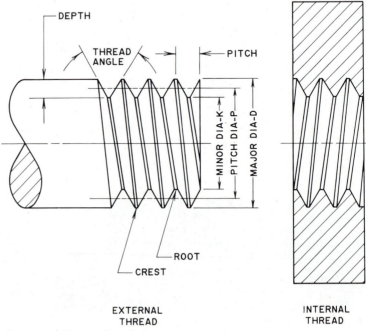

Figure 9.2 Screw-thread terms.

Other important terms are:

Axis   Longitudinal center line of screw or threaded cavity.

Crest   Outer surface of thread helix.

Root   Inner surface of thread helix.

Major Diameter   Largest diameter of screw thread, measured crest to crest (external and internal).

Minor Diameter   Smallest diameter of screw thread, measured root to root (external and internal).

Pitch   Distance between similar points on adjacent threads, measured parallel to the axis.

Pitch Diameter   Diameter of an imaginary cylinder that cuts the threads so that the distance to the crest equals the distance to the root.

Depth of Thread   Distance between crests and roots, measured normal to the axis.

Side   Surface of the helix connecting crest to root.

Thread Angle   Angle included between sides of thread, measured in a plane through axis of thread.

9.3   *OTHER PROPERTIES OF THREADS*

There are three other properties of threads with which you should be familiar. They include hand, form, and multiple threads. You will find a sketch for each property and an explanation in the following paragraphs.

9.3.1   Right-hand and Left-hand Threads

Threads can be made to advance when the screw is turned to the right (clockwise) or to the left (counterclockwise) (Figure 9.3). Common fasteners are all made right hand; if you need a left-hand fastener, you must specify LH. One common example of a left-hand thread is the outlet of a propane tank. Safety regulations require that this thread be different from standard threads, that is, left hand. The old adage "Turn it right to make it tight" does not apply when you attach a regulator to a propane tank. Look at the threads on your bow compass—one side of the screw is right hand, the other is left hand.

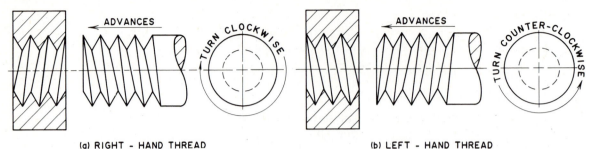

(a) RIGHT - HAND THREAD   (b) LEFT - HAND THREAD

Figure 9.3 Right-hand and left-hand threads.

9.3.2 Screw-Thread Forms

Not all screw threads look like V's when sectioned. The form of a thread is also called the profile. There are several types of threads, developed at different times for different purposes. The sharp V thread with a 60° thread angle was the U.S. standard thread for many years (Figure 9.4). It was replaced by the American National form, which has flattened crests and roots and is stronger than the sharp V. While the United States flattened the crests and roots, the British rounded them. Their standard thread form for many years was the Whitworth standard. Cooperation

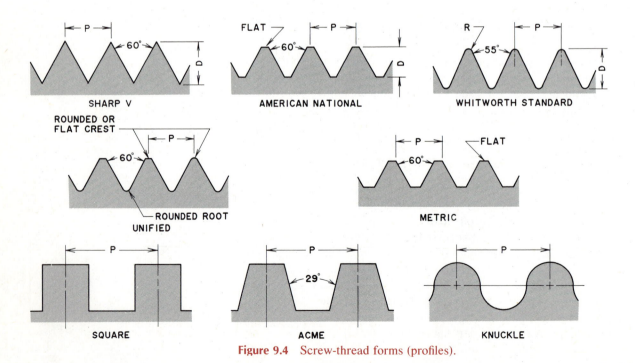

Figure 9.4 Screw-thread forms (profiles).

during World War II led the United States, Great Britain, and Canada to adopt a common thread form, the Unified, which includes features of the Whitworth and the American National. (This was the ABC accord.) The angle and proportions are from the American National, and the rounded roots and optional rounded crests on the external thread are from the Whitworth.

International commerce includes metric measurements and metric fasteners. Metric threads are similar in form to the American Standard or Unified; however, they are not cut as deep as ABC threads. Metric threads may have flat or rounded crests and roots, and they are measured in millimeters rather than in inches.

Threads may be used for fasteners, adjustment, or power transmission. Two forms used for power transmission are the square thread and the Acme thread. The square thread is theoretically ideal for power transmission, but it is hard to produce. A stronger thread form that is easier to manufacture is the Acme thread. An automobile jack or the lead screw on a lathe uses an Acme thread.

Most threads are cut with tools called taps and dies or rolled into the metal with hard metal-rolling tools powered by hydraulic cylinders. Another form of thread, which can be cast or rolled into the material, is the knuckle thread. If you examine a glass jar with a threaded top, you will see a form of the knuckle thread. It can be cast with fully rounded crests and roots. It can also be rolled into thin metal parts such as lids for glass jars and bases for light bulbs.

9.3.3 Multiple Threads

So far, you have been reading about single threads in which a single helix is wrapped around the cylinder. Sometimes we need to have the thread advance farther for each turn; in this case we wrap more than one helix about the cylinder and have multiple threads (Figure 9.5). The lead of a thread is the distance it advances in one revolution. The pitch is equal to the lead for a single

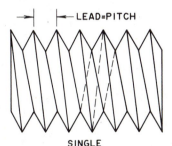

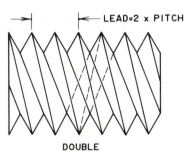

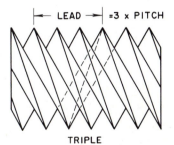

Figure 9.5 Single and multiple threads.

thread. With a double thread (two parallel helices on the shaft), the lead is twice the pitch; with a triple thread (three parallel helices), the lead is three times the pitch.

Multiple threads advance more rapidly than single threads but are not as strong. They are used for speed and convenience. If you have a large bow compass among your drawing instruments, examine the threads. They will be multiple threads for quick operation.

9.3.4 Thread Series

Series indicates the number of threads per unit of length for a given diameter. When threads were handmade, more than a century ago, each craftsman made the number of threads on a shaft that he felt were appropriate for the job. Over the years the number of threads per unit length has been standardized for different applications. There are several different series in the Unified system and in the metric or the International Standards Organization (ISO) system.

The two most common series in the inch system are the Coarse series and the Fine series. The Coarse series is used for most fasteners; when greater strength is needed or the length of thread engagement is short, the Fine series would be used. For the same diameter the Fine series has about 50% more threads per inch than the Coarse. The pitch is smaller for the fine thread for there are more threads per unit length, which puts them closer to each other. The profile, too, is smaller (Figure 9.6).

The National thread form is also available in the Extra Fine series (UNEF or NEF). This thread is used for fine adjustments or very short length of thread engagement (Figure 9.7). There are also three fixed-pitch series available: 8-pitch (8UN or 8N), 12-pitch (12UN or 12N), and 16-pitch (16UN or 16N). Fixed pitch means that the number of threads per inch is the same for a group of fasteners regardless of diameter. Fixed-pitch threads are used mostly on large-diameter fasteners and for adjustment devices.

Threads have been discussed so far as applied to cylindrical surfaces. They also apply to tapered surfaces—frustums of cones.

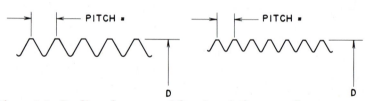

Figure 9.6 Profiles of coarse and fine threads for same diameter.

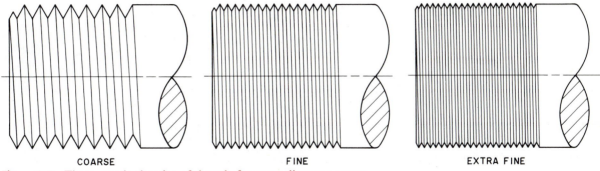

COARSE   FINE   EXTRA FINE

Figure 9.7 Three standard series of threads for same-diameter screw.

The National Pipe Thread (NPT) series is used for connecting pipes. The tapered threads seal securely to prevent leakage in piping systems. These threads are described in ANSI B2.1, which also includes straight pipe threads (Figure 9.8).

9.3.5 Specifying a Thread

When you specify a thread on a drawing or on a purchase order, you do not need to write out a complete description of all the features you need for that thread. There is an accepted short-hand way to describe what is needed. Unified (inch system) threads are specified as shown in Figure 9.9. The first number is the nominal or major diameter. Current standards call for the diameter to be expressed as a decimal inch figure; most fastener catalogs and older drawings will have a fractional inch size. The next number (after the dash) is the number of threads per inch. This will always be an integer. It is followed by letters that specify the thread form (UN or N) and the series (C for coarse in this example).

After the next dash you will find a number and, sometimes, a

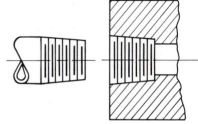

Figure 9.8 Schematic symbol for tapered pipe threads.

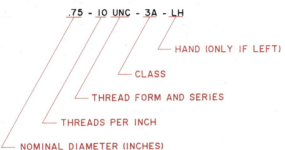

.75 - 10 UNC - 3A - LH

       — HAND (ONLY IF LEFT)

     — CLASS

    — THREAD FORM AND SERIES

   — THREADS PER INCH

— NOMINAL DIAMETER (INCHES)

Figure 9.9 Unified thread specification.

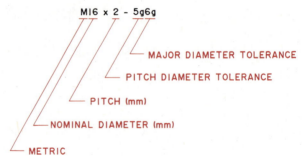

Figure 9.10 Metric thread specification (ISO).

letter. The number indicates the class of fit. You do not need to specify tolerances on standard fasteners—the one number covers a paragraph of specifications in ANSI standard B1.1, which describes the tolerance to which the thread should be made. Class 2 is most common; class 1 is for quick assembly of noncritical parts, and class 3 is used for precision threads and where close adjustment may be needed. The letter following the number (A or B) specifies an external thread (A) or an internal thread (B). Standard B1.1 describes both internal and external specifications for each class.

If the thread is not right hand, the class would be followed by a dash and the letters LH. In the case of a multiple thread the word DOUBLE or TRIPLE would be added to the thread specification.

Metric (ISO) threads are specified in a different manner (Figure 9.10). They always start with the letter M to indicate immediately that this is a metric thread conforming to the standards published by the ISO. The next number is the major diameter in millimeters. Mostly these are integers, but in smaller sizes they may be a one-place decimal such as 1.6 or 2.5. The diameter is usually followed by a multiplication sign and another number that is the pitch of the thread in millimeters. Since coarse threads are most common, the pitch may be omitted for coarse threads. If the fastener is a general-purpose one, the total specification could be M16.

Classes of tolerance may be specified after a dash. The first number refers to the tolerance grade for the pitch diameter—the smaller the number, the tighter the tolerance. Recommended tolerance grades are 4, 6, and 8. A tolerance grade of 5 represents good commercial practice. The lowercase letter following this number is the tolerance position or allowance for the pitch diameter. The next number is the tolerance for the major diameter, followed by the allowance for the major diameter. Tolerance positions are specified by lowercase letters for external threads and by capital letters for internal threads. The codes follow:

External threads   e, large allowance; g, small allowance; h, zero allowance.

Internal threads   G, small allowance; H, zero allowance.

9.4 THREADED FASTENERS

The concept of screw threads becomes useful when it is applied to fasteners. Three of the most common threaded fasteners are the bolt, the cap screw, and the stud as drawn in Figure 9.11. Photographs of actual fasteners are shown in Figure 9.12. These

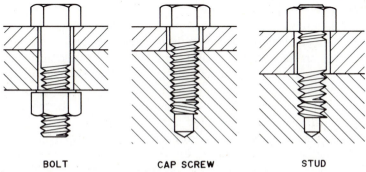

BOLT   CAP SCREW   STUD

Figure 9.11 Common threaded fasteners—bolt, cap screw, and stud—as they appear on drawings.

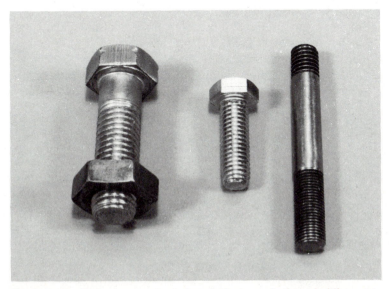

Figure 9.12 Bolt, cap screw, and stud photographed not holding parts together.

are the most common fasteners: many other types exist, and hundreds of new, special fasteners are designed each year.

9.4.1 Screws

Screws with a major diameter of $\frac{1}{4}$ in. or more are called cap screws. They may have several types of heads as shown in Figure 9.13. Standard diameters in the U.S. system go from $\frac{1}{4}$ up to 4 in. Standard diameters in the ISO system go from 1.6 to 100 mm. There are many lengths available in each diameter: The larger the diameter, the longer the lengths you can get. Screws are available in coarse, fine, or extra fine threads with the coarse series most readily available. Useful information about cap screws is given in the Appendices.

Small screws are called machine screws. Sometimes they are used with small nuts and act as little bolts. The types of heads available are pictured in Figure 9.14. They are available in the U.S. system from size 0 (0.060) to size 12 (0.216) and overlap cap screw sizes in some head styles from $\frac{1}{4}$ up to $\frac{3}{4}$. Where they have similar heads and the same diameters as cap screws, the machine screws will usually be threaded almost all the way to the head while cap screws are threaded for a length of a little over twice the diameter. Metric machine screws are available as small as 1 mm diameter and go up to 20 mm diameter. Lengths are available proportional to diameters, with the smallest screws produced as short as 2 mm and the largest as long as 100 mm. Metric machine screws mostly have coarse threads. You will find more data about machine screws in the Appendices.

Set screws are used to hold parts in place on a shaft. They pass through a threaded hole in the part and bear firmly on the shaft (Figure 9.15). Sometimes a slot or dimple is formed in the shaft to make a receptacle for the end of the set screw. Common set screw heads and points are shown in Figure 9.16. The headless

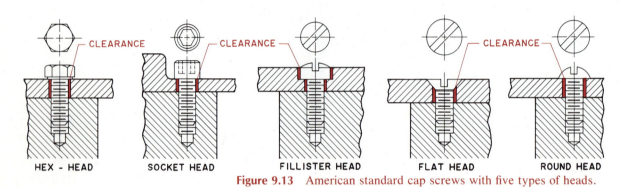

HEX - HEAD   SOCKET HEAD   FILLISTER HEAD   FLAT HEAD   ROUND HEAD

Figure 9.13 American standard cap screws with five types of heads.

screws are produced with diameters from #0 (0.060) to 2 in. Square-head set screws are available from $\frac{1}{4}$ to $1\frac{1}{2}$ in. diameter. Lengths run from $\frac{1}{16}$ to 6 in. for the headless varieties up to 12 in. for square-head screws. They can be ordered with coarse or fine threads.

Metric set screws range in diameter from M1 to M20 in lengths as small as 2 mm for the small-diameter screws to 80 mm for the larger diameters. Commercial standard threads are coarse threads.

Set screws are available with ends of soft material such as nylon or brass to bear on the shaft. This feature reduces damage to the shaft—desirable if the shaft is threaded or polished. They may also have locking devices built into the threads so they do not become loose from vibration. Nylon plugs imbedded in the thread are often used.

Another useful type of screw is the shoulder screw, or stripper bolt (Figure 9.17). Originally developed for holding movable parts of metal-forming dies together, it has a smooth section of constant diameter next to the head and a threaded section on the end. The smooth section is precisely machined with a close tolerance on

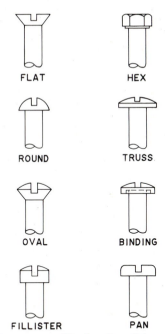

Figure 9.14 Head styles of American standard machine screws.

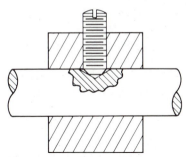

Figure 9.15 Set screws prevent relative motion between shaft and part mounted on it.

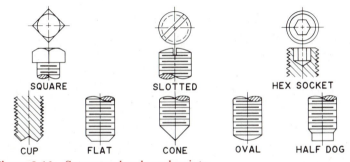

Figure 9.16 Set screw heads and points.

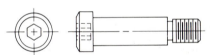

Figure 9.17 Shoulder screw (stripper bolt) has smooth, precisely finished shoulder.

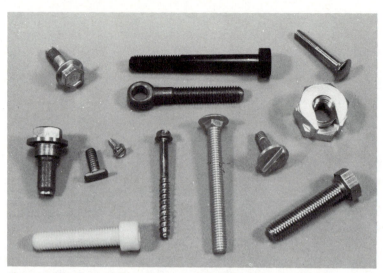

Figure 9.18 Some of the many specialized fasteners available.

the diameter. These fasteners can be used as guide pins or as axles in mechanisms. They are relatively inexpensive and can save time and cost over custom designs. Shoulder screws are available in inch and metric sizes.

Several other varieties of screws have been standardized. Wood screws are similar to machine screws, but the ends are tapered so that they start easily into wood and form their own threads in the wood. Lag screws are large wood screws with square or hex heads. Sheet metal screws are also tapered on the starting end. They are of hardened metal and have large heads for easy driving. Other threaded fasteners made of hardened metal can cut or form their own threads as they are installed.

There is an infinite variety of specialized fasteners. They are made for various industries having specialized needs, to combine functions, and to meet special fastening requirements. Screws are available that cut their own threads: they are tapered on the end, hardened, and act like a tapping tool to form threads. Others have integral locking devices such as plastic inserts in the thread, deformed threads, or one-way drives in the heads that allow the fastener to be installed but require special tools for removal. A few such fasteners are pictured in Figure 9.18.

9.4.2 Bolts, Nuts, Studs, and Washers

Bolts look like cap screws (Figure 9.19). They may have hexagonal or square heads or special-purpose heads. The length of the thread on bolts may be less than the thread length on cap screws of the same size. Specifications for cap screws and finished hex head bolts are combined. An unfinished fastener is machined only in the threads. Most hexagonal-head bolts and **nuts** have a smooth bearing face, called a washer face, machined on the surface that contacts the parts being fastened. These are known as semifinished. Finished nuts have a washer face on both sides. The tolerances for finished fasteners are more strict than those for semifinished. Square-head bolts and nuts are not finished except in the threads.

Bolts and nuts may be grouped according to use: Regular are for normal service; heavy are for harder service or for easier wrenching because the nuts and bolt heads are larger than regular heads and nuts. Square-head bolts are only available as regular. A special type of nut may be used with a standard nut to provide locking action. These thin nuts are called jam nuts.

Inch-size bolts and nuts are produced with diameters from $\frac{1}{4}$ to 4. Nuts are available down to size 0 for use with small machine screws. U.S. bolts and nuts are produced with coarse or fine threads.

Metric bolts and nuts are hexagonal. They can have coarse or fine threads extending the full length of the shaft or partly threaded. They are all semifinished or finished. Metric bolts and cap screws are considered to be the same and are known as metric hexagonal-head screws. They come in diameters from 5 to 100 mm. Nuts are available in sizes from 1 to 100 mm. A special series of metric fasteners is standardized for structural applications. These are called high-strength structural bolts and nuts and are produced in diameters from 16 to 36 mm appropriate for fastening the framing of buildings.

Threaded fasteners hold parts together by squeezing or gripping. Washers are used to increase the gripping area, to provide a smooth surface for wrenching the screw heads and nuts, to act

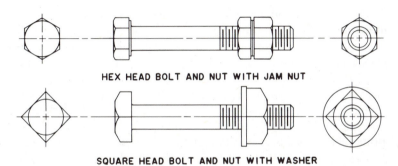

HEX HEAD BOLT AND NUT WITH JAM NUT

SQUARE HEAD BOLT AND NUT WITH WASHER

Figure 9.19 Bolts are designed to be used with nuts.

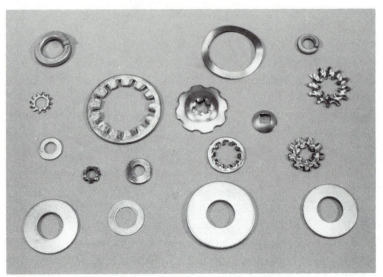

Figure 9.20 Washers come in different thicknesses and may be smooth or have locking features.

as a locking agent, and to provide thickness adjustments. The simplest form of washer is a circular disc with a hole through it. Some are not simple circles: They may have cuts or teeth to provide a locking action (Figure 9.20).

Washers are specified by the diameter of the fastener with which they are used. U.S. washers have two designations in many sizes: narrow, based on specifications of the Society of Automotive Engineers, and wide, based on structural needs. These flat washers are specified as type A, plain, N, or W. The tolerances on these washers are very broad for they are stamped from metal sheets at high speed.

9.4.3 Materials for Fasteners

Steel is the most common material for fasteners. Steel is a generic name that includes many alloys of iron and carbon with small quantities of other materials that add special properties. Controlled heating and cooling of steels is called heat treatment and can significantly change physical properties. Carbon steels are often heat treated to improve the tensile strength of the material.

Heads of U.S. hex-head bolts and screws are code marked to indicate their grade (strength). They may bear a number indicating grade or the specification number that they meet. Radius marks are also used to show the grade: no radius marks, tensile strength = 60,000 psi; three radius marks = 120,000 psi; and six radius marks = 150,000 psi. Metric screws with 5 mm or larger diameter,

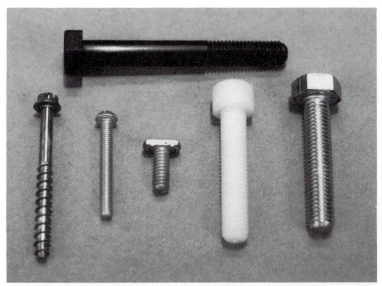

Figure 9.21   Many different materials are used to make fasteners.

both hex head and socket head, are marked with a property class number that indicates tensile strength and the ratio of yield strength to tensile strength. These numbers appear as 8.8, 9.8, 10.9, or 12.9. The higher integer number denotes a higher tensile strength. The higher decimal number denotes a higher ratio of yield strength to tensile strength.

Fasteners for corrosive environments may be made of stainless steel, brass, bronze, Monel metal, copper, nickel, or plastics. Nylon is often used for fasteners in chemical equipment. Aircraft fasteners may be made of high-strength aluminum or titanium. Fasteners of different materials are shown in Figure 9.21.

9.5   *REPRESENTING THREADS AND FASTENERS ON DRAWINGS*

Screw threads are helices on positive, negative, or tapered cylinders. Drawing them as continuous curves requires careful plotting of points located in two views of the screw thread. We seldom do this, except for presentation or catalog illustrations, because it is very time consuming.

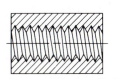

9.5.1   Detailed Thread Symbols

The representation that looks most like the true helices is the detailed symbol (Figure 9.22). The diameter is the actual diameter

Figure 9.22   Detailed thread representation.

of the thread. The pitch can be read from thread tables taken from ANSI or ISO standards (reproduced in the Appendix) and plotted to an approximate, convenient value. Remember, this is a symbol, and 11 threads per inch can be plotted as 10 threads per inch with no loss in meaning (and 10 per inch can be measured directly with a standard scale). The actual value of the pitch will be given in a note. The pitch should be laid out on the major-diameter line on the side of the thread and offset one-half the pitch distance when laid out on the opposite side. You locate the minor-diameter or root lines by drawing 60° angles between adjacent crests. Sharp V's are drawn for each pitch interval on each side of the thread. Crests should be connected with medium-weight (0.5-mm) lines. The crest lines will not be perpendicular to the axis but will slope. A right-hand, external thread will slope to the left. (Think: Turn it clockwise to make it advance.) A right-hand, internal thread will slope to the right; you are looking at the far side of the thread. You should also connect the roots with medium-weight lines. *Note:* Root lines will *not* be parallel to crest lines. If you are drawing a left-hand thread, the external thread lines will slope to the right and the internal ones to the left. Crest and root lines may be omitted on detailed internal thread symbols. The ends of external threads are usually tapered or chamfered for easier starting—the chamfer is normally drawn at 45° (Figure 9.23a).

Internally threaded holes that are through holes are drawn the same from one end to the other. Figure 9.24 shows a part cut open to reveal a tapped (threaded) hole with the tap above it, a drilled blind hole with the conical bottom matching the conical tip of the drill, and a drilled-through hole. Other features shown include a counterbored hole and a countersunk hole with the countersink above it. The countersink can provide a chamfered

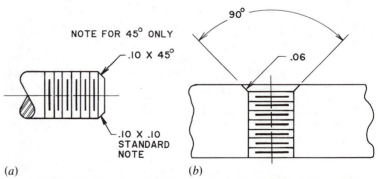

(a) (b)

Figure 9.23 (a) Chamfered shaft with note. (b) Chamfered hole with note.

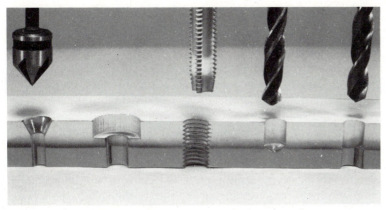

Figure 9.24 Transparent block showing drilled, threaded, countersunk, and counterbored holes. Note tools above drilled and tapped holes.

entrance for an internal thread. Both the countersink and the counterbore can provide a pocket for a screw or bolt head. Dimensioning these features was described in Figures 7.45 and 7.46. Some threaded holes are blind holes, which do not go all the way through. To produce such a threaded hole requires that the tap drill that removes material to the diameter of the root of the thread go further into the part than the desired depth of the thread. The tap (thread-cutting tool) is tapered on the end so that it can start cutting in a hole the size of the minor diameter. As it continues to cut, the full diameter of the tap does the cutting and produces full threads, but the tapered end never cuts full threads at the bottom of the hole. Full threads can be cut to the bottom of the hole with special "bottom" taps—this process requires a second tapping operation and more time. For this reason the tap drill hole is made a little deeper than the needed depth for the threaded hole. The end of the drilled hole is drawn with a 120° angle to show the conical shape made by the conical end of the drill. This is shown in one of the internal threads in Figures 9.25–9.26. Blind holes should be avoided when possible.

The end view of a detailed symbol for an external thread shows a solid circle for the major-diameter distance and a dashed circle for the minor diameter (you can't see the roots looking at the end of an external thread). If the end of the thread is chamfered, the minor-diameter circle will be shown as a solid line and the tapered entrance into the threads will be shown as pictured in Figure 9.23b. The end view of an internal thread will have a solid circle for the minor diameter and a dashed circle for the major diameter. This part of the thread is hidden inside the part, but you can see a hole the size of the minor diameter.

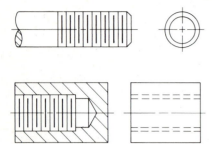

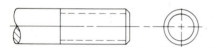

Figure 9.25 Schematic thread representation—pitch is approximate. Use convenient valve such as 0.1 in.

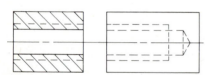

Figure 9.26 Simplified thread representation—the minor diameter is approximate. Major- and minor-diameter lines should be clearly separated.

9.5.2 Schematic Thread Symbols

Detailed symbols are seldom used on drawings for production. Although they are quicker and easier to draw than true helices, they are time consuming. Schematic symbols have been developed that convey the appearance of screw threads; they can be drawn more quickly than detailed symbols (Figure 9.25).

The major diameter of the thread is drawn and an approximate pitch scaled for an external symbol. A chamfer is usually shown at the end of the thread. Crest lines are drawn without slope as medium-weight lines (0.5 mm). One sharp V should be constructed to locate the root lines. Light construction lines should be drawn for the root diameter and the root lines drawn as thick lines (0.7 mm).

Internal threads, in section, are drawn similar to external threads. Internal threads, not in section, appear as four parallel dashed lines—two major-diameter lines and two minor-diameter lines. The circular view is the same as a detailed symbol.

9.5.3 Simplified Thread Symbols

The exact requirements for a thread are specified in the thread note that gives the diameter, the pitch, the thread form, and the tolerances. Simplified symbols that can be drawn quickly indicate that a thread is present but leave all specifications to the note (Figure 9.26). They do not show pitch or thread form. The external symbol is simply the outline of the threaded shaft, with chamfer, and a pair of dashed lines indicating the root diameter of the thread. If the thread runs less than full length, the symbol terminates at a visible line across the shaft.

The internal symbol is the same as the schematic symbol if the thread is not sectioned. If it is sectioned, the root lines are shown as visible lines, and the thread cavity between the root lines is not crosshatched. The major-diameter lines are shown as dashed lines on the crosshatching and extend as far as the length of thread. The termination of threads is a line perpendicular to the axis of the thread, hidden in the sectioned portion and visible in the unsectioned cavity. The circular view is the same for all three types of symbols.

9.5.4 Drawing Screw Heads and Nuts

You have learned how to represent screw threads on drawings with detailed, schematic, and simplified symbols. Threads by themselves are not of great value—they are useful as part of a fastener. The proportions of several types of fasteners are given in the Appendices. These are taken directly from the standards

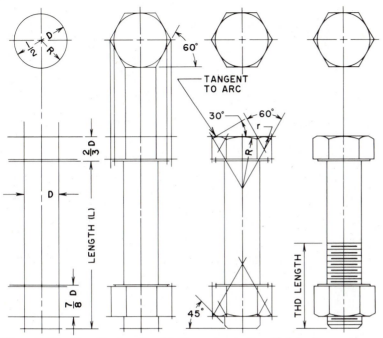

Figure 9.27   Drawing hex-head bolts and nuts (semifinished).

for such fasteners. If you buy a hex-head cap screw of a given size, it will always have the same proportions unless you request a special type of screw.

To draw a bolt and nut, you need to know where the center line of the fastener will be. You need to know where the contact surfaces are—where does the head of the bolt touch the parts— where does the nut touch? You need to know the size (diameter) of the fastener. Is it a hex-head bolt or a square-head bolt? When you have this information and the specification tables for standard bolts, you can draw the fastener. Figure 9.27 outlines the steps for drawing a hex-head bolt and nut. Bolt and screw heads and nuts are usually drawn across corners to show the maximum room that the fastener will take as it is turned to tighten or loosen it. If you don't have the standards table available, these proportions for regular, finished bolts and nuts will be helpful:

Diameter of head or nut	$1\frac{1}{2}$ × thread diameter
Thickness of bolt head	$\frac{2}{3}$ × thread diameter
Thickness of nut	$\frac{7}{8}$ × thread diameter
Length of thread	
< 6 in.	2 × thread diameter + $\frac{1}{4}$ in.
≥ 6 in.	2 × thread diameter + $\frac{1}{2}$ in.

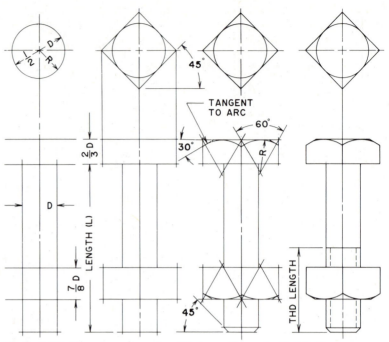

Figure 9.28   Drawing square-head bolts and nuts (unfinished).

The steps for drawing a square-head bolt and nut are similar and are outlined in Figure 9.28. The proportions for other fasteners are given in the standards copied in the Appendices.

9.5.5   Drawing Fasteners with a Computer

Computer graphics is a great tool for drawing fasteners. Many advanced CADD packages have a library of symbols and patterns that can be called from memory and displayed on the screen. A threaded fastener symbol, once brought to your screen, can be enlarged, reduced, shortened, lengthened, moved, mirrored, copied, or otherwise modified to meet your needs. If you do not have an advanced CADD package available, you can create your own pattern or "block." You create a "generic" threaded fastener by zooming in to draw to a large scale, so you can get the details right. You would use the same procedure on the screen as shown in Figure 9.27 or 9.28. Perhaps you would create a hex-head bolt and nut and store it in memory at some convenient scale. If you need a hex cap screw, call up your bolt and nut and delete the nut (Figure 9.29). If you need a headless set screw, delete the nut and the bolt head and modify the tip. Imagine two circular

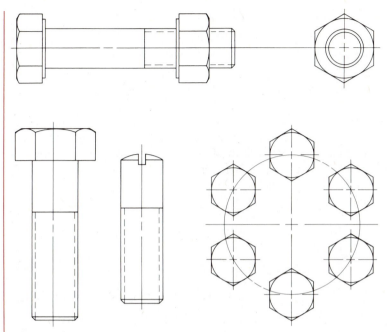

Figure 9.29   Using computer graphics to generate fastener drawings. (*Computer-generated image*)

disks to be bolted together—you retrieve your end view of the bolt head and duplicate it wherever there is a center for a bolt.

The power in a computer system lies in its ability to make changes rapidly. With a few fastener patterns in memory you can modify or combine them to produce almost any kind of fastener in a very short time. When making a pictorial drawing, you can enter the three-dimensional information—the data base for a fastener. You can then call the data base and show the fastener in any position or orientation on your drawing. You may also modify it as you would a two-dimensional drawing.

9.5.6   Specifying Fasteners

The symbol for the fastener alerts the reader of your drawing that a fastener is used at that location. The symbol by itself does not give the reader enough information to obtain a specific fastener. You need to provide the following information:

1. Thread specification (discussed in Section 9.3.5)
2. Fastener length
3. Fastener series (only for bolts: regular or heavy)
4. Type of finish (unfinished, semifinished, or finished)

5. Material (other than steel)
6. Head style or shape
7. Point style (only for set screws)
8. Name of fastener family
9. Special requirements (coatings, specifications to meet)

The information is grouped differently for U.S. and ISO fasteners. U.S. fasteners are described with the thread specification first and the length, followed by the other information in the order already listed. Metric fasteners are described by the head style, family name, thread specification, and material. Some examples follow:

$\frac{3}{4}$-10UNC-2A $\times$ $2\frac{1}{2}$ REG FINISHED HEX BOLT, STAINLESS STEEL

HEX SCREW, M16 $\times$ 1.5 $\times$ 110

SLOTTED SET SCREW, CONE POINT, M4 $\times$ 12, BRASS

#10-32UNF-3A $\times$ $1\frac{1}{2}$ OVAL HD MACH SCREW

1″-8UNC-2B FINISHED JAM NUT

Installation information is often added to assembly drawings. The torque, or turning force, necessary to install the fastener tightly but without damaging it is often specified. If the fasteners are installed with automatic machinery, the torque may be preset. If the fasteners are installed manually, a special wrench with a dial gauge built into it (a torque wrench) would be used.

9.6 OTHER MECHANICAL FASTENERS

There are other types of mechanical fasteners that are not threaded. These include rivets, pins, keys, splines, and retaining rings. There are also special-purpose clips and latches that are usually designed for specific applications. Fastener catalogs will provide you with a wealth of information on special fasteners.

9.6.1 Rivets

Rivets function like a bolt and a nut. They fasten by gripping or compression. A rivet has a smooth shaft and a large head on one end (Figure 9.30). The shaft is inserted through holes in the parts to be joined, and a new head is formed, in place, on the end without a head. This process may be accomplished with pressure only or with heat and pressure.

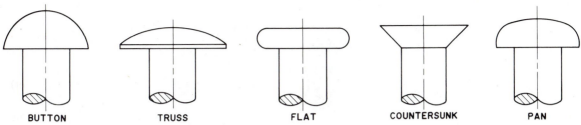

Figure 9.30 Head forms for American standard rivets.

Have you ever seen steelworkers high on a building under construction? One worker lines up the holes in the steel shapes to be joined while another uses a bucket to catch a red hot rivet heated by a third worker. The hot rivet is inserted in the aligned holes and held in place with a "bucking bar" while the other worker forms a new head on the smooth end with an air hammer. This is hot, noisy, and dangerous work. Many new buildings are joined with high-strength bolts assembled with powered wrenches.

Rivets are still used for shop assembly of steel shapes where the workers are not walking on steel beams hundreds of feet in the air. Until recently the outer covering, or "skin," of aircraft was riveted to the framing with thousands of rivets. Many small parts are assembled with rivets. Rivets are less expensive than screws, and the installation process has been automated in many cases.

One very useful kind of rivet is the "blind rivet," originally created to be used where the back side of the work was not exposed. A hollow rivet, with a long rod (mandrel) in the core is inserted in the holes in pieces to be joined (Figure 9.31a). Force on the head of the rivet holds it in place while the mandrel is pulled through the rivet, all from the front side. The mandrel deforms the far side of the rivet, creating a head on that side (Figure 9.31b). Blind rivets are made of a ductile material such as aluminum that will deform easily. The mandrel may be pulled through the rivet or may be designed to break off in the rivet to seal the core.

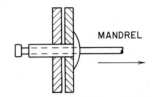

Figure 9.31 (a) Pull-through blind rivet with mandrel in place. Mandrel will move in direction of arrow; large end of mandrel will deform end of rivet, compress work pieces, and expand body of rivet.

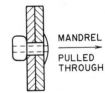

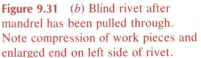

Figure 9.31 (b) Blind rivet after mandrel has been pulled through. Note compression of work pieces and enlarged end on left side of rivet.

9.6.2 Pins

Pins are cylinders that pass through preformed holes to join parts. When pins are used to fasten a part to a shaft, the holes will be perpendicular to the axis of the shaft. They have no threads and usually no heads, but some may have special shapes. Pins hold parts by interference or friction or the pin may have a split end that can be deformed to lock it in place. Some pins have a hole in the end and a locking wire or another, smaller pin is inserted

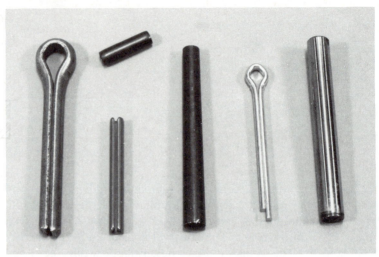

Figure 9.32   Some of the many varieties of pins.

in this hole to keep the pin in place. Several types of pins are shown in Figure 9.32.

The taper pin has been used for many years to assemble parts to a round shaft. As its name suggests, it is not a cylinder but is slightly tapered (a frustum of a long, slim cone). To fasten parts with a taper pin, the parts are located with respect to each other and a hole drilled through both of them. The hole is then enlarged and tapered with a tool that removes only a small amount of metal—a taper reamer. Then the carefully machined taper pin is driven into the tapered hole. The parts are locked firmly in place. Today, spring pins are replacing taper pins in many applications. They do not require a tapered hole, just a drilled one. The pin itself is rolled up from a sheet of spring steel—the round, rolled shape retains some ''springiness,'' so it adapts to irregularities in the hole. The hole is made a little smaller than the pin; the pin is driven in the hole and presses tightly against the sides of the hole, locking the parts together.

Groove pins are used much like spring pins. They are solid, round pins (dowel pins) with a longitudinal groove formed in them. This process deforms the pin so that it is no longer round. When the groove pin is driven into a hole, the groove edges tend to grip the sides of the hole and the deformed shape is forced back to a round shape. These actions lock the pin in the holes.

Cotter pins are formed from a single piece of D-shaped wire. They have two parallel legs, and the loop is usually opened up to form a head. The pin is inserted in a hole and the legs bent apart to keep it in place.

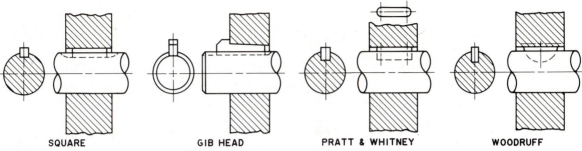

SQUARE   **GIB HEAD**   **PRATT & WHITNEY**   **WOODRUFF**

Figure 9.33 Four common types of keys.

9.6.3 Keys

Keys are metal shapes placed in holes or slots cut longitudinally in a shaft and mating part to lock them together so that they rotate together (Figure 9.33). Keys can be square in cross section or rectangular. They may be straight or tapered. Some may have a head to facilitiate removal (gib head key), and others may have rounded ends to fit a specially shaped keyseat. The keyseat is the slot or pocket cut into the shaft to receive part of the key. The other part of the key rests in a slot cut into the wheel, pulley, gear, or other device attached to the shaft. This slot is called a keyway.

One type of key that allows some freedom of movement to compensate for misalignment is the Woodruff key. It is shaped like a half moon, and the keyseat is a semicylinder. This shape allows the key to ''rock'' in the keyseat.

The size of the key is selected to meet the load on the shaft and the other part fastened to it. Proper selection of a key for specific design conditions is a design problem requiring some engineering analysis. Keys are seldom used that are wider than one-fourth the diameter of the shaft because cutting a larger keyseat would materially weaken the shaft.

If a shaft is formed with several key shapes formed on the shaft and the wheel-type part has several keyways to match, this is a splined connection (Figure 9.34). Splined shafts and holes require careful fabrication to get all the projections and cavities to match within close tolerances. Splines are used where loads are heavy, precision is needed, and the wheel-type part must move axially on the shaft while retaining the same radial relationships. If you have the opportunity to look inside an automobile transmission, you will see the gears riding on splined shafts.

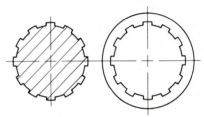

Figure 9.34 Splined shaft with mating wheel.

9.6.4 Retaining Rings

Retaining rings hold parts on a shaft by acting as an enlargement of the shaft (Figure 9.35). Most are open rings that fit into a groove

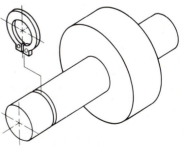

Figure 9.35 Application of external retaining ring.

around the shaft. They are held in the groove by spring action or by squeezing them into the groove. Some are complete circles with angled spring teeth that go on a shaft easily but "bite" into the shaft if a force attempts to move them in the other direction.

Retaining rings may be stamped from flat sheets or formed from square or round wire. Some require special tools for installation and removal.

9.7 OTHER MACHINE ELEMENTS

There are some other types of parts that are represented frequently on machine drawings. You will meet them here and will probably encounter them frequently in design courses and in your professional life. Springs are common devices for storing energy. They sometimes act as pure storage devices to collect energy in one part of a cycle and return it in another part of a cycle (valve springs in an automobile engine). They may also be used to even out movements by storing energy when moved in one direction and returning it to move in the other direction (body springs on an automobile).

A common form of spring is a helix wound from high-strength wire. Three common types are pictured in Figure 9.36, and the symbols for them are shown in Figure 9.37. They are named for the kinds of load they are designed to accept:

Compression   Accept an axial load that compresses them; then they try to resume their normal length when the load is removed.

Tension   Accept an axial load that stretches them; then they try to resume their normal length when the load is removed.

Torsion   Accept a radial twisting load (clockwise or counterclockwise); then they twist in the other direction when the load is removed.

There are also flat springs (the leaf springs on auto and truck bodies are stacked, flat springs) and power springs that look like a spiral of flat wire. Energy is stored in them by winding and released through a controlled movement, as in a mechanical clock or a wind-up toy.

Bearings and bushings are used to line a hole through which a shaft passes. Electric motors always have some type of bearings so that the motor shaft can spin easily. If the shaft turns or rotates in the hole, the liner is usually referred to as a bearing. If the shaft slides back and forth (reciprocates) in the hole, the liner is

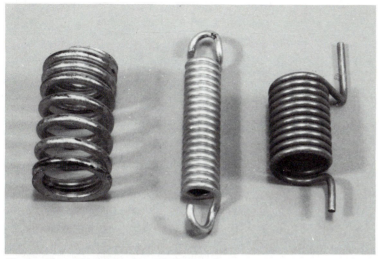

Figure 9.36 Three simple springs.

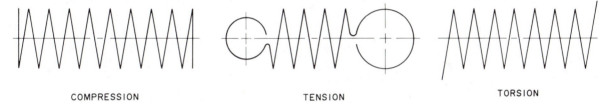

COMPRESSION   TENSION   TORSION

Figure 9.37 Symbols for three simple springs.

called a bushing. The term *bushing* is also used to describe some bearings and liners used to restore the diameter of holes that have increased in size through wear.

Bearings and bushings are used to decrease the friction between a shaft and the part through which it passes—some materials rubbing together have less friction than others. If the bearing or bushing wears, it is easier and cheaper to replace the small part than to replace an entire machine.

Bearings may be plain, just a sleeve of material inserted in the hole; it is this type that may also be referred to as a bushing. A greater reduction in friction occurs if balls or rollers are incorporated in the bearing—rolling friction is less than sliding friction. A ball bearing is shown in Figure 9.38. The rolling elements may be balls, cylindrical rollers, or tapered rollers. Tapered rollers are used for very heavy loads and to accept thrust loads in an axial direction (car and truck axles). Rollers of very small diameter may be used in tight spaces and for light loads; they are called needle bearings.

Figure 9.38 Ball bearing.

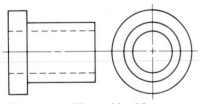

Figure 9.39 Flanged bushing.

Bushings are usually just a hollow cylinder. Sometimes they may have a flange on the end to make it easier to position or remove them (Figure 9.39). Materials for bushings and plain bearings are often made of brass or bronze. The basic material may have a lubricant such as oil, Teflon, or molybdenum disulfide impregnated in it.

Bearings and bushings are usually shown in outline form on the drawing and a reference given for ordering the part from a firm that makes them. Plain bearings and bushings may be fabricated for a small machine and shown with the drawings for that machine; only a bearing manufacturer would produce a ball or roller bearing because special materials and machines are required.

9.8   *THERMAL PROCESSES*

Threaded fasteners are generally reusable. Several processes are used for joining materials that result in a permanent bond. Some processes require heating the materials. Welding involves actually melting the parts to be joined at the point of contact; brazing and soldering require melting a filler material but not the parts to be joined.

9.8.1   Welding Processes

In **welding** the parts are joined by the application of heat or pressure, often with the addition of a filler material. The parts are

Figure 9.40 Arc welding generates heat through electric arc between electrode and work. (Courtesy of Ohio State University Engineering Publications)

usually similar metals. Heat is supplied by a gas flame, electrical arc, or electrical resistance. The filler material is a metal of a composition similar to the parts being welded. A flux is used to improve the quality of the weld by combining with unwanted products of the heating process, such as metallic oxides. It may also generate a gas cloud that protects the heated area from oxidation. In some processes an inert gas shield (carbon dioxide or argon) is used to prevent contamination of the heated area.

Electrical arc welding creates heat with an electrical arc drawn between an electrode and the work to be welded (Figure 9.40). There are several variations of the process. The electrode itself may be consumable—it supplies the filler material. The electrode may be nonconsumable, and a separate filler material is added in the arc. The electrode may be attached to a gas nozzle that supplies an inert gas shield. The operation may be controlled manually or automatically.

Gas welding uses a flame, commonly from burning acetylene with oxygen, to heat the parts being welded (Figure 9.41). The filler material, coated with a flux, is melted in the flame and added to the joint. Gas welding is slower than arc welding; it is often used for repair work and in small workshops. Generally it is done manually.

Resistance welding uses both heat and pressure to form a joint. The heat is generated by passing a large electric current through

Figure 9.41 Gas welding heats work and melts filler rod with high-temperature gas flame. (Courtesy of Ohio State University Engineering Publications)

the work pieces while they are clamped tightly together. The weld may be continuous or intermittent (spot welding). Because the process requires high currents and clamping devices, it is usually considered a mass production process.

Other exotic processes for welding coming into greater use include laser beam welding, electron beam welding, and plasma arc welding. These processes require expensive equipment and careful control. Laser beam and plasma arc produce clean welds on thin materials; they are suitable for high-performance materials—stainless steel, titanium, and aluminum alloys. Electron beam welding can produce a deep, clean weld and is useful for welding refractory metals or joining dissimilar metals.

9.8.2 Specifying Welds on Drawings

Welding is specified on drawings by symbols, a type of shorthand standardized by the American Welding Society. The various methods for making a joint in metals are defined, and the parts of the weld symbol describe the conditions of the weld (Figure 9.42).

The basic symbol is a reference line with an arrow pointing to the joint. The area below the reference line describes the *arrow side* of the weld—where the arrow touches the joint. The area

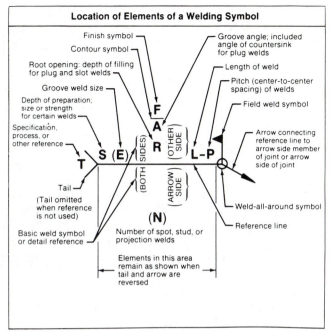

Location of Elements of a Welding Symbol

Finish symbol
Contour symbol
Root opening: depth of filling for plug and slot welds
Groove weld size
Depth of preparation; size or strength for certain welds
Specification, process, or other reference
Groove angle; included angle of countersink for plug welds
Length of weld
Pitch (center-to-center spacing) of welds
Field weld symbol
Arrow connecting reference line to arrow side member of joint or arrow side of joint
Tail
(Tail omitted when reference is not used)
Basic weld symbol or detail reference
Number of spot, stud, or projection welds
Weld-all-around symbol
Reference line
Elements in this area remain as shown when tail and arrow are reversed

F A R S (E) L-P T (N)
(BOTH SIDES) (ARROW SIDE) (OTHER SIDE)

Figure 9.42 Standard weld symbol from American Welding Society standards. (Courtesy of American Welding Society)

above the line describes the *other* side—opposite from where the arrow touches the joint. The type of joint is described by symbols; Figure 9.43 shows several common joints and their symbols. The sizes of the welds are given around the joint symbol. If the weld is a groove weld to be ground smooth after completion, a flush symbol is added. If a part is circular and is to be welded all around, a circle is drawn around the junction of the reference line and the

FILLET	PLUG AND SLOT	GROOVE WELDS				
		SQUARE	V	BEVEL	U	J

Figure 9.43 Types of welded joints and their symbols.

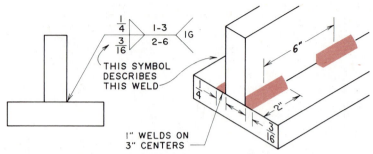

Figure 9.44 Weld symbol and welds it describes.

arrow line; a weld to be done in the field (not in the factory) gets a solid black circle at this point. Detailed specifications for welding are called for by a reference letter in the tail of the arrow. Figure 9.44 illustrates a weld symbol and the work you would expect from this symbol.

9.8.3 Soldering and Brazing

Soldering is a process for joining metals with a low-melting-temperature alloy (below 840°F). The parts to be joined are not melted; the solder alloy flows over the joint and adheres to the component parts. Soldering is used for electrical assemblies and for sealing metal cans. The heat can be supplied from a flame or an iron heated by electrical resistance. Most solder alloys are a mixture of lead and tin. Other alloys include tin and antimony, tin and silver, and lead and silver.

There are no standard symbols for soldered joints. Process conditions, solder alloy, and joint design are selected based on experience and research.

Brazing requires heating the parts to more than 840°F, but the work pieces are not melted. A filler metal is used that has a melting point lower than the metal in the parts to be joined. Copper, aluminum, and silver alloys are commonly used as filler materials. Heat may be applied in a furnace, in a bath of molten flux material, or by a torch, electrical induction, or electrical resistance. Brazing can join most all metals and is useful for joining dissimilar metals. Cast metals are difficult to weld but are readily brazed. Design of a brazing process requires knowledge of the base metals, the filler metal, the process equipment, and the service demands on the finished part. Brazed joints are noted with symbols similar to welding symbols (Figure 9.44). These are also specified by the American Welding Society.

9.9 *ADHESIVES*

Adhesives do not fasten parts mechanically, and they do not require high temperatures to join the parts. They do form a bond that transmits forces uniformly across the adhesive joint. The strength of the bond, is affected by many factors—bond strengths of 3000 psi are common. Some systems exceed 5000 psi.

First developed to solve difficult aerospace joining problems, they are now used in automobiles, appliances, and other common devices. Adhesives of some types have been used for centuries, but structural adhesives are a product of the last half of the twentieth century. They are all organic polymers, a part of the big family of plastics. Some are *thermoplastic:* They soften or melt as heat is applied. Other are *thermosetting:* Once formed, they do not soften with heat but hold their form. And some adhesives are plastic alloys, that is, mixtures of different types of polymers.

Adhesives can join thin materials and dissimilar materials (metals to nonmetals). They do not damage or mar the work surfaces; they can seal a joint as well as bond it. Because there are so many possible combinations of materials and adhesives, selection should be made carefully using information supplied by adhesive manufacturers. There are no industrywide standards as in welding, but every manufacturer provides data sheets and technical help.

Operating temperatures for the finished joint are determined by the type of adhesive used: silicone adhesives are useful up to 500°F; urethanes are good for low temperatures. Surface preparation is very important for some systems such as epoxies. The surfaces must be clean and free from impurities and may need to be roughened. Acrylics are more tolerant of impurities.

Some adhesives require low heat, ultraviolet light, the absence of air, or time (up to 24 hours) for curing. Several systems come in two parts—a resin and an activator—which can only be combined at the point of use. Adhesives can be applied as liquids by brushing, spraying, or dipping. Many types are available as films or tapes that can be laid between mating surface. Others come as powders, and some can be microencapsulated (in tiny beads).

One interesting application of adhesives is in combination with threaded fasteners. The threads can be coated with an adhesive, often in the form of microencapsulated beads, and when the fastener is installed, the adhesive acts as a locking agent to prevent loosening. Because not all the tiny beads are broken when the fastener is first installed, some can be reused up to four times with little loss in locking power. The adhesive can also act as a sealant to prevent fluid from escaping around the threads. One manufacturer has specified various colors of adhesives for different types of fasteners: Metric fasteners are coated with one color, and inch-size fasteners with another color.

9.9.1 Principal Types of Adhesives

Acrylics are thermosetting resins that cure quickly (less than 1 minute) at room temperature. They are two-component systems—a resin and an accelerator—although some one-part systems are available. Acrylics tolerate oily surfaces and are excellent for joining plastics. They also bond metals, glass, ceramics, and wood.

Anaerobics cure in the absence of oxygen; they are used in enclosed areas such as the screw thread sealants. Anaerobics are thermosetting resins such as dimethacrylates that cure in a few hours at room temperature or a few minutes at 250°F. Easy to apply, they are single-component resins. Surface preparation is required for joining metals, glass, ceramics, and some plastics.

Cyanoacrylates may be thermosetting or alloyed to become thermoplastic. A well-known thermoset is the household family of ''Superglues,'' which bond quickly; these are high in cost and hazardous to the skin. The thermoplastic varieties cure in $\frac{1}{2}$–5 hours at room temperature and will bond most any material.

Epoxies are the most widely used structural adhesives. They will bond dissimilar materials. The joint resists most chemicals. Epoxies are two-component systems that require careful mixing and surface preparation. Heat curing at about 300°F is desirable, but they will cure at room temperature in about one day.

Hot-melts are thermoplastic. They are low cost, easily applied, and produce flexible bonds. Because they are thermoplastic, they should not be used for service above 200°F. Hot-melts will bond most materials.

Silicones are thermosetting and are among the most expensive adhesives. They are one-component rubberlike resins that will bond most surfaces. Careful surface preparation is required. Curing is at room temperature and can be accelerated with ultraviolet light. Service temperatures are up to 500°F.

Urethane adhesives may be one- or two-component systems and are thermosetting plastics. They provide flexible bonds and cure at room temperature. Urethanes will bond most materials and are an excellent choice for low-temperature service.

Selection, application, surface preparation, curing conditions, operating environments, and cost are all important in designing an adhesive joint. Manufacturers of adhesives all provide design information and technical service to assist in the design. There are no industry-wide specifications for materials or procedures.

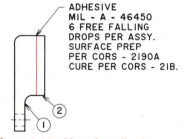

ADHESIVE
MIL - A - 46450
6 FREE FALLING
DROPS PER ASSY.
SURFACE PREP
PER CORS - 2190A
CURE PER CORS - 21B.

Figure 9.45   Note for adhesive application.

However, some types of adhesives are covered by military specifications (MIL specs). See Figure 9.45.

9.10 *THE FUTURE OF FASTENING AND JOINING*

Pins and threaded fasteners have been with us for many years. The taper pin, which requires careful manufacture and installation, is being replaced in some applications by spring pins. Three-part fasteners are being replaced by specially designed one-piece fasteners that perform all the functions of the three pieces for less cost. Stapling and stitching with wire are replacing separate fasteners in thin and soft materials. You will need to read, observe, and learn in order to be aware of trends and developments in fasteners.

A few areas you may want to consider as you advance in the design profession are noted. Joining of plastics can be done by mechanical fasteners, solvent bonding, welding of thermoplastics, and fusion bonding—new methods are reported each year. Welding processes are in use today using lasers, ultrasonics, and the electron beam that a generation ago were unknown. Fasteners made of materials other than steel are finding increasing applications—nylon, polyvinyl chloride, and titanium. Combinations of elements and materials such as adhesive-coated screws, nuts with integral sealing washers to seal in fluids, quick-operating fasteners, and explosive rivets will continue to be developed. If you become involved in manufacturing, look for faster, more reliable ways to install fasteners or new ideas for accomplishing the fastening and joining process.

9.11 *SAMPLE PROBLEM: TABLE SAW FIXTURE*

Our fixture needs to be fastened to the table of the saw. This will be done with a cap screw; a wide washer should be provided to cover the slot and increase the bearing area of the screw head.

We have the drawing in a CADD file. We can recall it and add the fastener parts that must be used with it. The drawing shows how they go together. (In the next chapter you will learn more about drawings that show how parts are assembled.) Symbols are again the key to quick and versatile solutions. Make a drawing of a wide washer, store it, and call it back in any size you need. Make a drawing of a bolt and nut; store it. Call it back

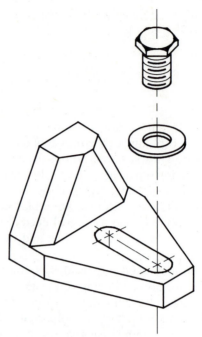

Figure 9.46 Fasteners can be stored in CADD files and called into a drawing as needed. These symbols or patterns can be changed in size, orientation, or number or have features modified to meet needs of many different drawings. *(Computer-generated image)*

and delete the nut or store a cap screw and nut separately. Change the scale to fit your drawing, and you can have screws anywhere you need them and in any size. When you recall symbols, remember where the reference point, or "handle," is for your symbol. You can store hundreds of handy symbols. An advanced CADD program may have many types of symbols available or you can purchase a package of symbols for your type of design (Figure 9.46).

9.12 SUMMARY

Fasteners hold our industrial society together. Threaded fasteners are the most common; other mechanical fasteners such as pins, keys, rivets, and retaining rings serve specific needs. Many types of threaded fasteners have been standardized by the ABC agreements and the American National Standards Institute for inch-size fasteners and by the ISO for metric fasteners. Fasteners are seldom drawn exactly as they look; standard symbols have been developed to represent fasteners on drawings. This chapter also included a brief review of other mechanical elements including springs, bushings, and bearings.

Joining processes provide permanent bonds. Welding is a high-technology discipline with many new processes available. The specification of welds on drawings is standardized through the specifications of the American Welding Society. Other joining processes that use heat to create a bond are soldering and brazing. Adhesives are the frontier of joining technology—the design of an adhesive-bonded joint requires much information and careful design. Fastening and joining technology is moving rapidly—you will need to be aware of advances in this area to do a good job of design.

PROBLEMS

These problems may be done on your computer or manually as specified by your instructor.

Problem 9.1
Copy this figure as a freehand or computer sketch. On your sketch label these features:

Major diameter	Depth of thread
Minor diameter	Thread angle
Pitch	

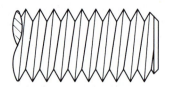

Problem 9.2.
Sketch the profile of these threads with the noted pitches:

Sharp V	Pitch $\frac{1}{4}$
Unified	Pitch 0.50
Square	Pitch 20 mm
Knuckle	Pitch 1 in.

Problem 9.3
A threaded shaft with a specification of 0.50-20UNF-3A DOUBLE is turned five revolutions in a mating nut. How far does it advance?

Problem 9.4
Copy and complete the following table:
Which of the threads per inch that you entered has the smallest pitch? Which has the largest pitch?

Major Diameter	Series		
	Coarse	Fine	Extra Fine
#6	UNC-32	UNF-?	—
?	UNC-24	UNF-32	—
.375	UNC-?	UNF-24	UNEF-32
.50	?	?	?
?	UNC-10	?	?
1.00	?	?	UNEF-20

Problem 9.5.
Name a tapered thread (noncylindrical). Why is it tapered?

Problem 9.6.
Write specifications for the following threads:
- (a) $\frac{1}{2}$ major diameter, fine external threads with precise tolerances
- (b) 0.75 major diameter, coarse internal threads with very loose tolerances
- (c) 12 mm major diameter with coarse threads

Problem 9.7.
Name the illustrated fasteners:

Problem 9.8.
Pick up two large hex screws and examine the heads. One looks like (*a*) and the other looks like (*b*). What do you know about each?

Problem 9.9.
Schematic thread symbols are used on assembly drawings; simplified symbols are usually used on detail drawings. Copy and complete the following assembly and detail drawings.

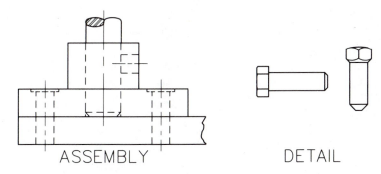

ASSEMBLY   DETAIL

Problem 9.10.
Write the specifications for the following fasteners:
- (a) 1 in. regular semifinished hex-head bolt, $3\frac{1}{2}$ long with a coarse thread
- (b) Metric cap screw with a 24-mm fine thread
- (c) 0.50 finished jam nut made of Monel metal and having precisely made fine threads

Problem 9.11.

Complete the assembly drawing two times the size shown in your text. Add a socket head set screw at point *A* and a Woodruff key at point *B*.

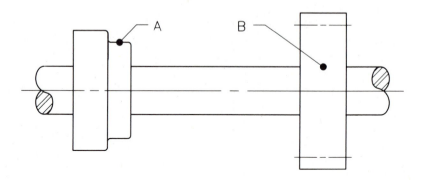

Problem 9.12.

Prepare the weld symbol for welding the end of a steel pipe to a plate using a $\frac{1}{4}$ fillet weld and welding all around.

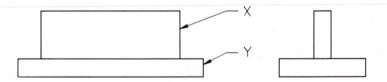

Problem 9.13.

Name three fastening processes that require heat and a filler material.

Problem 9.14.

Indicate that part *x* is to be welded to part *y* with an intermittent fillet weld on both sides. A $\frac{3}{8}$ fillet, $1\frac{1}{2}$ long, spaced at three intervals is required.

Problem 9.15.

You need to fasten two plastic spheres together as shown. List at least three ways to accomplish this.

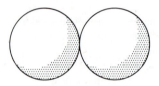

In Problems 9.16–9.20 prepare drawings for the fasteners noted. If you are using a CADD system, recall previous fasteners as starting views. If you do not have fastener symbols stored, you may wish to store these drawings as symbols for future work. Use schematic representation for your threads unless another type is requested.

Problem 9.16
Bolt head end and front views of a $\frac{3}{4}$-in. square-head bolt by 4 in. long with nut (coarse threads); letter a complete description for your bolt and nut. *References:* Appendix tables for square-head bolts and square nuts.

Problem 9.17
Thread end and front views of a hex screw, M16 × 110.

Problem 9.18
One view of a 5-in.-diameter bolt circle with heads of $5\frac{5}{8}$-in. hex bolt heads showing; indicate that these bolts have a tensile strength of 120,000 psi.

Problem 9.19
Two views of a 1-in. shaft, 6 in. long, with a coarse thread $1\frac{1}{2}$ in. long on one end and a fine thread 1 in. long on the other end; include thread notes for both ends.

Problem 9.20
Top and front views of a rectangular prism 200 mm wide, 38 mm high, and 76 mm deep; show the following holes centered in the 76-mm face: at station 50 mm, 1/2-13NC-2B through (section in front view); at station 100 mm, .75-16UNF-3B × 1.00 deep (show tap drill) (section in front view); and at station 150 mm, M12 through with 2-mm chamfered entrance (*not* in section).

TERMS YOU WILL SEE IN THIS CHAPTER

BILL OF MATERIAL   List of all the parts necessary to make up an assembly and some details about these parts.

PICTORIAL ASSEMBLY   Three-dimensional drawing showing how parts fit together.

EXPLODED VIEWS   Three-dimensional drawings showing the parts separated from each other but along predetermined axes to show how parts are assembled together.

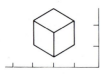

SECTIONED PICTORIAL DRAWINGS   Similar to exterior assemblies except some portion of the assembly has been "removed" so the viewer can see the interior features of the assembly.

ZONES   Coordinate grid used to identify areas of a drawing.

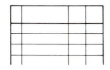

WIRE FRAME DRAWINGS   Traditional pictorial drawings generated from a computer data base. Hidden elements may be included or deleted.

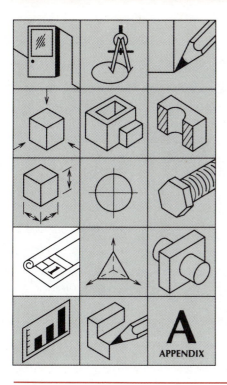

10

PRODUCTION DRAWINGS

APPENDIX A

10.1 INTRODUCTION

Production drawings are used to transmit and communicate information for the production of objects and assemblies (Figure 10.1). Production drawings are classified in two major categories, detail drawings and assembly drawings.

Detail drawings are drawings of single parts. They may be drawn one part per sheet or there may be several parts detailed on a large sheet. These detailed drawings include such additional information as dimensions and notes that relate to

Figure 10.1   Flow of production drawings.

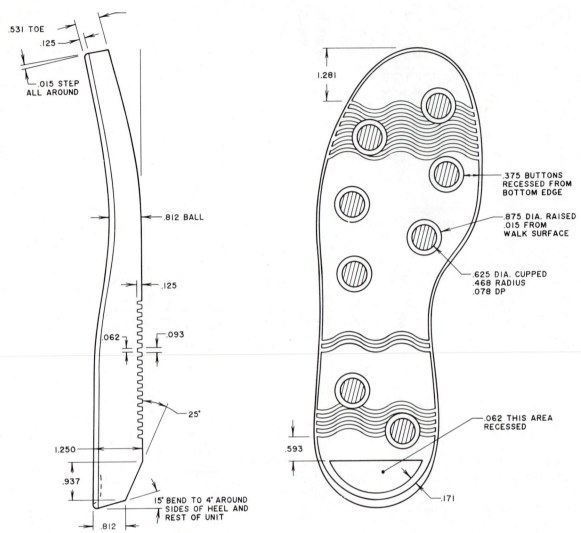

.531 TOE
.125
.015 STEP
ALL AROUND

.812 BALL

.125

.062 .093

25°

1.250

.937

.812

15° BEND TO 4° AROUND
SIDES OF HEEL AND
REST OF UNIT

1.281

.375 BUTTONS
RECESSED FROM
BOTTOM EDGE

.875 DIA. RAISED
.015 FROM
WALK SURFACE

.625 DIA. CUPPED
.468 RADIUS
.078 DP

.062 THIS AREA
RECESSED

.593

.171

Figure 10.2 Single detail drawing. (Courtesy of Jones and Vining, Inc.)

material, finish, weight, or standard tolerances (Figures 10.2 and 10.3).

Assembly drawings show how parts fit together or are functionally related. Dimensions are often found on assembly drawings. However, the dimensions refer to relationships among the parts, not to the size of the individual objects. Assembly drawings often include a listing of all parts necessary to make up the total assembly. This list is called a **bill of material** (Figure 10.4h). Large products requiring many parts

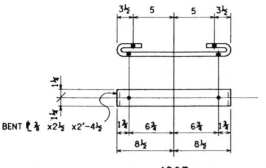

4803

1 – REQ'D. FOR STANDARD BOX MOUNTING
FURNISH 2 BOLTS ½"ø x 1¼ LG.
2 – REQ'D. FOR MOUNTING 4806
FURNISH 2 BOLTS ½"ø x 1¼ LG./DET.

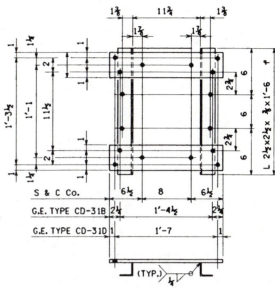

4805

USE FOR (CD-31B & D & S&C POTENTIAL DEVICE)
FURNISH 4 BOLTS ½"ø x 1¼" LG./DET.

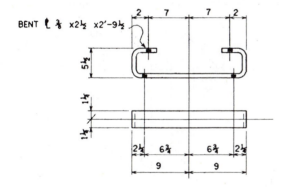

4804

2 – REQ'D. FOR LINE TUNING CABINET
& ADJUST UNIT MOUNTING
FURNISH 2 BOLTS ½"ø x 1¼ LG./DET.

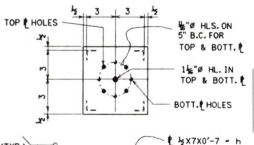

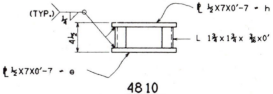

4810

2 – REQ'D. FOR B.B. TRAPS
FURNISH 1 BOLT 1"ø x 2" LG.

Figure 10.3 Several detail drawings on one sheet. (*Computer-generated image*) (Courtesy of American Electric Power Service Corp.)

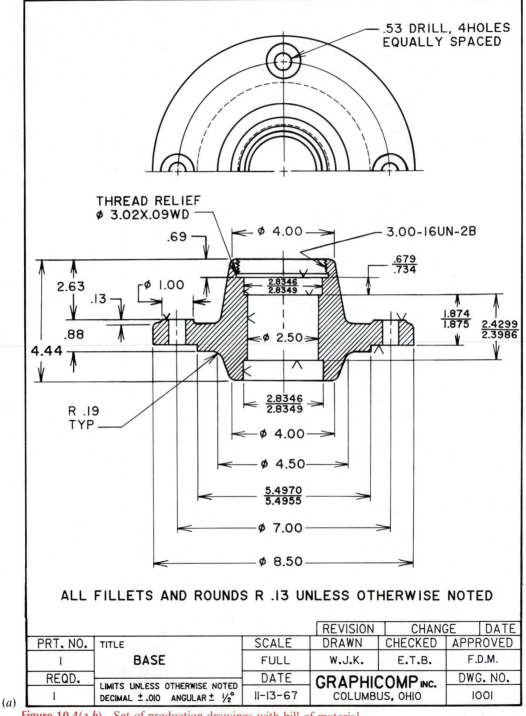

Figure 10.4(*a-h*) Set of production drawings with bill of material.
(Courtesy of OSU, Department of Engineering Graphics)

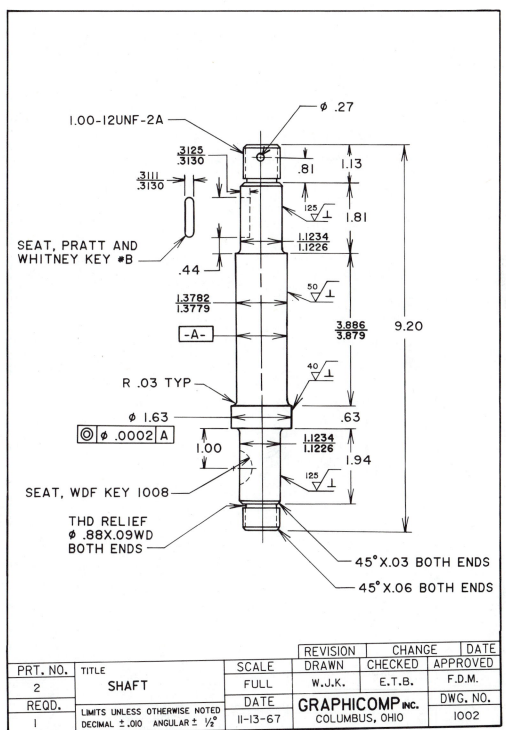

(b)

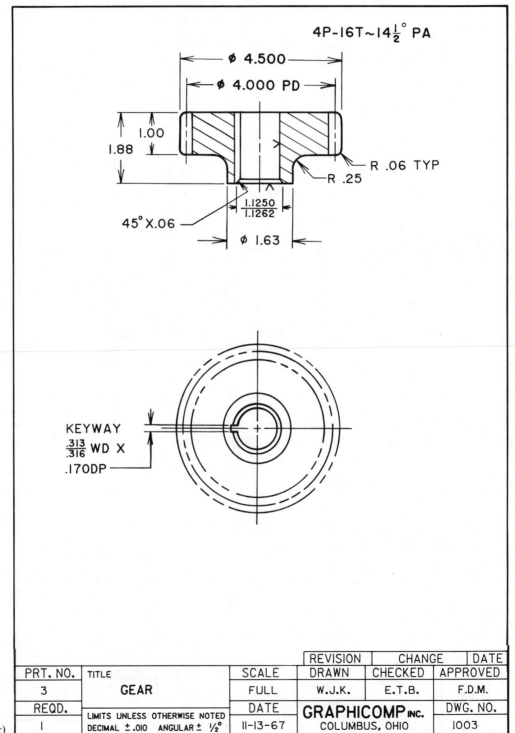

$4P-16T \sim 14\frac{1}{2}° PA$

$\phi 4.500$

$\phi 4.000$ PD

1.00

1.88

R .06 TYP

R .25

$\frac{1.1250}{1.1262}$

45° X.06

$\phi 1.63$

KEYWAY
$\frac{.313}{.316}$ WD X
.170DP

PRT. NO.	TITLE		REVISION	CHANGE	DATE
3	**GEAR**	SCALE	DRAWN	CHECKED	APPROVED
REQD.		FULL	W.J.K.	E.T.B.	F.D.M.
1	LIMITS UNLESS OTHERWISE NOTED DECIMAL ±.010 ANGULAR ± ½°	DATE	**GRAPHICOMP** INC.		DWG. NO.
		11-13-67	COLUMBUS, OHIO		1003

(c)

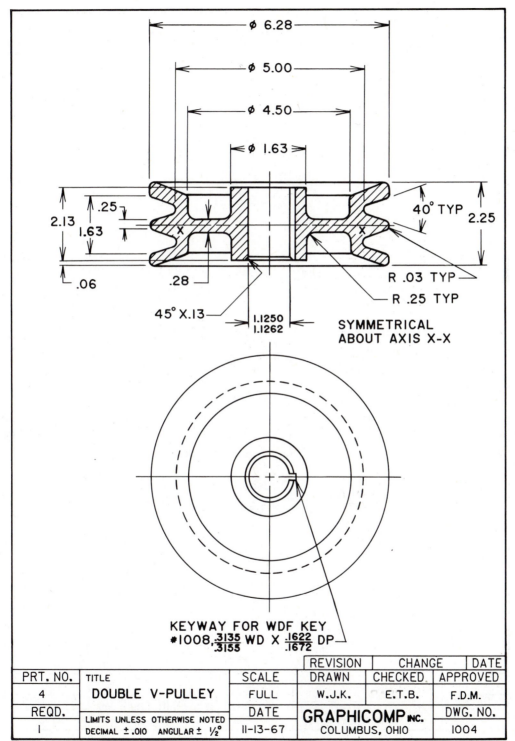

(d)

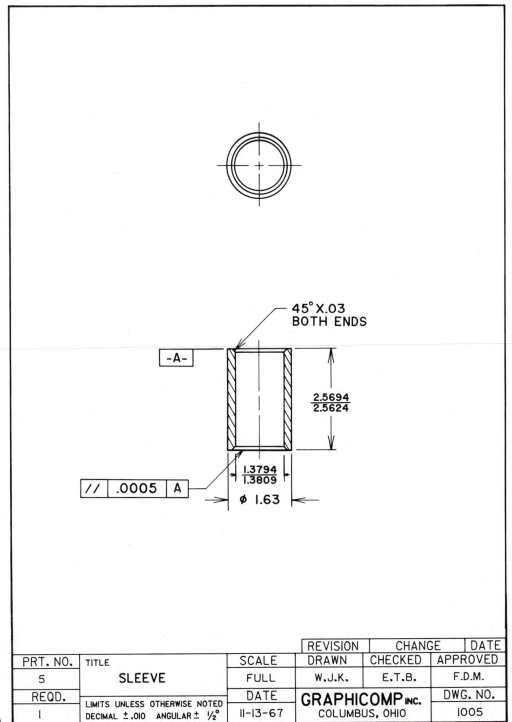

45° X .03
BOTH ENDS

-A-

2.5694
2.5624

// | .0005 | A

1.3794
1.3809

∅ 1.63

PRT. NO.	TITLE		SCALE	REVISION	CHANGE	DATE
5	SLEEVE		FULL	DRAWN	CHECKED	APPROVED
REQD.			DATE	W.J.K.	E.T.B.	F.D.M.
1	LIMITS UNLESS OTHERWISE NOTED DECIMAL ± .010 ANGULAR ± ½°		11-13-67	GRAPHICOMP INC. COLUMBUS, OHIO	DWG. NO. 1005	

(e)

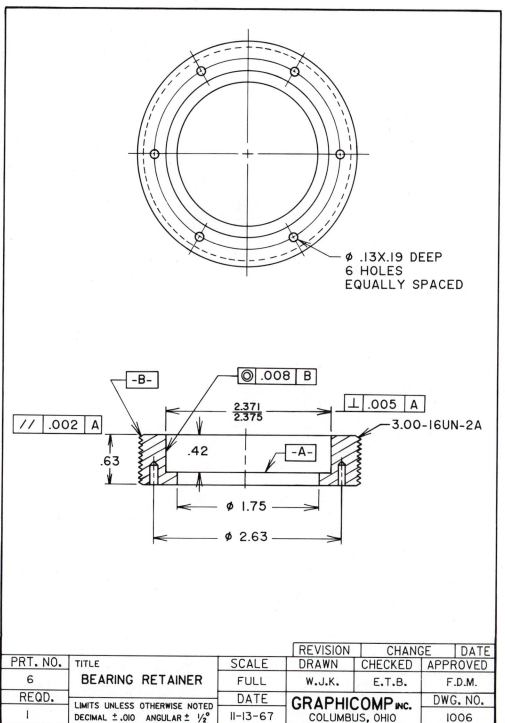

Ø .13X.19 DEEP
6 HOLES
EQUALLY SPACED

PRT. NO.	TITLE		SCALE	REVISION	CHANGE	DATE
6	BEARING RETAINER		FULL	DRAWN	CHECKED	APPROVED
REQD.			DATE	W.J.K.	E.T.B.	F.D.M.
1	LIMITS UNLESS OTHERWISE NOTED DECIMAL ± .010 ANGULAR ± ½°		11-13-67	GRAPHICOMP INC. COLUMBUS, OHIO		DWG. NO. 1006

(f)

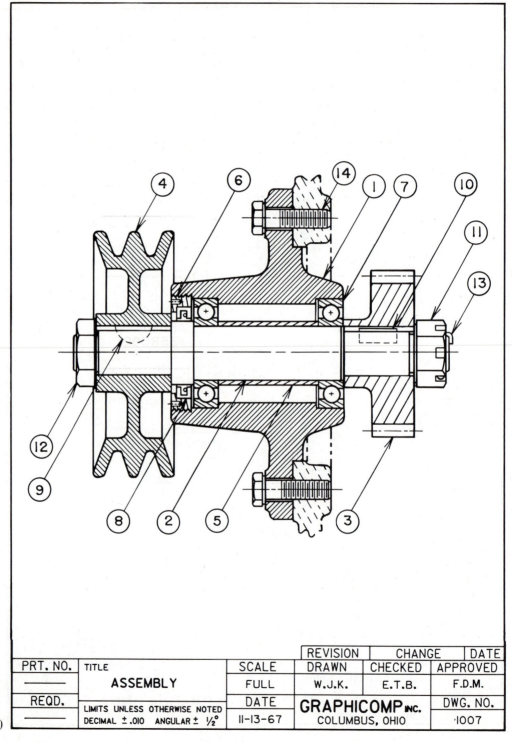

PRT. NO.	TITLE	SCALE	REVISION		CHANGE	DATE
————	**ASSEMBLY**	FULL	DRAWN	CHECKED		APPROVED
REQD.		DATE	W.J.K.	E.T.B.		F.D.M.
————	LIMITS UNLESS OTHERWISE NOTED DECIMAL ±.010 ANGULAR ± ½°	11-13-67	**GRAPHICOMP** INC. COLUMBUS, OHIO		DWG. NO.	1007

(g)

BILL OF MATERIAL

PART NO	NAME	MTL	REQD	NOTES
1	BASE	CI-CL25	1	
2	SHAFT	1030STL	1	
3	GEAR	3120STL	1	
4	DOUBLE V-PULLEY	85 AL	1	
5	SLEEVE	1020STL	1	
6	BEARING RETAINER	1020STL	1	
7	BEARING	——	2	AFBMA #35BCO2
8	GEAR SEAL	——	1	PERFECT OIL SEAL #235120
9	WOODRUFF KEY	——	1	#1008
10	PRATT 8 WHITNEY KEY	——	1	#B
11	HEX SLOTTED NUT	——	1	1.00-12NF-2B SEM-FN REG
12	HEX JAM NUT	——	1	1.00-12NF-2B SEM-FN REG
13	COTTER PIN	——	1	.25 X 2.00
14	HEX HD CAP SCR	——	4	.50-13UNC-2Ax2LG. SEMI-FIN

PRT. NO.	TITLE	SCALE	REVISION	CHANGE	DATE
———	**BILL OF MATERIAL**	FULL	DRAWN	CHECKED	APPROVED
REQD.		DATE	W.J.K.	E.T.B.	F.D.M.
———	LIMITS UNLESS OTHERWISE NOTED DECIMAL ± .010 ANGULAR ± ½°	11-13-67	**GRAPHICOMP** INC. COLUMBUS, OHIO		DWG. NO. 1008

(h)

may require several sets of subassembly drawings in order for the product to be manufactured and assembled.

10.2 APPLICATIONS OF PRODUCTION DRAWINGS

Production drawings are used in the design process. The first step usually includes sketches (freehand or computer generated) that describe the concept in general and provide some size information. Purchased parts to be used may be noted on sketches. From the sketches, designers prepare a layout drawing that shows the relationships of the parts and defines key dimensions. After the layout is approved by the responsible engineer, designers begin the detail drawings of parts that will meet the functional needs of the device and fit the key dimensions.

The principal reason for making assembly drawings is to show how pieces fit together. An assembly drawing is made by putting the detailed parts together on paper or on a computer screen to learn if the parts fit properly into the assembly. This is an important stage in the design process because the parts must fit together to be able to function properly. They may be used in the shop to guide the assembly process. Drawings of this type are also used for maintenance purposes. They are a great tool for the service repair person since they help the service person identify parts that need repair or replacement. Assembly drawings help identify the individual items within the assembly by part number to speed up the process of locating a replacement.

Production drawings affect many people. Once these drawings are prepared, different professional people will refer to them during the manufacturing and assembly process. The components and assemblies will be analyzed for strength, producibility, and cost. Also, you need to remember that the person who designed the product or the person who made the drawings will not be present to answer questions. For this reason, standard graphic and written descriptions must be given on the drawings to ensure that all parties who see the drawings will interpret the information in the same way.

Pictorial assembly and **exploded views** are often used in technical manuals and sales literature. These types of drawings are three dimensional and make it easier for a person not familiar with a device to understand the technical information.

10.3 TYPES OF PRODUCTION DRAWINGS

There are several types of drawings that can be considered as production drawings. You have read about details, assemblies, and pictorial drawings. More information about each type is given in the following paragraphs.

10.3.1   Orthographic Drawings

Orthographic drawings are very commonly used as production drawings. These are the easiest to dimension and are easy to interpret for the person trained to read technical drawings. You learned about orthographic drawings in Chapter 4. By now, you should be proficient in making and reading them.

An orthographic detail drawing may contain a single part or several parts. If threads are included on a detail drawing, they are usually shown with simplified symbols and totally described in the thread specification. If assembly drawings are sectioned, the section line patterns for each part will be the same, but adjacent parts will have a pattern of a different direction or spacing to clearly identify each part, as shown in Figure 10.4a-h. You learned about this in Chapter 6. If threaded parts that have been detailed or threaded fasteners are included on an assembly drawing, they are usually shown with schematic symbols. The threads on detailed parts will have been described on the detail drawings; threaded fasteners will be described in the bill of material.

A tabular drawing is a drawing of a part that includes dimensions identified by letter. The viewer refers to an adjoining table to read the actual dimensions. You first learned about these drawings in Chapter 7 (Figure 10.5). Standard drawings are drawings of parts an organization uses frequently. They are drawn once, and copies are included in any set of production drawings where they are needed (Figure 10.6).

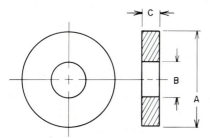

NOMINAL	A	B	C
.250	.625	.281	.065
.375	.812	.406	.065
.500	1.062	.531	.095
.625	1.312	.656	.095
.750	1.469	.812	.134
.875	1.750	.938	.134
1.000	2.000	1.062	.134
1.250	2.500	1.375	.165
1.375	2.750	1.500	.165

Figure 10.5   Tabular detail drawing.

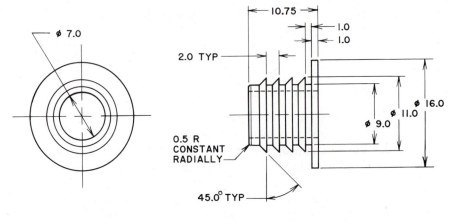

Figure 10.6 Standard detail drawing of often-used part. (*Computer-generated image*) (Courtesy of Ford Motor Company.)

10.3.2 Pictorial Drawings and Diagrams

Exterior assembly drawings show several parts assembled together. These drawings show only the surfaces and edges that are visible to the eye and exclude hidden surfaces and edges. Pictorial representations are easy for anyone to visualize. These drawings may or may not be dimensioned (Figure 10.7).

Sectioned pictorial drawings and diagrams are similar to exterior assemblies since they show many parts assembled together (Figure 10.8). The difference is that some portion of the assembly has been "removed" so the viewer can see certain interior features of the assembly. These drawings usually employ section lines to indicate the surfaces actually cut by the cutting plane. They may be full sections, half sections, or broken-out sections, as described in Chapter 6 (Figure 10.9).

Exploded pictorial drawings represent several parts assembled according to the axes of their assembly. However, the parts are not shown assembled but are moved apart along the principal axes of the product. This type of drawing typically is not dimensioned but is very helpful for assembly purposes on the production line. The worker can easily visualize the order and arrangement of assembly of parts that make up an assembly (Figure 10.10).

An installation assembly shows how one assembly attaches to another assembly. These drawings may later be used for maintenance purposes when disassembly is required (Figure 10.11).

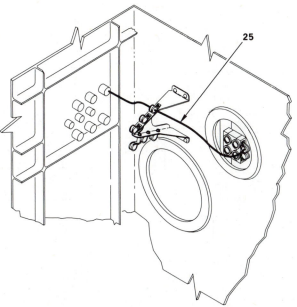

25

Figure 10.7 Exterior assembly shows outside of product in easy-to-read pictorial form. (*Computer-generated image*)

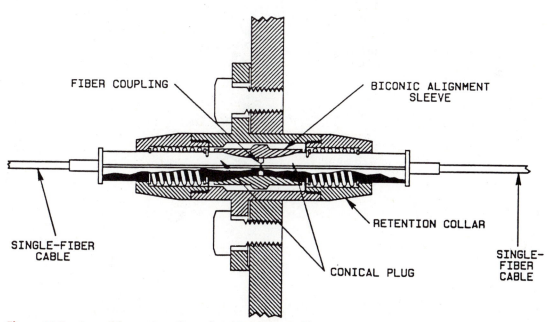

FIBER COUPLING

BICONIC ALIGNMENT SLEEVE

RETENTION COLLAR

SINGLE-FIBER CABLE

CONICAL PLUG

SINGLE-FIBER CABLE

Figure 10.8 Assembly section: Cross hatching is same for any one part but different for adjacent parts. (*Computer-generated image*) (Courtesy of AT&T)

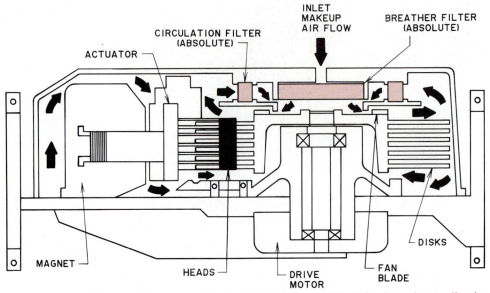

Figure 10.9 Sectional pictorial assembly helps in understanding inner workings of product. (*Computer-generated image*) (Courtesy of AT&T)

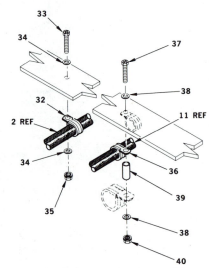

Figure 10.10 Exploded assembly shows how parts relate along principal axes of product and its components. (*Computer-generated image*)

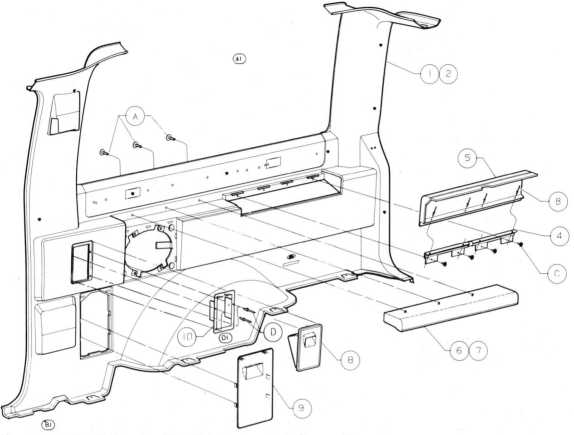

Figure 10.11 Installation assembly tells how components fit assembly. (*Computer-generated image*) (Courtesy of Ford Motor Company)

10.3.3 Drawings for Other Disciplines

The drawings described as required for production in the preceding paragraphs apply primarily to the manufacture of mechanical devices (appliances, machinery, and vehicles). Some engineering disciplines are not involved in the production of mechanical devices: Civil engineers design highways, bridges, and waterways; chemical engineers create process systems; architects design buildings; and electrical engineers often deal only with circuit design and electrical components (although these components may end up in a mechanical assembly). Each of these disciplines may have a slightly different system for naming various types of drawings or may need other types of drawings for production.

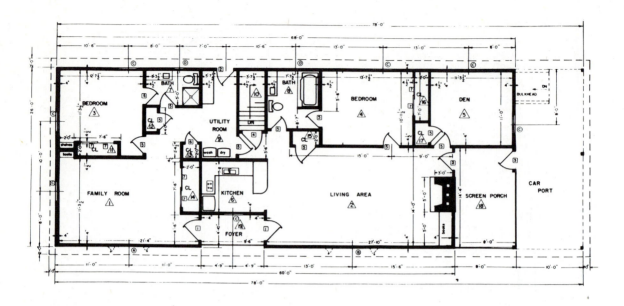

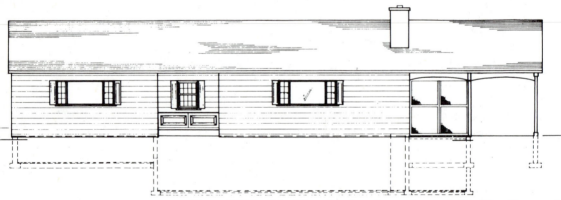

Figure 10.12 A building drawing shows plan and elevation views and may show section views. (Courtesy of Timothy C. Whiting)

Architects' drawings are commonly assembly drawings. Architects make plan-and-elevation drawings of buildings (Figure 10.12). These are really assemblies of the parts that make up the structure. They also prepare section drawings where they open up a wall to show how the components fit together inside the wall. These section drawings may not be crosshatched. Architects also make detail drawings of special parts that need to be built to fit the structure.

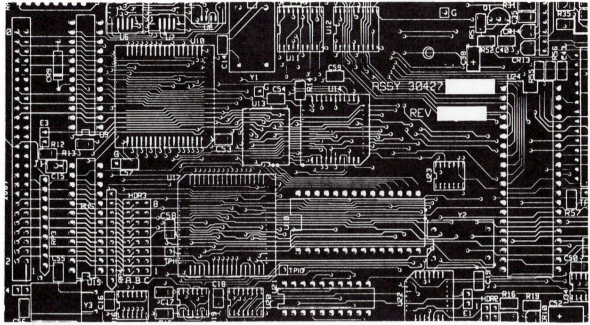

Figure 10.13 Printed circuit drawing for electronic component was produced with computer graphics and printed. (*Computer-generated image*)

Electrical designers may prepare schematic wiring plans showing where wires are run, leaving the exact locations and distances to the field electricians. On the other hand, small electronic components such as circuit boards may be detailed very carefully to scales of 100:1 or larger to be certain that elements of the circuit are the right size and in the correct relationship (Figure 10.13).

Civil and structural engineers often design assemblies such as steel trusses and leave it to the fabricator of the structure to prepare the detail drawings of the individual pieces of steel that go into the assembly. This is often done because different fabricators have different equipment for production. Each organization will design to take advantage of its own facilities.

Chemical engineers often make schematic assemblies of a process concept, including the major components, piping, and instruments for process control. These are called *process and instrumentation diagrams* (Figure 10.14). Chemical engineers, nuclear engineers, and others assembling large quantities of pipe and large components often have three-dimensional assemblies made. The finished job is built to scale with miniature components and wires

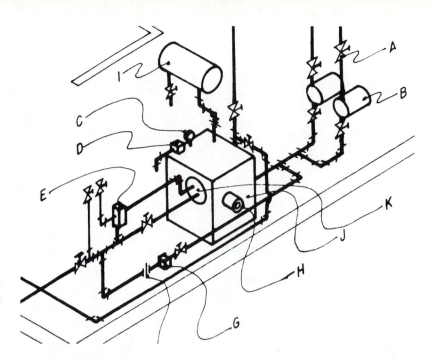

Figure 10.14 Chemical engineers prepare process and instrumentation diagrams such as this drawing. (Courtesy of Timothy C. Whiting)

Figure 10.15 Three-dimensional model of part of chemical plant. (Courtesy of O.S.U. Engineering Publications)

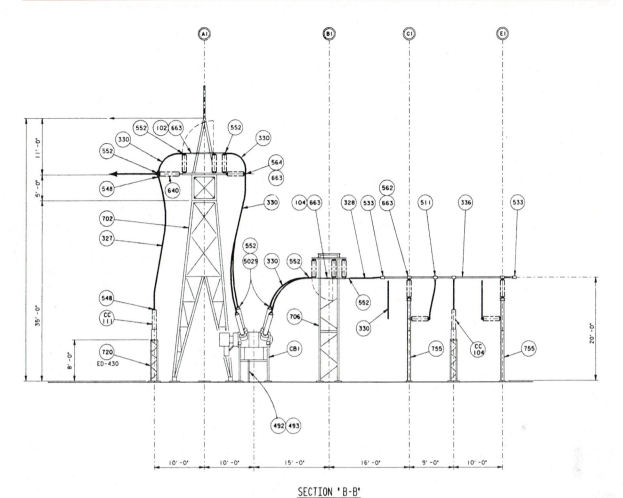

SECTION 'B-B'

Figure 10.16 Callouts on body of drawing referenced to bill of material. (*Computer-generated image*) (Courtesy of American Electric Power Service Corp.)

or rods for the piping. Sometimes this model becomes the principal assembly, and the construction workers obtain their information by "reading" the model (Figure 10.15).

10.4 *CALLOUTS*

Callouts are identifiers placed at the end of a leader line that direct the eye of the viewer from the part to the identifying note. The callouts identify a part by name, number, or letter and number. These help the viewer relate that particular part to a list of parts or details, as in a bill of material. Figure 10.16 shows compounds identified by number.

Callouts are placed around the perimeter of the assembly and are usually identified in sequence by letter or number. These identifiers may be placed in clockwise order or may be assigned according to the importance or size of the parts.

10.5 *METRIC IDENTIFICATION*

When the dimensions or notes of a production drawing are given in the metric system, the word millimeters should appear within a box in an open area of the drawing or near the title block.

If metric dimensions are given on the drawing, the standard unit of measure for small-parts production is the millimeter. The dimensions given in millimeters need not be followed with mm since this is understood if the note "millimeters" were labeled within the box (Figure 10.17). Conversely, if your organization makes both metric and inch drawings, you should identify

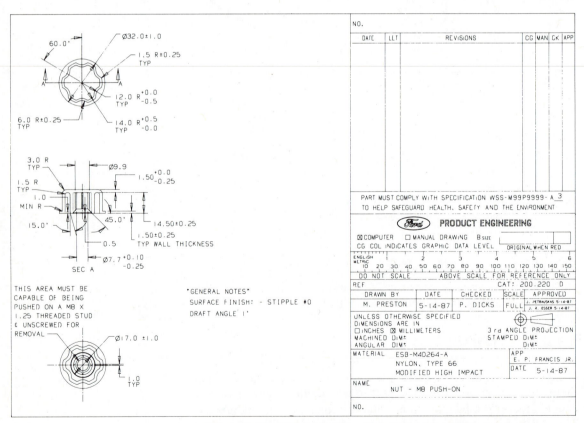

Figure 10.17 Metric drawing dimensioned in millimeters and identified in title block. (*Computer-generated image*) (Courtesy of Ford Motor Company)

the English system drawings with "inch" in a box near the title block.

10.6 *PAPER SIZES, FOLDS, AND BORDERS*

There are three basic sizes from which standard drawing papers are derived. They are 9×12 and $8\frac{1}{2} \times 11$ in. in the English system and 210×297 mm in the metric system.

The most common sizes in the English measure are the multiples of $8\frac{1}{2} \times 11$. The A size sheet can be copied in an ordinary office copy machine using standard $8\frac{1}{2} \times 11$ paper. Standard B, C, D, and E sheets are the multiples of $8\frac{1}{2} \times 11$.

An A size sheet will fit into standard file folders without covering the raised identification tab. The larger size sheets of the 9×12 multiples will not fit into standard office file folders and will become torn, or "dog eared," because the edges of the paper are exposed to wear. These and larger size sheets are usually filed in shallow, flat file drawers.

There is a standard method for folding the larger letter size sheets. This procedure provides for the exposure of the identifying information (such as the drawing number) on the drawing in a clearly visible location on the outside of the folded sheet (Figure 10.18). Some drawing sheets have fold marks printed along the

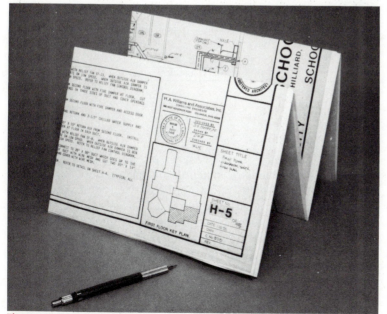

Figure 10.18 Copies of large drawings are folded with "accordion" folds so drawing number is visible. (Courtesy of McDonald-Cassel & Bassett, Inc.)

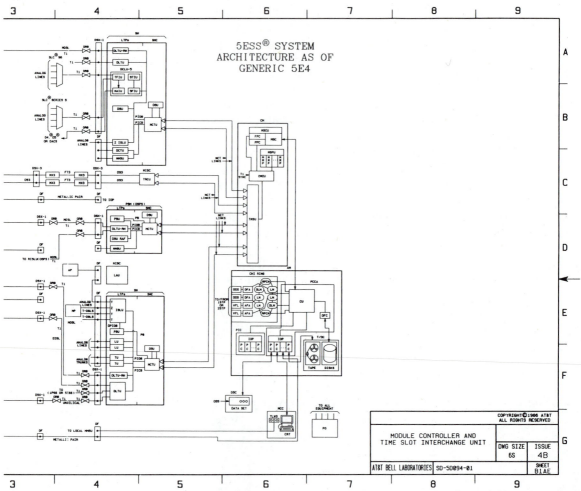

Figure 10.19 Zones help locate areas on drawing quickly. (*Computer-generated image*) (Courtesy of AT&T)

edge of the sheet. Industry standards are specific about how larger sheets should be folded. The accordion style fold makes it easy to locate particular identifying information and allows the sheet to be easily opened for reading, similar to reading a road map.

Border lines may be drawn or printed on standard size drawing sheets. These improve the appearance of the sheet for presentation purposes. Also, the borders are printed a short distance from the edge of the paper to prevent the drawing from extending to the edge of the paper. Since well-used drawings become damaged on the edges, the border space provides a safety area to prevent the loss of important information.

Letters or numbers are often printed to identify **zones.** If the lines between zones were extended across the working area of the paper in a coordinate grid, then identification of various areas of the drawing can be made more quickly and accurately. You can relate to the zones on a road map when visualizing the grid on a drawing. The zoned grid is a particularly important aid when talking on the telephone. The speaker can help orient the listener to a specific area of the drawing quickly if proper use of the zone identifiers is made (Figure 10.19).

10.7 TITLE BLOCKS

The title block is usually found at the bottom of the drawing. The title block is an area within a box that contains a concentrated amount of information. Although title blocks vary according to the information required by the industry, some items are standard (Figure 10.20). Items usually included within title blocks are as follows:

Drawing number	Identifying number of the drawing. The numerals are the largest numerals in the lettering block.
Sheet number	Sheet _____ of _____ . This notation means the drawing sheet number followed by the total number of sheets in the entire set of plans. It is very important to know the total number of drawings in order to establish that you have a complete set of drawings.
Title	Name of the part or assembly. This identifier usually consists of larger letters than other items in the lettering block.
Company name	Name of the company for whom the drawing was created.
Company logo	For quick identification, the company logo or other simplified graphic may appear near the company name.
Scale	Scale at which the drawing was made. This shows the relationship of the size of the drawing to the size of the actual part or assembly.
Date	Date the drawing was created. This date may be very important for establishing the date the work was created for patent or other protection purposes.

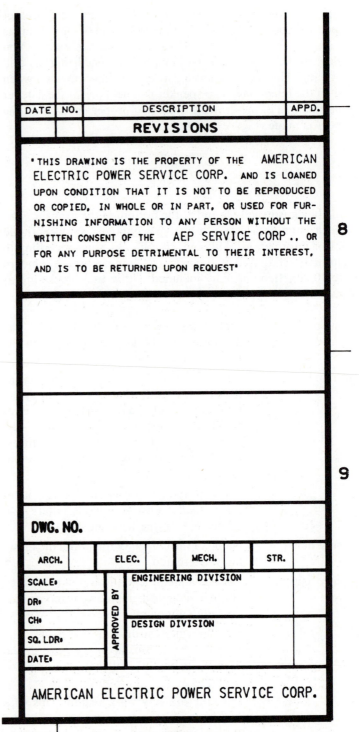

Figure 10.20   Typical title block.
(*Computer-generated image*)
(Courtesy of American Electric Power
Service Corp.)

Tolerances	Standard tolerances for the company or industry. They may be entered in decimals, fractions, millimeters, or degrees. These tolerances usually state a range of acceptable limits that can be applied to dimensions on the drawing. Tolerances are given as fractional, decimal (to a given number of places), or angular for angles and bends. This is described in detail in Chapter 7.
Drafter	Name or initials of the person who created the drawing. With this entry, questions or comments related to that drawing can be addressed to the person who created the drawing.
Checker	Name or initials of the person who checked the drawing for drafting accuracy and compliance with standards.
Supervisor	Name or initials of the supervisor who approved the drawing. The supervisor is the person ultimately responsible for the quality and accuracy of the drawing.

Optional items that might appear in the lettering block could include such things as the following:

Weight	This may be very important for shipping reasons. This will allow someone to calculate the gross weight of a single part or the weight of a total shipment as well as the total weight of an assembly.
Material	If only a single part is detailed on the drawing, then it is appropriate to identify the material used to make the part. However, if there are details of several parts on the same sheet, especially if the parts are made of different materials, then the material should be listed under the part being detailed, not in the lettering block. When this condition arises, the term *as noted* should be lettered in the lettering block box reserved for identifying the material.
Number required	Number of parts to be produced.
Surface	Surface roughness, texture, or other such characteristics as may be required.

Hardness	Hardness of the material may be listed. This value is given for a standard hardness test such as a Rockwell hardness test.
Heat treatment	If the part is to be heat treated, there may be an entry to note the type or degree of heat treatment required.
Sheet size	Seldom will it be necessary to identify the size of the drawing sheet, but given within some title blocks is a place to check which corresponds to the standard letter size sheets A–E.

10.8 REVISIONS

All revisions made to an original drawing must be recorded. For this purpose a revision chart or schedule is shown on the drawing (Figure 10.21). The date is important for legal or patent purposes and to inform any reader of the time of the change. A brief description of the change is given as well as the zone, if on a large drawing. The changes must also be made on the specific parts or

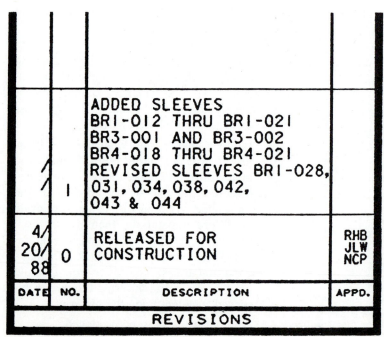

Figure 10.21 Revision schedule. (*Computer generated image*) (Courtesy of American Electric Power Service Corp.)

specifications affected by the change. A change letter or other code is noted beside the change to identify it.

Since information concerning revisions is relatively standard, it is common to list the revisions near the title block. If the revision schedule starts at the title block, notice that revision number 1 is the closest to the title block. This allows for additional revisions to be added to the "open-ended" schedule.

10.9 PARTS LIST

A bill of material may be listed along the side of an assembly drawing or on a separate sheet. This list typically includes all items necessary to make an assembly. This includes standard items such as bolts, nuts, and washers and other purchased parts. The listing of these standard items eliminates the necessity for detailing many standard parts (Figure 10.22).

The parts list typically includes but is not limited to the part number, part name, number required, and a place to enter sizes or other details.

Parts lists, revisions, and other notes are primarily words rather

BILL OF MATERIAL

| NO. REQ'D. | | MARK | DESCRIPTION | LENGTH | | WEIGHT |
ITEMS SHIPPED	SHOP ASSEMBLY			FT.	IN.	
1		4801	8'-0 HIGH FRAME			
	4	4801A	WELDED ASSEMBLY			
	4	a	℄ 1½ x 4	0	4	
	4	b	L 2½ x 2½ x $\frac{3}{16}$	7	11	
	4	4801B	WELDED ASSEMBLY			
	4	c	L 2½ x 2½ x ¼	1	3½	
	20	4801C	L 1¾ x 1¾ x $\frac{3}{16}$	1	10$\frac{3}{16}$	
1		4802	10'-0 HIGH FRAME			
	4	4802A	WELDED ASSEMBLY			

Figure 10.22 Typical bill of material (parts list). (*Computer-generated image*). (Courtesy of American Electric Power Service Corp.)

than lines and symbols. There may be pages of description defining how various operations will be performed, how testing will be done, packaging and shipping requirements, and other legal requirements. These descriptions are called specifications. Because of the complexity of specifications and the time needed to prepare them, word-processing programs are commonly used on the computer to help generate specifications. Also, additions and deletions can be made to the master set of specifications very quickly using word-processing programs.

The specifications will normally be produced on $8\frac{1}{2} \times 11$ paper. Notes, bills of material, and revision schedules appear on the face of the drawings, which may be much larger than $8\frac{1}{2} \times 11$. Some organizations use wide-carriage printers that allow the information to be printed directly on the drawing. In other cases the information is printed on adhesive-coated transparent sheets that can be attached to the drawing.

10.10 *GENERAL NOTES*

Some general notes may be appropriate. These should be neatly lettered and placed near but outside the lettering block. Occasionally, these notes are lettered at an angle to attract the viewer's attention. Also, the size of the lettering is generally larger than other lettering on the drawing. They might include such notes as:

All dimensions in millimeters

All undimensioned radii = _____

FAO (finish all over)

Not approved

For estimate only

For quotation purposes only

Not to scale

Sometimes notes are placed for security purposes. This is particularly appropriate on government drawings. Examples of such security notes might include:

Top secret

Security clearance required

Restricted drawing

''In house'' drawing only

10.11 *COMPUTER-GENERATED PRODUCTION DRAWINGS*

There are many advantages to making production drawings on the computer using available graphics packages. For example, if the graphics package has the capacity to generate a data base, then after some views have been entered, others can be generated without having to draw them individually. The points can simply be generated from the data base.

Various types of views may be generated from the data base. For instance, once the points that define a building have been entered, orthographic views and perspective pictorials can be generated from this data base. Perspective pictorials can be generated after noting such variables as the distance the eye is from the object, the height of the eye above ground level, and whether a one-, two-, or three-point type of perspective drawing is desired (Figures 10.23 and 10.24). The same techniques apply to mechanical parts: A pictorial drawing can be generated from the data base created by preparing the orthographic drawings.

The same data base can be used in auxiliary programs that do stress analysis (finite-element analysis programs). A few large manufacturing organizations also analyze the part data base for

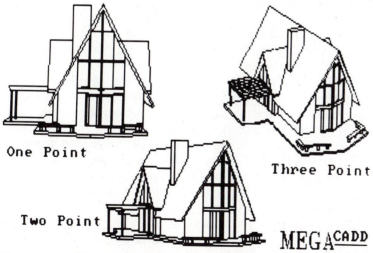

Figure 10.23 Computer-plotted perspective generated from three-dimensional data base. (*Computer generated image*) (Courtesy of MEGA-CADD)

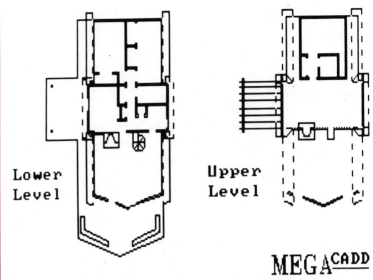

Figure 10.24 Elevation views from same data base. (*Computer-generated image*) (Courtesy of MEGA-CADD)

producibility and cost to manufacture. One manufacturer has publicized a program that performs economic optimization—changing material thicknesses and contours to give a part maximum performance using minimum material.

Wire frame drawings are the traditional pictorials generated from the computer data base (Figure 10.25). These drawings indicate the edges, points, and surfaces that make up the object. You generally see the lines for all the surfaces, both front and back with no regard for hidden lines. These drawings may be further refined by solid modeling. However, some programs allow wire frame diagrams to be viewed showing only visible surfaces and edges (Figure 10.26).

Solid modeling programs use the data base information to produce a diagram that displays the object as a three-dimensional pictorial clearly depicting the visible surfaces of the object (Figure 10.27). Hidden surfaces are seldom pictured. Solid modeling is very effective when displaying industrial parts and assemblies. The object can be examined from many points of view, and clearances and relationships among adjacent parts can be checked.

The architectural application of solid modeling is illustrated by generating the visible surfaces of a building from the data base (Figure 10.28). Relationships with adjacent structures as viewed from different points can also be evaluated.

Zoomed portions of detail drawings allows the CADD operator to produce large images of detailed areas of production drawings. This can be very effective when dimensions, notes, or details

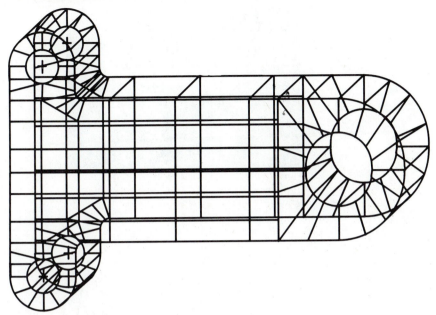

Figure 10.25 Ordinary wire frame drawing from computer data base.
(*Computer-generated image*)

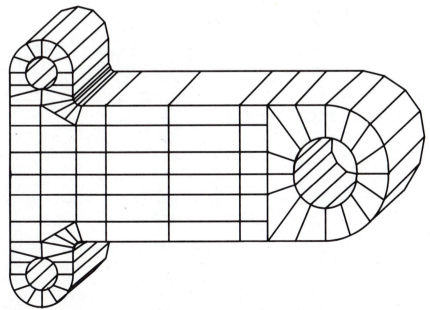

Figure 10.26 Wire frame drawing with hidden lines deleted.
(*Computer-generated image*)

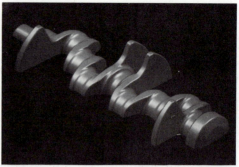

Figure 10.27 Solid model from computer data base. (*Computer generated image*) (Courtesy of O.S.U. Engineering Publications)

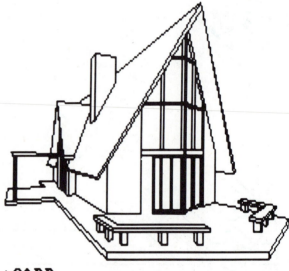

Figure 10.28 Solid model for building from architectural data base. (*Computer-generated image*) (Courtesy of MEGA-CADD)

must be added to a complicated portion of the drawing. Zoomed images are also appropriate for showing enlarged views of complex sectioned details (Figure 10.29).

Layered drawings were discussed in Chapter 4. They have many advantages and can easily be obtained from the data base (especially if you plan ahead and assign information to different layers). For instance, plans for a chemical processing plant might be developed with drawings with limited information for certain contractors. A floor plan with only the electrical details added

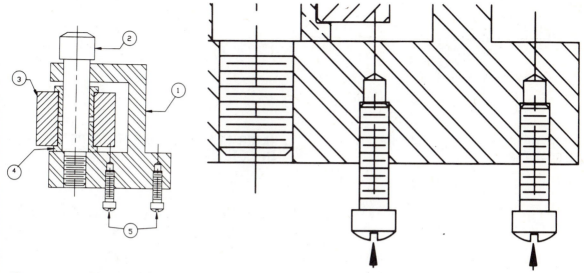

Figure 10.29   Complex drawing magnified by zooming. (*Computer-generated image*)

would be appropriate for the electrical contractors (Figure 10.30). A plan with the heating and air-conditioning ducts would be appropriate for the heating, ventilation, and air-conditioning contractors (Figure 10.31). Another plan with the location of the equipment superimposed on the floor plan would be appropriate for the general contractor installing the equipment (Figure 10.32). Once the data base has been entered, it is easy to develop drawings with only the required information for particular applications.

Large companies that must provide many sets of drawings for the construction of facilities can save time and money by producing only the plans necessary for each of the bidders. In the past, entire sets of drawings would be provided to all parties. This was very expensive and time consuming. However, once a computer data base has been entered, then it is a relatively simple process to rearrange the data in a form to produce only the needed information for specific bidders.

The use of the computer to manipulate data bases can also be useful in the transmittal of information over the telephone lines. Information for drawings as well as written material can be transmitted from one location to another by using a telephone modem (Figure 10.33).

The same data base used to produce drawings can be used to drive computer-aided manufacturing (CAM) machines such as milling machines, lathes, or profilers. There are various acronyms for this type of integration. It has been referred to as CAD/CAM

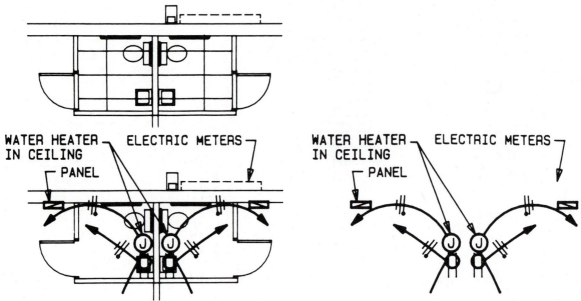

Figure 10.30 Layered drawing: Electrical plan only. (*Computer-generated image*) (Courtesy of Consulting Engineers, Mechanical/Electrical/Energy Consultants, Inc.)

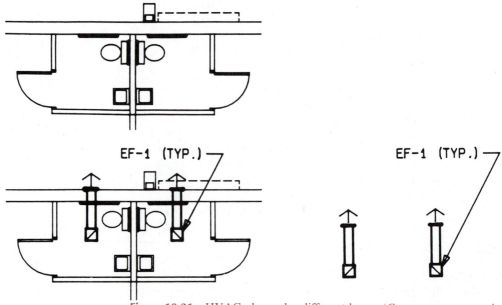

Figure 10.31 HVAC plan only; different layer. (*Computer-generated image*) (Courtesy of Consulting Engineers, Mechanical/Electrical/Energy Consultants, Inc.)

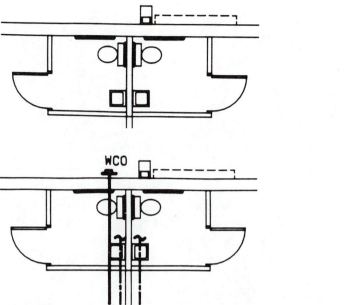

Figure 10.32 Equipment plan; only equipment shows on this layer. (*Computer-generated image*) (Courtesy of Consulting Engineers, Mechanical/Electrical/Energy Consultants, Inc.)

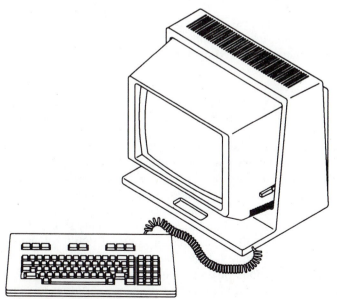

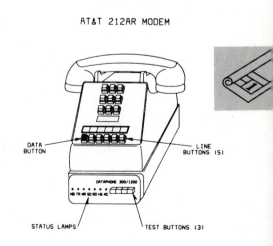

Figure 10.33 Computer work station with Modem for telephone transmission of data. (*Computer-generated image*.) (Courtesy of AT&T)

Figure 10.34 Machine tool with integrated numerical control. (Courtesy of O.S.U. Engineering Publications)

because of the progress achieved in getting all types of manufacturing facilities to communicate electronically. It has also been called computer-integrated manufacturing (CIM) (Figure 10.34).

10.12 *DRAWING MEDIA*

In our contemporary industrial world, two media are used almost exclusively for making production drawings. They are vellum and film. Vellum is a translucent paper that may be used to make large copies. It is semitransparent, which allows for tracing; this can save the drafter a considerable amount of time.

Film is nearly transparent and is very dimensionally stable. This means the drawings will not expand or contract due to changes in temperature and humidity. It also ages very slowly (deteriorates slowly over long periods of time). This can be important for archival storage (many years) of master drawings. Another advantage of film is that it is so tough that it may be run through a copy machine thousands of times with little wear. If vellum is used for a large number of copies, the wear and tear on the vellum will degrade the quality of the copies.

Film masters are dimensionally stable. This stability makes film quite attractive for those industries making drawings of electronic circuitry that will be reduced photographically for the making of printed circuits. Because of the accuracy required, the complexity of the drawings, and the need for excellent line quality, these drawings can best be produced using a computer-driven plotter. For working purposes, some computer-generated prints can be printed on opaque paper.

Sepias are intermediate copies. They are brown line drawings on a semitranslucent paper or film base. You may draw on a sepia to alter an original, use it as a print, or make copies from it.

10.13 *REPRODUCTION OF DRAWINGS*

Original drawings never leave the engineering office. Only copies of the originals are sent to other professionals or the production or construction site. Because these original drawings must be used to make many copies, they must be very clear and have legible lettering. Generally speaking, if the original drawing is difficult to read, the copy will be worse.

Small drawings, details, or specifications made on $8\frac{1}{2} \times 11$-in. opaque paper may be copied on a common office copy machine. Drawings produced on vellum or film may also be copied on these office copy machines, but a backing sheet of white paper may have to be placed behind the vellum or film before the copy is made. Some of these machines will copy larger sheets, but most business copiers are limited to small sheets (Figure 10.35).

Copy machines are now readily available for copying larger sheets. These machines can easily handle D and E size master drawings (Figure 10.36).

Large drawing sheets have commonly been reproduced by the diazo process. This is a two-stage process involving exposure to light and then exposure to ammonia fumes. In the first stage, the vellum or film is placed on a sheet of copy paper that has been specially treated with diazo salts. When the two sheets are inserted into the diazo machine, the ultraviolet light passes through the master and "burns off" the salt crystals on the emulsion surface of the copy paper. The pencil or ink lines prevent the light from affecting the salts, so the salts remain under the areas covered by the lines.

In the second stage, the exposed copy paper is subjected to ammonia fumes. The areas exposed to the light will not be affected by the fumes, but the remaining salt crystals that were under the

Figure 10.35 Standard office copier accepts originals $8\frac{1}{2} \times 11$ in. and may take originals up to 11×17 in. (Courtesy of Xerox Corp.)

Figure 10.36 Copier used to copy large-size drawings. (Courtesy of Xerox Corp.)

lines of the master during the exposure stage will turn to a deep blue (or other color if special paper is ordered) (Figure 10.37).

Contemporary reproduction methods include many products and methods that add greater flexibility for the user. Clear film appliques are often applied to master drawings where repetitious shapes must appear on the drawing. This avoids the time-consuming process of redrawing common shapes. Appliqués are made on clear film and have a clear, light adhesive on the back.

Large architectural, engineering, and design firms now use overlay reproduction on complicated sets of drawings of major projects. The master drawings are actually drawn on separate layers of film. The film layers are aligned through a pin registration system. Each sheet is accurately punched with a series of up to seven holes that are spaced to fit a standard pin registration bar.

Once the main registration drawing has been completed (say, the floor plan wall layout), other professionals can each begin on a different overlay (electrical, plumbing, dimensions, etc.). This helps reduce drawing time. This process also makes each layer much less complicated, and thus changes can be made easily to any layer without disturbing the information that is now on another layer.

These layers can be easily copied on clear film with the lines on each layer being of a different color. These are used for design and checking purposes. When the set is complete, it will be sent to the printer, and the final copies will be printed in color by offset printing. One might first think this is the most expensive route to follow, but this process really saves a great amount of time and money on large projects. For instance, those bidding or contracting for a particular segment of a project (say, the electrical portion) only need a limited number of drawings. This process results in

Figure 10.37   Diazo process machine for reducing engineering drawings. (Courtesy of Bruning, Division of American International, Inc.)

more accurate and quicker reading of the set of drawings. This overlay process also aids in the bidding process by reducing time in cost estimation.

10.14 *FLOW OF INFORMATION*

You need to understand the role of production drawings in the manufacturing process. Many people's activities are affected by the information on the drawings, so they must communicate a very clear message. This message takes into account graphic, numeric, and written information. The flow of engineering information from the designer to the final product that is ready for the user may take several paths. Two possible paths within an organization are shown in Figure 10.38. Many variations of these paths are possible. Sometimes this information travels on paper and sometimes it may travel electronically. A delay in the flow because of misunderstanding, missing information, or unclear line work costs time and money.

10.15 *ACCOMPANYING WRITTEN SPECIFICATIONS*

If all the needed information cannot be placed on the detail or assembly drawings, then this additional information is supplied in written form. General standards for testing and quality control, types of finish, and packaging requirements are written in specifications rather than lettered directly on the drawing (Figure 10.39).

Assembly drawings related to the assembly at the job site (large machinery, conveyors, chemical process equipment, etc.) will require many pages of specifications to describe the needs for packaging, shipping, protection of parts, storage conditions, and procedures for assembling the parts together at the job site. Testing and checking procedures are also included in the written specifications. The specifications for a large-machine installation may be the size of several large city telephone books.

10.16 *SAMPLE PROBLEM: TABLE SAW FIXTURE*

Production drawings of assemblies show how parts fit together or are functionally related. Many production drawings are pictorial in nature. In Figure 10.40, the table saw fixture is shown in pictorial view. The relationship of the casting, the washer, and the bolt are shown in relationship to each other along the center line.

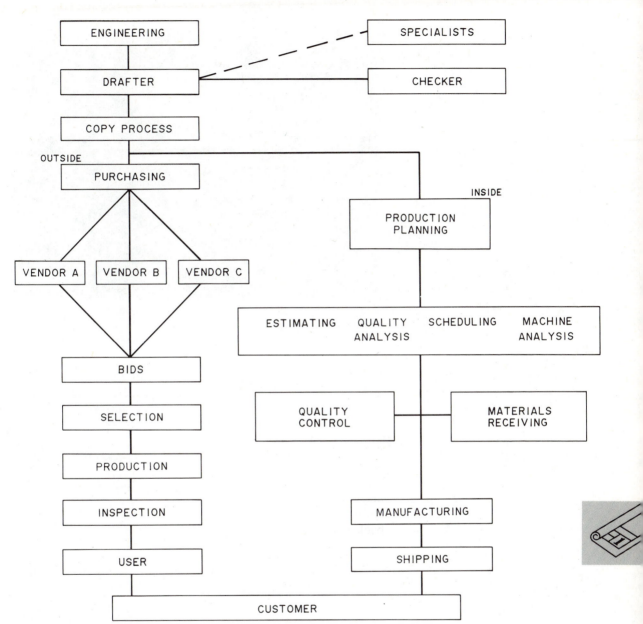

Figure 10.38 Flow of engineering information and production drawings may take several paths.

Figure 10.39 Specifications include engineering information too lengthy to include on drawings. (*Computer-generated image*) (Courtesy of American Electric Power Service Corp.)

NO.	NO REQ'D	NAME	MAT'L	DESCRIPTION
1	1	BODY	C.I.	
2	1	HX HD CAP SCREW	STL	.88-9UNC-2A X 1.5 LG
3	1	WASHER	STL	.88 STD WIDE

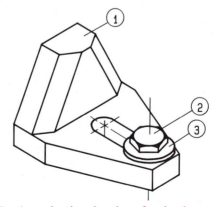

Figure 10.40 A production drawing of a simple assembly showing the relationship of parts along a center line.

10.17   *SUMMARY*

Production drawings carry information from the engineering office to the sites where it will be used to plan, buy, produce, and assemble the components and final form of useful devices. These drawings may also be used for maintenance and repair. The information may be transmitted on paper or electronically.

The two most common types of production drawings are detail drawings of individual parts and assembly drawings, which show how parts are installed relative to each other. Drawings may carry information in orthographic or pictorial form; they may be produced manually or with a computer. Computer-generated drawings create a data base that may be useful for producing other types of drawings, engineering analysis, or direct transmission to computer-aided manufacturing.

The notes, bills of material, title blocks, and accompanying specifications are important parts of the total set of production drawings. The total package needs to be clear, accurate, and correct in order to avoid delays, which cost time and money.

PROBLEMS

Problems 10.1–10.5

Problems for this chapter consist of several assembly problems. Any individual part from these assemblies may be drawn and detailed as a single unit or may be drawn in isometric as a pictorial view.

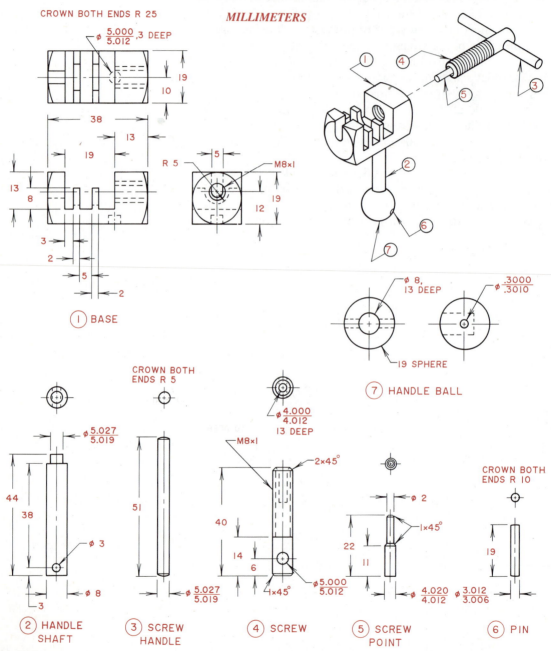

MILLIMETERS

CROWN BOTH ENDS R 25

$\phi \frac{5.000}{5.012}$, 3 DEEP

19
10
38
13
19
R 5
5
M8×1
13
8
19
12
3
2
5
2

① BASE

ϕ 8, 13 DEEP
$\phi \frac{.3000}{.3010}$
19 SPHERE

⑦ HANDLE BALL

$\phi \frac{5.027}{5.019}$
44
38
ϕ 3
ϕ 8
3

② HANDLE SHAFT

CROWN BOTH ENDS R 5
51
$\phi \frac{5.027}{5.019}$

③ SCREW HANDLE

$\phi \frac{4.000}{4.012}$ 13 DEEP
M8×1
2×45°
40
14
6
1×45°
$\phi \frac{5.000}{5.012}$

④ SCREW

ϕ 2
1×45°
22
11
$\phi \frac{4.020}{4.012}$

⑤ SCREW POINT

CROWN BOTH ENDS R 10
19
$\phi \frac{3.012}{3.006}$

⑥ PIN

Problem 10.1

The instructor may elect to assign the assemblies to be drawn as exploded views, assemblies, sectioned assemblies, or assembled pictorials (which may or may not be rendered).

INCHES

CHAIN RIVET
EXTRACTOR

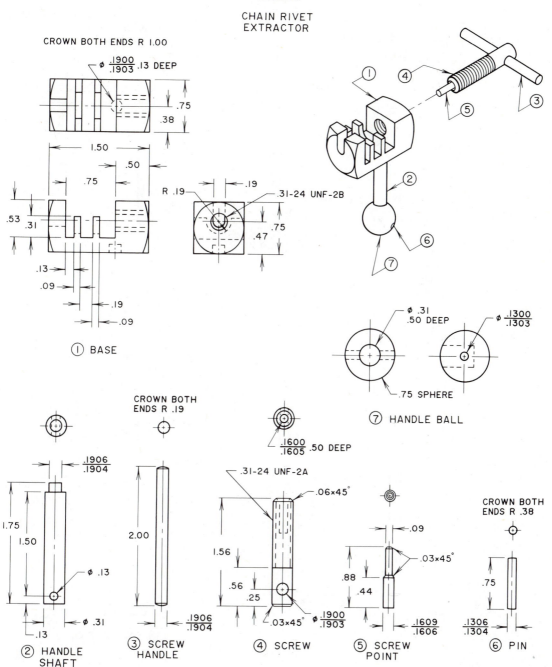

CROWN BOTH ENDS R 1.00

$\phi \frac{.1900}{.1903}$ .13 DEEP

.75
.38
1.50
.50
.75
R .19
.19
.31-24 UNF-2B
.53 .31
.75
.47
.13
.09
.19
.09

① BASE

ϕ .31 .50 DEEP
$\phi \frac{.1300}{.1303}$
.75 SPHERE

⑦ HANDLE BALL

CROWN BOTH ENDS R .19

$\frac{.1600}{.1605}$ .50 DEEP

.31-24 UNF-2A
.06×45°

$\frac{.1906}{.1904}$
1.75
1.50
ϕ .13
ϕ .31
.13

② HANDLE SHAFT

2.00
$\frac{.1906}{.1904}$

③ SCREW HANDLE

1.56
.56
.25
.03×45° $\phi \frac{.1900}{.1903}$

④ SCREW

.09
.03×45°
.88
.44
$\frac{.1609}{.1606}$

⑤ SCREW POINT

CROWN BOTH ENDS R .38

.75
$\frac{.1306}{.1304}$

⑥ PIN

Problem 10.1

MILLIMETERS

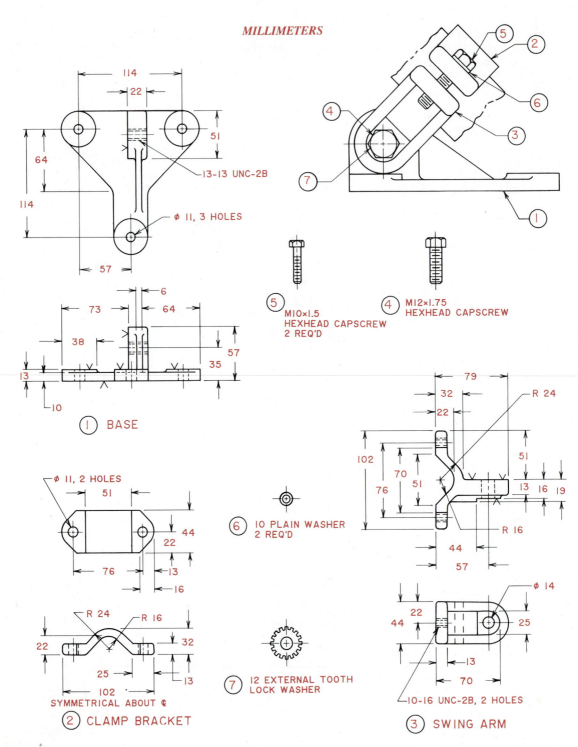

① BASE

② CLAMP BRACKET

③ SWING ARM

④ M12×1.75 HEXHEAD CAPSCREW

⑤ M10×1.5 HEXHEAD CAPSCREW 2 REQ'D

⑥ 10 PLAIN WASHER 2 REQ'D

⑦ 12 EXTERNAL TOOTH LOCK WASHER

Problem 10.2

INCHES

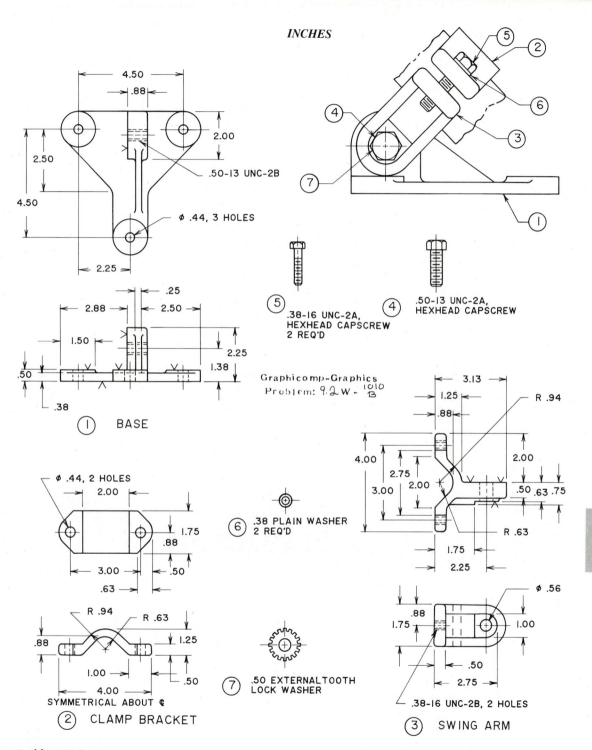

.50-13 UNC-2B

Ø .44, 3 HOLES

.38-16 UNC-2A,
HEXHEAD CAPSCREW
2 REQ'D

.50-13 UNC-2A,
HEXHEAD CAPSCREW

Graphicomp-Graphics
Problem: 9.2W - 1010 B

① BASE

Ø .44, 2 HOLES

.38 PLAIN WASHER
2 REQ'D

R .94

R .63

R .94 R .63

.50 EXTERNAL TOOTH
LOCK WASHER

Ø .56

.38-16 UNC-2B, 2 HOLES

SYMMETRICAL ABOUT ₵

② CLAMP BRACKET

③ SWING ARM

Problem 10.2

MILLIMETERS

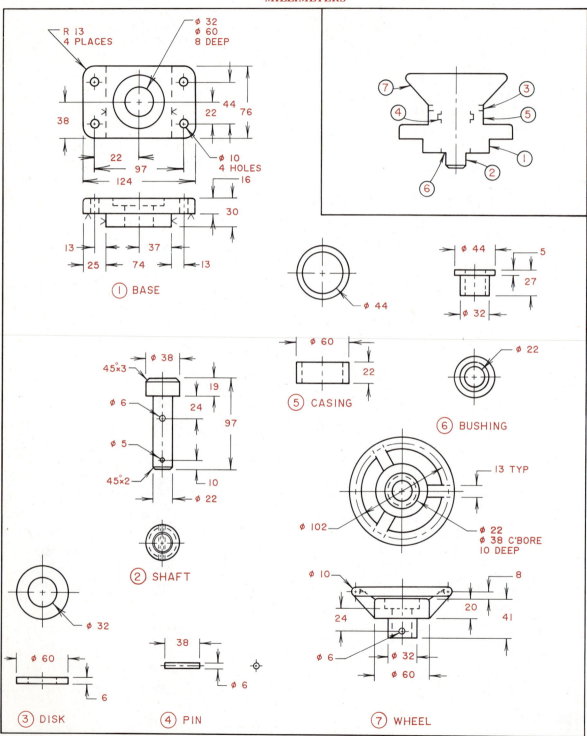

R 13
4 PLACES

Ø 32
Ø 60
8 DEEP

44
76
22
38
22
97
124

Ø 10
4 HOLES

16
30
13
25
37
74
13

① BASE

Ø 44

Ø 44
Ø 32
5
27

7
4
3
5
1
2
6

Ø 60
22

⑤ CASING

Ø 22

⑥ BUSHING

45°×3
Ø 38
19
24
97
Ø 6
Ø 5
45°×2
10
Ø 22

② SHAFT

13 TYP

Ø 102
Ø 22
Ø 38 C'BORE
10 DEEP

Ø 32

Ø 60
6

③ DISK

38
Ø 6

④ PIN

Ø 10
8
A
A
20
41
24
Ø 6
Ø 32
Ø 60

⑦ WHEEL

Problem 10.3

INCHES

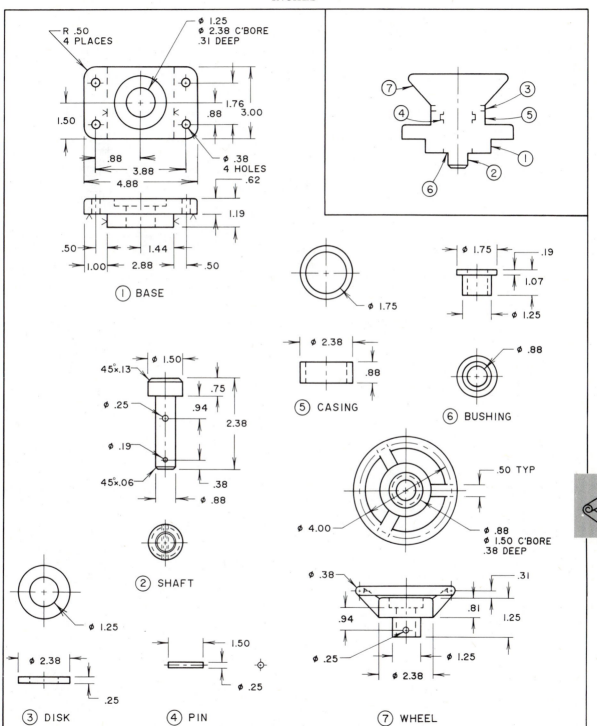

R .50
4 PLACES

Ø 1.25
Ø 2.38 C'BORE
.31 DEEP

1.76
.88
3.00

1.50

.88

Ø .38
4 HOLES

3.88
4.88

.62

1.19

.50
1.00
1.44
2.88
.50

① BASE

Ø 1.75

Ø 1.75
.19
1.07
1.25

Ø 1.50
45°x.13
Ø .25
.75
.94
2.38
Ø .19
45°x.06
.38
Ø .88

Ø 2.38
.88

⑤ CASING

Ø .88

⑥ BUSHING

② SHAFT

Ø 1.25

Ø 2.38
.25

③ DISK

1.50
Ø .25

④ PIN

.50 TYP

Ø 4.00

Ø .88
Ø 1.50 C'BORE
.38 DEEP

Ø .38
.31
.94
.81
1.25
Ø .25
Ø 1.25
Ø 2.38

⑦ WHEEL

Problem 10.3

MILLIMETERS

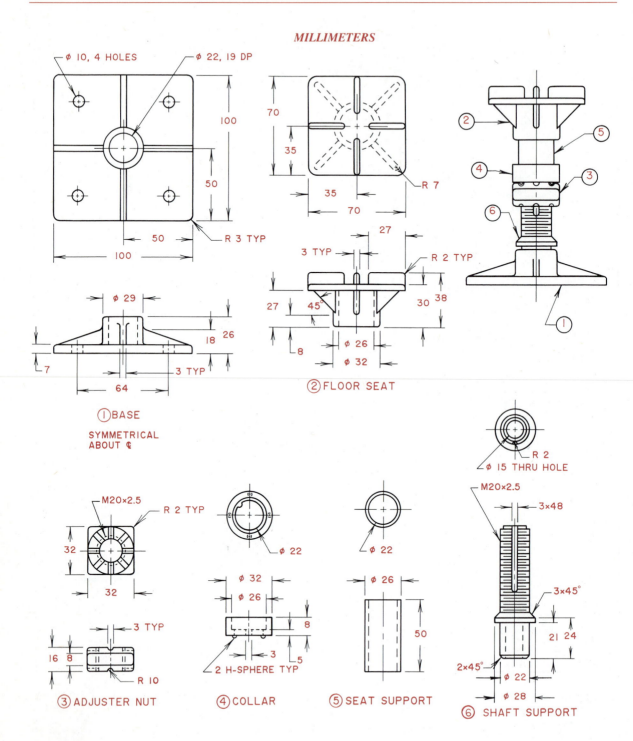

Ø 10, 4 HOLES

Ø 22, 19 DP

100

35

50

50

100

R 3 TYP

70

35

35

70

R 7

27

3 TYP

R 2 TYP

27 45°

30 38

8

Ø 26

Ø 32

②FLOOR SEAT

Ø 29

18 26

7

3 TYP

64

①BASE

SYMMETRICAL
ABOUT ₵

②

⑤

④

③

⑥

①

R 2
Ø 15 THRU HOLE

M20×2.5

R 2 TYP

32

32

3 TYP

16 8

R 10

③ ADJUSTER NUT

Ø 22

Ø 32

Ø 26

8

3

5

2 H-SPHERE TYP

④COLLAR

Ø 22

Ø 26

50

⑤SEAT SUPPORT

M20×2.5

3×48

3×45°

21 24

2×45°

Ø 22

Ø 28

⑥ SHAFT SUPPORT

Problem 10.4

INCHES

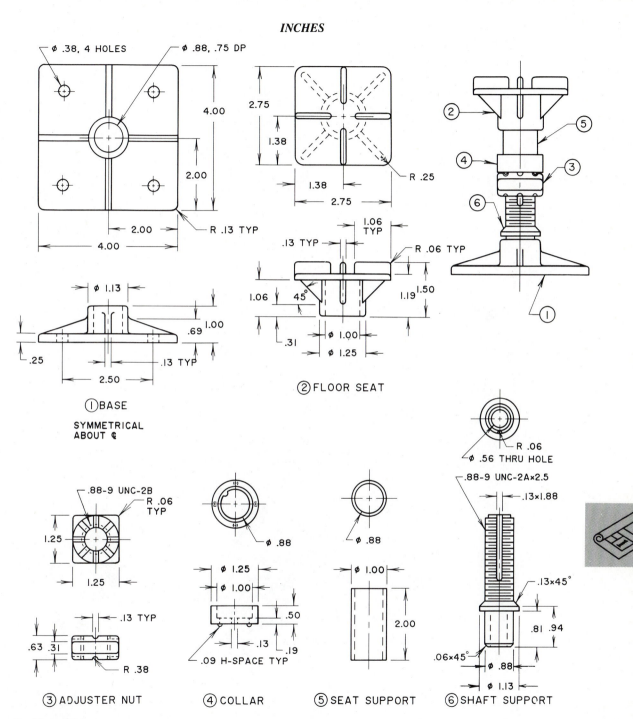

Ø .38, 4 HOLES Ø .88, .75 DP

4.00 2.00 2.00 4.00 R .13 TYP

2.75 1.38 1.38 2.75 R .25

Ø 1.13 1.00 .69 .25 .13 TYP 2.50

① BASE

SYMMETRICAL
ABOUT ℄

1.06 TYP .13 TYP R .06 TYP 1.06 45° 1.19 1.50 .31 Ø 1.00 Ø 1.25

② FLOOR SEAT

R .06 Ø .56 THRU HOLE

.88-9 UNC-2A×2.5 .13×1.88

.88-9 UNC-2B R .06 TYP 1.25 1.25

.13 TYP .63 .31 R .38

③ ADJUSTER NUT

Ø .88 Ø 1.25 Ø 1.00 .50 .13 .19 .09 H-SPACE TYP

④ COLLAR

Ø .88 Ø 1.00 2.00

⑤ SEAT SUPPORT

.13×45° .81 .94 .06×45° Ø .88 Ø 1.13

⑥ SHAFT SUPPORT

Problem 10.4

INCHES

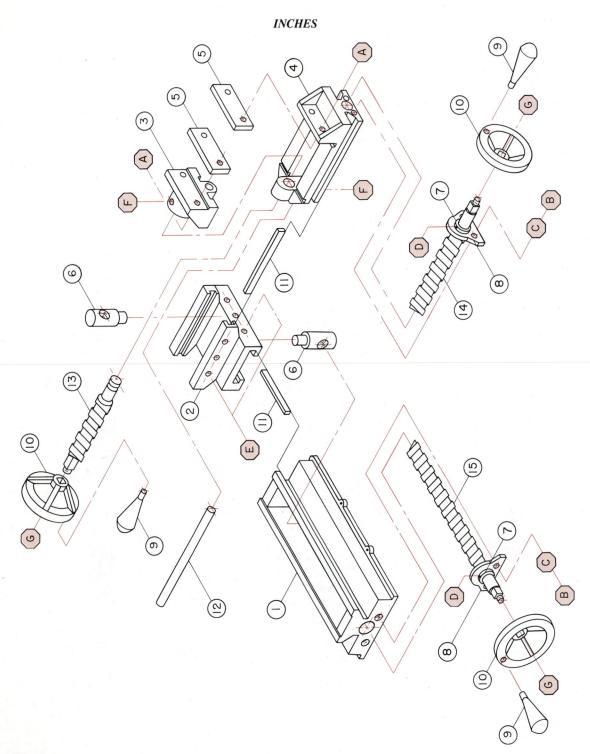

Problem 10.5

INCHES

HARDWARE SCHEDULE

PART	NO. REQ'D	DESCRIPTION
Ⓐ	4	#10-24 UNC-3A x .50 LG, CUPPOINT, HEADLESS, SETSCREW
Ⓑ	4	#10-24 UNC-3A x .50 LG, RHMACH SCREW
Ⓒ	4	#10 N WASHER
Ⓓ	2	#5-40 UNC-3A x .13 LG, CUPPOINT, HEADLESS, SETSCREW
Ⓔ	6	#5-40 UNC-3A x .25 LG, CONEPOINT, HEADLESS, SETSCREW
Ⓕ	2	#5-40 UNC-3A x .38 LG, CUPPOINT, HEADLESS, SETSCREW
Ⓖ	3	#10-24 UNC-3B HEXNUT
ALL HARDWARE-STEEL		

BILL OF MATERIALS

PART	NO. REQ'D	MAT'L	DESCRIPTION
①	1	CI	BASE
②	1	CI	INTERMEDIATE SLIDE
③	1	CI	JAW
④	1	CI	SLIDE BASE
⑤	2	3140 STL	JAW SPACER
⑥	2	1040 STL	PIN
⑦	2	1040, STL	COLLAR
⑧	2	BRZ	COLLAR BRACE
⑨	3	1020 STL	HANDLE
⑩	3	1020 STL	HAND WHEEL
⑪	2	BRASS	PRESSURE SPACER
⑫	1	1020 STL	ALIGNMENT ROD
⑬	1	1040 STL	SPINDLE A
⑭	1	1040 STL	SPINDLE B
⑮	1	1040 STL	SPINDLE C

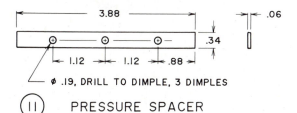

ⓘⓘ PRESSURE SPACER

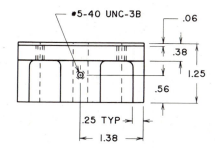

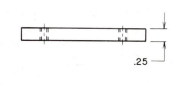

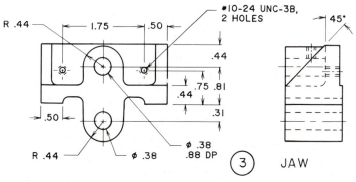

③ JAW

⑤ JAW SPACER

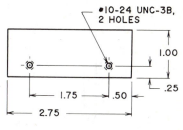

Problem 10.5

INCHES

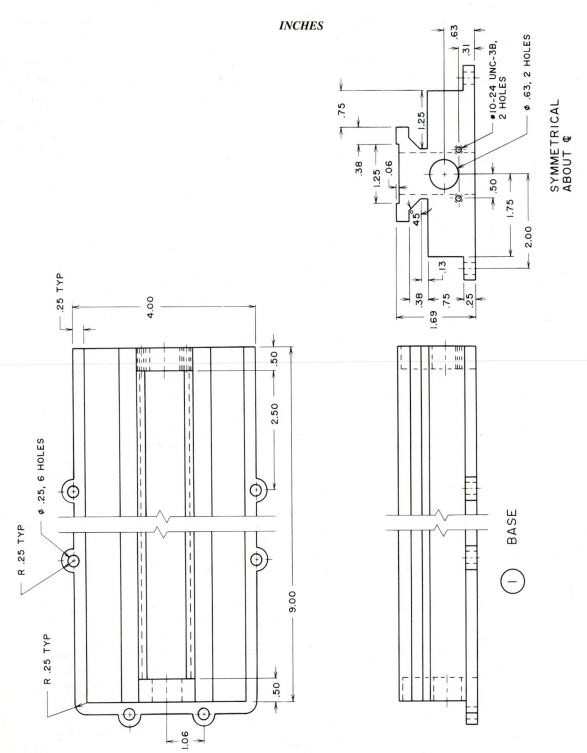

SYMMETRICAL ABOUT ℄

.63

.31

#10-24 UNC-3B, 2 HOLES

⌀ .63, 2 HOLES

1.25

.75

.38

1.25

.06

.50

1.75

4.5°

2.00

.13

.38

.75

.25

1.69

.25 TYP

4.00

.50

2.50

⌀ .25, 6 HOLES

R .25 TYP

R .25 TYP

9.00

.50

1.06

BASE

①

Problem 10.5

INCHES

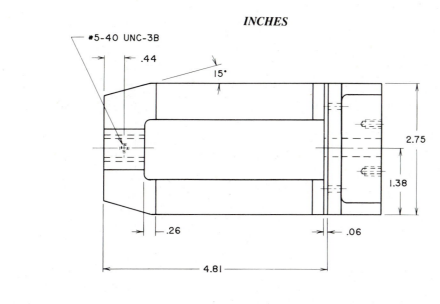

#5-40 UNC-3B

.44

15°

2.75

1.38

.26

.06

4.81

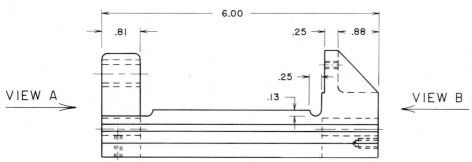

6.00

.81

.25

.88

.25

.13

VIEW A →

← VIEW B

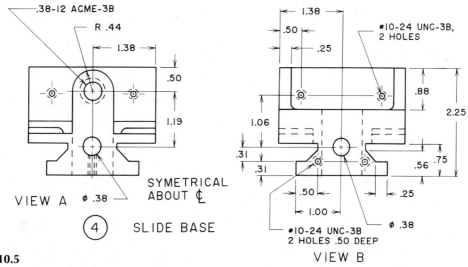

.38-12 ACME-3B

R .44

1.38

.50

1.19

VIEW A   φ .38

SYMETRICAL
ABOUT ℄

④   SLIDE BASE

1.38

.50

.25

#10-24 UNC-3B,
2 HOLES

.88

2.25

1.06

.31

.31

.75

.56

.50

.25

1.00

#10-24 UNC-3B
2 HOLES .50 DEEP

φ .38

VIEW B

Problem 10.5

INCHES

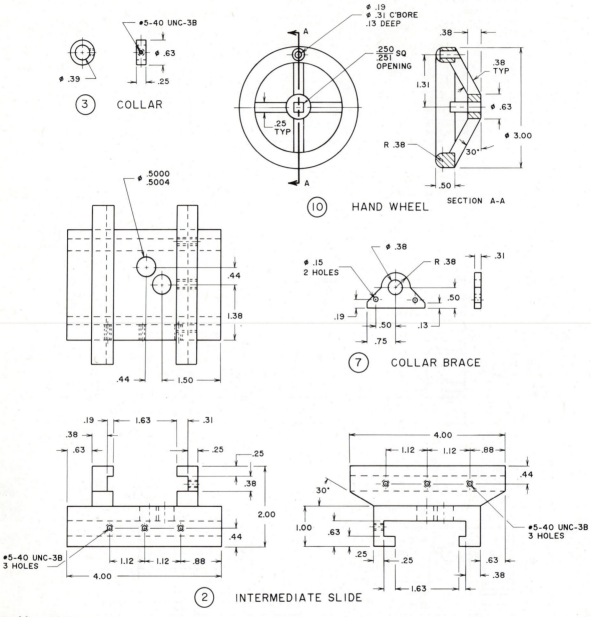

#5-40 UNC-3B

Ø .63

.25

Ø .39

③ COLLAR

Ø .19
Ø .31 C'BORE
.13 DEEP

A

.250 SQ
.251
OPENING

.25
TYP

.25
TYP

A

⑩ HAND WHEEL

.38

.38
TYP

1.31

Ø .63

Ø 3.00

R .38

30°

.50

SECTION A-A

Ø .5000
.5004

.44

1.38

.44

1.50

Ø .38

Ø .15
2 HOLES

R .38

.31

.50

.19

.50

.13

.75

⑦ COLLAR BRACE

.19

1.63

.31

.38

.63

.25

.25

.38

2.00

.44

#5-40 UNC-3B
3 HOLES

1.12

1.12

.88

4.00

② INTERMEDIATE SLIDE

4.00

1.12

1.12

.88

.44

30°

#5-40 UNC-3B
3 HOLES

1.00

.63

.25

.25

.63

.38

1.63

Problem 10.5

INCHES

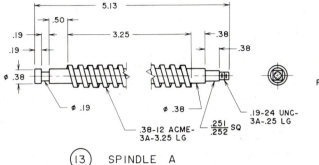

.19
.19
.50
5.13
3.25
.38
.38
Ø .38
Ø .19
.38-12 ACME-
3A-3.25 LG
$\frac{.251}{.252}$ SQ
.19-24 UNC-
3A-.25 LG

⑬ SPINDLE A

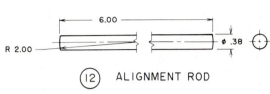

6.00
R 2.00
Ø .38

⑫ ALIGNMENT ROD

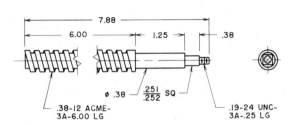

7.88
6.00
1.25
.38
Ø .38
$\frac{.251}{.252}$ SQ
.38-12 ACME-
3A-6.00 LG
.19-24 UNC-
3A-.25 LG

⑭ SPINDLE B

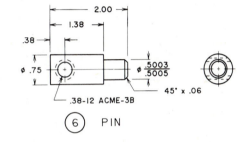

2.00
1.38
.38
Ø .75
$\frac{.5003}{.5005}$
.38-12 ACME-3B
45° x .06

⑥ PIN

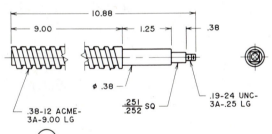

10.88
9.00
1.25
.38
Ø .38
$\frac{.251}{.252}$ SQ
.38-12 ACME-
3A-9.00 LG
.19-24 UNC-
3A-.25 LG

⑮ SPINDLE C

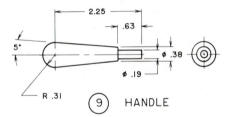

2.25
.63
5°
Ø .38
Ø .19
R .31

⑨ HANDLE

Problem 10.5

TERMS YOU WILL SEE IN THIS CHAPTER

AUXILIARY VIEW An orthographic view, other than a principal view, to provide additional information.

ROTATION Method of solving descriptive geometry problems in which all the points of the object revolve about an axis.

TRUE-LENGTH LINE View of a line in which it is parallel to the picture plane.

PARALLEL LINES Two or more lines that are nonintersecting where all the points of one line are separated from the other line(s) by the same distance.

PERPENDICULAR LINES Two or more lines that cross each other at 90°.

SKEW LINES Two or more lines that are nonintersecting and are not parallel.

EDGE VIEW View of a plane that appears as a line.

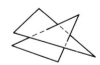

LINE OF INTERSECTION Points that are common between two planes that intersect.

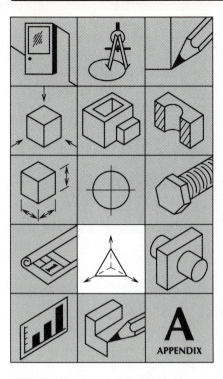

11

THREE-DIMENSIONAL GEOMETRY CONCEPTS

A
APPENDIX

11.1 *INTRODUCTION*

We live in a three-dimensional world and must solve three-dimensional problems. Designers must think in three dimensions and be able to visualize their three-dimensional problems (Figure 11.1). For many years engineers and scientists sought ways to describe these three-dimensional designs and problems without having to build models. They wanted to be able to use pencils or pens and paper to draw representations of their problems. Although orthographic projection and pictorials as presented in Chapters 4 and 5 did much to provide the information required, they did not help solve the problems of finding true distances, true angles, and true areas.

Descriptive geometry was developed by Gaspard Monge for solving design problems for military fortifications while he was a student in France. His graphical solutions to these complex problems were deemed so valuable that they were considered to be a defense secret for the French. He published his work in

Figure 11.1 Designers must be able to visualize three-dimensional shapes.

a book entitled *Descriptive Geometry* in 1795. Monge found that any view could provide two of the three dimensions, and thus two adjacent views could provide all three dimensions.

Descriptive geometry is a key part of a technical student's graphic and problem-solving tool kit. It allows you to represent three-dimensional problems in two dimensions. Studying descriptive geometry will also improve your ability to visualize. Today, two dimensions refers to a drawing either on paper or on a CRT screen.

Fundamental concepts will be introduced using a cut block. Figure 11.2 shows pictorial and wire frame views of the block. The wire frame view also shows the axis system to be used and the coordinates of the block's vertices based on the axis

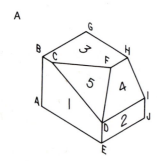

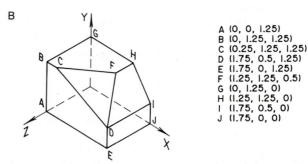

A (0, 0, 1.25)
B (0, 1.25, 1.25)
C (0.25, 1.25, 1.25)
D (1.75, 0.5, 1.25)
E (1.75, 0, 1.25)
F (1.25, 1.25, 0.5)
G (0, 1.25, 0)
H (1.25, 1.25, 0)
I (1.75, 0.5, 0)
J (1.75, 0, 0)

Figure 11.2 (*a*) Pictorial of cut block. (*b*) Wire frame pictorial drawing showing axis system and vertice coordinates.

(a)

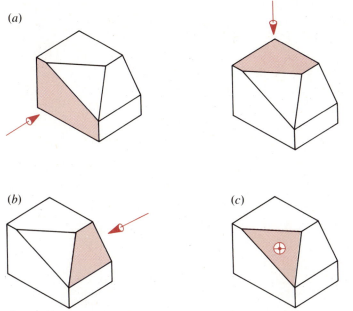

(b)   (c)

Figure 11.3 Pictorial showing (a) principal planes highlighted; (b) inclined plane highlighted; (c) oblique plane highlighted.

system. This block contains principal, inclined, and oblique planes. These planes are labeled and highlighted in Figure 11.3. This figure also shows the line of sight, or normal vector, for each of the designated planes. The block will be used to present the physical location in space of the planes and lines for the basic problem-solving concepts.

The flat pattern for the block has been printed on heavier paper on the inside back cover of this text. The three-dimensional coordinate information for this block, shown in Figure 11.2, can be used to create a data file for CADD software that has three-dimensional capability.

This chapter uses three methods to solve three-dimensional problems: (1) the direct method using reference planes, (2) the glass box with fold lines, and (3) **rotation** or revolution. The direct method and fold line method work well using manual drawing techniques. Normally, rotation using manual methods is very difficult; however, a computer does rotation quickly and is better suited for rotation or revolution methods.

Each demonstration problem for the fundamental line and plane problems is solved using the three methods where possible. When these three methods are presented, all three

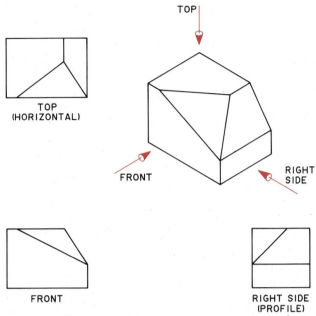

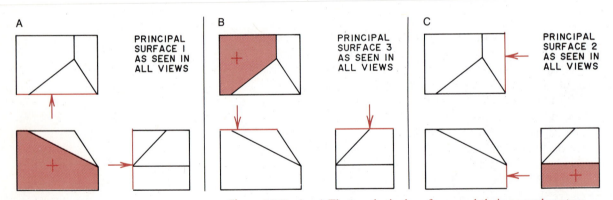

Figure 11.4 Pictorial and orthographic views of cut block.

Figure 11.5 (*a–c*) Three principal surfaces and their normal vectors.

problem solutions will be shown on the same or adjacent pages so that you can compare them. Later in the chapter, problems will be solved using the most commonly used method(s).

Figure 11.4 shows the pictorial and orthographic views of the cut block. Figure 11.5 shows the block with principal surfaces highlighted and their normal vectors as they appear in the three

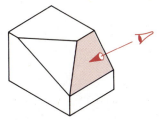

Figure 11.6 Pictorial with inclined plane highlighted.

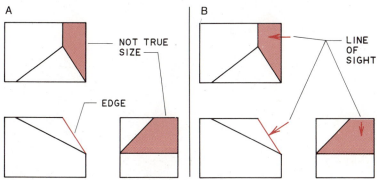

Figure 11.7 (*a*) Three views with inclined plane highlighted. (*b*) Line of sight or normal vector added in each view.

views. Normal vectors are the vectors that are perpendicular to the surfaces. Figure 11.6 shows the pictorial view of the block with the inclined surface highlighted and the inclined surface's normal vector appearing as a point. Figure 11.7 shows the three orthographic views with the inclined surface highlighted and the arrows representing the line of sight. The transparent reference planes are added in Figures 11.8 and 11.9 to show the relative location of the block and the reference planes. The latter of these illustration sets shows which planes would be visible in individual views. Figure 11.10 shows how combinations of reference planes are then used to make up a glass box that surrounds the object. The glass box is unfolded to describe the three-dimensional object using two-dimensional views. The two-dimensional views are called principal views, and the particular planes are called principal planes. Refer to Chapter 4 for more information on principal planes. Also look ahead to Figure 11.25.

(*a*)

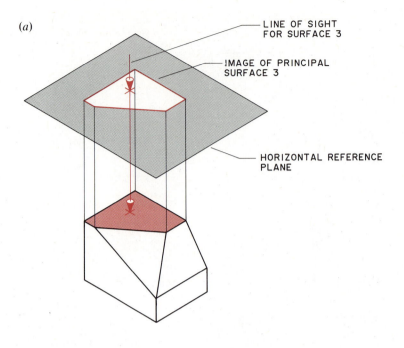

LINE OF SIGHT
FOR SURFACE 3

IMAGE OF PRINCIPAL
SURFACE 3

HORIZONTAL REFERENCE
PLANE

(*b*)

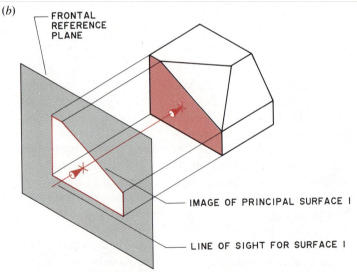

FRONTAL
REFERENCE
PLANE

IMAGE OF PRINCIPAL SURFACE I

LINE OF SIGHT FOR SURFACE I

Figure 11.8 (*a*) Pictorial of block with horizontal reference plane added. Principal plane highlighted. (*b*) Pictorial of block with frontal reference plane added. Principal plane highlighted. (*c*) Pictorial of block with profile reference plane added. Principal plane highlighted.

(c)

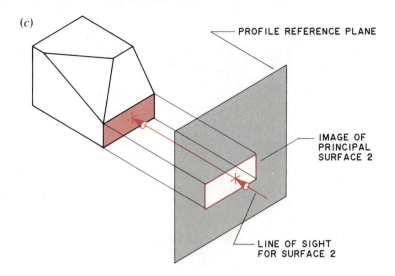

PROFILE REFERENCE PLANE

IMAGE OF
PRINCIPAL
SURFACE 2

LINE OF SIGHT
FOR SURFACE 2

Rotations will always be done around the X, Y, and Z axes where the X and Y axes are in the plane of the paper or CRT screen for the front view and the Z axis is coming out of the paper or the CRT screen toward you. The principal views of the cut block are created by rotating the block about the axes. Lines that are true length and planes that are true size will appear in another color or shade of gray in the figures.

Your first step is to cut out the pattern on the inside back cover of this book and assemble the model of the cut block. Once assembled, rotate the finished model so that it matches the examples shown in Figures 11.2–11.10 so that you can see how each face looks in each of the principal views. The faces have been numbered, and they will be referred to by their numbers. You should note whether the faces are visible or hidden and whether they are true size, foreshortened (smaller than true size), or an edge.

If you are using three-dimensional software on your computer, load your current CADD package. You will need to create the files for the cut block using the data shown in Figure 11.2. As you study the material in this chapter, it is helpful to have both the physical model and the computer available for understanding the solutions being presented. The manipulation practice will increase your visualization skills, which are vital in communicating your ideas.

(a)

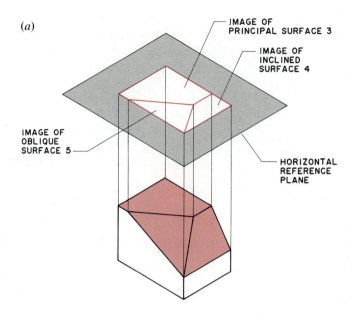

IMAGE OF PRINCIPAL SURFACE 3

IMAGE OF INCLINED SURFACE 4

IMAGE OF OBLIQUE SURFACE 5

HORIZONTAL REFERENCE PLANE

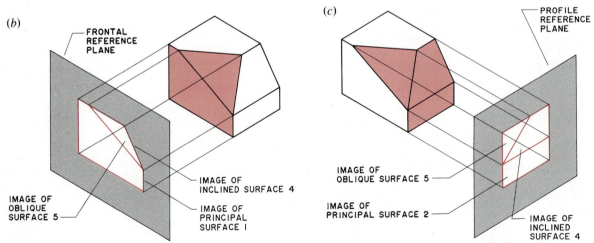

(b)

FRONTAL REFERENCE PLANE

IMAGE OF INCLINED SURFACE 4

IMAGE OF PRINCIPAL SURFACE I

IMAGE OF OBLIQUE SURFACE 5

(c)

PROFILE REFERENCE PLANE

IMAGE OF OBLIQUE SURFACE 5

IMAGE OF PRINCIPAL SURFACE 2

IMAGE OF INCLINED SURFACE 4

Figure 11.9 (a) Pictorial of block with horizontal reference plane added. All visible surfaces highlighted. (b) Pictorial of block with frontal reference plane added. All visible surfaces highlighted. (c) Pictorial of block with profile reference plane added. All visible surfaces highlighted.

(a)

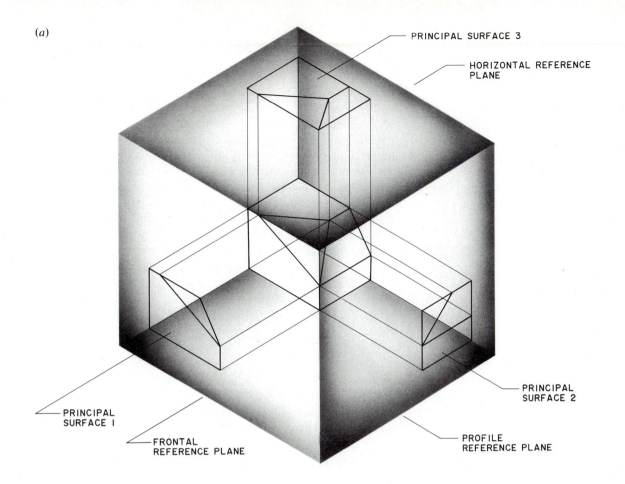

PRINCIPAL SURFACE 3

HORIZONTAL REFERENCE PLANE

PRINCIPAL SURFACE 2

PROFILE REFERENCE PLANE

PRINCIPAL SURFACE 1

FRONTAL REFERENCE PLANE

(b)

PRINCIPAL VIEWS OF SOLID

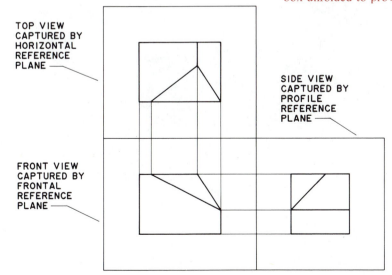

TOP VIEW CAPTURED BY HORIZONTAL REFERENCE PLANE

SIDE VIEW CAPTURED BY PROFILE REFERENCE PLANE

FRONT VIEW CAPTURED BY FRONTAL REFERENCE PLANE

Figure 11.10 (a) Pictorial of block with all three reference planes added to form glass box. (b) Glass box unfolded to provide three orthographic views.

11.2 *APPLICATIONS OF DESCRIPTIVE GEOMETRY*

Houses, other buildings, chairs, cars, and tables are objects that we use that have combinations of planes and lines that intersect. The questions are: Where do they intersect? How long is the **line of intersection?** If the line of intersection is created by two planes of a toolbox, we need to know how many spot welds, how many rivets, or the length of weld bead is required to join the two surfaces. Figure 11.11 shows a photo of a spot-welded tool box.

Another aspect of complex objects such as cartons is how to create the three-dimensional object from a flat sheet. This requires a pattern. In the pattern development process, the object is drawn first as a three-dimensional object, usually a pictorial sketch, and then as three or more orthographic views; then the development is done by laying off true-length lines. Figure 11.12 shows the orthographic views, the pictorial, and the flat pattern for the example object. Developments are important in areas other than packaging: Applications are in heating and air-conditioning duct work, aircraft design, building construction, luggage, and clothing, to name a few.

Another important application of descriptive geometry lies in the area of mapping. Mapping includes subdivision layouts, con-

Figure 11.11 Tool box.

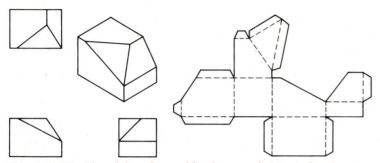

Figure 11.12 Pictorial, orthographic views, and pattern development for cut block.

tour maps, and construction plans for roads, water mains, sewers, and dams. Directions on maps are important. Chapter 12 will define bearings and azimuths, which are two methods used in descriptive geometry to define directions. Contour lines on a map define the location of constant elevations. See Figure 11.13. Descriptive geometry methods can be used to transform these contour lines into cross sections through the earth to show better

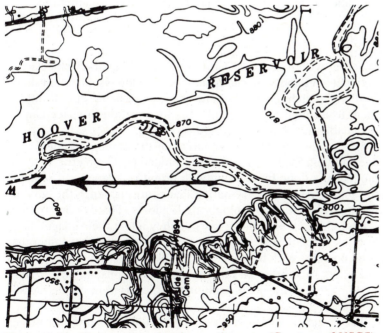

Figure 11.13 Contour map showing north arrow. (Courtesy of USGS)

detail. Determining slopes and scaling are important problem-solving methods when applying descriptive geometry to mapping. Three-dimensional visualization is needed in creating multilayer printed circuit boards and plans for chemical processing plants.

11.2.1 Problem-solving Skills

The only way to solve a complicated problem is to reduce the complicated problem to a series of simple problems. You then determine which problem to solve first and proceed to solve the simple problems until the complicated problem is solved. In most engineering problems there are a variety of correct solutions. For example, count the number of different makes of cars, bicycles, chairs, television sets, and so on; all of the products in each category solve the same problem.

In descriptive geometry you will be given a statement of what you are required to find and enough information to solve the problem. This text provides a set of rules and techniques to get from the given information to the solution. There will usually be two possible solutions that will provide the desired information. Thus, descriptive geometry provides experience in basic problem-solving skills and helps you learn how to attack any problem. However, in the real world you may find that not all of the information is given and you may have to make assumptions.

11.2.2 Developing Visualization Skills

Working with a series of views of real and abstract objects (individual lines and planes in space) can develop visualization skills. Looking at individual lines and planes and being able to attach them to an object or a set of axes for location in space can be invaluable in solving real-world problems.

One of the fastest growing areas in the computer-aided engineering field is that of geometric modeling. Geometric or solid modeling is done by working with primitive shapes and requires the ability to visualize. Primitive shapes are chosen as the building blocks to create more complicated objects. This is done by preschool children on a daily basis as they allow their imaginations to think of castles, roads, houses, buildings, airplanes, and other solid objects. With building blocks, all that can be done is to add one block to another and the blocks are of constant size. Figure 11.14 shows some of the primitive shapes used to create models of real objects.

The designer working with solid modeling has some advantages over the preschool child. Computer-based modeling systems provide the ability to select a primitive and resize it. A set of prim-

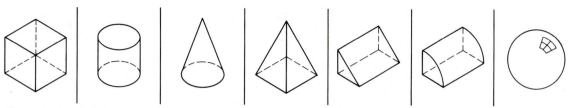

Figure 11.14 Primitive shapes for solid modeling.

A

B

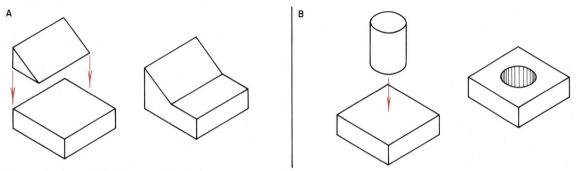

Figure 11.15 (*a*) Adding two blocks to form new solid. (*b*) Differencing block and cylinder to produce hole.

itives can be chosen and modified for use by the designer. Then he or she can add them together to create a more complicated object or an assembly of complicated objects. However, the designer can also subtract (difference) objects. For example, the difference between a rectangular solid and a cylindrical solid can be a hole in the rectangular solid. Figures 11.15*a,b* show the adding and differencing operations. Other mathematical operations are also possible, but these simple examples give an idea of the power of such tools.

In the coming years, solid modeling will expand to cover much of what we now think of as descriptive geometry. In the meantime, a good understanding of three dimensions through descriptive geometry will pave the way for efficient learning and use of solid modeling systems.

11.3 *OPERATIONS*

11.3.1 The Direct Method

The direct method as used in solving descriptive geometry problems is a manual method. This means that problem solutions are

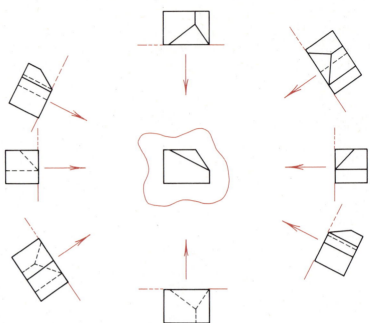

Figure 11.16 Figure showing front view with reference plane installed and several adjacent views.

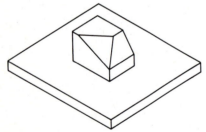

Figure 11.17 Drawing of cut block sitting on desk.

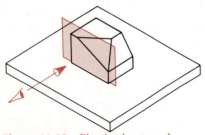

Figure 11.18 Plastic sheet used as frontal reference plane.

more easily found using a pencil, paper, and drawing tools rather than using a computer. The basic concept of the direct method is that an infinite reference plane placed anywhere in a given view will appear true size in that view and as an edge in any adjacent view. Figure 11.16 shows the front view of the cut block with the reference plane installed and several adjacent views of the block that show the reference plane as an edge. You may find it useful to use a sheet of clear plastic to model the reference plane with the cut block.

Set your model on the table in front of you. See Figure 11.17. As you sit at a desk, you can see several of the planes that make up the object. You are seeing a pictorial view of the object. Orient the block so that the sides are lined up with the desk and plane 1 is facing you, plane 2 is to the right, and plane 3 is on top. Drop your eye to desk level and look at the block. You are seeing the front view of the object. If you have a clear plastic sheet, hold it so that it is between you and the object and is vertical and parallel to the front of the object. See Figure 11.18. If you were to trace the outline and edges of the object's planes on the clear plastic sheet, you would have the front view of the object and plane 1 is seen true size. The clear plastic sheet represents the picture plane or reference plane.

Without moving the plastic sheet or the object, crouch down by the side of the desk and look at the object again. See Figure 11.19. This is the side view and plane 2 is seen true size. Notice that the plastic sheet has now become an edge when you look at the side view. Therefore, the reference plane for one orthographic view becomes an edge for the adjacent orthographic views. Now stand up and look directly down on the object while continuing to hold the plastic sheet in the location where it was when viewing the front view. See Figure 11.20. This is the top view and plane 3 appears true size. Again, note that the reference plane for the front view appears as an edge when viewing the object from the top.

Auxiliary views are orthographic views. This was discussed in Chapter 4. They are not principal views. There are an infinite number of auxiliary views possible for any given view; however, only six principal views are available, as described in Chapter 4. In this next series of steps you will be positioning yourself so that you can see auxiliary views. Continue to hold the reference plane so that it is a frontal reference plane and do not move the object. Now, slowly move from your position when viewing the object from the top and crouch beside the desk again. Figure 11.21 shows various positions of a viewer and some of the possible lines of sight from the top to the right side views. In this process, note

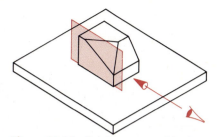

Figure 11.19 Looking at cut block to see plane 2 true size and reference plane as edge.

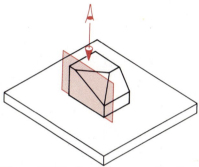

Figure 11.20 Looking at cut block to see plane 3 true size and reference plane as edge.

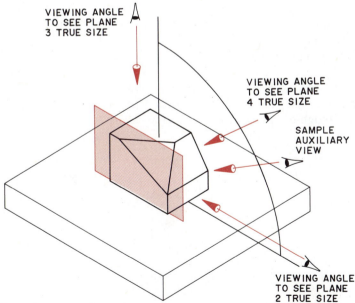

VIEWING ANGLE TO SEE PLANE 3 TRUE SIZE

VIEWING ANGLE TO SEE PLANE 4 TRUE SIZE

SAMPLE AUXILIARY VIEW

VIEWING ANGLE TO SEE PLANE 2 TRUE SIZE

Figure 11.21 Various viewing positions to see top and right side views and some of auxiliary views between top and right side views.

that the reference plane continues to appear as an edge in all of the positions. Any position between the top view and the side view is considered to be an auxiliary view.

When you moved from the top view to the side view, you moved through a position where the inclined plane, plane 4, appeared true size. This occurred when your line of sight was perpendicular to plane 4. See Figure 11.21.

Finding the true size of surfaces is one use of descriptive geometry. The process provides us with ways to find the true size and shape of inclined and oblique planes. When we do this, we also get **true-length lines** and true-size angles.

If you had desired to draw any of the auxiliary views on paper, you could have had another reference plane to hold between you and the object. See Figure 11.22. On that reference plane you could trace the object as you see it. What is the relationship between the two reference planes? They are perpendicular. It is important to point out that you saw the frontal reference plane in the same position relative to both the top and side views as well as to any of the auxiliary views in between. This allows us to measure from the frontal reference plane in one view to get the depth of a point (distance from the reference plane) and be able to transfer it to another view where we can see the frontal reference plane as an edge, as shown in Figure 11.23. This illustrates an important concept. In any situation where we see the reference plane true size in one view, it appears as an edge in *any* of the adjacent views. Hold the reference plane parallel to surface 3 in the top view and then look at the edge of the plane in the front and right side views. See Figure 11.24.

The direct method uses a systematic method of drawing and labeling reference planes. There are two approaches. One uses a standard black pencil for drawing the representations and labels. The other uses colored pencils. Both approaches require sketching a rough-cut plane around the view where the reference plane

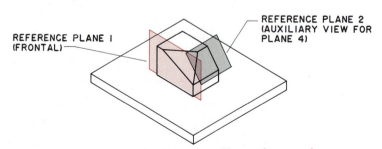

Figure 11.22 Block with frontal and auxiliary reference planes.

REFERENCE PLANE 1 (FRONTAL)

REFERENCE PLANE 2 (AUXILIARY VIEW FOR PLANE 4)

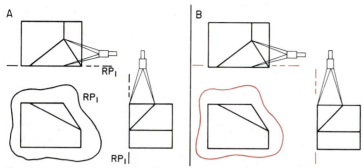

Figure 11.23   (*a*) Frontal reference plane used for transferring distances. Reference plane, RP1. (*b*) Frontal reference plane used for transferring distances. Reference plane colored.

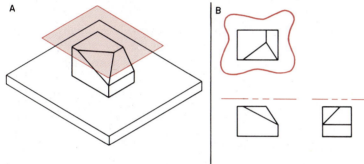

Figure 11.24   (*a*) Horizontal reference plane (HRP) installed on block. (*b*) Orthographic views showing HRP.

appears true size. This rough-cut plane is called the "cloud" view. The reference plane is installed in the adjacent views as an edge or a line. Labels are attached to both the cloud view and edge views of the reference plane. Examples of these labels are RP1 or RP2 or, in general, RPn. The other approach is to use colored pencils to draw the reference planes, and each plane is drawn in a different color. Then the adjacent views have a line drawn in the same color representing the edge view of the reference plane. The colored planes can also be labeled, but the labels may be superfluous unless copies are to be made of the drawing. See Figures 11.23*a,b*.

Measurements for construction are then made either away from the given view in both adjacent views or toward the given view in both adjacent views. When a reference plane is attached to a principal plane, it is often designated the horizontal, frontal, or profile reference plane (HRP, FRP, or PRP) rather than RPn.

11.3.2 The Glass Box: Fold Line Method

Another way of looking at the reference plane and objects is the fold line method. Leave the reference planes located in the position for each of the three principal views. If three more planes were added, a box would be created around the object. Refer again to Figures 11.10a,b. Look through each of the three reference planes and draw what you see inside the box. You have drawn three of the principal views. This can be a very convenient approach for beginners because it fixes the reference planes relative to each other. If we were to put hinges on the box between the frontal and top and the frontal and right sides, we could unfold the box and we would have the orientation of three of the six principal views. This glass box approach for the six principal planes is also presented in Chapter 4.

In addition, Chapter 4 presents the concept of the auxiliary view. With the three principal views in place, it is possible to add a fourth plane that is parallel to plane 4. The hinges would be placed between the frontal plane and the seventh, or auxiliary, plane. When the box is unfolded, it is easy to see the location of the auxiliary view relative to the principal views. Figure 11.25a shows the complete glass box constructed for six principal views and one auxiliary view. Figure 11.25b shows the box open to show the relative positions of the views.

You need to know at this point that while creating the glass

A   B

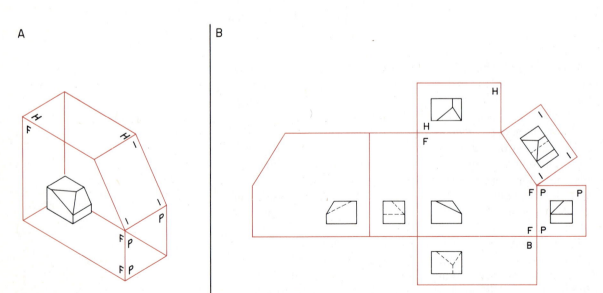

Figure 11.25 (a) Auxiliary plane added to glass box. (b) Seven planes opened to show relative position.

box for a simple figure with an inclined surface is relatively easy, it can be much more difficult to think of the glass box for oblique planes such as surface 5.

In the glass box construction, the edge views of the reference planes are called fold lines because this is where the box is folded or hinged. When working on paper, fold lines, the edge views of the reference planes, are labeled by the planes that they are between. For example, the fold line between the front and top views is labeled F/T or F/H and the fold line between the front and side (or profile), views is labeled F/P.

At this point in the discussion, it is important to point out that the principal views are known by more than one name. The top view is sometimes called the plan or horizontal view. The front view is sometimes called an elevation, and the right and left side views can be called profile views or elevations. These names are used interchangeably in this text, but the use of a particular name, for example, "plan view," is used by civil engineers to designate the top view.

The fold lines have traditionally been labeled F/T or F/H between the top, or horizontal, and front views. The fold line between the front and right side, or right profile, view is labeled the F/R, or F/P, line. It is not important which letters are used, but they should be used consistently.

The designation for the fold lines between auxiliary and principal views normally uses a letter and a number. For example, H/1 indicates that the fold line is between the horizontal and first auxiliary view. When the fold line is between two successive auxiliary views, it might be designated by 1/2 or 2/3.

In Figure 11.25*b*, the glass box with seven planes is unfolded so that you can see the reference planes and fold lines with the proper labeling. Note that the fold line method uses a 1 for the number representing the first auxiliary views. Subsequent views are labeled by 2, 3, and so on.

11.3.3 Rotation

When working with objects, it is important to know the frame of reference within which you are working. *Frame of reference* means the set of three axes to define an origin and three mutually perpendicular axes in space. In this case we shall discuss **rotation,** and it is important to know what is rotating and what is considered to be the frame of reference. We consider that the frame of reference stays fixed in space and the objects move (rotate or translate).

When you rotate something, there must be an axis about which the rotation takes place. There are three dimensions and therefore

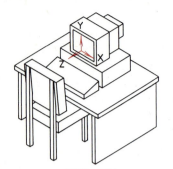

Figure 11.26 CRT with axes.

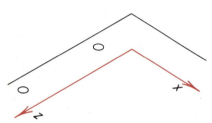

Figure 11.27 Axes shown on sheet of paper.

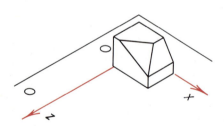

Figure 11.28 Cut block placed on paper.

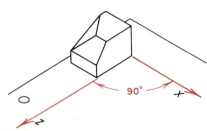

Figure 11.29 Block rotated 90° clockwise about Y axis.

we have three axes. For this text, as was discussed in Chapter 5, we use a set of axes where the X axis goes to the right, the Y axis is vertical, and the Z axis points toward the person looking at the object.

For the previous discussion on reference planes and glass boxes, we placed the cut block on the desk so that the top surface of the block was parallel to the desk surface and the front surface was perpendicular to the desk surface and was facing you. This would mean that the X axis points to the right, the Y axis points toward the ceiling, and the Z axis toward you. It may be easier to think of this orientation as a CRT sitting on a desk with the Z axis coming out of the CRT. Figure 11.26 is a copy of Figure 5.32 showing the CRT and axes.

When we start this rotation process, the width of the object is along the X axis, the height is along the Y axis, and the depth is along the Z axis. This is called the "home" position. In order to preserve the frame of reference, take out a clean sheet of paper and draw the X and Z axes on it. See Figure 11.27. Put the sheet on your desk with the axes aligned as shown in the figure. This will be your frame of reference and will not move. Set your object on the paper so that it is in the home position as shown in Figure 11.28.

First, we rotate the object around the Y axis. When we start, the object is oriented to give the object's front view to the person seated at the desk. If we rotate the object 90° clockwise with the axis through the back left corner, then we see the right side view. See Figure 11.29. Note that in this case you are rotating the object with the reference plane fixed between you and the object and perpendicular to the Z axis of the frame of reference. Figure 11.30 shows the viewing plane added to Figure 11.29. Now return the object to the home position.

As shown in Figure 11.31a, rotate the object about the Y axis 45° clockwise. This allows you to see both the front and right side although both are foreshortened. We want to rotate about the X axis approximately 35° counterclockwise. If the desk were made of sand, we could push the front edge of the object down into the sand and see a pictorial view. The desk is not made of sand. Therefore, you will have to raise the back of the object so that the base of the object forms an approximate 35° angle with the desk top. You have now positioned the object as an isometric view, as shown in Figure 11.31b. Once you have done this, return the object to the home position.

The purpose of these rotations is not to find a way to display an isometric pictorial view of the object. The purpose is to prepare you to find the true size of planes or the true length of lines using rotation as a method.

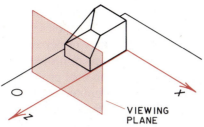

Figure 11.30 Viewing plane added to Figure 11.29.

Two rotations are normally sufficient to find the answer to most problems. The following two operations will display the plane 4 true size. Note that plane 4 makes an angle of approximately 55° with the horizontal. First rotate about the *Y* axis 90° clockwise as shown in Figure 11.32*a*. Next rotate about the *X* axis approximately 35° counterclockwise, as shown in Figure 11.32*b*. This set of rotations shows you, the viewer, the true size of plane 4 just as drawing the auxiliary view showed the true size of plane 4 in Figure 11.25*b*.

You now have seen the three methods for solving descriptive geometry problems. Each has its applications and conventions.

A

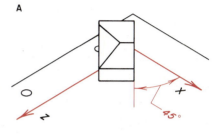

B

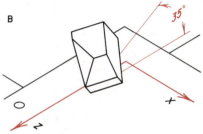

Figure 11.31 (*a*) Rotation of block 45° clockwise about *Y* axis. (*b*) Second rotation of block approximately 35° counterclockwise about *X* axis.

A

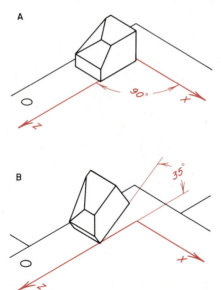

B

Figure 11.32 (*a*) Rotation of block 90° clockwise about *Y* axis. (*b*) Second rotation of block approximately 35° counterclockwise about *X* axis.

As you use them, you will develop a set of tools that gives you confidence to solve problems. The choice of the direct method or the fold line method is a matter of personal preference. Both can be used to solve the same type of problems. Rotation, on the other hand, is required in development problems. Rotation is also a very natural approach for computer displays, and you should be comfortable with the concepts of rotation.

11.4 LINE RELATIONSHIPS

In technical drawings, figures are composed of collections of lines, planes, and curved surfaces. In order to work effectively with technical drawings, you will need to know how lines appear. In this section we look at lines as part of solid objects and lines as individual items.

11.4.1 Principal Lines

In Chapter 4, you learned about principal lines. These are lines that appear true length in one of the principal views and are referred to as frontal, horizontal, or profile lines. Figure 11.33a shows some principal lines highlighted in the three given views of the block. Figure 11.33b shows these same lines in space. Note that, first, the true-length view is the longest of the given views of the lines and, second, the line has to be parallel to the reference plane for the view in which it appears true length. This means that it appears as a line parallel to the edge view of the reference plane or the fold line in the adjacent views. As you look at a principal line, your line of sight is perpendicular to it.

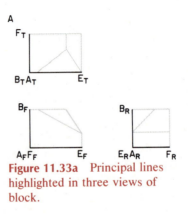

Figure 11.33a Principal lines highlighted in three views of block.

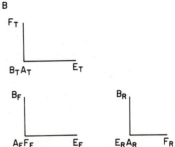

Figure 11.33b Principal lines highlighted in Figure 11.33a shown in three views without block.

11.4.2 Inclined Lines

Inclined lines are lines that appear true length in one of the three principal views and foreshortened in the other two principal views. Figure 11.34 shows three views of one inclined line as part of the block and the line in space. Note that the foreshortened views of inclined lines are parallel to the edge view of the reference plane or fold line between the views. When the fold line or the line representing the edge view of the reference plane is parallel to a view of a line, the line will appear true length in the adjacent view.

11.4.3 Oblique Lines

The term *oblique* refers to lines that are not parallel to any of the principal planes. Figure 11.35 shows three views of an oblique line as part of the block and in space. This presents the problem of determining the length of that line. Why do we need the true length of lines? In engineering applications, a line can represent the wire bracing used on a utility pole to keep it vertical, the structural component that stiffens a framework, the pipeline that leads from a pump in a refinery to the top of a storage tank, or the overflow pipe on an earthen dam. Figures 11.36–11.38 show some of these applications.

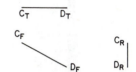

Figure 11.34 (*a*) Inclined line highlighted on block. (*b*) Inclined line shown in three views without block.

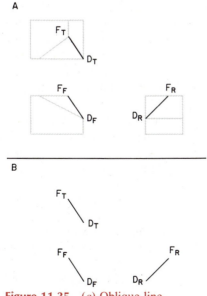

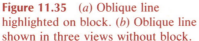

Figure 11.35 (*a*) Oblique line highlighted on block. (*b*) Oblique line shown in three views without block.

Figure 11.36 Utility pole with wire bracing.

Figure 11.37 Building structural components. (Courtesy of Ohio State University Engineering Publications)

Figure 11.38 Radio telescope. (Courtesy of Ohio State University Engineering Publications)

Descriptive geometry provides two solutions for the problem of finding the true length of a line. One way is using auxiliary views. This means that the viewer's line of sight is perpendicular to the given view of the line but from a direction other than a normal view of the principal plane. Figures 11.39 and 11.40 show the auxiliary view for a line as part of a three-dimensional figure and the line isolated in space, respectively. Figure 11.41 shows the steps required for construction of the true length of the line using the direct method. Remember that when an auxiliary view is being projected, there are three views involved: the auxiliary view that is being constructed and two given adjacent views. The projection is done from one of the given views, and measurements for the construction must be made in the other given view. Figure 11.41c shows where the measurements are taken in the front view and where they are positioned in the auxiliary view.

The measurement to point D in the front view is made *away* from the top view and the location of point D in the auxiliary view is made *away* from the top view as well. Figure 11.42 shows the same construction using the fold line method and nomenclature.

It is also possible to find the true length of the line by rotating the line until it is parallel to one of the principal planes. This approach requires two basic operations: a rotation in one view and a translation in an adjacent view. Translation is a movement

FOLD LINE

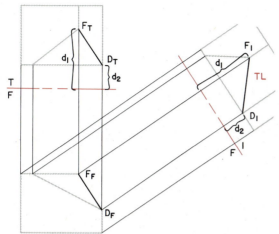

Figure 11.39 Auxiliary view to get true length of line on block.

FOLD LINE

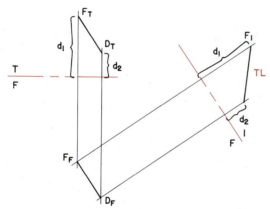

Figure 11.40 Auxiliary view to get true length of same line not attached to block.

of one end of a line to new coordinates with that end at the same distance from the reference plane. In this case, you choose one of two views for the rotation and a point (one end of the line) in this view to locate the axis of rotation.

The axis is installed as a point in this view and as a true-length line in the adjacent view. In the case shown in Figures 11.41 and 11.42 the front and top views were provided. Figure 11.43 shows the rotation in a pictorial. Figure 11.44*a* shows the given infor-

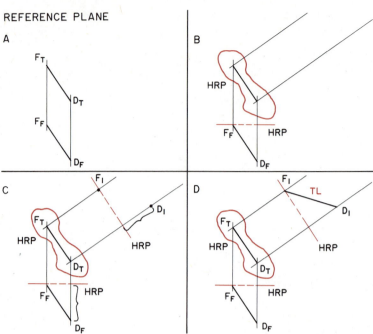

Figure 11.41 Steps to solve for true-length line using direct method.

mation. We shall choose the top view for rotation and choose the end F as the pivot point. The axis of rotation is installed through F in both views. In the top view, the axis of rotation appears as a point, and in the front view it appears as a center line. In Figure 11.44b, D is rotated in the top view until FD is parallel to the frontal reference plane. A projector is then drawn from D in the top view perpendicular to the reference plane in the front view. Then D in the front view is translated parallel to the reference

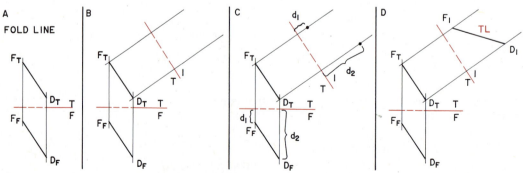

Figure 11.42 Steps to solve for true-length line using fold line method.

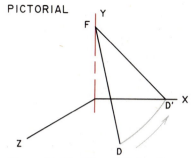

Figure 11.43 Pictorial showing rotation of oblique line to find true length.

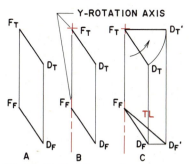

Figure 11.44 Steps of rotation method to solve for true length of oblique line.

plane until it crosses the projector. This locates D in the front view. When F is connected to D in the front view, the problem has been solved, and FD in the front view is the true length, as seen in Figure 11.44c. Compare the true length found by rotation with the true length found by projection. These true lengths are identical.

11.4.4 Parallel Lines

Parallel lines appear parallel in all views. Therefore, if lines do not appear parallel in one view, they are not parallel. Figure 11.45 shows two lines RS and UV that are parallel in the front and top views. When an auxiliary view is drawn to determine the true length of one of the lines, the lines still appear parallel and both appear true length. Figure 11.46 shows another auxiliary view where the line of sight is drawn at an arbitrary angle. This new view shows the two lines as being parallel. Therefore, if two lines are parallel, they are parallel in any view.

The distance between the lines may also be important. If the two lines represent center lines of two pipes, it is possible for the two pipes or the insulation around the two pipes to interfere. The distance between parallel lines can be determined with two auxiliary views. Figure 11.47 shows the steps to find the distance between two parallel lines. The first auxiliary view shows both lines true length. This second auxiliary view is drawn with the line of sight parallel to the true-length lines. This makes them appear to be points, and the space between points is the true-length distance. This procedure is similar for both the first method and the fold line method.

Whenever a second auxiliary view is required and you are using the rotation method, a second rotation is required. While this can

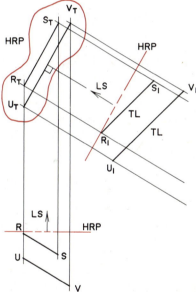

Figure 11.45 Parallel lines in top, front, and auxiliary views. Line of sight 90° to top view gives auxiliary view that shows true length of lines.

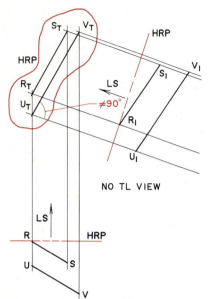

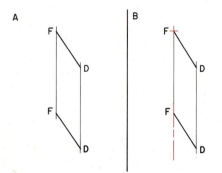

Figure 11.46 Parallel lines in top, front, and auxiliary views. Line of sight at arbitrary angle to top view gives auxiliary view that shows lines are still parallel but not true length.

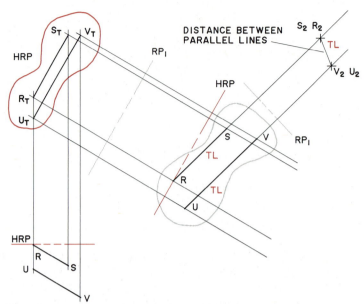

Figure 11.47 Distance between parallel lines.

be confusing, it is a little easier if the given views are in black, the intermediate step (first rotation) is done in one color, and the second rotation is done in another color. Where this cannot be done, an alternative is to label each point *carefully* so that first rotations are primes and second rotations are double primes (*A′* and *A″*). Figure 11.48 shows the double rotation to find the point view of a line. More than two rotations are difficult to show with a two-dimensional drawing on paper; however, a computer system that allows rotation does not retain previous images and can continue to rotate without confusion as long as you define the axis and the angle of rotation.

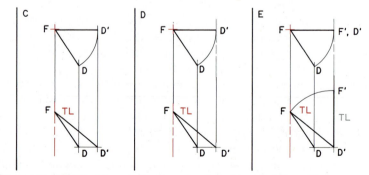

Figure 11.48 Double rotation to get point view of oblique line.

11.4.5 Perpendicular Lines

Figure 11.49 shows **perpendicular line** relationships. Two lines representing the edges of a box have been highlighted with color in the front and right side views. It is easy to see that they are perpendicular when looking at the front view. What makes you sure of this fact? Both lines appear true length in the front view. When both lines are true length and the angle between them is 90°, they are perpendicular. Is it necessary for both lines to appear true length to be able to determine perpendicularity? Do the lines have to intersect in order to be perpendicular? When the relationship between lines is unknown, you must find a view where *one* of the lines appears true length. If the two lines then appear to be perpendicular, they, in fact, are perpendicular.

Figure 11.50*a* shows two **skew lines** (lines that do not intersect in space) and an auxiliary view with one of the lines true length. Figure 11.50*b* shows the second auxiliary view. It also shows the shortest perpendicular distance from one line to the other. The second auxiliary view is drawn with the line of sight parallel to the true-length line. A perpendicular line is drawn from the point view of one line to the other line. Do lines have to be perpendicular to be able to determine the shortest distance between them? If

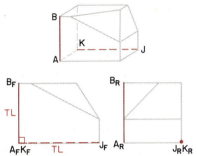

Figure 11.49 Pictorial, front, and right side views of block showing two lines highlighted that are perpendicular.

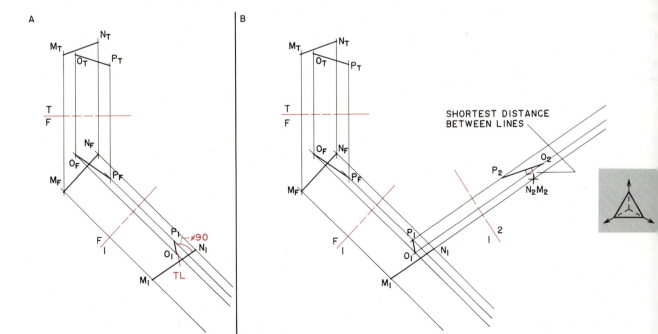

Figure 11.50 Two auxiliary views used to find shortest distance between skew lines.

you think about this question, the answer is no. Skew lines are lines that are neither perpendicular nor parallel to one another; however, there is a shortest distance line between two skew lines that can be determined.

11.4.6 Angular Relationships

Angular relationships apply to lines whether or not they intersect or whether or not they are in the same plane. To find the angular relationship between lines, the lines have to be true length in the same view. This requires a three-step procedure for the general case.

Figure 11.51 shows the steps to find the angle, or apparent angle, between two lines *AB* and *CD*. The first step is to find the true length of one line *CD* and project the other line *AB* into the same view. The second step is to find a point view of line *CD* and again project *AB* into this view. The third step requires a view taken off the point view of *CD* such that the line of sight is perpendicular to line *AB*. This view will show line *AB* true length. It will also show *CD* true length because any view projected from the point view of a line will show its true length.

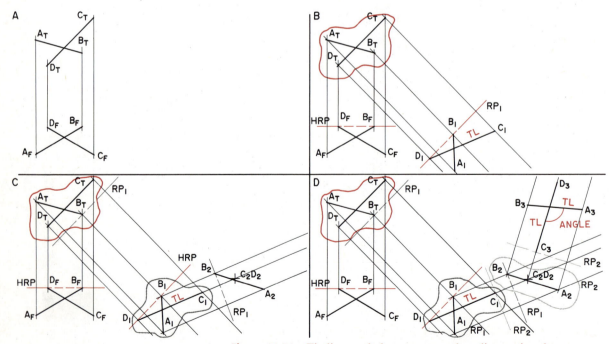

Figure 11.51 Finding angle between two skew lines using three auxiliary views.

Lines that intersect form planes. Planes are discussed later in this chapter.

11.4.7 Applications

There are a variety of applications for line relationships. Some were mentioned in the preceding. One of the most important is the application to graphical vectors.

A vector has a magnitude and a direction. The magnitude is determined by the length of the line and the direction is determined by the arrowhead on one end of the line, as shown in Figure 11.52. Two-dimensional vectors require only one view because they are all true length. See Figure 11.53. Three-dimensional vectors require at least two views, as shown in Figure 11.54. The techniques discussed here, single and multiple auxiliary views or single or multiple rotations, can be used to solve vector problems. Vector problems are discussed in Chapter 12.

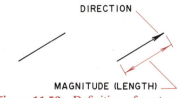

Figure 11.52 Definition of vector.

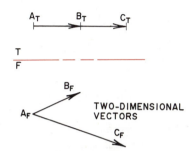

Figure 11.53 Two-dimensional vectors. Two views.

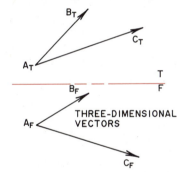

Figure 11.54 Three-dimensional vectors. Two views.

11.5 PLANE AND LINE RELATIONSHIPS

11.5.1 Principal Planes

The cut block that you have been studying has six surfaces that are principal planes. They are numbered 1–3 and 6–8. It also has two surfaces, 4 and 5, that are inclined and oblique, respectively. We shall study these surfaces individually, with some lines that can be represented by your pencil, and in pairs to learn the relationships necessary to solve problems. See Figure 11.55.

The principal planes are those that are parallel to the surfaces of the glass box. Pick up the cut block and hold it in the home position so that you can see plane 1 true size. Then study each of the other planes in turn. Rotate the block so that you can see

Figure 11.55 Block with surfaces numbered and vertices identified with letters.

planes 2, 3, and 6–8. In this process start with the block in the home position and make the X or Y rotations in order to see each of the principal planes. Make some notes for yourself either mentally or on paper as to whether the rotations were clockwise or counterclockwise and the number of degrees required. In each of these cases also note that when you see one principal plane true size, you will see the adjacent principal planes as edges. Keep in mind that each plane's normal vector, the unique vector perpendicular to the plane's surface, will appear as a point when the plane is viewed true size.

Your computer software can be used for this exercise as well.

11.5.2   Inclined Planes

Repeat the studying process outlined in the preceding with plane 4, the inclined plane. First, how did this plane appear when you were studying each of the principal planes? When you can see plane 1 true size, 4 appears as an edge. See Figure 11.56. However, when you rotate the block 90° clockwise about the Y axis to see plane 2 true size, you see 4 foreshortened. See Figure 11.57. In order to see 4 true size, you also have to rotate the block about the Z axis clockwise by an angle of arctan $\frac{2}{3}$, or approximately 34°. See Figure 11.58.

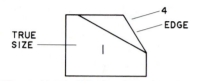

Figure 11.56   Block oriented to show plane 1 true size.

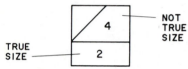

Figure 11.57   A 90° Y-axis rotation to show plane 2 true size.

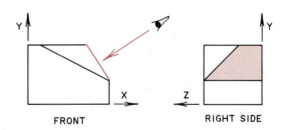

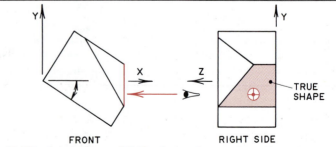

Figure 11.58   Arctan $\frac{2}{3}$ or ~34° Z-axis rotation to show plane 4 true size in right side view.

As an alternative to rotating the block, set it on the desk in the home position. See Figure 11.59. Get up from your chair, walk around to the right side of the desk, and stand in a position so that the line of sight from your eye is perpendicular to plane 4. Place a pencil on 4 so that it appears as a point. See Figure 11.60. Hold it in that position and sit down again. Make your line of sight perpendicular to plane 1. See Figure 11.61. What angle does the pencil form with the surface of the desk? You must use this angle to project a view of the cut block that will show plane 4 true size.

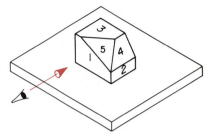

Figure 11.59 Block on desk in home position.

How does this projection process take place? There are some principles involved that must be reviewed. First, any orthographic view of an object only shows two dimensions. In the case of the front view, these are the height and the width. The depth is missing. You know from previous discussions in this text that you can find depth dimensions in the top and right side views. Therefore, if you are going to project from one view, the front in this case, you must have an adjacent view, the top view, either side or profile view, or the bottom view. You also need a reference from which to measure the depths for each feature of the block.

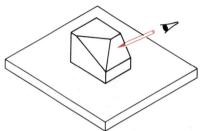

Figure 11.60 Pencil perpendicular to plane 4.

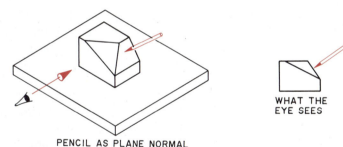

PENCIL AS PLANE NORMAL

WHAT THE EYE SEES

Figure 11.61 Viewer's line of sight perpendicular to plane 1; pencil perpendicular to plane 4.

In the previous sections, we have talked about using a clear plastic sheet as a picture plane, and in the preceding section, we used it as a reference plane. Here, again, we use the plastic sheet as a reference plane. Put the sheet against plane 1 on the cut block. See Figure 11.62. The distance from that reference plane to any individual point on the block remains constant. We use this property to solve the problem of an auxiliary view that gives the true shape of the inclined plane 4.

Study Figure 11.63a, which shows the front and top views of the block. Figure 11.63b shows the reference plane attached to the front view as a cloud and to the top view where it appears as

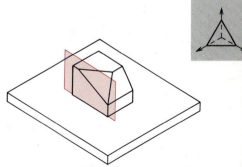

Figure 11.62 Pictorial of block with reference plane installed.

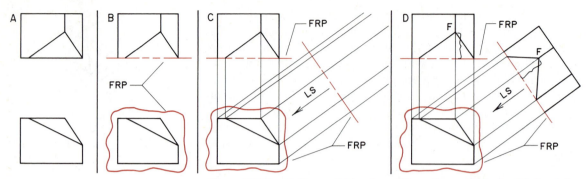

Figure 11.63 Orthographic and auxiliary views of block with frontal reference plane.

an edge. Figure 11.63c shows the edge view of the reference plane drawn perpendicular to the line of sight. The front view locates the starting points for the projection lines, which are drawn parallel to the line of sight in Figure 11.63c. Measurements are made in the top view, as shown in Figure 11.63d, and transferred to the auxiliary view. The completed auxiliary view is shown in Figure 11.63d.

Compare the auxiliary view in Figure 11.63d with the rotated view of the block shown in Figure 11.58. They should be the same view because the line of sight of the viewer is parallel to the normal vector of plane 4. Therefore, you have two ways to solve for the true size of an inclined plane assuming that you work with a three-dimensional CADD system that allows you to manipulate a cut block, have an exact model of the block, or are willing to use the reference plane method to solve the problem.

There are two more paper methods to solve the same problem. Figure 11.64 shows the fold line method of solving the problem. Figure 11.65a shows the cut block inside the glass box for reference. Note that it is very similar, again, to the direct method. The exception is the labeling of the fold lines instead of the edge views of the reference plane. It is more difficult to think of the fold lines as being attached to the object as we did with the reference plane. Figures 11.65 and 11.66 show both of these solutions with just the plane rather than the plane as part of the block. Figures 11.65a and 11.66a are pictorials that provide a reference for the two methods shown in Figures 11.64–11.66.

The final method is to solve the problem by rotation on paper. The steps are the same as they were in the revolution of the block by hand or rotating on the computer. However, your work has to be done more neatly, and you have to keep track of labels

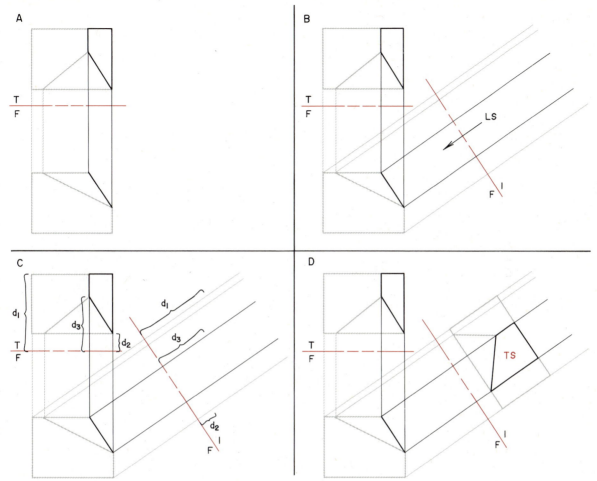

Figure 11.64 Auxiliary view of block using fold line method.

because the first and second rotations are drawn directly over the given views of the object.

Two adjacent views with one showing the inclined plane as an edge are sufficient to solve for the true size of the plane. Figure 11.67*a* shows top, front, and right side views as the given information. However, only the front and right side views are necessary to solve the problem. In this case, a single rotation about the *Z* axis of arctan $\frac{2}{3}$ degrees, or 33.7°, is sufficient to solve the problem.

If you are restricted to displaying the true size of the plane in the front view, as with some computer software, then three views are required. This method takes longer and requires a 90° rotation

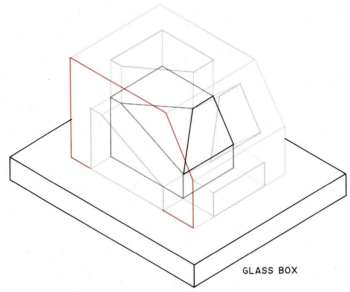

Figure 11.65a Pictorial showing cut block inside glass box.

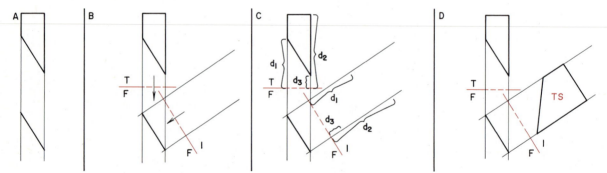

Figure 11.65b Auxiliary view of plane 4 using fold line method.

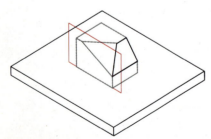

Figure 11.66a Pictorial showing the reference plane.

about the Y axis first and a 33.7° counterclockwise (arctan $\frac{2}{3}$) rotation about the X axis second. Figure 11.67*b* shows the top, front, and side views during the rotation processes.

The available tools will dictate how you solve your problems. The goal here has been to get you to think about the possible methods that are available and to strengthen your visualization skills.

11.5.3 Oblique Planes

Finding the true size of oblique planes requires the same basic steps and information. First, you have to find a view that shows

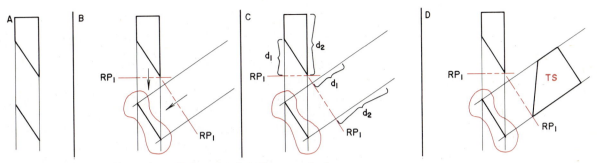

Figure 11.66b Auxiliary view of plane 4 using direct method.

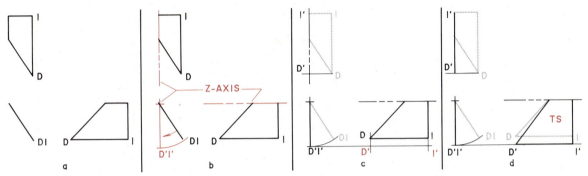

Figure 11.67a Auxiliary view of plane 4 with single rotation.

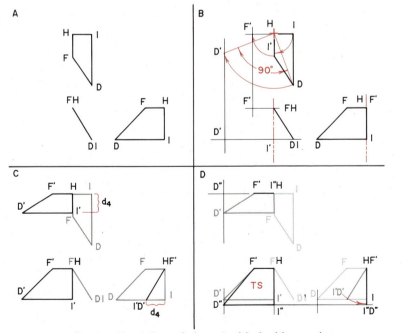

Figure 11.67b Auxiliary view of plane 4 with double rotation.

the **edge view** of the oblique plane and then a view that shows the normal vector of the plane parallel to your line of sight.

In this case, the direct method will be shown first. The fold line method will be shown next. The third method will be to rotate the block with the correct angle of rotation about the *Y* and *X* axes to display the true size of plane 5. The last will be to perform the same operations on a computer.

The key to many descriptive geometry problems involving planes is to find a true-length line in the plane. In this case we have one true-length line in each of the given views. See Figure 11.68*a*. Lines *CD* and *CF* are true length in the front and top views, respectively. This means that either view can be used to start the construction. When there is no true-length line in the plane, you must construct one.

In this case we choose the top view to start the construction. The first auxiliary view will be drawn to the right and above the top view. See Figure 11.68*b*. Put in all the projection lines and the reference plane between the front and top views and touching the front view so that measurements can be made in the front view. Show the cloud view of the reference plane in the top view. Put in the projectors for the auxiliary view parallel to line *CF* in the top view and install the reference plane far enough away to clear any of the given views. Label this reference plane HRP in all views. See Figure 11.68*b*. Make measurements in the front view, transfer them to the auxiliary view, and connect the points with visible lines. See Figure 11.68*c*, which shows the completed first auxiliary view. Figure 11.68*d* shows the projectors from the first auxiliary view, which are drawn perpendicular to the edge view of plane 5.

A second reference plane, RP1, is drawn in between the first auxiliary view and the top view so that it touches the top view. The cloud view of this reference plane is attached to the first auxiliary view, and the edge view of RP1 for the second auxiliary view is drawn perpendicular to the projection lines. See Figure 11.68*d*. Measurements are made in the top view from RP1 to the specific points on the top view and laid off from RP1 in the second auxiliary view. Figure 11.68*d* shows the finished views with plane 5 labeled true size. Figures 11.69*a–d* show the fold line solution to the same problem.

The key to solving the problem by revolution is the same as for solving the problem by projection. You must find the true-length line. In this case, the same true-length line will be used. See Figure 11.70*a* for the given views. Note that in this case we need three views—the top, the front, and the right side views—and that both end points on the *V* line have been labeled with letters. The first step here is to rotate the object in the top view

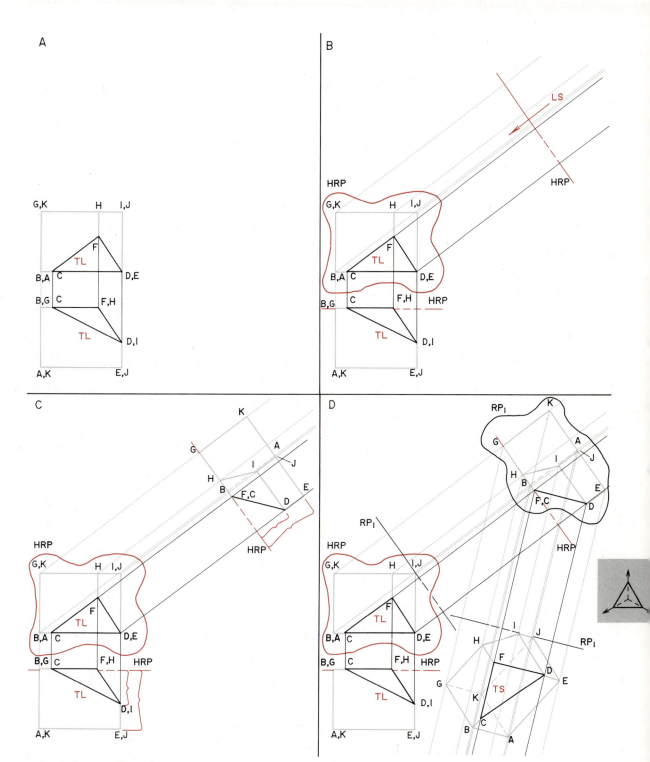

Figure 11.68 True size of plane 5 by direct method.

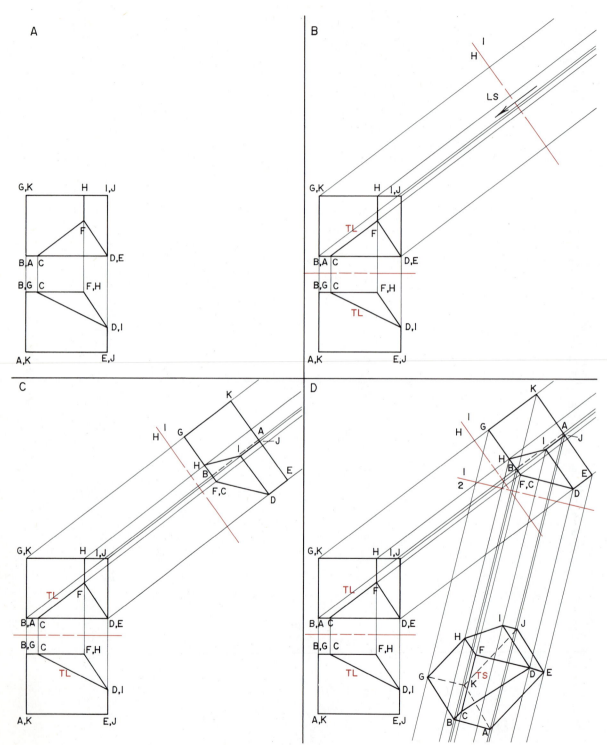

Figure 11.69 True size of plane 5 determined by fold line method.

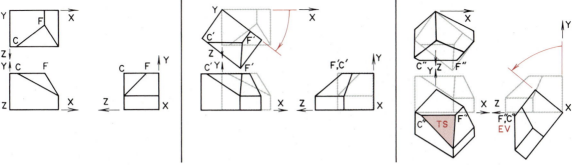

Figure 11.70 True size of plane 5 by rotation.

clockwise about the Y axis until the true-length line CF is aligned with the X axis. You can do this with the model and with the drawing. When the top view has been rotated, you must also show the location of all the points in the front and right side views by labeling them with their letter and a prime. In this case, Figure 11.70b shows the views in the block's new position as dark lines and the original position as gray lines. The right side view shows plane 5 as an edge. The next step in the solution is to rotate the block in the right side view counterclockwise about the X axis until the edge view is perpendicular to the Z axis. Figure 11.70c shows this rotation with the resultant views where the front view shows plane 5 true size.

This particular series of steps was chosen so that the same or similar operations could be done on a computer when all three views are shown simultaneously. In this example, as in the manual example, the front view provides the finished view.

Extra steps will probably be required if you are working on a computer system that displays only one view at a time and allows only X and Y rotations. The steps might be as follows. You are looking at the front view or down the Z axis. First, rotate the block counterclockwise about the Y axis until you can see plane 5 as an edge. Then rotate 90° counterclockwise about the X axis so line CF appears true length and parallel to the Y axis. The third step is a counterclockwise rotation about Y so that the plane appears as an edge perpendicular to the X axis. Fourth, rotate about the Y axis 90° clockwise so that the plane appears true size. Figure 11.71 shows these steps.

When you have only one view available at any one time and can rotate about all three axes, the required steps are as follows: Rotate about Y counterclockwise until line CF appears as a point and plane 5 appears as an edge. Rotate the block about Z clockwise until plane 5 is perpendicular to the X axis. Finally, rotate 90° clockwise about Y again, and you will see the true size of the plane in the front view. See Figure 11.72.

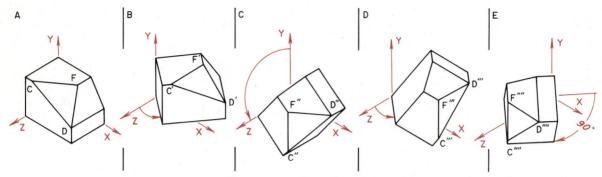

Figure 11.71   True size of plane 5 by rotation. Alternate method.

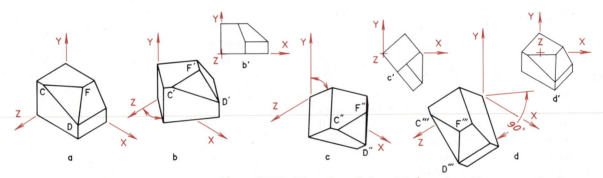

Figure 11.72   True size of plane 5 by rotation. Alternate method.

One more axis of rotation provides you with the ability to do one less rotation.

If your system also provides a normal to each plane, you can reduce the rotations to two. First, rotate the block about the Y axis until the plane 5 normal is parallel to the Y axis and then rotate about the X axis until the normal vector appears as a point in the front view. When the normal appears as a point in the front view, you see the plane true size. See Figure 11.73a–c.

A review of the procedures points out the following facts. When you are solving problems involving planes, no matter which method you use, you will need to know how to find the true length of a line in a plane. The next step is the *point view* of the line, which produces the edge view of the plane. The last step is either viewing the plane so that you see the true size or rotating the block so that you see the true size. Assuming that you can either construct

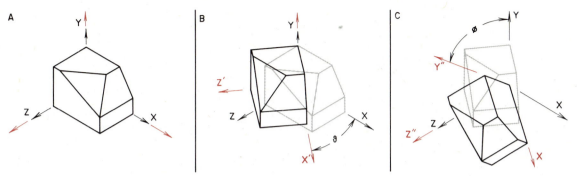

Figure 11.73 True size of plane 5 by rotation. Alternate method.

or have been given the direction normal, the plane will appear true size when the normal vector appears as a point.

11.5.4 Lines Parallel to Planes

A line is parallel to a plane when it is parallel to any line in the plane. If plane 4 is used as an example, we can see that lines *EJ*, *BG*, and *AK* are parallel to the plane because each is parallel to both the lower and upper edges of the plane. When looking down the *Z* axis, you see the edge view of plane 4, and you see lines *EJ* and *BG* as points. With a pencil, draw a line parallel to line *CF* on plane 3 on your block. See Figure 11.74. Observe what happens to that line and to plane 5 as the block is rotated or when you walk around the block.

If you wish to determine whether a given line is parallel to a given plane, draw a line on the plane in one view that is parallel to the line in that view. Project that line to the other view and check to see whether it is still parallel to the constructed line. If it is not, then the line is not parallel to the plane, as shown in Figure 11.75.

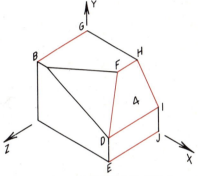

PICTORIAL SHOWING PLANE 4
AND LINES BG AND EJ.

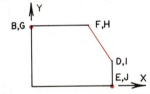

VIEW DOWN Z AXIS SHOWING PLANE
4 AS AN EDGE AND LINES BG AND
EJ AS POINTS

Figure 11.74a Pictorial and front views of block highlighting plane 4 and lines *EJ* and *BG*.

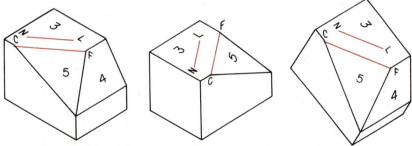

Figure 11.74b Line *NL* parallel to plane 5 and line *CF* in plane 5.

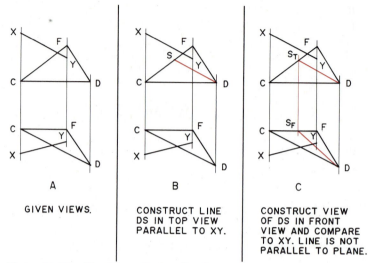

Figure 11.75 Checking whether line is parallel to plane.

11.5.5 Lines Perpendicular to Planes

In the cut block there are a series of lines that are perpendicular to planes. Line *BG* is perpendicular to plane 1. Rotate the model and study the relationships between *BG* and plane 1. When *BG* is a point, you can see the plane true size. When *BG* is true length, you can see plane 1 as an edge. See Figure 11.76.

How do you determine whether a line is perpendicular to a plane? Think about the last two statements in the last paragraph. If you know that the line is true length and the plane is an edge, then you have a line perpendicular to the plane. If the plane is an edge and the line appears perpendicular to the plane, do you necessarily have perpendicularity? Study Figures 11.77a,b. In Figure 11.77a, the line appears perpendicular to the edge view of the plane in the frontal view. However, the top view shows that

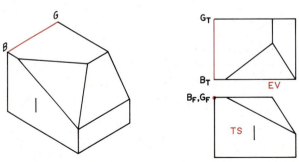

Figure 11.76 Lines perpendicular to planes.

it is not perpendicular because the line *WZ* is not perpendicular to *DI* and *FH*, which are true length in the top view. The frontal view in Figure 11.77b appears just the same as the frontal view in Figure 11.77a, but the top view shows line *WZ* is perpendicular to the true-length lines.

On your model, points *P* and *Q* are located with crosses on planes 2 and 5. Push a straightened paper clip or toothpick through these points. See Figure 11.78, a drawing of the cut block. In the drawing, the line *ST* is shown in color. Is the line perpendicular to either plane? Does it appear to be perpendicular to either plane from any view?

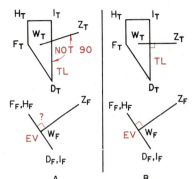

Figure 11.77 Determining whether line is perpendicular to plane.

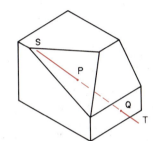

Figure 11.78 Checking perpendicularity with block and colored line *ST* through *PQ*.

11.5.6 Angles between Lines and Planes

What angle does line *ST* make with plane 5? How do you go about measuring the angle? There are projection methods to accomplish this task. One is to find the plane's edge view, then the true-size view, and then the edge view of the plane in a direction that shows the line true length. Figures 11.79a–d show the steps using the direct method. Figures 11.80a–d show the same steps using the fold line method. Figures 11.81a–d show the rotation steps required on paper. The first step is to rotate plane 5 counterclockwise until it appears as an edge in the front view. The rotation is determined by making sure that line *CF* is parallel to the *Z* axis in the top view. Rotate the block clockwise around the *Z* axis until the edge view appears perpendicular to the *X* axis in the front view. The plane should now appear true size in the right side view. Rotate the block clockwise about the *X* axis until the line *ST* appears parallel to the *Y* axis in the right side view and the line will appear true length in the front view. Plane 1 still appears as an edge and *ST* is true length. In each of these cases the last step in the solution process is to measure the angle.

When you are holding the sample block in your hands, it is

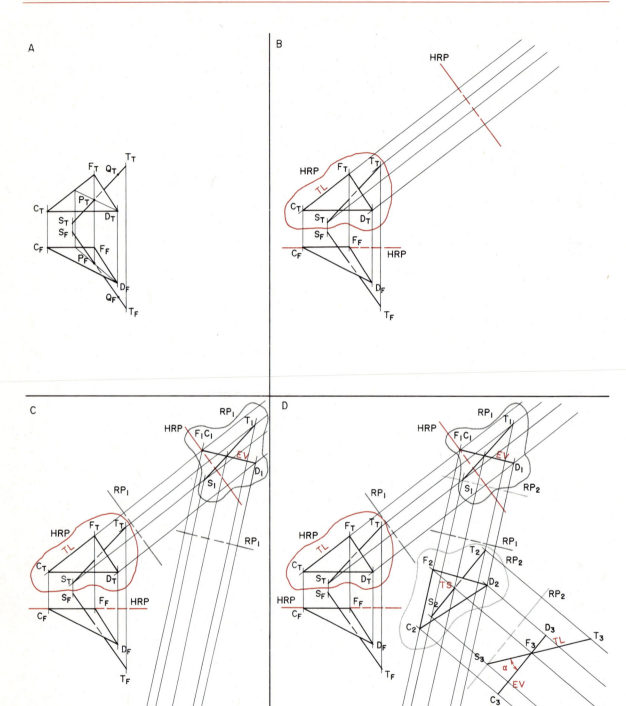

Figure 11.79 Direct method to determine angle between line and plane.

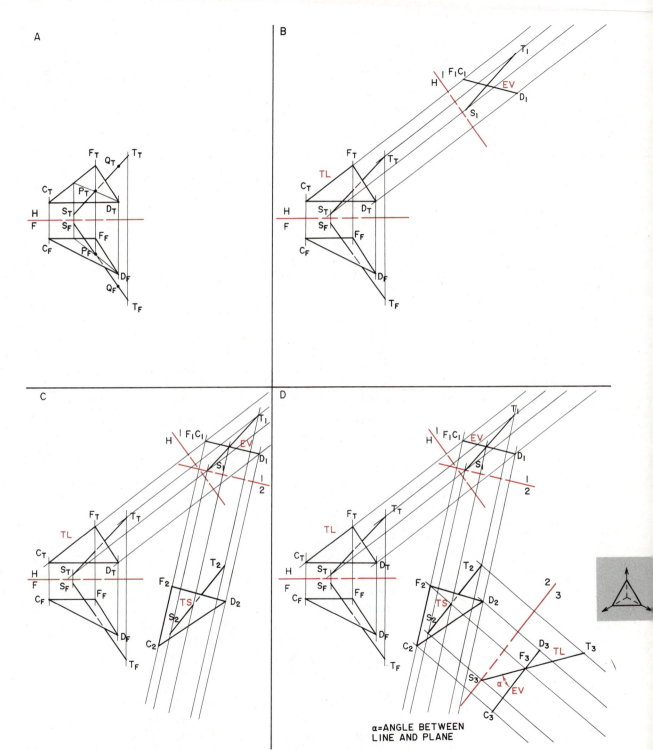

Figure 11.80 Fold line method to determine angle between line and plane.

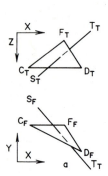
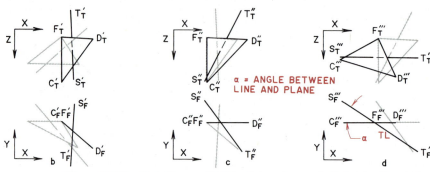

Figure 11.81 Rotation method to determine angle between line and plane.

much easier to find the true angle because you are able to rotate the block about any axis.

11.5.7   Lines Intersecting Planes

One of the key solutions for descriptive geometry problems is finding where a line intersects a plane. There are two methods. If you find the edge view of a plane, it is easy to determine the intersection point on the line. Once you have found the intersection point on the line, that point can be transferred into the other views to locate the point on the plane. Figure 11.82 shows the given views and the auxiliary view using the direct method with the intersection point located in each view. Either the fold line method or rotation can be used to solve this problem.

You need to know what part of the line is visible in each view. In the auxiliary view the entire line is visible. It can be determined which part of the line is visible in the top view. The part of the line "above" the plane (closest to the top view) in the auxiliary view is visible in the top and the part of the line "below" the edge view of the plane is hidden.

The remaining problem is to determine visibility in the front view. Where is the line visible and where is the plane visible? You can determine this by checking the apparent intersections between line *ST* and the edges of the plane in the front view. Trace projectors from these points up to the top view. Which does each projector hit first, the line or the plane? If they hit the line first, as the projector on the left does in this case, then this portion of line *ST* is visible in the front view. This means that the line will not be visible on the other side of the point of intersection. To check this, use the projector on the right side. This time your projector hits the plane first, and therefore the plane is visible on the other side of the point of intersection and the line must be represented by a dashed (hidden) line. If you could not see the

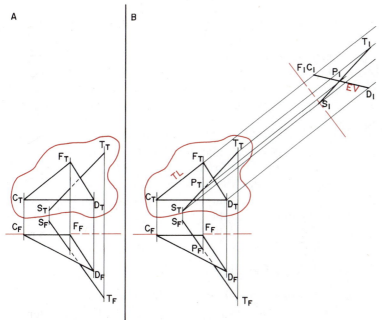

Figure 11.82   Intersection point between line and plane using direct method and edge view of plane.

visibility between the top view and the auxiliary view, this method would also find the visibility for that case. See Figure 11.82 for the steps for determining visibility.

There is another method that can be used to find the intersection point between a line *ST* and plane 5, or *CDF*: the cutting plane method. This method allows you to find the solution for this problem using only the two given views. Therefore, it saves time. More important is the fact that this concept can be used effectively in all intersection problems, as will be shown later in this chapter. The steps are shown in Figure 11.83. The key point is that the edge view of an imaginary cutting plane is constructed that con-

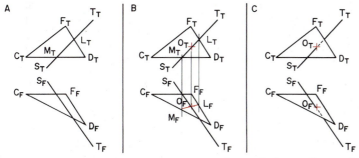

Figure 11.83   Intersection point between line and plane using cutting plane method.

tains the given line in one view where the edge view of the cutting plane is shown in color or labeled CPn. In the example problem, the edge view of the cutting plane is constructed in the top view. The points where this cutting plane crosses the edges of the plane are marked and labeled *L* and *M*. These points are projected to the front view and connected by a line called the trace. The point of intersection, *O*, is where line *LM* in the front view crosses *ST*. Point *O* is projected back to the top view.

The remaining problem is to determine visibility. The method described for determining the visibility in the front view must be used twice: once for the front view and once for the top view. The procedure was described for the front view. For visibility in the top view, you must start in the top view with the apparent intersections and look to the front view to find whether you see the line *ST* or the plane 5 first.

11.5.8 Planes Parallel to Planes

At times you have enough information to define two planes and yet do not know whether they are parallel. You have to find the edge view of one plane and see if the other one also appears as an edge *and* is parallel. Figure 11.84*a* shows two planes that are not parallel while Figure 11.84*b* show two planes that are parallel.

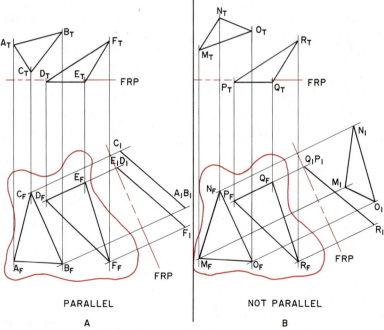

Figure 11.84 Parallel and nonparallel planes.

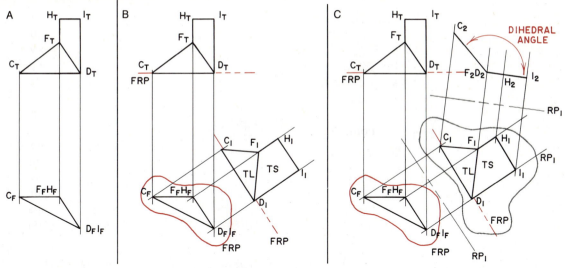

Figure 11.85 Determining dihedral angle using direct method.

11.5.9 Dihedral Angles

Another important problem is finding the angle between planes. This is termed the *dihedral angle*. The key to this particular problem is identifying the line of intersection between planes and then getting the true length and the point views of that line. When the point view has been constructed, both planes will appear as edges and the angle can be measured.

In the example block, the oblique and inclined planes, planes 5 and 4, respectively, have a line of intersection that is labeled *FD*. Figure 11.85 shows the steps to get the true angle between the planes. In this case only the planes have been shown for the reference plane solution. Because the direct and fold line solutions are so similar, the fold line solution has been omitted.

To solve the problem by rotation, you start with the front view by rotating clockwise around the Z axis until FD appears parallel to the Y axis. This will make $F'D'$ true length in the right side view. The next rotation occurs in the right side view by rotating the planes around the X axis until line $F''D''$ again appears parallel to the Y axis. The top view now shows $F''D''$ as a point and the angle between the planes can be measured. Figure 11.86 shows the required steps.

If you have only Y and X rotations available and the front view, then rotate the block clockwise about Y until $F'D'$ appears parallel to the Y axis. Now rotate the block counterclockwise about the X axis until you see $F''D''$ as a point.

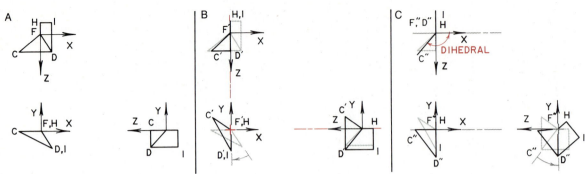

Figure 11.86   Determining dihedral angle by rotation.

11.5.10   Planar Intersections

When the line of intersection between two planes is not known, another projection method becomes important. In this case we think of passing two cutting planes, labeled *ZW* and *XY*, through both of the given planes. If need be, return to the section on intersection of lines and planes for a review of cutting planes. Figure 11.87*a* shows the given planes. Figure 11.87*b* shows the cutting planes being installed in the top view. The traces of the cutting planes are shown in the front view in Figure 11.87*c*, which

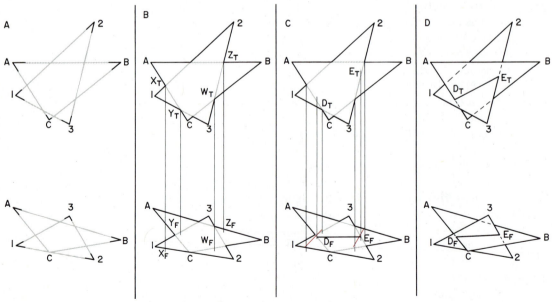

Figure 11.87   Finding line of intersection between two planes using cutting planes.

also shows the line of intersection, *DE*. The line of intersection, *DE*, is transferred to the top view in Figure 11.87*d*. Figure 11.87*d* also shows the visibility of both planes. Visibility must be checked as before.

The rotation solution requires the revolution of the planes as a pair to get one of the planes perpendicular to one of the axes, for example, the Y axis, and both planes as edge views. In order to do this, we first must construct a true-length line in the top view of one of the planes (the black plane or plane 1 was chosen). Rotate the planes about the Y axis until the true-length line is parallel to either the X axis or the Z axis. In this case the planes are rotated clockwise so that the true-length line is parallel to the X axis. We now have the edge view of plane 1 in the right side view. Next, rotate the planes clockwise about the X axis until the edge view of plane 1 is coincident with the XZ plane. This produces the true size of the plane in the top view.

Now construct a true-length line in plane 2 (gray) in the top view. Rotate the planes until the new true-length line is parallel to either the X or Z axis. For this example, the planes are rotated clockwise about the Y axis so that the second true-length line is parallel to the Z axis. The front view shows the edge view of both planes and the point view of the line of intersection. If the intersection does not occur, extend the edge views to find the point view of the line of intersection so that the dihedral angle can be measured. Any adjacent view will show the line of intersection true length. See Figure 11.88.

11.5.11   Plane–Solid Intersections

When planes and solids intersect and the intersection between the two is sought, it becomes a matter of applying the principles of line–plane intersections. We shall deal with two different plane–solid intersections. The first is the intersection of a plane with a prism and the second is the intersection of a plane with a pyramid. In order to make it easy to see the principles, the prism will have a triangular cross section and the pyramid will have three sides and a base (tetrahedron).

If the plane is an edge in the front or any of the given views and we can see the sides of the prism as edges in the other view (top in this case), then the solution is easy. See Figure 11.89. The edge view of the plane appears as a line, and the places where the line cuts across the edges of the prism mark the intersections on the surface. The line of intersection is simply marked as a line in the front view and assumes the cross-sectional shape of the prism in the top view.

A

B

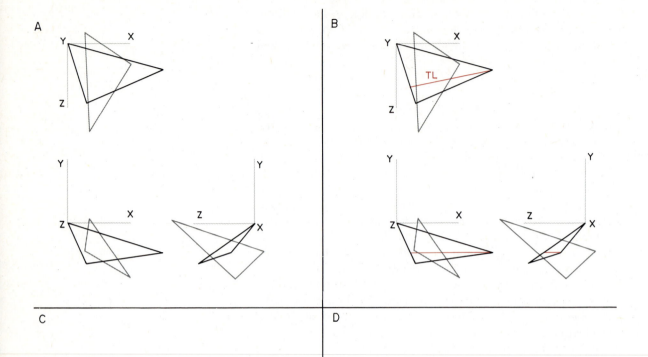

C

D

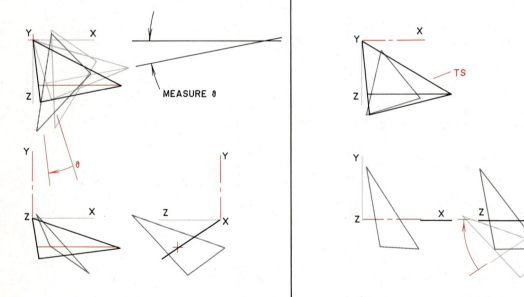

Figure 11.88 Finding line of intersection by rotation. (*Computer-generated image*)

E

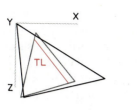

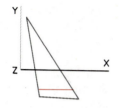

TL

F

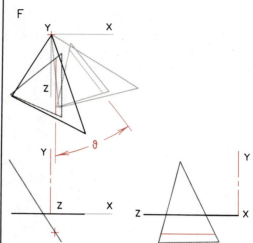

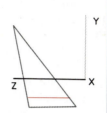

ϑ

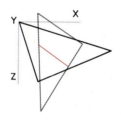

G

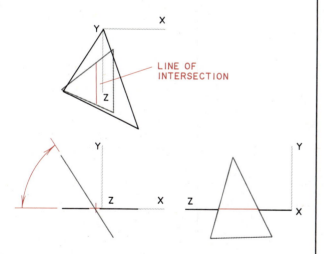

LINE OF
INTERSECTION

H

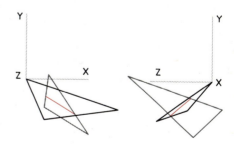

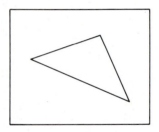

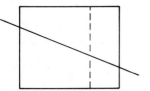

Figure 11.89 Plane intersecting three-sided prism. Front view shows plane as edge.

If the plane does not appear as an edge in any of the given views, then there are two possible solutions. The first is to construct an edge view of the plane. See Figure 11.90. Label each of the intersection points where the prism goes through the plane. Do this in both the top and auxiliary views. Then project the intersection points to the front view by measuring from the reference plane or fold line in the auxiliary view and transferring those measurements to the front view.

Once the points that form the line of intersection have been located in the front view, it is time to determine the visibility. Here, with the help of the auxiliary views, it is relatively easy to see which part of the prism is above and which part is below the plane.

The same problem is presented in Figure 11.91. This time we use the cutting plane method. Starting in the top view, three cutting planes are installed so that they contain the edges of the prism. This creates three traces on the plane. The three traces are then located in the front view, and the points where the traces cross the constructed edges of the prism locate the line of intersection. However, without the edge view provided by the auxiliary view, it is a little more difficult to determine whether the

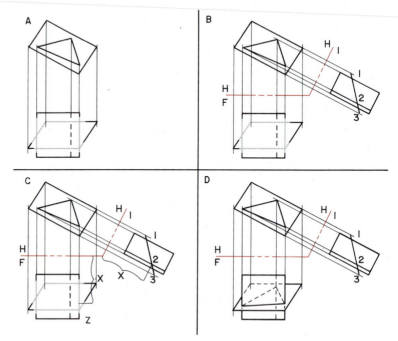

Figure 11.90 Plane intersecting three-sided prism. Top view shows end view of prism. Solve by constructing edge view as auxiliary from top view.

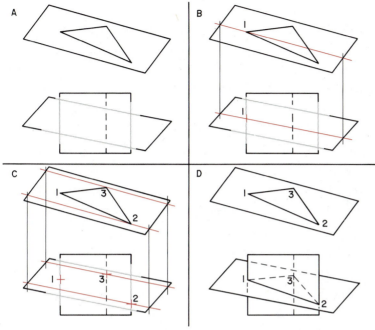

Figure 11.91 Plane intersecting three-sided prism. Top view shows end of prism. Problem solved by cutting plane method.

plane or the prism is visible. You will have to observe in both views whether the traces are located above or below in the front view and are in front of the prism, touch, intersect, or are behind the prism in the top view. You can always check visibility by constructing the edge view. Another method for checking your problems is to take your hexagonal pencil and sketch the plane on a piece of paper. Then push the pencil through the constructed plane and compare your model with your drawings.

Although there are modeling programs that will determine the intersections of two- and three-dimensional objects, many CADD programs do not yet provide the capability. Therefore, the methods described in the preceding paragraphs will have to be used. If the prism happens to have its axis aligned with one of the display axes on the system, then many systems will allow you to rotate the figure until the plane appears as an edge. Do not forget to construct a true-length line in the view where the rotation is to take place. This is the key to solving these problems because it can be used to estimate or determine the required angle of rotation. Figure 11.92 shows a three-sided prism intersected by a plane. The CADD system allows you to see only one view at a time, and this figure shows the given top and front view, the

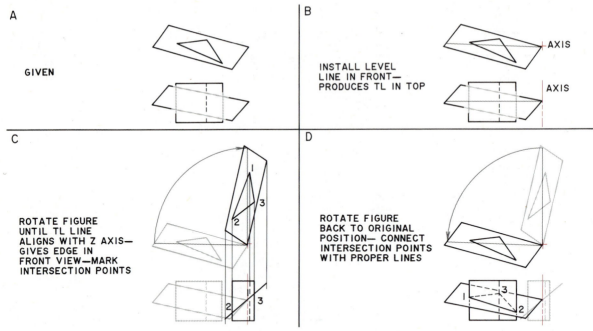

A

GIVEN

B

INSTALL LEVEL
LINE IN FRONT—
PRODUCES TL IN TOP

AXIS

AXIS

C

ROTATE FIGURE
UNTIL TL LINE
ALIGNS WITH Z AXIS—
GIVES EDGE IN
FRONT VIEW—MARK
INTERSECTION POINTS

D

ROTATE FIGURE
BACK TO ORIGINAL
POSITION— CONNECT
INTERSECTION POINTS
WITH PROPER LINES

Figure 11.92 Three-sided prism intersected by plane, solved by rotation. (*Computer-generated image*)

construction of the true-length line, and the rotation so that the edge view of the plane can be seen. Once the edge view is seen in the front view, the points of intersection can be marked on the edges of the prism and the figure can be rotated back to the given position. Now the points can be connected with lines to form the line of intersection.

The intersection of a plane and a pyramid is treated in much the same way as the intersection of a plane and a prism. The techniques for determining the intersection using a cutting plane are slightly different.

When given two adjacent views of a plane intersecting a prism where the line of intersection must be determined, it is always possible to construct an auxiliary view that shows the edge view of the plane. The intersection of the edges of the pyramid with the edge view of the plane are easy to determine. These points are then located in the front view and the line of intersection can be drawn. See Figure 11.93.

You can solve this problem using cutting planes in the two given views. In this case, the key is to install the cutting planes through the vertex of the pyramid coincident with the corners of the pyramid. See Figure 11.94. The traces from the cutting planes

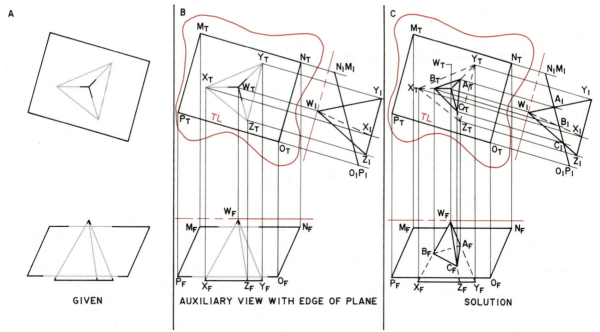

Figure 11.93 Three-sided pyramid intersected by plane. Solved by direct method using an edge view of the plane.

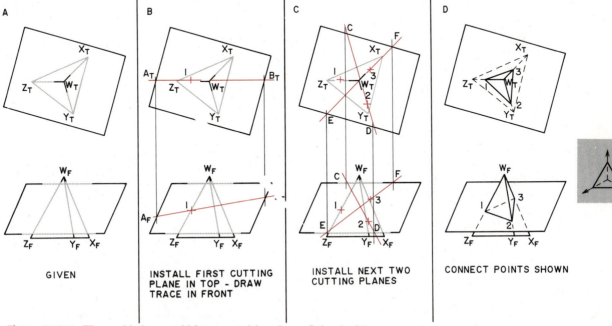

Figure 11.94 Three-sided pyramid intersected by plane. Solved with cutting planes.

are then drawn in the front view where they provide the intersection points. These points can be connected, creating the line of intersection.

11.5.12 Solid–Solid Intersections

The problem to be dealt with in this section on planar figures is the intersection of two solids. Here, the key is to remember the elements already covered: line–plane, plane–plane, and plane–solid intersections. Solving for the intersection of two solids is nothing more than breaking down the solids into manageable pieces. As a demonstration we solve for the intersection of a triangular prism with a three-sided pyramid. See Figure 11.95.

In this case, two of the pyramid edges intersect the upper and lower surfaces of the prism. Because these surfaces appear as edges, it is easy to mark the four intersection points in the front view and project them to the top view. The final two intersection points must be located by passing a cutting plane from the vertex of the pyramid through the point view of the line forming the prism edge and extending through the edge view of the pyramid base. This cutting plane creates two traces, one on the front

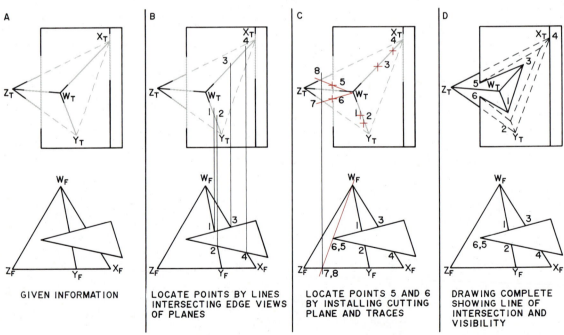

GIVEN INFORMATION

LOCATE POINTS BY LINES INTERSECTING EDGE VIEWS OF PLANES

LOCATE POINTS 5 AND 6 BY INSTALLING CUTTING PLANE AND TRACES

DRAWING COMPLETE SHOWING LINE OF INTERSECTION AND VISIBILITY

Figure 11.95 Three-sided pyramid intersected by three-sided prism. Front view shows end view of prism so that sides of prism are edge views.

surface of the pyramid and one on the rear surface of the pyramid. The intersection points are located where the traces intersect the line forming the edge of the prism in the top view. The six points are connected with lines to give the line of intersection. Note that part of the intersection is visible on the upper surface of the prism and part is hidden on the lower surface in the top view.

In situations where the end view of a prism is not provided, it is easiest to solve the problem by constructing an end view of the prism. The methods shown here work for the intersections of prism with prism and pyramid with pyramid as well.

The key points in this section are the relationships between lines and planes. By knowing how to determine angles between lines and planes or planes and planes, we can see the three-dimensional relationships that are important in design. You should now know when lines are perpendicular or parallel to each other and when a line is perpendicular or parallel to a plane. You should also know how to solve for the intersections between lines and planes through solids intersecting with other solids. All are solved by breaking an apparently difficult problem into a series of simple problems.

11.6 CURVED SURFACES

The previous sections have dealt with lines and planes. You also need to learn about curved surfaces. Curved surfaces include cylinders, cones, spheres, and warped surfaces (Figure 11.96). The material presented here deals mostly with right-circular cyl-

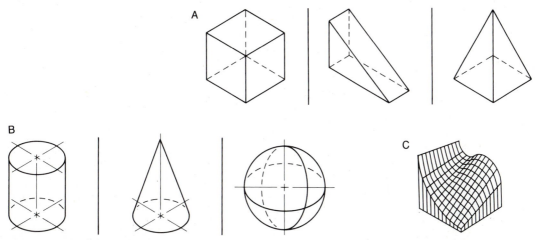

Figure 11.96 (a) Planar primitives, (b) curved surface primitives, (c) warped surface. (*Computer-generated image*)

inders and cones. *Right circular* means that the cross section of the cylinder or cone taken perpendicular to the cylinder or cone axis appears to be a circle.

Although cylinders and cones have been shown in preceding chapters, it is important to review this material again and look at the possible orientations for these primitive shapes. A series of views are shown that display the cylinder and cone in a variety of positions. The series begins with the axis vertical so that the top view appears circular. The subsequent views show the primitives rotated about the base. See Figures 11.97 and 11.98. Note that the front view did not change its shape while the top view shows the circular portion of the object first becoming an ellipse and then a line. When the bases of the primitive shapes are seen as lines in the top view, the top view and front views become identical.

This series of top views could have been drawn by using a series of auxiliary views (changing lines of sight), as shown in Figures 11.99 and 11.100.

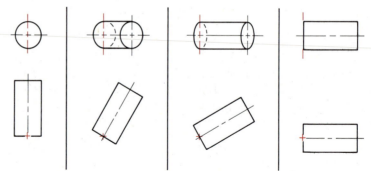

Figure 11.97   Successive views of rotated cylinder. (*Computer-generated image*)

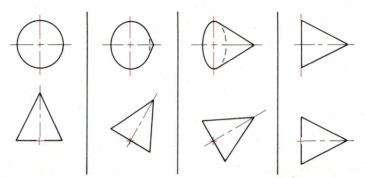

Figure 11.98   Successive views of rotated cone. (*Computer-generated image*)

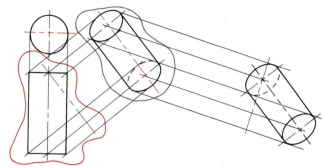

Figure 11.99 Multiple auxiliary views of cylinder.

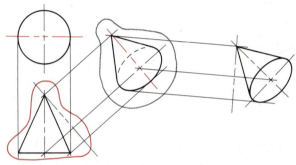

Figure 11.100 Multiple auxiliary views of cone.

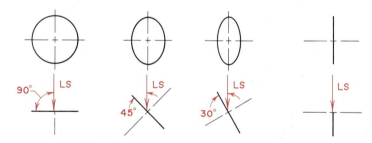

Figure 11.101 Edge and top views of circle with line of sight.

The construction of the ellipses can be accomplished in several ways. First, we must define the relationship between the line of sight and the edge view of the plane. See Figure 11.101. The angle measured is used to specify the proper ellipse template. Templates are available in 5° increments. See Figure 11.102. If the measured angle is a multiple of 5°, the proper template can be chosen and the ellipse completed. Practically, if the angle is other than in 5° increments, the closest angle is used. Figure 11.103 shows how to use an ellipse template to draw a partial ellipse.

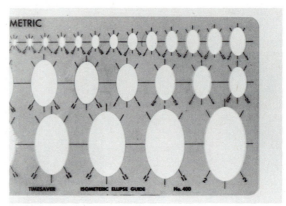

Figure 11.102 Ellipse template.

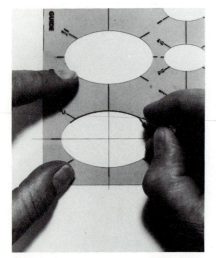

Figure 11.103 Ellipse partially drawn with template.

A second method is known as the trammel method. This method allows construction to proceed if both the major and minor ellipse diameters are known. Figure 11.104 shows this construction method. It is important here to note and remember that the major diameter of the ellipse is always perpendicular to the axis of the cone or cylinder.

A third method is always possible when a true-size view of the circular base can be drawn. Auxiliary views constructed with either a reference plane or rotation can provide the true-size view of the circular base. In this case both the circular view and the elliptical view of the base are projected from the edge view. The measurements are made in the circular view and transferred to the elliptical view. See Figure 11.105. Note that when using the reference plane, it is convenient to locate the plane along the axis so that only one set of measurements needs to be made for all four quarters of the circle.

Although it seems obvious from studying a spherical shape, it is important to remember that we always show the sphere as a circle on paper. We use this fact as we learn about the intersections between various lines, planes, and primitive figures.

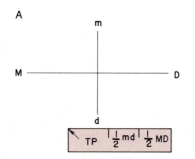

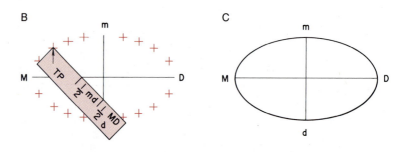

Figure 11.104 Trammel construction.

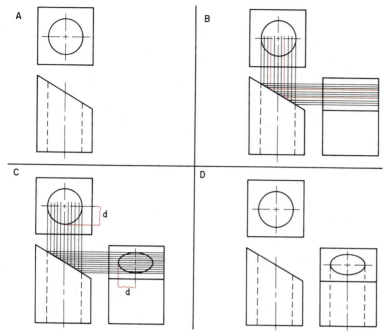

Figure 11.105 Ellipse construction. Edge true-size method.

You will need to know how to find the intersections between lines, planes, and solid primitives. These principles can be applied to complicated shapes. The two techniques for line–plane intersections are useful here. The first was to find the edge view of a plane and the second was to use the cutting plane method. If the problem is simple, the cutting plane is convenient because it requires no extra views. The methods described here apply both to manual and CADD drawings where the CADD package does not have the ability to determine intersections.

11.6.1 Line–Cylinder Intersections

We start with the equivalent of the edge view of the plane. This is the end view of a cylinder, which gives the true cross section of the cylinder. If we do not have the end view of the cylinder (point view of the cylinder axis), it is possible to obtain it by getting the true length of the axis and then the point view of the axis. This can be done by auxiliary views with reference plane–fold line techniques or by rotation. Once the end view is constructed, the intersection point(s) are visible and can be projected back to the given views. See Figure 11.106 for an example using one of the auxiliary view methods. This provides a very sure approach to solving the problem but is not necessarily the most efficient.

The second method requires an edge view of the base through the axis and an adjacent view with this base shown as a circle or an ellipse. If an appropriate base is not available, you can choose the type of cross section needed and install a cutting plane to produce that section. In this case, install a cutting plane through the cylinder so that it is parallel to the cylinder axis and contains the line. The cutting plane is defined by colored construction lines drawn through points on the intersecting line and parallel to the cylinder axis. (Two parallel lines define the cutting plane.) This is done in both of the given views. In the view that shows the

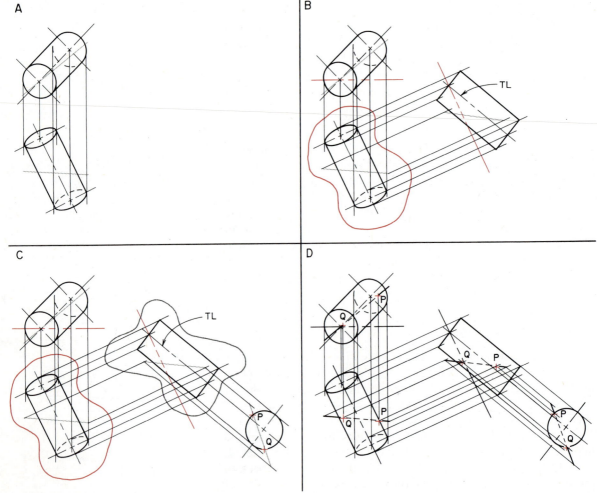

Figure 11.106 Intersection by auxiliary view method.

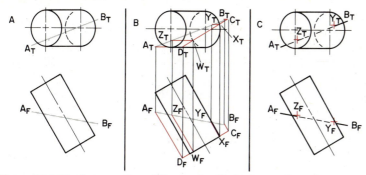

Figure 11.107   Intersection of line and cylinder by cutting plane method.

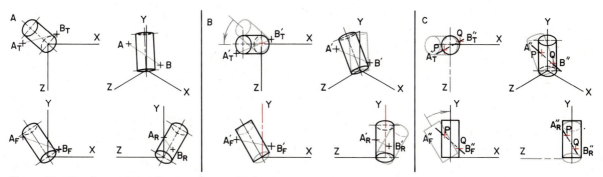

Figure 11.108   Intersection by rotation method. (*Computer-generated image*)

edge of the cylinder cross section, a colored line is drawn that extends the cross section until it intersects the two construction lines. These points are marked in this view and projected into the adjacent view and connected. This line should cross the elliptical view of the cross section at two points. (If it does not, the line does not intersect the cylinder.) These are two points on the trace of the cutting plane and the feet of elements on the surface of the cylinder. Element lines are drawn from these two points parallel to the cylinder axis, and these lines are on the surface of the cylinder. The points at which these elements cross the intersecting line are the intersection points between the line and the cylinder. They are on the surface of the cylinder *and* on the line. These points can be projected to any views. See Figure 11.107.

The intersection of a line and a cylinder can be solved by rotation. The key to the solution is to get a point view of the axis. In Figure 11.108, the axis cannot be seen as a point in either of the given views. The first step is to rotate the cylinder axis about the Y axis until it appears parallel to the X axis. The cylinder axis now appears true length in the front view. The second step is to rotate the cylinder axis about the Z axis in the front view until it coincides with (or is parallel to) the Y axis. The cylinder now appears as a circle in the top view, and the intersection points can be determined as they were in the auxiliary view method.

11.6.2   Line–Cone Intersections

The cutting plane method for cones requires an edge view of a cross section or base through the axis and an adjacent view with this cross section shown as a circle or an ellipse. In this case, the cutting plane is drawn through the cone so that it goes through the tip of the cone axis and contains the line DE. The cutting plane is defined by colored construction lines drawn through points on the intersecting line (the points can be the ends of the line) and the tip of the cone. Two intersecting lines define the cutting plane. This is done in both of the given views.

In the view that shows the edge of the cone cross section, a colored line is drawn that extends the cross section until it intersects the two construction lines. This line is in the cutting plane. These points are marked in this view and projected into the adjacent view and connected. This colored line should cross the elliptical view of the cross section at two points. These points are the feet of elements on the cone and are on the trace of the cutting plane. Element lines are drawn from these two points to the tip of the cone. These lines are traces of the cutting plane on the surface of the cone. The points at which these element lines cross the intersecting line are the intersection points between the line and the cone. These points can be projected to any views. See Figure 11.109.

11.6.3   Line–Sphere Intersections

When a line intersects a sphere, it is possible to find the intersection by constructing a cutting plane that contains the line and then finding the edge view and true-size view of the plane. This provides a true-size view of the portion of the sphere and the cutting plane that coincide. Once this view is obtained, it is possible to see the intersection and project the intersection points to the other views. See Figure 11.110.

11.6.4   Plane–Cylinder Intersections

Finding the intersection of a plane with a cylinder requires the components of the intersection methods described in the preced-

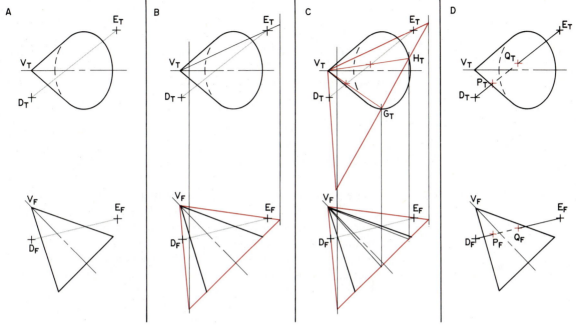

Figure 11.109 Intersection of line and cone by cutting plane method.

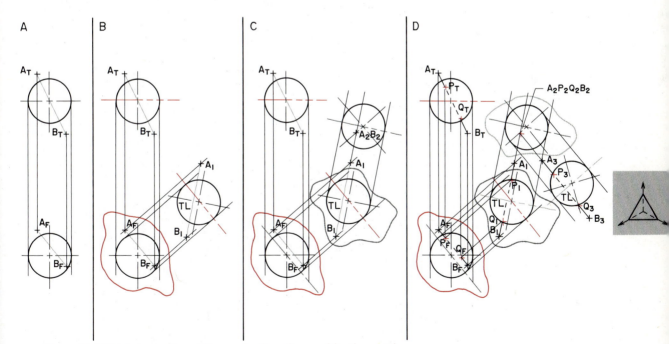

Figure 11.110 Intersection of line and sphere by combination cutting plane and auxiliary views.

ing. The intersection will be a circle when the cylinder is right circular and the plane is perpendicular to the cylinder axis. It can also be an ellipse if the plane is not perpendicular to the axis. The method for solving this problem utilizes cutting planes and a point view of the cylinder axis. The example is shown where the point view of the cylinder is provided. If there is no point view of the cylinder axis, then auxiliary views or rotation must be used to get a point view of the cylinder axis.

The cutting planes are installed in the top view through the cylinder axis. They are evenly spaced to provide a sufficient number of intersection points on the cylinder to produce a smooth line of intersection. The cutting planes are extended to the edges of the plane in the top view. The cutting planes produce traces on the plane and elements on the cylinder in the front view. The traces intersect the elements to provide intersection points that are connected to show the line of intersection. When doing this construction, you should use either numbers to label all points or a series of colored lines to represent the edge views and traces of the cutting planes. See Figure 11.111.

11.6.5 Plane–Cone Intersections

When solving for the intersection of a plane and a cone, it is important to have a point view of the cone axis. A series of cutting

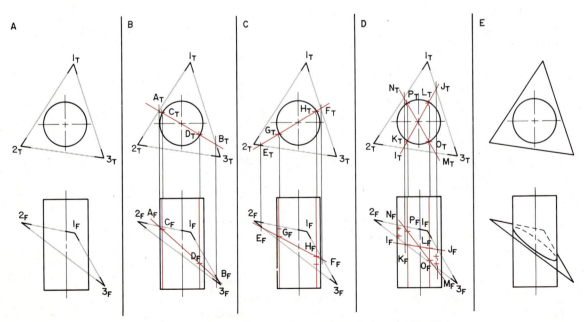

Figure 11.111 Intersection of plane and cylinder.

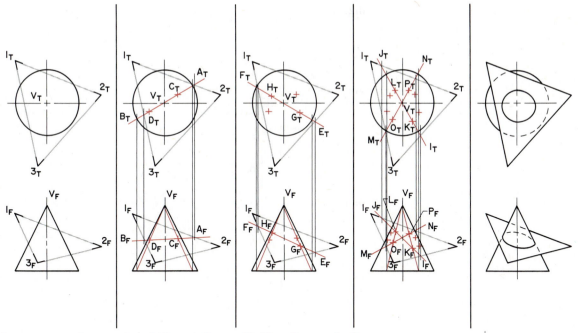

Figure 11.112 Intersection of plane and cone. Cutting plane method.

planes can then be constructed through the vertex of the cone so that they intersect both the plane and the base of the cone. The cutting planes produce traces on the plane and elements on the surface of the cone. The intersection points can be transferred to the other views and connected to form the line of intersection. In Figure 11.112, the cutting planes were installed in the top view and the resulting traces and elements constructed in the front view. The intersection of the traces and elements produce intersection points on the line of intersection. These points are transferred back to the top view to produce the line of intersection in the top view. When the point view of the cone axis is not provided, it must be constructed using either auxiliary views or rotation.

11.6.6 Cylinder–Cylinder Intersections

Cylinder–cylinder intersections have complex lines of intersection. A combination of auxiliary views and cutting planes is used to find the line of intersection. The example shown in Figure 11.113 shows the cylinders oriented so that the axis of one appears as a point in one view. The other cylinder is oriented so that the axis appears true length in the other view. This allows a series of cutting planes to be constructed in the top view. When working with several cutting planes, it is easier to keep track if the planes

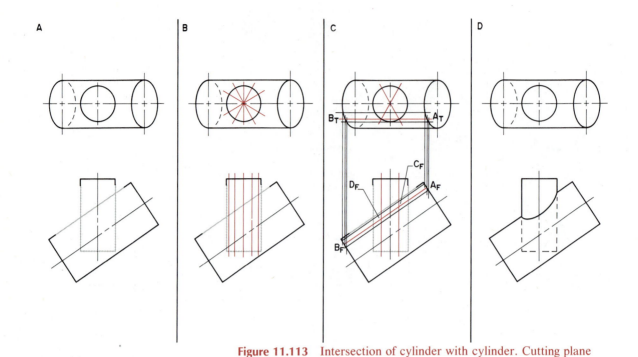

Figure 11.113 Intersection of cylinder with cylinder. Cutting plane method.

are colored or numbered. The traces of the cutting planes in the front view are found by projection from the top view. When the traces are plotted on each of the cylinders, the intersection points are determined by the locations where the traces cross. Once the points have been plotted, they are connected using an irregular curve. On the computer, they might be connected using a spline. It is easier to solve the problem if one cutting plane is constructed, the trace projected, and the points of intersection found before starting on the next one.

In real problems of intersecting cylinders you may not have the convenience of the point view of one axis and the true-length view of the other axis. This will require the construction of sufficient views so that both the point view of one axis and the true-length view of the other axis are in adjacent views.

11.6.7 Cylinder–Right Circular Cone Intersection

The key for determining any solid–cone intersection is to have a point view of the cone's axis in one view, in this case the top view, and a point view of the cylinder's axis in an adjacent view. This allows cutting planes to be constructed perpendicular to the cone axis in the front view and parallel to the cylinder axis. Then

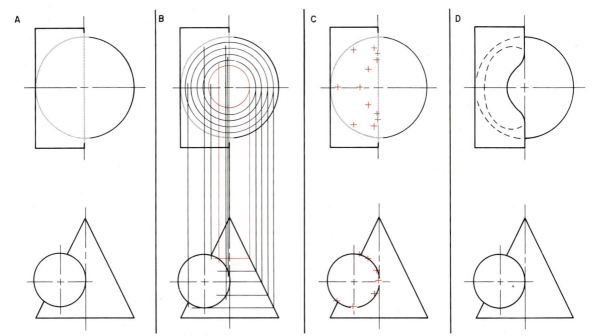

Figure 11.114 Intersection of cylinder with cone. Special case with axes perpendicular.

in the top view with the cone's axis as a point the traces of the cutting plane on the cone are circles while the traces on the cylinder are lines parallel to the cylinder axis. This approach works well for situations when the cylinder axis is 90° from the cone axis. This is a special case and is Figure 11.114. The general case where the axes are not perpendicular is described next.

When the cylinder and cone axes are not perpendicular, an auxiliary view must be constructed that gives the point view of the cylinder axis. Cutting planes are constructed as edges in the auxiliary view through the vertex of the cone. The traces are then constructed in both the front and top views on both the cone and the cylinder. The intersection points are located and then connected either with the irregular curve by hand or with an irregular curve (in some cases a spline) function if you are working on a computer. See Figure 11.115.

11.7 COMMON PROBLEM

In the previous chapters, we have seen the common problem, the saw table fixture, in dimensioned orthographic views. If we wanted to paint or plate the surface area to protect it from cor-

A B C D

B_T
A_T

A_I
A_F
B_I

B_F

Figure 11.115 Intersection of cylinder with cone. Combination of auxiliary view and cutting plane method.

rosion and make it more attractive, we would need the total surface area of the guide. The surface areas of most of the planar surfaces are relatively easy to determine because they are seen true size in one of the principal views. However, the surface area of the oblique plane labeled *ABCD* cannot be readily determined without auxiliary views. We will solve the problem by both projection and rotation. Computer software is used for both solutions.

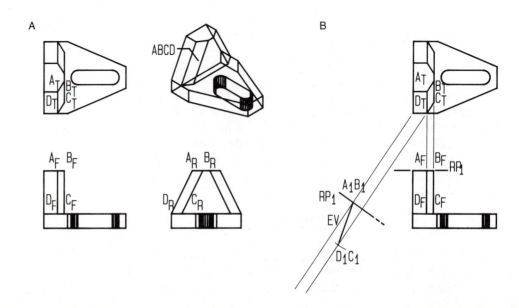

A   B

ABCD

A_T B_T
D_T C_T

A_F B_F
D_F C_F

A_R B_R
D_R C_R

A_T B_T
D_T C_T

RP_1
A_1B_1
EV
D_1C_1

A_F B_F RP_1
D_F C_F

First, we will solve for the true size of the surface by projecting two auxiliary views. Figure 11.116a shows the pictorial of the saw fixture with plane labeled *ABCD*. It also shows the three orthographic views (top, front, and right side). The first step is to find or create a true-length line in one of the views. In this case, line *AB* is true length in the top view. The projectors for the first auxiliary view are drawn parallel to *AB* in the top view. Auxiliary view 1 is located down and to the left, as shown in Figure 11.116b. This view shows the edge view of plane *ABCD*. The projectors for the second auxiliary view are drawn perpendicular to the edge view. The true size of surface *ABCD* is shown in the second auxiliary view.

Second, we will solve for the true size of the plane *ABCD* by rotation. Figure 11.117a shows the three orthographic views of the fixture and an isometric pictorial. The first, step is to find or create a true-length line on surface *ABCD* in one of the views. Both *AB* and *DC* are true length in the top view. We will rotate the fixture in the top view using an axis through *X*. Figure 11.117b shows the same four views after a two-dimensional rotation in the top view. The angle of rotation is 55.5° clockwise. This rotation produces an edge view of plane *ABCD* in the right side view. Figure 11.117c shows the second rotation in the right side view about *X*. The axis of rotation is located at the same corner (labeled *X*), and the angle of rotation is 16.07° counterclockwise. This produces the true size of *ABCD* in the front view.

Figure 11.118a shows just the plane *ABCD* in the three orthographic views and the pictorial. The lower left, rear corner from the fixture has been retained to serve as coordinate axes to

c

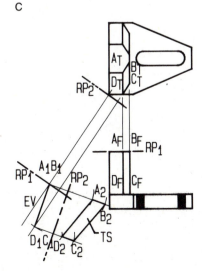

Figure 11.116 (*a*) Orthographic and pictorial views of the table saw fixture. (*b*) Construction of the edge view, first auxiliary view of the plane *ABCD*. (*c*) Construction of the true-size view, second auxiliary view of plane *ABCD*.

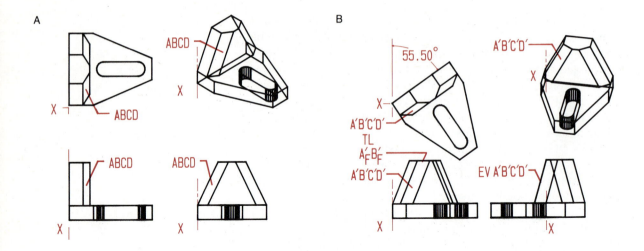

show the location and orientation in space. Again, both AB and DC are true length in the top view. We will rotate the fixture in the top view using an axis through D. The view is rotated clockwise until DC is parallel to the X axis. This is 55.5°. Figure 11.118b shows the orthographic views plus the pictorial view after the first rotation. Plane ABCD is seen as an edge view in the right side view. The next rotation is in the right side view about the point view of line CD. This rotation is counterclockwise until the edge view of ABCD is vertical. This is 16.07°. Figure 11.118c shows the top, front, and right side views of plane ABCD after the two rotations. The front view shows the true size of ABCD while ABCD now is seen as an edge in both the top and side views.

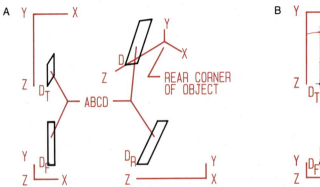

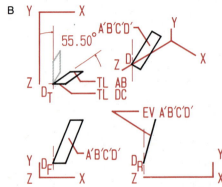

c

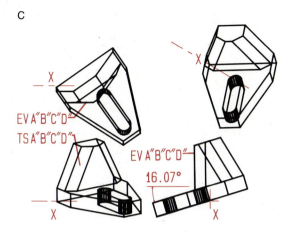

EV A"B"C"D"
TS A"B"C"D"

EV A"B"C"D"

16.07°

Figure 11.117 (*a*) Orthographic and pictorial views of the three-dimensional model of the table saw fixture. (*b*) Two-dimensional rotation in the top view produces an edge view of plane *ABCD* in the right side view. (*c*) Two-dimensional rotation in the side view produces a true-size view of plane *ABCD* in the front view.

11.8 SUMMARY

In this chapter, we have studied lines, planes, and objects as they appear in different views. We have studied the relationships between these elements, which include parallelism, perpendicularity, and finding the angle between elements. We have also looked at intersections of elements and the resulting visibility. These presentations should have given you the picture of how elements appear for various situations and should have improved your ability to visualize. Chapter 12 will make use of this knowledge with some practical applications. The following problems will give you a chance to test your knowledge and skill.

c

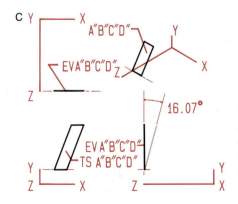

Y   X
    A"B"C"D"   Y

EV A"B"C"D"
              Z   X

Z

16.07°

EV A"B"C"D"
TS A"B"C"D"

Y   Y
Z   X   Z ———— X

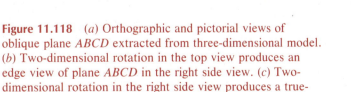

Figure 11.118 (*a*) Orthographic and pictorial views of oblique plane *ABCD* extracted from three-dimensional model. (*b*) Two-dimensional rotation in the top view produces an edge view of plane *ABCD* in the right side view. (*c*) Two-dimensional rotation in the right side view produces a true-size view of plane *ABCD* in the front view.

PROBLEMS

In the following problems, the grid divisions represented by the calibrated axes associated with each figure may be assigned any value per grid division. Suggestions: 1 grid unit = 0.20 in., 0.25 in., 5 mm, 10 mm. Label any true length lines, TL, edge views of planes, EV, and true size views of planes, TS.

Problems 11.1–11.6
Sketch the given orthographic views and the right profile view of each figure on grid paper, and color or crosshatch the principal planes in green, inclined planes in blue, and oblique planes in red. Use a double-width colored line for the edge views of planes.

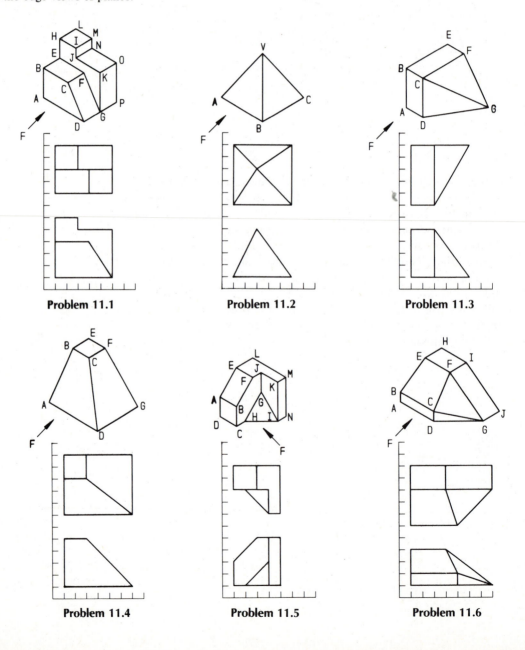

Problem 11.1   Problem 11.2   Problem 11.3

Problem 11.4   Problem 11.5   Problem 11.6

Problem 11.7

Referring to the figure of problem 11.2, draw the given orthographic views and find (or construct the necessary views to find) the true length, of:

    (a) Line *VA*.

    (b) Line *VB*.

    (c) Line *VC*.

Problem 11.8

Referring to the figure of problem 11.3, draw the given orthographic views and find (or construct the necessary views to find) the true length of:

    (a) Line *DG*.

    (b) Line *CG*.

Problem 11.9

Referring to the figure of problem 11.4, draw the given orthographic views and find (or construct the necessary views to find) the true length of:

    (a) Line *BA*.

    (b) Line *CD*.

Problem 11.10

Referring to the figure of problem 11.6, draw the given orthographic views and find (or construct the necessary views to find) the true length of:

    (a) Line *EB*.

    (b) Line *CG*.

    (c) Line *FC*.

Problem 11.11

Referring to the figure of problem 11.1, draw the given orthographic views and construct the necessary view(s) to find the true size of plane *CFGD* and label the edge view, EV, and true size view, TS.

Problem 11.12

Referring to the figure of problem 11.2, draw the given orthographic views and construct the necessary view(s) to find the true size of:

    (a) Plane *VBC*.

    (b) Plane *VAB*.

Problem 11.13

Referring to the figure of problem 11.3, draw the given orthographic views and construct the necessary view(s) to find the true size of:

    (a) Plane *CFG*.

    (b) Plane *CDG*.

Problem 11.14

Referring to the figure of problem 11.6, draw the given orthographic views and construct the necessary view(s) to find the true size of:

    (a) Plane *CDG*.

    (b) Plane *FGJI*.

    (c) Plane *BEFC*.

    (d) Plane *CFG*.

Problem 11.15
Referring to the figure of problem 11.5, draw the given orthographic views and construct the necessary view(s) to determine whether line *HG* is perpendicular to the line *GI*. If they are not perpendicular, determine the angle between them.

Problem 11.16
Referring to the figure of problem 11.3, draw the given orthographic views (you may draw just the given views of the points, lines, or planes specified rather than the entire figure in this and all succeeding problems) and construct the necessary view(s) to determine the shortest distance between:
- (a) *F* and line *CG*.
- (b) *E* and line *DG*.
- (c) *A* and line *CG*. Show the line representing the shortest distance in all views.

Problem 11.17
Referring to the figure of problem 11.6, draw the given orthographic views and construct the necessary view(s) to determine the shortest distance between:
- (a) *B* and line *DG*.
- (b) *B* and line *FG*.
- (c) *I* and line *CG*. Show the line representing the shortest distance in all views.

Problem 11.18
Referring to the figure, draw the given orthographic views and construct the necessary view(s) to determine the shortest distance between the following pairs of lines:
- (a) *AB* and *CD*.
- (b) *AB* and *HK*.
- (c) *EG* and *CD*.
- (d) *EG* and *HK*.
- (e) *CD* and *HK*.

(*Optional:* Show the line representing the shortest distance in all views.)

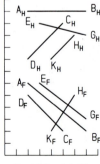

Problem 11.18

Problem 11.19
Referring to the figures, draw the orthographic views of the plane, the given view of the dashed line through the lettered point, and the lettered point in the adjacent view. Construct the adjacent view of the dashed line so that the line is parallel to the plane:

    (a) The first (*a*) figure and letter *A*.
    (b) The first (*a*) figure and letter *C*.
    (c) The second (*b*) figure and letter *E*.
    (d) The second (*b*) figure and letter *G*.

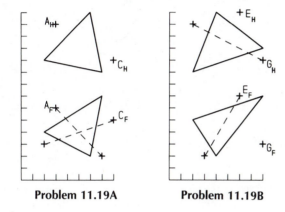

Problem 11.19A   Problem 11.19B

Problem 11.20
Draw the figure on linear grid paper and construct a plane *ABC* parallel to plane *XYZ*.

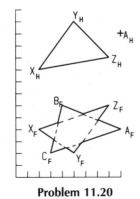

Problem 11.20

Problem 11.21
Draw the figure on linear grid paper and construct a plane *HJW*
parallel to plane *STU*.

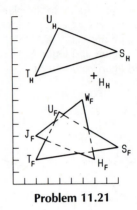

Problem 11.21

Problem 11.22
Referring to the figure, draw orthographic views of the plane shown
and the point specified on linear grid paper. Construct the necessary
view(s) to determine the shortest distance between the point and the
plane. Draw the line representing that distance in the true-length view
and label the point where the line strikes the plane:

    (a) Point *A*.
    (b) Point *B*.
    (c) Point *C*.

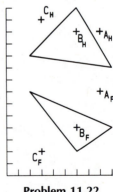

Problem 11.22

Problem 11.23
Referring to the figure of problem 11.5, draw plane *HGI* and point *F* in
two orthographic views and construct the necessary view(s) to find the
distance from point *F* to plane *HGI*.

Problem 11.24
Referring to the figure of problem 11.5, draw the two orthographic views of plane *GHI* and the line specified and construct the necessary view(s) to find the angle between the specified line and plane *GHI*:
    (a) Line *LM*.
    (b) Line *MN*.

Problem 11.25
Referring to the figure of problem 11.6, draw the two orthographic views of plane *CFG* and the line specified and construct the necessary view(s) to find the angle between the specified line and plane *CFG*:
    (a) Line *BF*.
    (b) Line *HF*.

Problem 11.26
Referring to the figure of problem 11.6, draw the two orthographic views of plane *CFG* and the plane specified and construct the necessary view(s) to find the dihedral angle between the specified plane and plane *GHI*:
    (a) Plane *GFIJ*.
    (b) Plane *CDG*.

Problem 11.27
Referring to the figure of problem 11.5, draw the two orthographic views of plane *GHI* and the plane specified and construct the necessary view(s) to find the dihedral angle between the specified plane and plane *GHI*:
    (a) Plane *GIKJ*.
    (b) Plane *ABCD*.

General Note: Label all points of intersection and show visibility in the following problems.

Problem 11.28
Referring to the figure, draw the two orthographic views of the prism *ABCDE* and the line *XY* and construct the necessary view(s) to find the intersection between line *XY* and the specified plane:
    (a) Plane *ABC*.
    (b) Plane *ADE*.

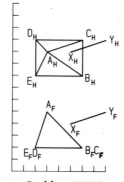

Problem 11.28

Problem 11.29

Referring to the figure, draw the two orthographic views of the prism *ABCDEF* and the line *ZW* and construct the necessary view(s) to find the intersection between line *ZW* and the specified plane:

    (a) Plane *ABCD*.

    (b) Plane *CDEF*.

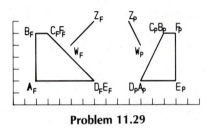

Problem 11.29

Problem 11.30

Referring to the figure, draw the two orthographic views of the prism containing plane *ABC* and the line *ST* and construct the necessary view(s) to find the intersection between line *ST* and the plane *ABC*.

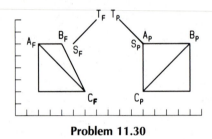

Problem 11.30

Problem 11.31

Referring to the figure, draw the two orthographic views of the prism containing *KLM* and the line *VR* and construct the necessary view(s) to find the intersection between line *VR* and the plane *KLM*.

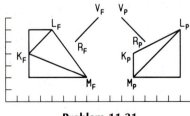

Problem 11.31

Problem 11.32
Referring to the figure, draw the two orthographic views of the prism
VABC and the plane *KLM* and construct the necessary view(s) to find
the intersection between plane *KLM* and the plane(s) specified:
    (a) Plane *VAB*.
    (b) Plane *VAC*.
    (c) Entire prism *VABC*.

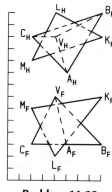

Problem 11.32

Problem 11.33
Referring to the figure, draw the two orthographic views of the prism
ABCDEGH and the plane *XYZ* and construct the necessary view(s) to
find the intersection between plane *XYZ* and the plane(s) specified:
    (a) Plane *ECG*.
    (b) Plane *ABC*.
    (c) Plane *BEC*.
    (d) Entire prism *ABCDEGH*.

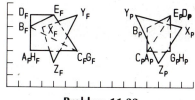

Problem 11.33

Problem 11.34

Referring to the figures, draw the two orthographic views of the prism and plane combination specified and construct the necessary view(s) to find the intersection between the prism and the plane specified:

    (a) Prism *VABCD* and plane *STU*.

    (b) Prism *VKLM* and plane *PQR*.

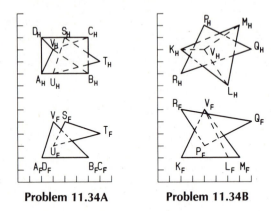

        Problem 11.34A   **Problem 11.34B**

Problem 11.35

Referring to the figures, draw the given orthographic views of the figure specified and find the intersections between the line and the cylinder or cone and label each intersection point. Show visibility.

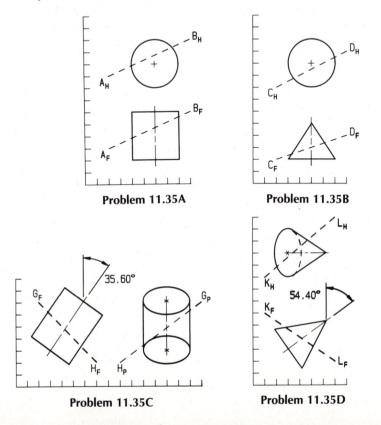

    Problem 11.35A   **Problem 11.35B**

    Problem 11.35C   **Problem 11.35D**

Problem 11.36

Referring to the figures, draw the given orthographic views of the figure specified and find the line of intersection between the plane and the cylinder or cone and label key intersection points.

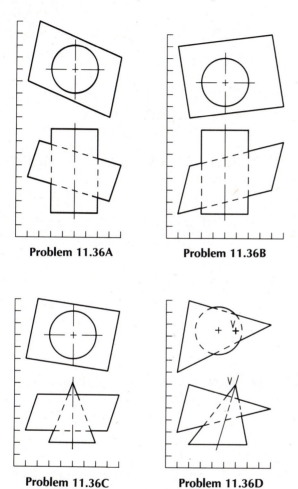

Problem 11.36A   **Problem 11.36B**

Problem 11.36C   **Problem 11.36D**

PATTERN DEVELOPMENT   Process by which a three-dimensional object can be made from flat stock.

MODEL   A three-dimensional representation of plans, created at scale, to show feasibility of ideas.

CONTOUR MAP   A map with lines of constant elevation to show hills and valleys.

BEARING   A direction measured from north or south toward the east or the west—N 60 E.

AZIMUTH   A direction measured from north—N 130.

VECTOR   An arrow representing a direction with the arrowhead and a magnitude with its length.

STANDARD LIGHT RAY   A light ray coming from above the viewer and over the viewer's left shoulder.

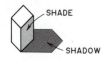

SHADE AND SHADOW   Shade is the part of the object not hit by light rays and shadow is the part of the surroundings not hit by light rays because the object is hit first.

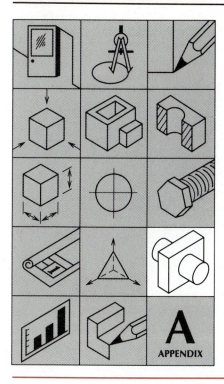

12

3-D GEOMETRY APPLICATIONS

A
APPENDIX

12.1 *INTRODUCTION*

In Chapter 11, you learned about lines, planes, and solid objects and the relationships among them. In this chapter we are going to use the knowledge and the skills that you developed and apply them to solving some practical problems.

We are going to study four applications: development of flat patterns and their use in models; contour maps and the associated symbols; vectors for solving distance, motion, and force problems; and shades and shadows cast by light sources shining on various objects. These are a few of the many applications.

12.2 *DEVELOPMENTS*

12.2.1 Principles and Examples

Designers are faced with many problems that require knowing the true size of all surfaces of an object. The term **pattern devel-**

Figure 12.1 Tool box.

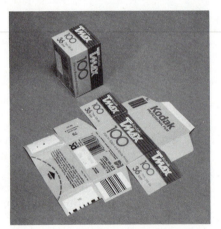

Figure 12.2 Two cardboard boxes, one assembled and one laid flat.

opment defines a process that takes two or three given views and creates the flat pattern or patterns for all of the necessary parts. There are relatively simple development problems associated with containers such as tool boxes, fiberboard milk cartons, and frozen juice cans. There are also some relatively difficult problems that are involved in bulk materials handling that require transition pieces between square or rectangular shapes and circular or elliptical shapes.

Figures 12.1–12.6 show some of the examples outlined in the

preceding. However, all of the items have several things in common. First, each began as a flat piece of a raw material called flat stock. Second, to cut the flat stock, you must know how to find the true shape of each of the surfaces that define the item that is to be made from the flat stock. Third, the only way to solve the assembly problem is to know how to efficiently cut the extra

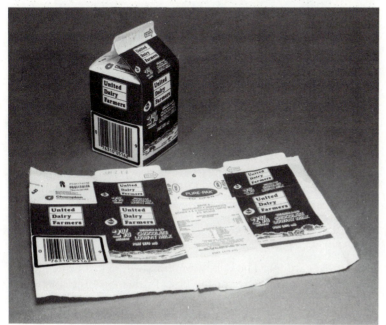

Figure 12.3 Two milk cartons, one assembled and one torn open and laid flat.

Figure 12.4 Funnel, conical-to-conical transition.

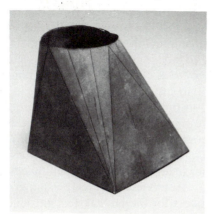

Figure 12.5 Transition piece, square to circular.

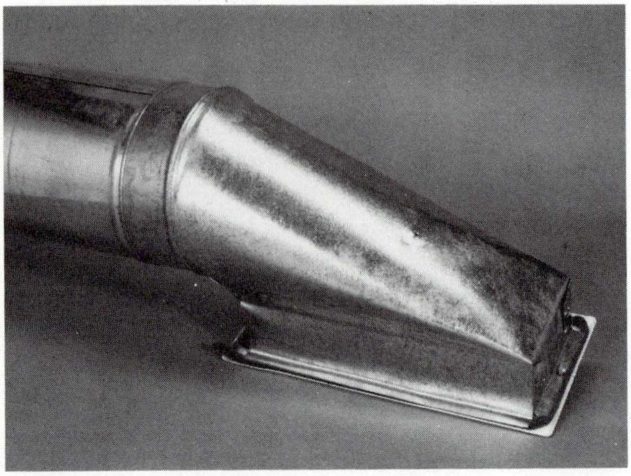

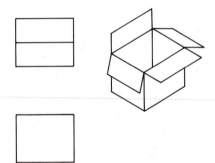

Figure 12.7   Top, front, and pictorial of box.

Figure 12.8   Box with string.

Figure 12.6   HVAC ductwork section.

material that is used to join the surfaces together (tabs). Joining processes include tape, screws, or rivets (mechanical processes); soldering, brazing, or welding (thermal processes); and adhesives (chemical processes). See Chapter 9 for further information on fastening and joining processes.

12.2.2   Right Solids

The simple cardboard box shown in Figure 12.3 can be developed from front and top views of the box. In this case the box opens at the top and bottom. We would like to connect as many of the equal-height surfaces as possible. In this case those would be the front, right side, left side, and back surfaces. See Figure 12.7.

If a string is wrapped all the way around the box, marked, and then pulled out straight, it provides the length of material needed for these surfaces, as shown in Figure 12.8. This is called the stretch-out line, and it is drawn next to the front view. For construction purposes, this line begins close to the front view and is drawn to the right or left of the front view an indeterminant length. It is a good idea to label the points in the top and front views so that you can keep track of the panels. When doing this, imagine that you are standing inside the box and looking at the side panels. Turn clockwise (to the right) and label the corners as you turn. The length of this line can be determined from the given views by taking dividers and measuring the perimeter in the top view. As the length of each side is measured, it is marked and labeled on the stretch-out line. The height of each panel or side can be projected from the front view. Figure 12.9 shows the steps in the development process for the box. In this particular example, all of the sides are the same height, and it does not make any dif-

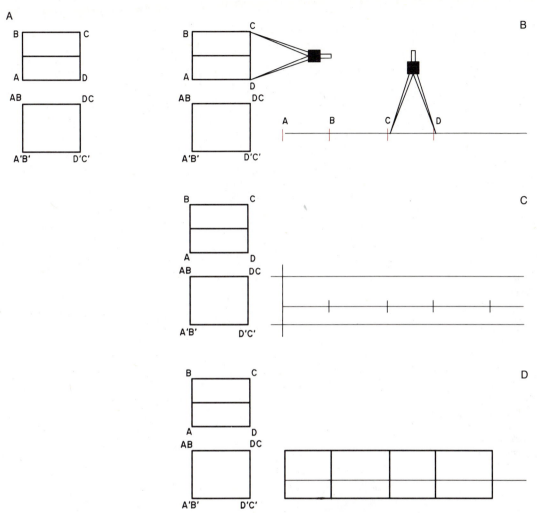

Figure 12.9 (*a*) Top and front views of box with stretch-out line and points labeled on given views. (*b*) Top and front views showing *A* and *B* marked and measuring the distance from *C* to *D*. (*c*) All points marked on stretch-out and first projection. (*d*) Side panels complete.

ference where you start and where you finish. This is not true for many other problems.

A complete box also requires a top and a bottom. For most packing boxes, the top is formed by having a flap on each panel above and below so that the box can be assembled with tape. Another flap is required at the beginning or end of the set of side panels so that tape or glue can be used to hold the side panels together to form the body of the box. See Figure 12.10.

Once the pattern has been drawn on paper, it is transferred to

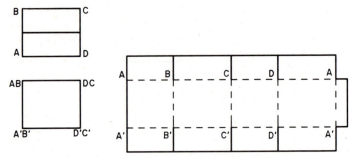

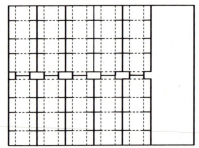

Figure 12.10   Figure 12.9 pattern with flaps added. (*Computer-generated image*)

Figure 12.11   Flat stock shown with more than one pattern nested. (*Computer-generated image*)

the flat stock so that it can be cut out. Most patterns are constructed as inside patterns so that the markings on the stock do not show once the material is assembled. In order to conserve material and avoid scrap, multiple patterns are arranged on a large sheet of stock material so that there is minimum scrap after the patterns are cut. See Figure 12.11.

The steps to solve a second example with a rectangular cross section are shown in what follows. In this case, both the heights of the corners and sides of the base are of unequal length. Notice that the greatest heights are used for folding the material while the shortest height is used for the location of the seam. This saves money because a smaller number of rivets or spot welds are required or less adhesive is required. This also promotes the integrity of the finished package. Notice also that the attachment for the ''lid'' is on one of the long sides and the ''flaps'' to create the seams are on the two short sides and only one long side. See Figure 12.12.

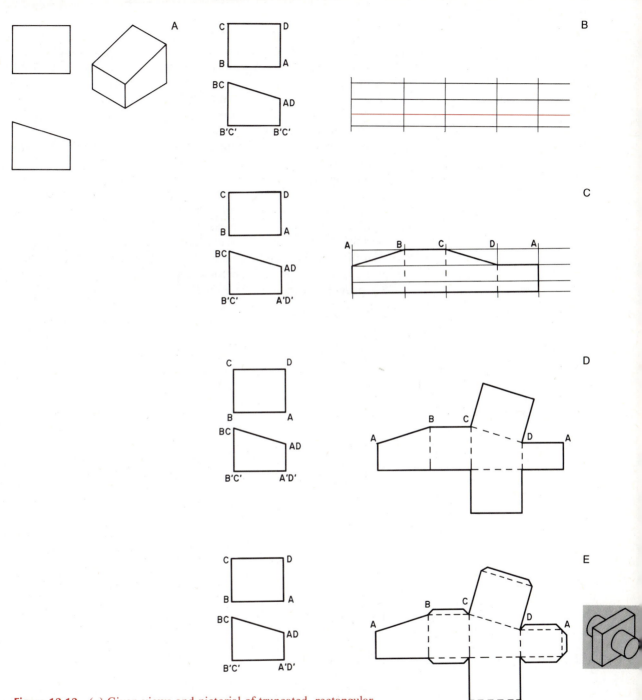

Figure 12.12 (*a*) Given views and pictorial of truncated, rectangular prism. (*b*) Stretch-out line and height projection. (*c*) Finished sides. (*d*) Top and bottom lids added. (*e*) Flaps added for gluing or riveting. (*Computer-generated image*)

12.2.3 Right Cylinders

Normally the objects with circular or elliptical cross section are drawn so that the top view shows the cross section. The front view shows the "rectangular" or longitudinal view of the object. The stretch-out line is drawn to the right or left of the front view. The height, either constant or variable, can be projected from the front view of the pattern view. See Figure 12.13.

When working with circular cross sections, there are two approaches that work reasonably well. In both cases, the circle is subdivided into a number of equal arcs and the subdivisions are labeled. In the first case, the circumference of the circle is calculated and then is divided by the number of arcs to give the arc length. The arc length is measured on the stretch-out line once, and the dividers are used to mark the remaining arc lengths. Each mark is labeled. The heights for the ends of each vertical section are projected from the front view and the flat pattern is completed by providing a tab at one end. If the height is constant and the

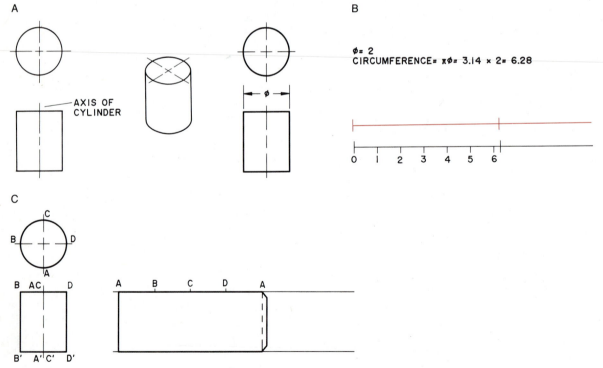

Figure 12.13 (a) Top, front, and pictorial views of cylinder. (b) Top and front views with stretch-out line. Calculations shown for arc length. Figure shows scale laying off arc length on stretch-out line. (c) Construction for projection of height and finished pattern. (*Computer-generated image*)

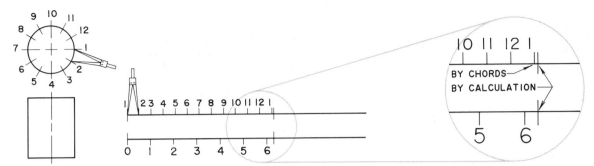

Figure 12.14 Figure 12.13 repeated but showing chord of arc being measured with dividers. (*Computer-generated image*)

resulting pattern rectangular, the construction is relatively simple. The second approach to creating the proper length on a stretch-out line assumes that relatively small arcs and the cords for the small arcs are close enough to the same size that a calculation does not have to be made. The first approach works all of the time. Figure 12.14 shows a comparison.

When a tube with a circular cross section is to be joined with another circular cross section tube or to a flat surface but where the plane of the joint is at an angle to the axis of the cylinder, the height will vary so that the pattern will be curved at the top. Figure 12.15 shows two intersecting tubes. Figure 12.16 shows a tube that has been cut at an angle. The top view appears as it did before, as a circle. The same processes take place except that labeling is very critical and the height projections must be done from the vertical line in the front view to the matching vertical line in the pattern. Here, again, it is important for the drafters to assume that they are standing inside the object in the top view and turning to their right (clockwise) to see the successive panels in the correct order.

Once each of the heights have been marked, an irregular curve can be used to create the correct shape of the top of the pattern. See Figure 12.17. Where the pattern is symmetrical, you can speed up your work by marking your irregular curve and flipping it over so that you can draw the same curved portion on the other part of the pattern. If you are uncertain if the irregular curve was drawn correctly, mark and project a few additional points on the circumference to check your work. Remember that the more segments you draw, the more accurate your construction will be.

Creating flat patterns with the computer is generally no more difficult than it is to do them by hand. A feature that is required of the CADD system is the ability to fit a curve to the points.

Figure 12.15 Two HVAC cylindrical tubes intersecting.

A

B

C

D

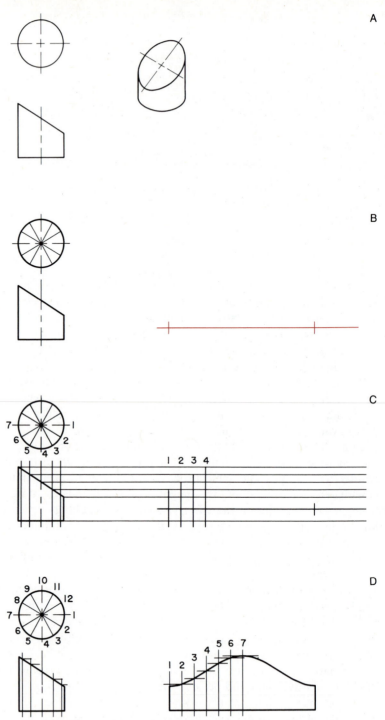

Figure 12.16 (*a*) Top and front view of truncated cylinder. (*b*) Stretch-out line added. (*c*) Construction (projection) of first two heights shown. (*d*) Finished construction. (*Computer-generated image*)

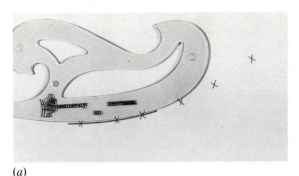
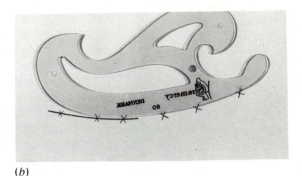

(a)   (b)

Figure 12.17 (a) Construction of ellipse by connecting three or more points with irregular curve. (b) Irregular curve flipped for mirrored points.

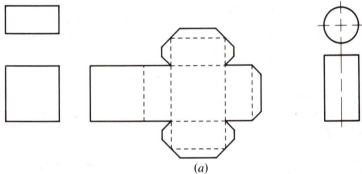
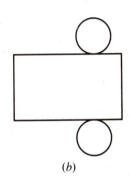

(a)   (b)

Figure 12.18 Pattern development done on CADD systems. (a) GraphiCAD and (b) AutoCAD. (*Computer-generated image*)

However, once the flat pattern is created with a computer system, the drafter can draw the stock size of the material from which the pattern(s) will be cut. The patterns can then be moved around on the stock size so that there is the least amount of waste. In addition, some CADD systems have been interfaced (connected) to laser on mechanical cutting systems so that no programming is required to set up the production system once the pattern has been drawn. Figure 12.18 shows patterns developed on two CADD systems. Refer back to Figure 12.11 for an example of multiple patterns closely packed to minimize waste.

12.2.4 Cones

When dealing with both cones and pyramids, it is important to keep in mind the necessity for true-length lines in order to be able to complete developments. It is easy to lose track of which side lengths are known and which are unknown. These shapes require a thorough understanding of the principles of revolution to obtain the true lengths of sides. As a reminder, if you rotate in one view, you translate in the other view and make the measurements in that view. Labeling is very important to keep track of the given views and the true-length segments.

Cones will be examined first because a view where the axis of a right-circular cone is true length provides a true-length edge or radial line. In Figure 12.19, the top view of the cone appears to be a circle and the front view, where the cone appears as a triangle, is the one that provides the true-length radial line from the top to the base of the cone. The radial line is highlighted in both views.

In Figure 12.19 the top and front views are given. The first step is to find a true-length line, which is shown highlighted in Figure 12.19*b*. The development drawing begins in Figure 12.19*c* by locating the apex of the cone and drawing the first radial line and the arc for the base. Two methods are available for finding the length of the base arc: the calculation method described in Section 12.2.3 and the chord method that is shown in Figure 12.19*d*, which shows the completed pattern.

Figure 12.20 shows the same right-circular cone with the top having been truncated at an angle to the cone axis. In this case much additional work must be done. The base circle in the top view is divided into an even number of equal arcs. These arcs are labeled in both the top and front view. In the front view, each mark has two labels. One for the front of the cone and one for the back. These marks are connected to what would be the apex of the cone in both views. You can see that the radial lines have

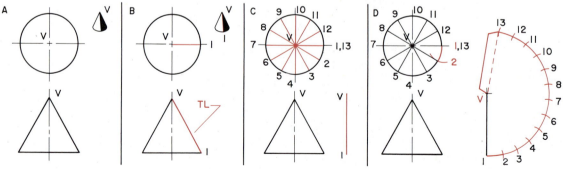

Figure 12.19 (*a*) Top, front, and pictorial views of a right-circular cone; (*b*) top, front, and pictorial views with *TL* line highlighted; (*c*) beginning of development with apex of cone and first radial line with arc for base; (*d*) chord system for laying off base arc and finished drawing.

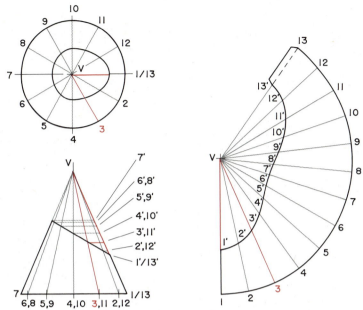

Figure 12.20 Development of truncated cone.

varying lengths because of the truncation. Each radial line must be laid off individually in order to complete the pattern development. In this case only half of the pattern need be drawn because the object is symmetrical. After tracing the half pattern on the stock piece, the pattern can be flipped over and the second half of the material to be cut can be traced.

12.2.5 Pyramids and Tetrahedrons

The pyramid shown in Figure 12.21 has none of its inclined edges (legs) shown true length in either view. The first step is to number or letter the vertices of the pyramid. Again work from the inside and number clockwise. The next step is to revolve one of the legs about the axis of the pyramid in the top view and translate the lower end in the front view to get the leg's true length. Lay the leg's true length off near the two given views and begin the development. Use this length as the radius of an arc and strike an arc to the right. The pyramid base is shown true size in the top view. Measure the length of the first edge of the base with your compass and strike an arc with the lower end of the first leg as center. This second arc will intersect the large arc locating the second corner of the base. Continue this process until you have done each of the edges of the pyramid. Construct a tab along the finishing edge. Attach a base with tabs to the first side and darken

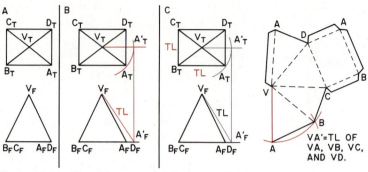

Figure 12.21   Development of full pattern for right pyramid.

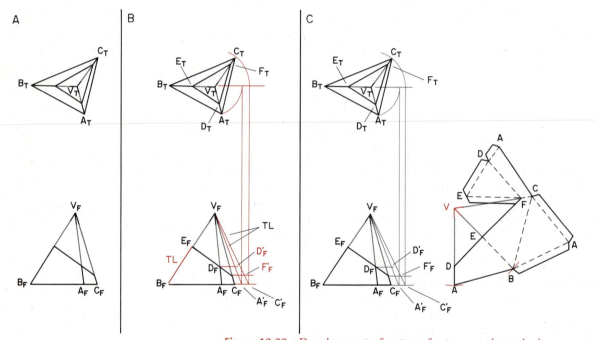

Figure 12.22   Development of pattern for truncated tetrahedron.

the edges of the pattern that will be cut. Use a dashed line for the location of the folds. In this case half of the pattern is sufficient if the proper starting point is chosen.

Figure 12.22 shows the development of a truncated tetrahedron where the axis is not perpendicular to the base. In this case the true length of every line must be determined either through rotation or auxiliary views.

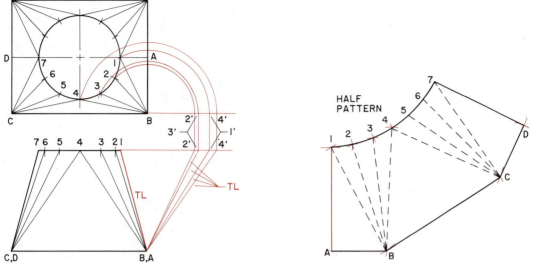

Figure 12.23 Half pattern development for transition piece.

12.2.6 Transition Pieces

There are occasions when one shape for tubing or ductwork must change to a different shape. One example of this is when a furnace's air outlet is round and the desired ductwork shape is rectangular. A piece of ductwork is needed to provide the transition from one shape to the other.

Figure 12.23 shows the steps for the development of a transition piece for a square to a circular cross section. The key here is to remember that corners are in effect part of a cone while the sides are triangles. A quarter of the pattern can be drawn and used four times to get the full pattern. In the example shown, half of the pattern was constructed. You can see the work that could be saved by clever planning.

12.2.7 Models from Intersecting Solids

In Section 11.5.12, a three-sided pyramid is intersected by a plane. Figure 11.94d is reprinted as Figure 12.24a. In order to build a **model** of this figure, the patterns (true size) of the plane and the pyramid must be constructed. Redraw the figure (top and front views) so that they occupy approximately two-thirds of an 8.5 × 11-in. sheet of paper. Use a sheet of 11 × 17-in. paper for the patterns. Develop the patterns by finding the true size of the plane and the true lengths of the edges of the pyramid. See Figure 12.24*b*. Figure 12.24*c* shows the patterns. Mark the points where the plane intersects the edges of the pyramid and label them points 1, 2, and 3. Locate these points on both patterns (see Figure 12.24*c*) and compare the distances 1–2, 2–3, and 3–1 on the patterns. If they are different, go back to the given views and check your construction. Also check the mating edges of the pyramid. Make xerographic copies of your patterns on cheap paper and cut them out. Do you really have inside patterns? Did you remember the flaps? If not, go back and reconstruct your patterns.

Assuming that there were no differences or that the patterns have been corrected, transfer the patterns to bristol board (no. 2 or 3) or other heavy paper. Patterns can be transferred by taping the pattern to the bristol board and poking holes at intersections with a pin or the point on your dividers through the pattern into the bristol board. When all the intersections have been located,

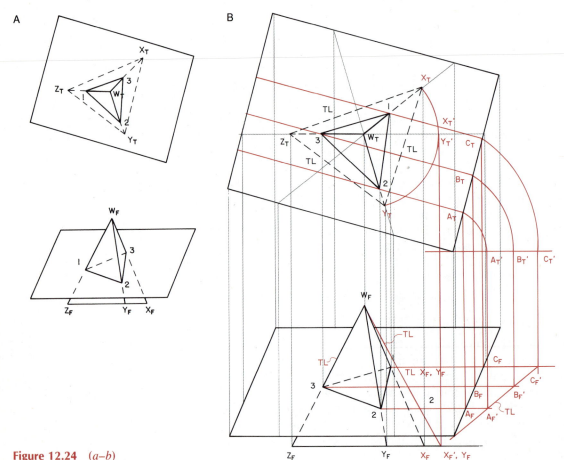

Figure 12.24 (*a–b*)

connect the holes on the bristol board using a very sharp, hard (6H or harder) pencil. If you press hard, you will crease the board, making it easier to fold your board into a model. Also locate points 1, 2, and 3 on the exterior of the surfaces of the pyramid. Draw lines 1–2, 2–3, and 3–1 on the outer surface of the pattern. Patterns can also be transferred by putting a piece of carbon paper between the pattern and the bristol board and tracing the pattern. You need to trace the carbon patterns with the hard pencil as described above.

The creases will help guide the scissors or knife when you cut out the model. Fold the pattern including the flaps. Hold your pyramid model together with your hand. Check for accuracy of mating edges. Some of the flaps may have to be trimmed before gluing. All lines except 1–2, 2–3, and 3–1 should be on the inside of the pyramid. If everything matches, glue the flaps and the model is complete. Take the plane and slip it over the pyramid and see if points 1, 2, and 3 and the lines 1–2, 2–3, and 3–1 on both the pyramid's surface and the plane match. If they do, then you have completed an accurate model and are ready to start a model of the intersection of two solids. Figure 12.24d shows a picture of the completed model.

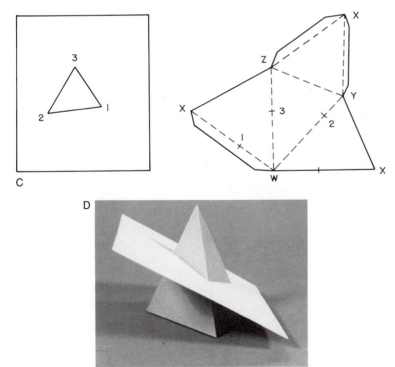

Figure 12.24   (a–c) Steps for building a 3-D model of a plane intersecting a pyramid using patterns. (d) A photo of the finished model.

Figure 12.25*a* is a reprint of Figure 11.95*d*. This shows a triangular pyramid intersecting a triangular prism. The steps in the process for developing the patterns and then the model are the same as outlined above. Figures 12.25*b,c* show the steps. However, the mating surfaces are much more critical for the model of two intersecting solids than they were for a plane intersecting the solid. Number or letter the lines of intersection. Develop the patterns and check each of these lines on both patterns. Any length difference in mating surfaces will result in a model with warped surfaces. Note that the flap that joins the sides of the prism is on the edge of the prism that is not cut by the triangular prism. Note also that in the example shown here the flaps alternate between the two solids. Again, assemble the model by hand and check the fit before gluing. Figure 12.25*d* shows a picture of the completed model.

A

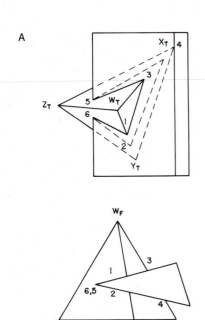

B

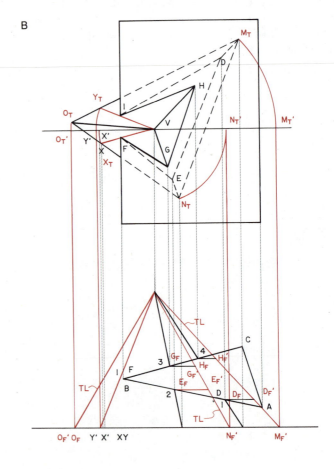

The techniques discussed above will help you when you have to create more complicated models, intersecting air ducts, machine parts, or complex containers.

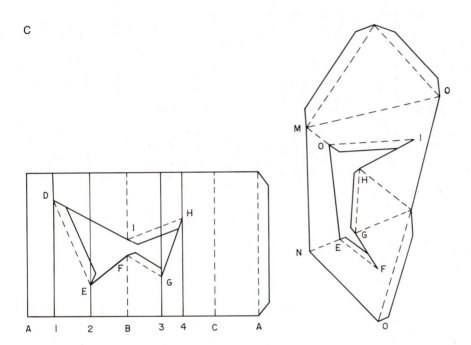

C

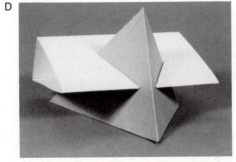

D

Figure 12.25 (*a–c*) Steps for building a 3-D model of a triangular prism intersecting a pyramid using patterns. (*d*) A photo of the finished model.

Figure 12.26 Container pattern. (Used by permission of McDonald's Corporation)

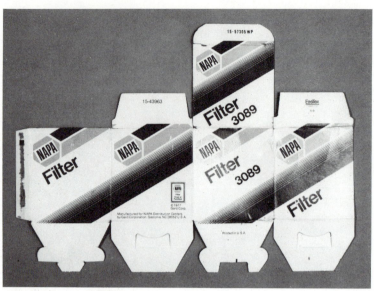

Figure 12.27 Container pattern.

12.2.8 Applications

Take time to observe the number of items that you use daily that are three-dimensional shapes created from patterns drawn on flat stock. Some examples are grocery sacks, french fry containers, waxed cardboard drink containers, ice cream cartons, and boxes for toys. The floor mats in cars, metal desks, bicycle fenders, wing section panels for airplanes, and some newspaper stands are also the product of development methods used in design (Figures 12.26 and 12.27).

12.3 MAPS AND CONTOUR PLOTS

One of the more important graphics topics for designers is the subject of maps and plans that depend on maps. Any construction project or transportation project involves looking at space or distances. If you have flown on an airplane or looked down at the ground for a high building or hill, you have seen what needs to be represented on paper. Figure 12.28 shows the view from an airplane. Your eyes have provided the information about what items were present and whether they had some height or altitude. The colors and textures also provided your brain with information about the presence of each of the items such as buildings, trees, streams, roads, bridges, and other items of interest.

Graphics provides the rules for representing the details of a particular section of terrain on paper and the symbology so that

Figure 12.28 View from an airplane. (Courtesy of OSU Engineering Publications)

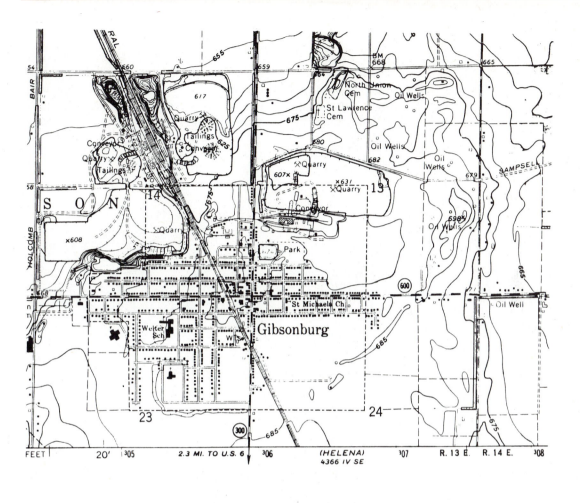

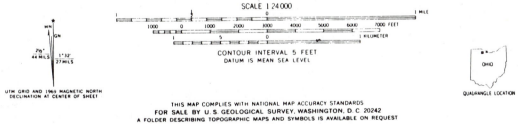

Figure 12.29   Section from geological survey map. (Courtesy of USGS)

all items present can be correctly interpreted. It also provides the conventions for direction, distance or scale, and altitude, elevation, or height. Most of the critical information needed is provided on the map or in the margin around the map and is called the legend. Figure 12.29 shows a section of a geological survey map.

12.3.1 Map Introduction

When we look down from a high point, everything appears to be smaller than it really is. A two-dimensional representation or drawing must be much smaller than real objects, and we have to make very small distances represent very large distances. This is called scaling. Typical scales for maps might be 1 in. on the paper equals 1000 ft on the ground or 1 in. equals 1 mile or several miles. Figure 12.30 shows three portions of maps or plans at different scales.

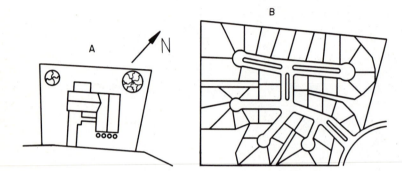

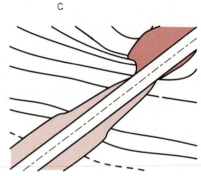

Figure 12.30 (*a*) Typical lot with house (1 in. = 20 ft); (*b*) typical subdivision (1 in. = 500 ft); (*c*) typical highway section (1 in. = 100 ft). (*Computer-generated image*)

Engineer's scales provide 1 in. divided into 10, 20, 30, 40, or 50 parts. The parts can represent 1 ft, 10 ft, 20 ft, 2000 ft, and so on. Metric scales are available and are used in Europe and Canada. Figure 12.29 shows a section of a contour map where the scale is 1:50,000. Figure 12.31 shows the ends of two scales. One is an engineer's scale based on the inch and the other is a metric scale.

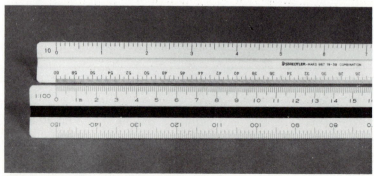

Figure 12.31 An engineer's scale and a metric scale.

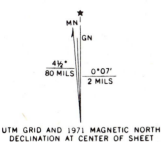

The 1976 MAGNETIC BEARING is 1°38' (29 mils) EAST of GRID NORTH.

ANNUAL CHANGE INCREASING 2.3'

GRID NORTH is 1°45 (31 mils) WEST of TRUE NORTH for centre of map.

Le REPÈRE MAGNÉTIQUE en 1976 est à 1°38' (29 mils) EST du NORD DU QUADRILLAGE.

VARIATION ANNUELLE CROISSANTE 2.3'

NORD DU QUADRILLAGE est 1°45' (31 mils) à l'ouest du NORD GÉOGRAPHIQUE au centre de la carte.

UTM GRID AND 1971 MAGNETIC NORTH
DECLINATION AT CENTER OF SHEET

Figure 12.32 Declination diagram from USGS map and written information from Canadian map. (Courtesy of USGS. This map is based on information taken from 52-I-14 © 1976. Her Majesty the Queen in Right of Canada with permission of Energy, Mines and Resources Canada)

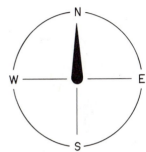

Figure 12.33 Compass points.

Unless the sun is shining and we know what time it is and therefore where shadows are located, we do not know whether a direction is north or south or some other compass point. With a compass, we can determine where north is. The convention or custom in drawing maps is to locate north at the top or near the top of the paper or map. North is indicated by an arrow with the letter *N* or the word *north*. Compasses can provide the location of magnetic north, but this is not the basis for map directions. We need to know where true north is, and this can be determined from magnetic north depending on where we are in the world. Figure 12.32 is a copy of a declination diagram from the U.S. Geological Society (USGS). This figure also includes the written declination information from a Canadian map. The diagram provides the information on the difference between magnetic north and true north.

South is the opposite of north, obviously, and at the bottom of most maps. East is to the right and west to the left. Directions, when expressed as **bearings,** are always measured from north or south and are designated by the angle and whether it is east or west. Examples might be N 30 E or S 85 W. Another convention, called **azimuth,** is to assume that all directions are measured clockwise from north. Examples would be N 30 or N 330. Figures 12.33–12.35 show the compass points and the conventions for giving directions.

There are several different conventions for indicating direction depending on whether you are working for industry or the military or just going for a hike. However, the key is always to know where to find north.

A map drawn on a flat sheet of paper must provide the information about a section of the earth. The earth is not flat, and in order to represent the differences in height or elevation, map-

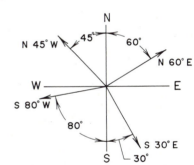

Figure 12.34 Bearing convention east or west of north or south.

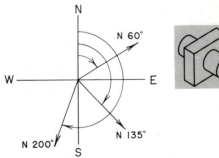

Figure 12.35 Azimuth convention angle to 360° clockwise of north.

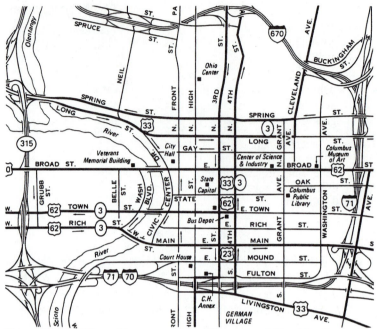

Figure 12.36 Map without contours. (Courtesy of Franklin County Engineer's Office)

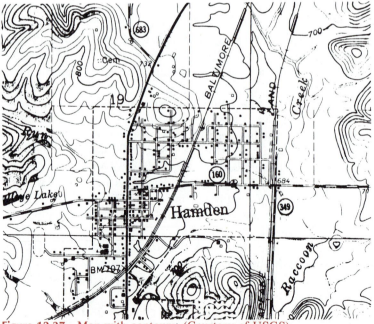

Figure 12.37 Map with contours. (Courtesy of USGS)

makers have chosen to show lines of constant elevation or contour lines. Some or all of these lines can be labeled with the numbers that represent the elevation above sea level or just the elevation above or below some known point. See Figures 12.36 and 12.37.

These lines of constant elevation are determined by surveyors, people who know how to determine direction and measure distances accurately. They have used their instruments to determine the heights of specific points on the ground. This information is plotted on paper and then the contours are drawn in at constant elevations. Figures 12.38*a*–*c* show the original area and preliminary drawing from a surveyor's notes used to create a **contour map**, and Figure 12.38*d* shows the map with the contour lines finished.

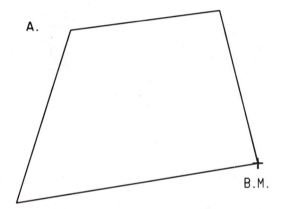

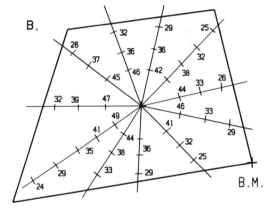

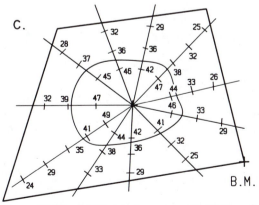

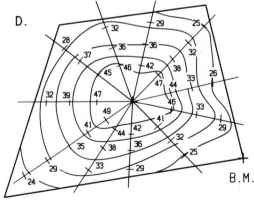

Figure 12.38 (*a*) Plot to be surveyed. (*b*) Drawing from surveyor's notes showing directions, distances, and elevations. (*c*) Single contour finished. (*d*) Completed contour map. (*Computer-generated image*)

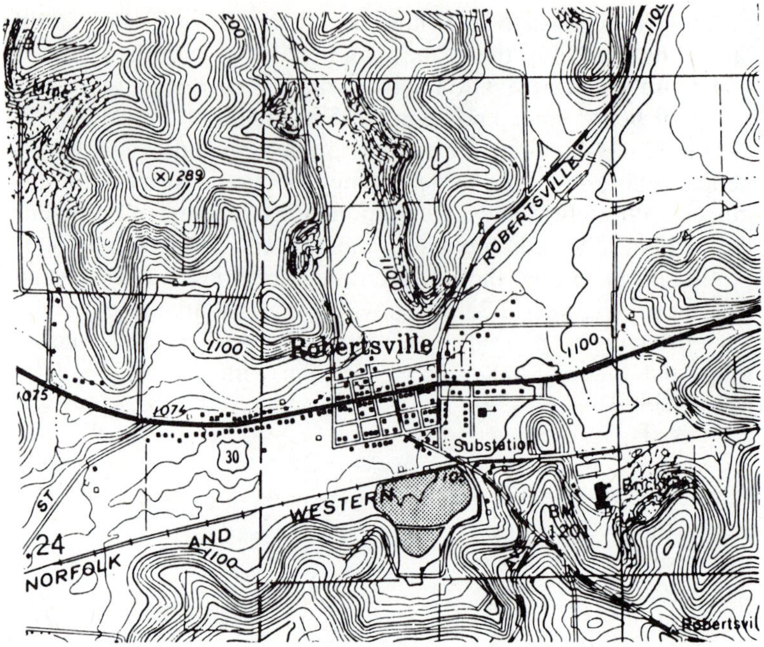

Figure 12.39 Section showing small town from USGS map. (Courtesy of USGS)

Symbols can be used to represent specific details. Rectangles can represent buildings. If the symbol for a building is modified with a flag, it represents a school. A cross on the rectangle symbolizes a church. Figure 12.39 shows the section of a map that represents a small town. Figure 12.40 shows a legend from a map

LANDMARK FEATURES	POINTS DE REPÈRE	
HOUSE, BARN	MAISON, GRANGE	▪ ▪
CHURCH, SCHOOL	ÉGLISE, ÉCOLE	⊹ ⊳
POST OFFICE	BUREAU DE POSTE	•P
HISTORICAL SITE	LIEU HISTORIQUE	⊕
TOWERS: FIRE, RADIO	TOURS: FEU, RADIO	⊚
WELL: OIL, GAS	PUITS: PÉTROLE, GAZ	o
TANK: OIL, GASOLINE, WATER	RÉSERVOIR: PÉTROLE, ESSENCE, EAU	●
TELEPHONE LINE	LIGNE TÉLÉPHONIQUE	⊥ ⊥ ⊥ ⊥ ⊥
POWER TRANSMISSION LINE	LIGNE DE TRANSPORT D'ÉNERGIE	- - - - - - - -
MINE	MINE	⚒
CUTTING, EMBANKMENT	DÉBLAI, REMBLAI	⟨⟨⟨⟨⟨ ⋁⋁⋁⋁⋁
GRAVEL PIT	GRAVIÈRE	⦂GP⦂

Figure 12.40 Legend from map. (This map is based on information taken from 52-I-14 © 1976. Her Majesty the Queen in Right of Canada with permission of Energy, Mines and Resources Canada)

giving the symbols for such features as a bridge, a dam, or rail-road, and so on.

When maps are colored, the color can be used to represent a variety of features. In some cases, color represents elevation on a map of a very large area. In other cases the colors represent land features. For example, blue may represent water; green, parks or vegetation; and black or gray, roads and bridges. Again, the legend on a map should describe the symbols or colors.

12.3.2 Concept of Representing Elevation

When discussing a map with contour lines in the preceding we talked about the fact that contour lines represent constant elevations. This would mean that you could follow the contour line around a hill and never go up or down. However, when you walk up the hill, if it is a steep hill, you get tired very rapidly. This means that you are crossing contour lines frequently or that they are close together. If, on the other hand, the walk up the hill does not tire you, the contour lines are farther apart and the slope is rather shallow.

Study the contour map shown in Figure 12.41. Note the places

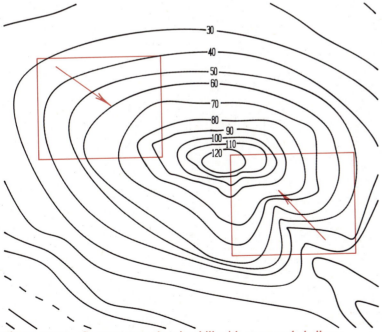

Figure 12.41 Contour map showing hill with steep and shallow slopes. (*Computer-generated image*)

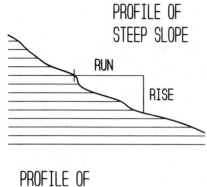

PROFILE OF STEEP SLOPE

PROFILE OF SHALLOW SLOPE

Figure 12.42 Diagrams for rise and run showing portion of contour map for run and profile showing rise and run. Includes calculations. (*Computer-generated image*)

that have contour lines close together and those that have contour lines far apart. These represent steep and shallow slopes. What does slope mean? It means that for every unit of distance that you go in a particular direction, you either go up or down a certain distance. In other words, you are changing elevation. Slope is the change in elevation (rise) divided by the change in horizontal distance traveled (run). Figure 12.42 shows two different slope calculations. The slope is represented as a ratio and can be converted to an angle measured from the horizontal.

12.3.3   Sections or Profiles and Applications

As in orthographic drawings, there are times when we may need to look underground. Or we may need to plan for a road, dam, water supply system, or sewer system. In these cases, we have to draw a cross section of the earth along a particular direction. This is called a sectional view or, many times, the profile. The contours become parallel horizontal lines in the profile or sectional views. Figure 12.43 shows a portion of a contour map and a cross section through the ground where the cutting plane is shown.

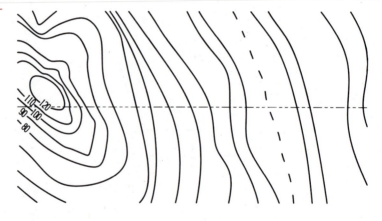

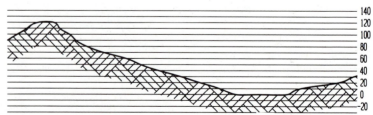

Figure 12.43   Section of contour map (plan) and profile. (*Computer-generated image*)

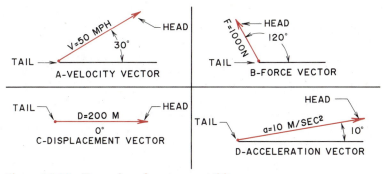

Figure 12.44 Examples of vector quantities.

12.4 *VECTOR GRAPHICS*

Many technical problems can be solved by using **vectors.** These problems can be solved with or without the graphical representation of the vectors; however, a picture of the problem makes it easier to understand. Typical problems that can be solved using vectors include the analysis of simple structures in statics, dynamics, strength of materials, machine design, and fluid mechanics. These subject areas will become familiar to you as you advance in your technical education.

Understanding descriptive geometry is required to solve three-dimensional vector problems. You may wish to review the auxiliary view material presented in Chapter 11 before proceeding with this section.

Vector quantities have a direction and a magnitude. To represent vector quantities graphically, we draw a line and put an arrowhead at one end. The end of the vector that has the arrowhead is called the "head," or "tip," of the vector. The end without the arrowhead is called the "tail" of the vector. The length of the line is proportional to the magnitude of the vector and the arrowhead indicates the vector's direction. Accuracy in drawing the vector is important. A good protractor, straight edge, and scale are required to solve vector problems.

Vectors can represent forces, velocities, accelerations, displacements, and other quantities that have both magnitude and direction. The vector is drawn from a location in a direction specified by an angle. A scale is used to mark the length and an arrowhead is placed on the end of the vector to indicate direction (Figures 12.44*a–d*).

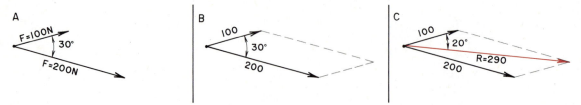

Figure 12.45   Vector addition: Parallelogram method.

12.4.1   Vector Addition and Subtraction

Vectors can be added or subtracted. For these processes, we assume that the vectors are concurrent–coplanar. This means that all the vectors (two or more) act at the same point and are contained in the same plane. Also, assume the plane on which the vectors are acting is the plane of the paper or the plane of the computer screen (Figure 12.45a). When adding two concurrent–coplanar vectors, first lay out the vectors to scale and in their respective directions. Then, using the two vectors as two sides of a parallelogram, lay out the full parallelogram (Figure 12.45b). Addition of two vectors is accomplished by drawing the diagonal of the parallelogram from the common point on which the two vectors act to the opposite end of the parallelogram (Figure 12.45c). This is called the resultant vector. Another approach with regard to vector addition is to visualize the tail of the second vector being located at the head of the first vector. The sum is then the vector drawn from the tail of the first vector to the head of the second vector. This is called the triangle method of vector addition (Figure 12.46). The triangle method can be expanded to the polygon method when more than two concurrent–coplanar vectors are to be added. Figure 12.47a shows four vectors that can be added to give a resultant vector. Simply connect the head of one vector to the tail of another until all vectors are connected. Then draw a resultant vector from the tail of the first vector to the head of the last vector to complete the polygon (Figure 12.47b).

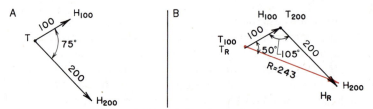

Figure 12.46   Vector addition. Triangle method.

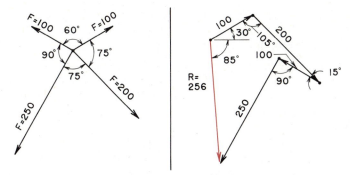

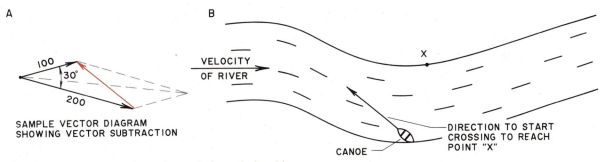

Figure 12.47 Vector addition. Polygon method.

A

B

SAMPLE VECTOR DIAGRAM
SHOWING VECTOR SUBTRACTION

VELOCITY
OF RIVER

X

DIRECTION TO START
CROSSING TO REACH
POINT "X"

CANOE

Figure 12.48 Vector subtraction: relative relationships.

The magnitude of the resultant vector can be determined by measuring with the scale used to lay out the parent vectors. Its direction can be determined using a protractor.

Vectors can also be subtracted using the same principles discussed for vector addition. Figure 12.48*a* shows vector subtraction. The opposite diagonal of the parallelogram from the diagonal used in vector addition is used in vector subtraction. Vector subtraction shows relative relationships between vectors. For example, Figure 12.48*b* shows a top view of a river flowing with a current. On one shore is a canoe the canoeist is going to use to cross the river and arrive at point X. The canoeist must start at an angle upstream in order to arrive at the designated point. This example illustrates the concepts of vector subtraction.

12.4.2 Applications

Statics: Two-Dimensional, Single-Joint Problem

When objects are said to be in static equilibrium (there is no movement), it means that the sum of all of the forces on an object

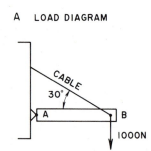

Figure 12.49 Vector diagram showing forces in equilibrium. Vector diagram closes when vectors are connected in head-to-tail fashion.

or any part of the object must be zero. The vector diagram of such an object is shown in Figure 12.49. Note that there is no resultant force shown in the vector diagram. When the vectors are connected in head-to-tail fashion, the vector diagram closes to form a polygon. This means that the resultant vector for equilibrium is zero. This makes it convenient to solve problems with no more than two unknown values. At this point, you need to remember that vectors have two properties: direction and magnitude. When solving problems with vectors, typical situations might be one vector (magnitude and direction) is unknown, the magnitudes of two vectors are unknown, or two of the directions are unknown.

Figure 12.50 illustrates a situation where the load on a connection point or joint is known and the direction of the forces provided by the rigid support and the cable are known. When solving such problems, you must remember that cables, ropes, and chains always pull on the joint and their forces are directed away from the joint. Rigid members can either pull or push on a joint, and their vectors can be represented as either toward the joint or away from the joint.

The solution step where the initial vector directions are assumed is called a free-body diagram. This diagram is drawn showing that the load, the chain, and the rigid member have all been replaced by their forces on a single point (Figure 12.51a).

The next step is to draw the known vector with the proper direction and magnitude (Figure 12.51b). Next from the head end of the known vector draw a construction line in the direction of the cable, and from the tail end of the known vector draw a second construction line in the direction of the rigid member (Figure 12.51c).

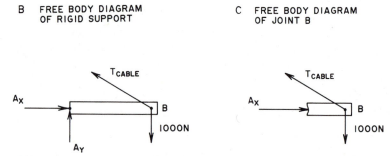

Figure 12.50 Two-dimensional force analysis of joint: (*a*) load diagram; (*b*) free-body diagram of rigid support; (*c*) free-body diagram of joint at *B*.

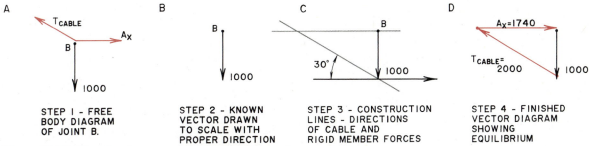

A

STEP 1 - FREE
BODY DIAGRAM
OF JOINT B.

STEP 2 - KNOWN
VECTOR DRAWN
TO SCALE WITH
PROPER DIRECTION

STEP 3 - CONSTRUCTION
LINES - DIRECTIONS
OF CABLE AND
RIGID MEMBER FORCES

STEP 4 - FINISHED
VECTOR DIAGRAM
SHOWING
EQUILIBRIUM

Figure 12.51 Step-by-step procedure to determine unknown quantities related to vectors using graphical means. (*a*) Draw free-body diagram of joint. (*b*) Lay out to scale the known vector quantities in proper directions. (*c*) Draw directions of forces in cable and in rigid member using light construction lines. (*d*) Complete vector diagram by connecting vectors in head-to-tail fashion until polygon closes indicating equilibrium. Measure unknown values at selected scale to determine their magnitudes.

The final step is to draw the arrowheads for the two vectors that were unknown and to measure their length using the scale chosen at the beginning of the problem. The completed solution is shown in Figure 12.51*d*. Recall that when a joint is in equilibrium, the vector diagram of the joint closes when the vectors are connected head to tail.

Now try to solve this problem first using your drawing instruments and then sketching on grid paper. When using grid paper, assign a value to each grid square and then use your pencil or the edge of a second piece of paper to determine the magnitudes of any vectors that do not lie on either the vertical or horizontal axes.

Statics: Two-Dimensional Truss Analysis

The forces in the structural members of complicated trusses (those with many joints) can be solved finding the joints that have only two unknown values. These unknowns can be magnitudes or directions. From the previous discussion, we know that these types of problems can be solved graphically using vectors. Once the unknowns for a joint have been solved, subsequent joints can be solved because the number of unknowns for that joint will be reduced to 2. Figures 12.52*a–d* show the steps involved in creating a vector diagram to aid in solving for the unknown forces acting at joint *A*. Figures 12.52*e–h* show the vector diagrams for the remainder of the truss structure. The black vectors are the known forces and the vectors shown in color are the unknown forces.

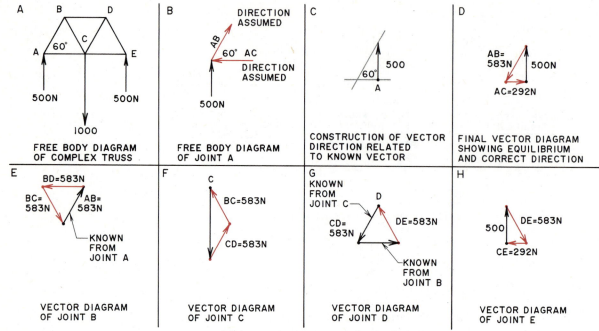

Figure 12.52   Two-dimensional force analysis of truss: (*a*) free-body diagram of truss with applied loads and reactions; (*b*) free-body diagram of joint *A*; (*c*) construction of vector diagram using directions of unknown vectors related to known vectors; (*d*) final vector diagram showing equilibrium and magnitudes of unknown vectors; (*e–h*) procedure repeated to get vector diagrams for each joint in truss. Magnitude of each unknown force vector is found.

Statics: Three-Dimensional Analysis

It is also possible to solve three-dimensional problems with vectors. To do this, you need to remember some of the information presented in Chapter 11. Specifically, you should know that by drawing any two adjacent orthographic views of a line, you can find the true length of any vector (line). Sometimes you will need an auxiliary view to determine the true length.

Figure 12.53*a* shows a top and front drawing of the rigging of a tow truck. The vector labeled as 1000 lb represents the cable of the tow truck that is loaded in the direction indicated. Assuming that all of the forces go through a single joint (joint *A*), a free-body diagram can be drawn (Figure 12.53*b*).

Figure 12.53*c* shows a vector diagram with a top and front view. The front view is solved first because the load (the car being towed) is known (the vector is true length) and there are only two unknowns. Again the black vectors represent the known forces

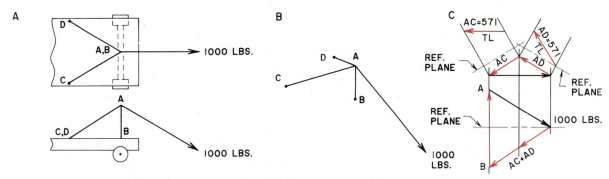

Figure 12.53 (*a*) Load diagram of three-dimensional force system. (*b*) Free-body diagram of tow truck rigging. (*c*) Top and front views of vector diagram of forces in rigging drawn to scale. Using methods of descriptive geometry, unknown force vectors in members are solved.

and the colored vectors represent the unknown forces. Once the front view has been solved, the auxiliary views can be used to determine the magnitudes of the unknown forces in the top view (*AC* and *AD*).

12.5   *SHADES AND SHADOWS*

People can understand pictorial drawings more easily when **shades and shadows** are added. Shade and shadow gives a pictorial drawing depth and makes it look more like a picture. In this section you will learn how to create shade and shadow for orthographic views, isometric pictorials, and to a limited extent, perspectives. Shade and shadow can be added reasonably quickly to both drawings and sketches using the methods that follow. Figure 12.54 shows an illustration of a house, a flag pole, and a wall with shade and shadow.

12.5.1   Shade, Shadows, and the Standard Light Ray

Shadow is the part of an object's surroundings that is hidden from the light by the object. Shade, on the other hand, is that part of the object that is hidden from the light by itself. See Figure 12.55. The boundaries of shades and shadows are determined by the last light ray that clears the object and strikes the ground. When working with orthographic views, a light ray coming from over the observer's left shoulder is defined as being the diagonal of a unit cube. Figure 12.56 not only shows the light ray but also shows the projection of the light ray on the front, top, and right side

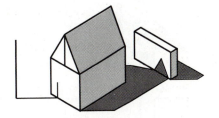

Figure 12.54   An illustration with shade and shadow.

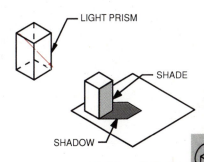

Figure 12.55   Definition of shade and shadow.

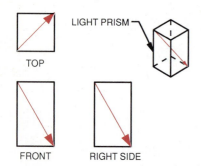

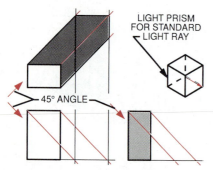

Figure 12.56 A light ray in isometric and orthographic views.

Figure 12.57 Shade and shadow for a block in three orthographic views.

(profile) surfaces of the cube. It is the projections of the light ray on the cube surfaces that is used to define the direction of the light ray in the front, top, and profile views. Figure 12.57 shows three views of a block and the projection of the light rays that touch the block's corners. We can pick light rays other than the standard; this discussion is based on the **standard light ray.** The light source is assumed to be large enough and far enough away to project parallel light rays rather than being a point source.

This same light ray can be used for isometric drawings. When it is used with pictorials and orthographics, it should define the same shade and shadow for both. This will be covered in more detail in the section on isometric shade and shadow.

The approach used in this section is to define the shadow for a vertical pole with the standard light ray. When the light ray clears the object, it travels until it strikes another surface. In other words, it is a situation where we have to determine where the line (light ray) strikes the surface of the earth or other surrounding objects (planes). Additional features will be added so that eventually solid objects will be defined. Any apparent discrepancies can be resolved by reducing a problem to a set of vertical poles and checking each pole with the light ray. This approach is used for both orthographic and pictorial drawings. Figure 12.58 shows the vertical pole with the light ray and shadow in three orthographic views and the isometric pictorial. Note that the light ray is in color, the edges of the object are relatively thick black lines, construction lines are thin black lines. The shadow is a darker gray pattern and the shade is a lighter gray pattern.

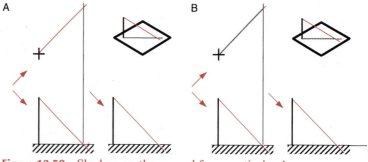

Figure 12.58 Shadow on the ground for a vertical pole.

12.5.2 Orthographic Views

We define a vertical pole as a vertical line of finite length. In the front and profile views, we see it as a true-length line while in the top view it appears as a point (plus sign). In the front and profile views, the surface of the earth (ground) appears as an edge.

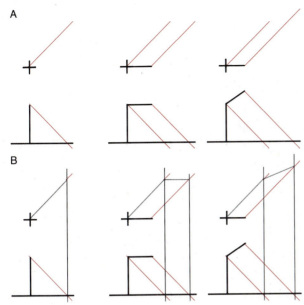

Figure 12.59 Shadows on the ground for a vertical pole and two additional vertical poles with extensions.

In the front view in Figure 12.59 it is easy to see where the standard light ray strikes the ground. That intersection point is projected to the top view so that the length of the shadow can be found. In the top view, the shadow is drawn from the base of the pole in the direction of the light ray to the point of intersection.

We cannot see the shadow in the front view. However, in the process of learning about shade and shadow, sometimes we draw the shadow as a line on the surface to help understand the method.

The shade is on the surface of the vertical pole away from the origin of the light ray.

When a vertical surface (wall) is placed behind the pole, the light ray may strike the vertical surface before it strikes the ground. The light ray has to be drawn in both views to find the point of intersection. When the light ray strikes the wall first, part of the shadow appears on the ground and part is on the wall. The part on the ground is drawn from the base of the pole as though there was no wall present but the shadow stops at the base of the wall. From that point, the shadow goes up the wall to the point of intersection. The part of the shadow on the wall can be seen in the front view unless the wall appears as an edge in the front view. See Figure 12.60a.

Figure 12.60b shows an inclined surface that appears as an edge in the profile view. In this case, the profile view shows that the light ray strikes the inclined surface before it strikes the ground.

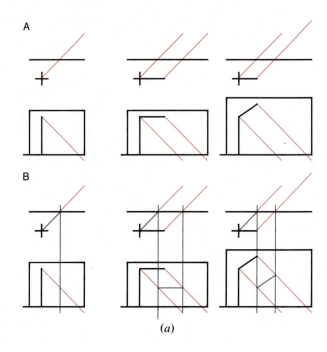

(a)

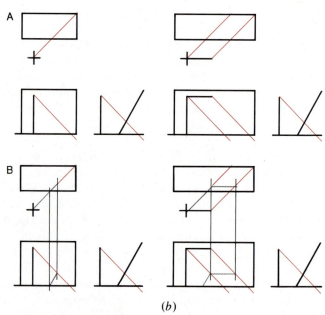

(b)

Figure 12.60 (*a*) Shadows on the ground and a vertical wall for a vertical pole and two additional vertical poles with extensions. (*b*) Shadows on the ground and an inclined wall for a vertical pole and two additional vertical poles with extensions.

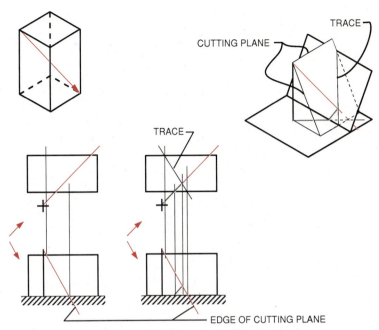

Figure 12.61   Cutting plane method for finding the shadow on an inclined wall of a vertical pole in two views.

As in the case of the vertical wall, the shadow starts at the base of the pole and extends along the direction of the light ray until it intersects the base of the inclined surface. From this point the shadow is drawn to the point of intersection. The shadow on the inclined surface is visible in both the top and front views. Therefore, when there is an inclined surface, each of the three views has to be checked to determine where the light ray intersects the surface. The third (profile) view may not be necessary. In Chapter 11, the cutting plane method was used to find the point of intersection for a line and a plane. This method eliminated the work of creating a third view (principal or auxiliary) in order to find the intersection point. Figure 12.61 shows the solution for Figure 12.60*b* without having the third view.

A horizontal pole can be attached to a wall of some sort. In Figure 12.62*a*, the pole is attached to a vertical wall. In the top and profile view, the pole appears true size. In the front view, it appears as a point. In order to determine where the light ray that touches the end of the pole strikes either of the surfaces, think of the end of the horizontal pole as though it was the top of a vertical pole. When you use this mental construction, the determination of the intersection point is the same as it was in the vertical pole.

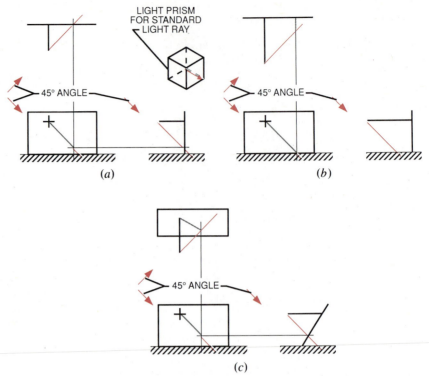

Figure 12.62 (*a*) Shadow for a horizontal pole attached to a vertical wall. (*b*) Shadow on the ground and a vertical wall for a horizontal pole attached to a vertical wall. (*c*) Shadow for a horizontal pole attached to an inclined wall.

The location of the shadow is obviously different. In the case illustrated by Figure 12.62*b*, the shadow starts at the base of the horizontal pole on the wall and ends at the intersection point. When the intersection point is on the wall, the shadow is a straight line from base to intersection point. When the intersection point is on the ground rather than the wall (see Figure 12.62*b*), then there are two parts to the shadow. The first part of the shadow is drawn parallel to the horizontal pole from the intersection point to the base of the wall. From that point it is drawn to the base of the horizontal pole.

When the wall is an inclined surface, the intersection point can be determined either by drawing the third view (Figure 12.62*c*) or by using the cutting plane method. In this case the third view is shown. There are two horizontal poles shown. One is short enough so that the end of the pole casts a shadow on the inclined surface while the other is long enough that part of the shadow is on the ground and part is on the inclined wall.

When a pole is neither vertical nor horizontal, its shadow is determined using the same method as the horizontal pole. Imagine that the end of the inclined pole is the top of a vertical pole. Draw the light rays through the top of the imaginary pole and find the intersection with the ground or walls. Figures 12.63a–c show the end of the inclined pole shadow located on the ground, on the vertical wall, and on an inclined surface. In Figure 12.63a, the entire shadow is on the ground and is drawn from the base of the inclined pole to the intersection point.

In Figure 12.63b, the intersection point is on the vertical wall. Part of the shadow is on the ground. This part is drawn by connecting the base of the pole to the point where the light ray drawn through the end of the pole would have struck the ground if the vertical wall was not there. From the point where the shadow

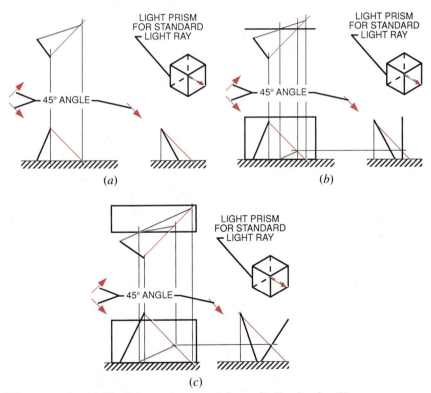

Figure 12.63 (a) Shadow on the ground for an inclined pole. (b) Shadow on the ground and a vertical wall for an inclined pole. (c) Shadow on the ground and an inclined wall for an inclined pole.

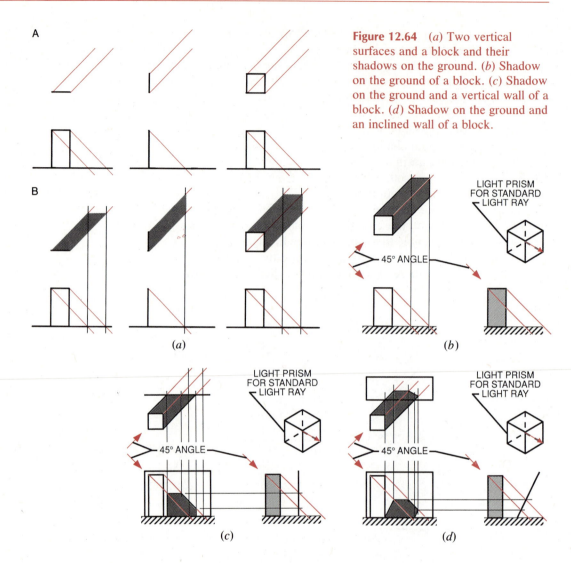

Figure 12.64 (*a*) Two vertical surfaces and a block and their shadows on the ground. (*b*) Shadow on the ground of a block. (*c*) Shadow on the ground and a vertical wall of a block. (*d*) Shadow on the ground and an inclined wall of a block.

strikes the base of the wall, draw a line to the intersection point where the limiting light ray strikes the wall. This same technique is used in the case of the inclined wall as well. See Figure 12.63*c*.

There are several primitive shapes that cast distinctive shadows and whose surfaces are partially shaded. The first one shown is a rectangular block (see Figure 12.64*a*). The solution to locating the shadow on the ground is to think of having four vertical poles (one at each of the corners). The one on the front left corner does not come into play because of the shadow of that pole would fall in the shadow of the upper surface. The shadows are drawn for the other three corners (poles) and the ends of the shadows are connected. These lines form the limits of the shadow cast by the block.

The part of the block away from the light source (two surfaces) is shaded. One of the shaded surfaces can be seen in the profile view. Figures 12.64b–d show the shadow cast on the ground, a vertical wall, and an inclined wall. Fill in the shadow in one color or pattern and the shade with a different color or pattern.

The shadow for a pyramid can be drawn by assuming that the vertex is a vertical pole and locating the shadow of the vertex either on the ground or on the walls. When the entire shadow is on the ground, it is drawn by connecting the corners of the pyramid base to the shadow of the vertex. Remember that the edges of the pyramid drawn from the corners of the base to the vertex are inclined poles.

When the shadow is partially on the ground and partially on the wall, the problem must be solved by constructing the shadow on the ground as though the wall was not present and finding where the shadow on the ground strikes the base of the wall. Connect these points to the location of the vertex shadow on the wall. The two surfaces away from the light source are shaded. See Figure 12.65.

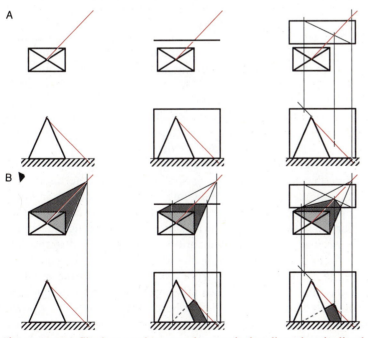

Figure 12.65 Shadow on the ground, a vertical wall, and an inclined wall for a pyramid.

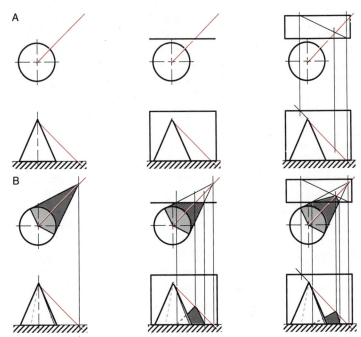

Figure 12.66 Shadow on the ground, a vertical wall, and an inclined wall for a cone.

A cone shadow is constructed in much the same manner as the pyramid shadow. The locations of the intersection(s) of the light ray that touches the vertex with the ground (and wall, if present) are determined by construction. The shadow of the cone is located by drawing tangents to the circular base through the location of the shadow of the vertex. The lines connecting the tangent points to the vertex mark the dividing line between the part in the shade and the part that is lighted. See Figure 12.66.

The top surface of a cylinder is circular and casts a circular shadow. The location of the circular shadow is determined by drawing a light ray through the center of the circular top in both the top and front views. The light ray through the center of the circular top in the front view strikes the ground locating the center of the circular shadow. This location can be projected to the top view. A circle can be drawn in the top view about the center to locate the shadow. Light rays are drawn tangent to the circular top and to the circular shadow in the top view. The tangent points are projected to the front view where vertical lines are drawn through them (vertical poles). Between the tangent points away from the light source the cylinder is shaded. Part of the shade is visible in the front view. The extent of the shade (for construction

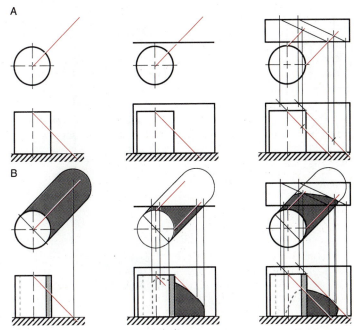

Figure 12.67 Shadow on the ground, a vertical wall, and an inclined wall for a cylinder.

purposes in these exercises) is shown by a dashed line on the rear side of the cylinder. While this would not normally be shown in an illustration, the location is important if additional views are to be drawn. See Figure 12.67.

The shadows for the cylinder on vertical and inclined walls are also shown in Figure 12.67. For accurate construction, it is advisable to construct two or more vertical poles on the edge of the cylinder to locate the extent of the shadow on the walls. This would also be true if the top surface of the cylinder was not parallel to the ground. The shadow can then be drawn by connecting the points with an irregular curve.

Complex objects are composed of primitive solids. Adding shade and shadow to drawings of complex objects is not difficult but requires finding the primitives. Once they are found, their shadows can be drawn. However, the shadow may fall on the ground or another part of the object. When the shadow is located partially on the object and partially on the ground, you must extend the primitive until it touches the ground in order to be able to determine the boundaries defining the shadow limits. This means that a pyramid attached to a rectangular block sitting on the ground must be extended to the ground. The shadow of the pyramid on

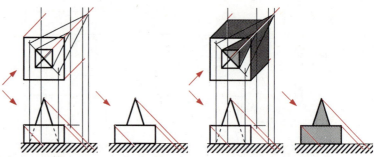

Figure 12.68   Shade and shadow for a pyramid and block combination.

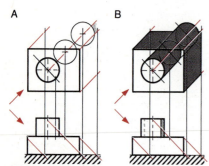

Figure 12.69   Shade and shadow for a cylinder and block combination.

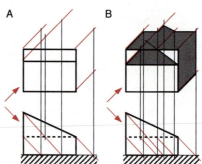

Figure 12.70   Shade and shadow for an inclined block on top of a rectangular block.

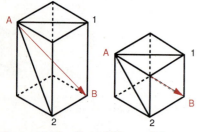

Figure 12.71   Two different light rays shown in isometric.

the ground is constructed. Then the shadow that the pyramid casts on the block must be drawn. See Figure 12.68. Figures 12.69 and 12.70 show additional combinations of primitive shapes.

12.5.3   Shades and Shadows in Isometric Pictorials

Shades and shadows are produced by the light rays striking an object. In pictorial drawings, the light rays can be installed in two ways. In the first and simplest case, light rays are installed so that they are in the picture plane. This makes it very simple to determine where the light ray strikes the ground. Figure 12.71 shows two light rays as diagonals in isometric boxes (light prisms) where the light ray is parallel to the picture plane. In this case the light ray (*AB*) can be at any angle to the horizontal. In most cases, it is somewhere between 30° and 60° with the horizontal. In these situations the shadow of a vertical pole appears to be a

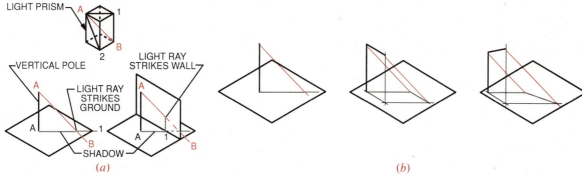

Figure 12.72   (*a*) Shadow in isometric on the ground and on a vertical wall for a vertical pole. (*b*) Vertical pole and two vertical poles with extensions and their isometric shadows.

horizontal line on the horizontal ground (A_1). See Figure 12.72*a* A vertical pole and two vertical poles with extensions and their isometric shadows are shown in Figure 12.72*b*.

Therefore, the limits of the shadow are relatively easy to determine. The light ray is drawn at the specified angle through the top or limiting feature of an object. Then a horizontal line is drawn from the foot of a vertical pole through the limiting feature. Where the horiziontal line crosses the light ray determines the location of the limit of the shadow. Figure 12.73 shows the shadow of a solid object.

Figure 12.74 shows several primitive shapes and their shaded portions and shadows. These were the same objects used to demonstrate shade and shadows in orthographic views.

If there are either vertical walls or sloped walls behind an ob-

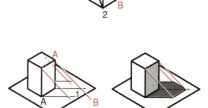

Figure 12.73   Construction of the shadow on the ground for an isometric pictorial of a block.

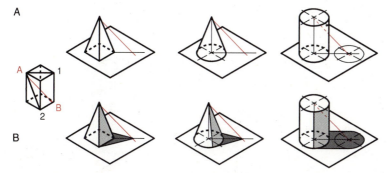

Figure 12.74   Shadow of three additional geometric primitive shapes in isometric pictorials.

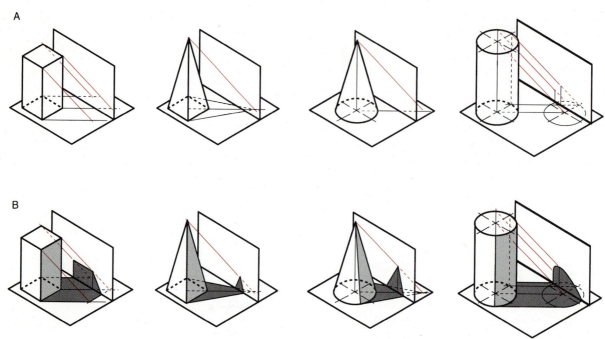

Figure 12.75 Construction of the isometric shadows on the ground and a vertical wall of four primitive shapes.

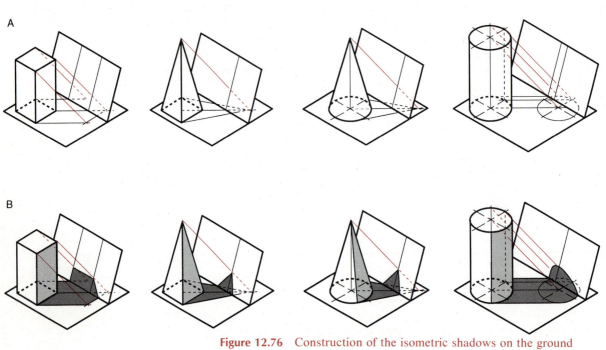

Figure 12.76 Construction of the isometric shadows on the ground and an inclined wall of four primitive shapes.

ject, the shadow appears to go across the ground and up the wall. Figure 12.75 shows the shadows of some of the same objects on vertical walls while Figure 12.76 shows the shadows on inclined walls. Shadows on steps that go up or down in the apparent path of the shadow can be constructed by changing the foot of the vertical pole to match that change in elevation. Figure 12.77 shows the shadow of a vertical pole on steps.

We have assumed that we are looking north at objects facing south, so that the light rays in the picture plane represent sunlight in the afternoon. It is possible to construct light rays for objects facing directions other than south or for morning hours. This is done by drawing a light prism with a nonsquare base or by drawing the light ray so that it enters from the upper right rather than the upper left.

Figure 12.78 shows a light ray that is not in the picture plane and enters from the upper right. A vertical pole with its shadow is shown for the specified light source. Note that the shadow (A_1) is drawn parallel to the diagonal (A_1) on the top surface of the light prism, while the light ray (AB) is drawn parallel to the light ray (AB) in the light prism.

12.5.4 Shades and Shadows in Perspective Pictorials

Shades and shadows in perspective drawings are also produced by two general types of light rays. As in the isometric drawings they can be installed so that they are in the picture plane. This makes it very simple to determine where the light ray strikes the ground. Figure 12.79 shows a triangle formed by the vertical pole, the light ray, and the shadow of the vertical pole. In this case the light ray can be at any angle to the horizon. In most cases, it is somewhere between 30° and 60° with the horizontal. When the light ray is parallel to the picture plane, the shadow of a vertical pole appears to be a horizontal line on the horizontal ground. Therefore this "light" triangle is a right triangle.

The limits of shadow are relatively easy to determine. The light ray is drawn at the specified angle through the top or limiting features of an object. Unlike the isometric, where the front and back of a rectangular solid are the same height, the heights vary from front to back in a perspective. Here you must construct a light triangle for every feature of the perspective drawing in order to determine the limits of the shadow. These limits are then connected to form the shadow. Figure 12.80 shows the shadow for a cut block.

Figure 12.81 shows a primitive shape with its shade and shadow. This same object will be used when the light ray is not in the picture plane so that you can compare the effect with different light rays.

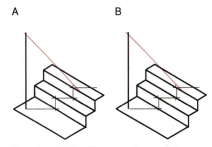

Figure 12.77 Construction of the isometric shadow of a vertical pole on steps.

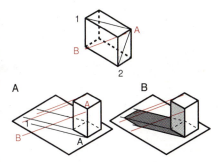

Figure 12.78 Shadow of a rectangular block with light coming over the right shoulder of the observer rather than the left shoulder. The light ray is not in the picture plane.

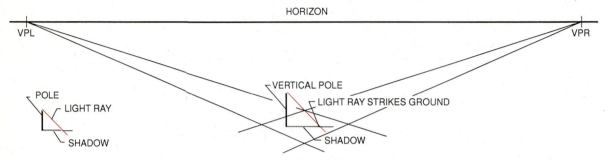

Figure 12.79 Shadow of a vertical pole shown in perspective. The light ray is in the picture plane.

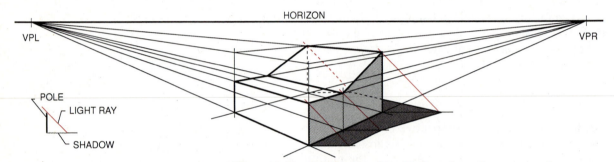

Figure 12.80 Cut block and its shadow shown in a perspective drawing.

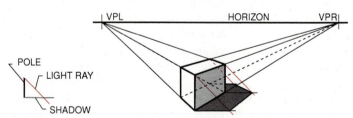

Figure 12.81 Rectangular block and its shadow shown in a perspective drawing. The light ray is in the picture plane.

Vertical walls or sloped walls behind an object require special construction similar to the construction shown in the section on isometric drawings.

If light rays are not in the picture plane, the light triangle is

A

B

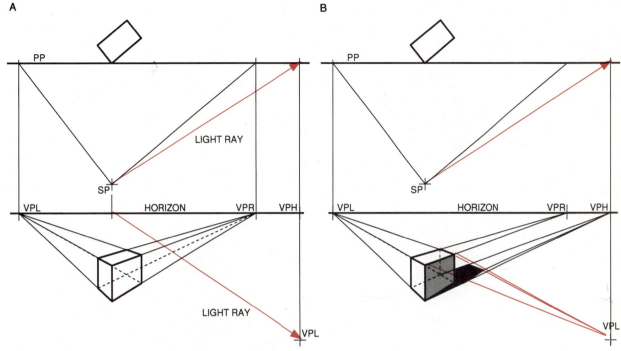

Figure 12.82 Rectangular block and its shadow shown in a perspective drawing. The light ray is not in the picture plane.

something other than a right triangle. The vertical pole is still vertical but the shadow is drawn at an angle to the horizontal. It can appear to be above the horizontal to represent the light coming from in front of the object away from the observer. It can also appear to be below the horizontal so that it represents light coming from behind the object and toward the observer.

Figure 12.82 shows a primitive shape with its shade and shadow. This same object was used when the light ray was in the picture plane so that you can compare the effect with different light rays.

12.6 COMMON PROBLEM

In order to be able to sell a product, the product needs to be presented so that nontechnical decision makers can visualize the product. In the case of the universal saw fixture, we are going to add shades and shadows to a pictorial. This will add depth and realism to the fixture that will make it easier for a nontechnical person to interpret. The light source will be in the picture plane as described in Section 12.5.1. Notice how the shade and shadow bring out the slot and shape of the object. The isometric pictorial is shown in Figure 12.83*a*. Figure 12.83*b* shows part of the con-

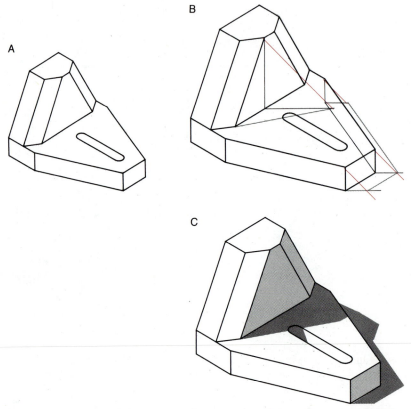

Figure 12.83 Universal saw fixture shown as (*a*) an isometric pictorial, (*b*) an isometric pictorial with construction for the shade and shadow partially finished, and (*c*) an isometric pictorial with shade and shadow.

struction to determine where to locate the shades and shadows. Figure 12.83*c* shows the finished illustration.

12.7 *SUMMARY*

In this chapter, we have studied four applications of three-dimensional geometry: developments, contour maps, vectors, and shade and shadow. Each one relies on the basic three-dimensional geometry fundamentals developed in Chapter 11. The first three will extend your abilities to solve problems. The study of these applications should have improved your visualization skill and improved your knowledge of technical graphics. The following problems will give you a chance to test your skill and knowledge.

PROBLEMS

Problems 12.1–12.15
For the following figures, develop the inside full, half, or quarter
patterns.

    (a) With the figure only with no tabs, bottom or top lids.
    (b) With tabs for assembly without bottom or top lids.
    (c) With tabs and the bottom and top lids.
    (d) With tabs and bottom and top lids. Then lay out the pattern
         on heavy paper, cut it out, and assemble it.

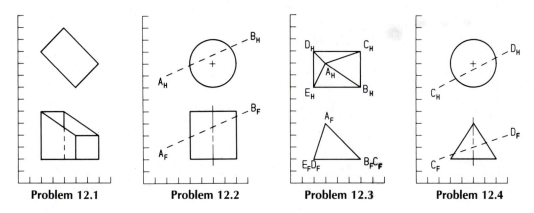

Problem 12.1   Problem 12.2   Problem 12.3   Problem 12.4

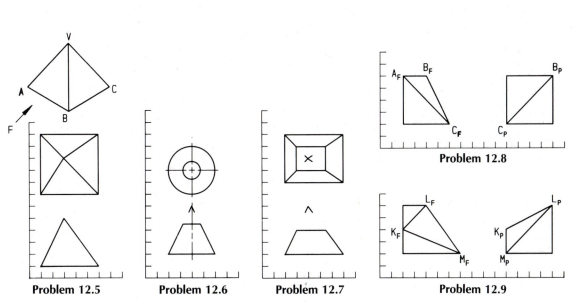

Problem 12.5   Problem 12.6   Problem 12.7   Problem 12.8

Problem 12.9

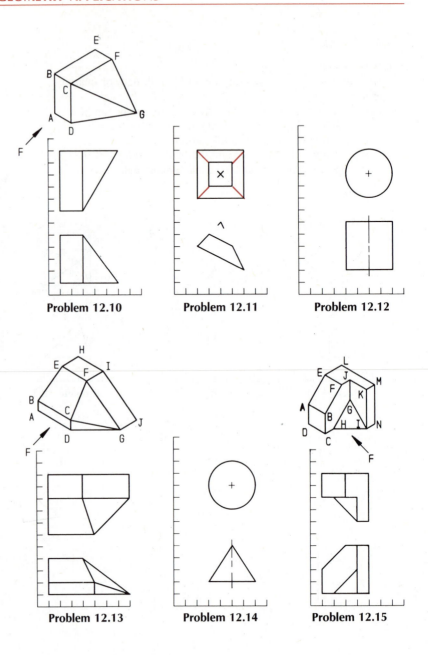

Problem 12.10

Problem 12.11

Problem 12.12

Problem 12.13

Problem 12.14

Problem 12.15

Problems 12.16–12.22

For the following figures, draw each set of views to fill an 8.5 × 11-in. page. Solve for the intersection of the solid and the plane. Construct a pattern for each solid object and plane. Cut slits along the line of the intersection on the solid to be assembled to the matching tabs on the plane. Assemble the model using rubber cement or an equivalent adhesive. Be sure that no flaps or construction lines show on the models.

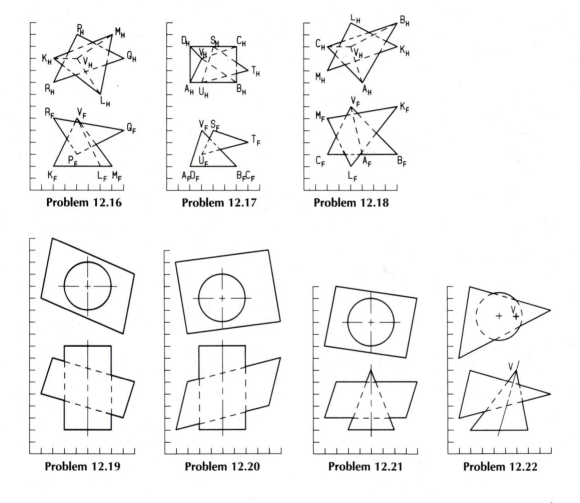

Problem 12.16

Problem 12.17

Problem 12.18

Problem 12.19

Problem 12.20

Problem 12.21

Problem 12.22

Problems 12.23–12.25
For the following figures, draw each set of views to fill an 8.5 × 11-in. page. Solve for the intersection of the two solids. Construct a pattern for each solid object. Assemble the model using rubber cement or an equivalent adhesive. Be sure that no flaps or construction lines show on the models.

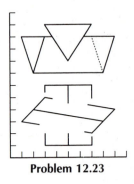

Problem 12.23

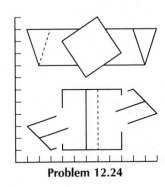

Problem 12.24

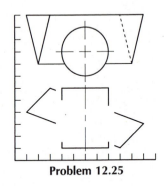

Problem 12.25

Problem 12.26
For the figure, measure the angle with a protractor then write out the bearing and the azimuth for the vectors assigned relative to the north arrow.

(a) *A, C, E, H.*
(b) *B, D, F, G.*

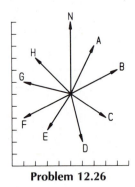

Problem 12.26

Problem 12.27
Copy the elevations in the figure to a piece of linear grid paper and draw the contours as 2-ft intervals.

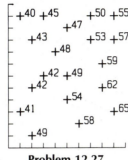

Problem 12.27

Problem 12.28
Copy the contour map shown in the figure to a piece of linear grid paper. Grid scale is 10 ft per division. Plot a 20-ft-wide level roadway along the given center line with a cut and fill of 1:1 at one of the following elevations as assigned:

(a) 20 ft.
(b) 30 ft.
(c) 40 ft.
(d) 50 ft.

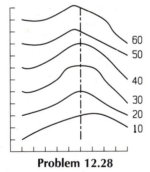

Problem 12.28

Problem 12.29

Copy the contour map shown in the figure to a piece of linear grid paper. Grid scale is 10 ft per division. Plot a 20-ft-wide level roadway along the given center line with a cut and fill of 1:1.5 at one of the following elevations as assigned:

    (a) 10 ft.
    (b) 20 ft.
    (c) 30 ft.
    (d) 40 ft.

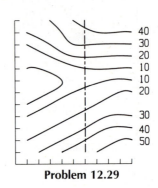

Problem 12.29

Problem 12.30

Referring to the figure of problem 12.26, determine the length of each vector shown using a scale matching the coordinate system. In addition, determine the length using the Pythagorean theorem and the coordinates of the ends of the vectors. Arrange your answers in a table.

$$L^2 = (X_2 - X_1)^2 + (Y_2 - Y_1)^2 \quad \text{(Pythagorean theorem)}$$

Problem 12.31

Determine the resultant force of the concurrent coplanar force system shown. Scale: 1 in. = 50 lb; 10 mm = 50 newtons (N).

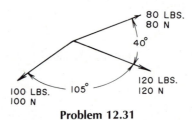

Problem 12.31

Problem 12.32

Determine the magnitude and direction of the force in each member of the frame shown. Direction can be indicated by a *T* for tension or a *C* for compression. Tension indicates the force is directed away from the

joint; compression indicates the force is directed toward the joint.
Scale: 1 in. = 5000 lb; 10 mm = 5000 N.

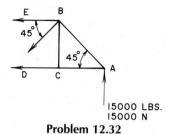

15000 LBS.
15000 N

Problem 12.32

Problem 12.33

For the three-dimensional force system shown, determine the
magnitude of the resultant force. Show the resultant force in the
horizontal and frontal views. Scale: 1 in. = 50 lb; 10 mm = 50 N.

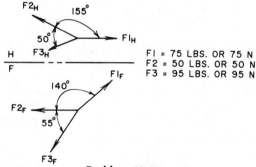

F1 = 75 LBS. OR 75 N
F2 = 50 LBS. OR 50 N
F3 = 95 LBS. OR 95 N

Problem 12.33

Problems 12.34–12.48

Draw, in the order shown, the orthographic views of the figure (a
minimum height of 2 in. or 50 mm for the front view). Using the
standard light ray, add the shade and shadow for the given views.

12.34	refer to	12.1	12.42	refer to	12.10
12.35	refer to	12.5	12.43	refer to	12.15
12.36	refer to	12.3	12.44	refer to	12.19
12.37	refer to	12.14	12.45	refer to	12.20
12.38	refer to	12.7	12.46	refer to	12.21
12.39	refer to	12.6	12.47	refer to	12.23
12.40	refer to	12.12	12.48	refer to	12.25
12.41	refer to	12.11			

Problems 12.49–12.63

In the order shown for problems 12.34–12.48, construct an isometric
drawing of the figure (use a minimum height of 2 in. or 50 mm). Using
the standard light ray, add the shade and shadow for the given
pictorial views.

TERMS YOU WILL SEE IN
THIS CHAPTER

DESKTOP PUBLISHING   Merging text and graphics by computer.

AXES   Scales along which data are plotted.

DATA POINT   Symbol used on a graph to show the location of a specific value relative to the axes.

GRAPH TITLE   Name of the chart or graph.

AXIS CALIBRATION   Units and markings on an axis that allows all data to be plotted.

AXIS TITLE   Name and units for the data to be plotted along the axis.

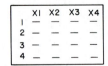

SPREADSHEET   Computer software that processes data using rows and columns.

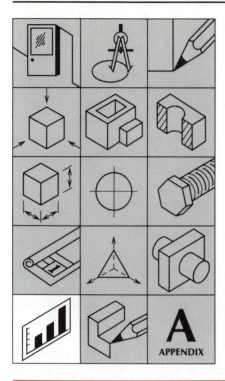

13

GRAPHICAL PRESENTATION OF DATA

13.1 *INTRODUCTION*

Graphical presentation of data is the creation and use of charts
and graphs to record and display data. Charts and graphs can
relate complex parameters as design features or they may
simply show trends over time. In the first part of this chapter
you will study several types of charts and graphs and the
methods used to create them. In the next part, you will learn
about using the computer to display information in a clear and
concise manner. Then in the third part, you will learn about the
concepts of merging text with graphics: **desktop publishing.**
Examples of charts and graphs are shown in Figure 13.1.

Figure 13.1 Sample illustrations for charts and graphs.

13.2 *CHARTS AND GRAPHS*

Designers use charts and graphs to present information in graphical or pictorial form. They know that information that is presented graphically is read and understood much more quickly by most people than the same information presented in words or tables of numbers. Either of the terms *chart* or *graph* can be used, although the former is generally used when referring to a map or presentation display and the latter when presenting technical information. This chapter will cover some typical types of charts and graphs and their uses.

13.2.1 Classes of Charts and Graphs

Charts and graphs can be classified according to their intended use. Some are used to provide technical information, such as the horsepower being developed by an automobile engine in a laboratory test, while others are used to provide general information, such as a report of weather conditions. Those used for technical purposes are generally prepared with more attention to detail and precision. Those used for general information purposes may provide less detail but more attention may be paid to readability and appearance. Additional detail can make the chart or graph more difficult to read.

13.2.2 Typical Uses of Charts and Graphs

One of the many ways in which charts and graphs are used is in the analysis of data. Figure 13.2 is a graph of U.S. foreign trade from 1965 to 1985. Both total exports and total imports are plotted at five-year intervals over the period. By looking at the graph, it is easy to determine (1) the years in which exports exceeded imports, (2) the years in which imports exceeded exports, (3) whether exports are increasing or decreasing relative to imports, and (4) other important information such as the amount by which exports and imports differed in any year. Because data are plotted at five-year intervals, it is not possible to tell the exact year when imports began to exceed exports. Plotting data points for every year would aid in the analysis of the data.

Figure 13.3 is a graph from a recorder attached to an environmental chamber. It shows the temperature and the humidity of the process on a continuous basis as a function of the time of day. This graph tells at a glance whether the chamber is maintaining the temperature and humidity within acceptable limits or whether adjustments need to be made.

We have all seen weather maps on television. These maps are

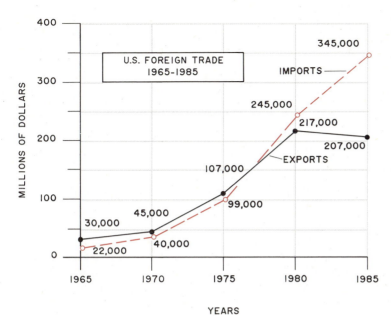

Figure 13.2 Example of line graph.

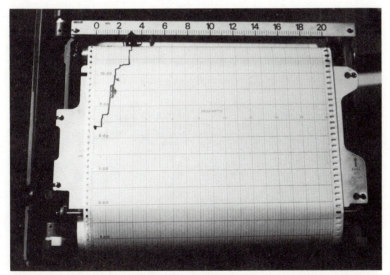

Figure 13.3 Strip chart from recording instrument. (Courtesy of Ohio State Engineering Publications)

a form of chart. They are used to present a wide variety of weather information that would be nearly impossible for the weather forecaster to present in words. Examples of other commonly used maps are road maps and survey maps showing land parcel locations and uses.

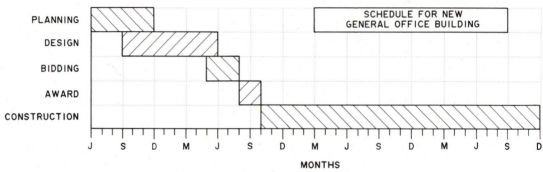

Figure 13.4 Construction schedule for new general office building.

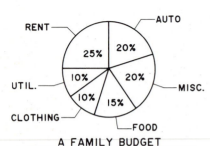

A FAMILY BUDGET
EXPENDITURES BY CATEGORY

Figure 13.5 Pie chart showing family budget.

Other ways in which charts and graphs are used to display information include using them to show schedules (Figure 13.4), to describe the makeup of something such as a budget (Figure 13.5), and to show trends such as in the stock market (Figure 13.6).

13.2.3 Types of Graphs and Their Uses

Line Charts

Line charts relate one quantity to another. For instance, a sales chart might show the sales of an item with respect to time (Figure 13.7). The type of line chart with which we may be most familiar

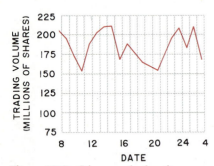

Figure 13.6 Line graph showing trend of stock market from July 8 through July 27, 1987.

Figure 13.7 Line chart: Sales of item over period of time.

has rectilinear or linear coordinate scales along its **axes,** that is, scales with even spacing between units, as represented by Figure 13.7. Some line charts used for data analysis or problem solving may use a logarithmic scale on one axis and a rectilinear scale on the other axis (Figure 13.8). These are often referred to as semilogarithmic, or semilog, scales. In other cases they may use a logarithmic scale on both axes (Figure 13.9). These are referred to as logarithmic, or log–log, scales. Applications where semilog or logarithmic scales would be used will be covered later in this chapter.

Multiple-Line Graphs

Many times it is desirable to plot two or more different relationships on the same line graph. In that case the plotted points are marked with different symbols for each relationship, and different line styles or colors are used for the lines (Figure 13.10).

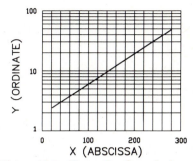

Figure 13.8 Example of graph that uses logarithmic scale along ordinate and arithmetic scale along abscissa: Semilogarithmic graph.

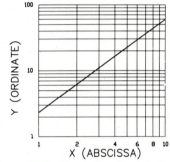

Figure 13.9 Example of graph that uses logarithmic scale along both axes: Logarithmic graph.

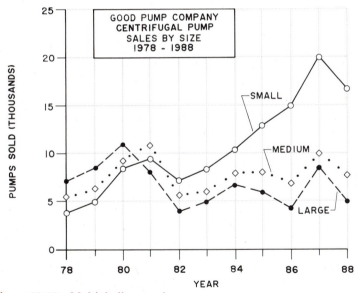

Figure 13.10 Multiple-line graph.

Multiple-Scale Graphs

Sometimes, when we wish to plot more than one relationship on a single graph, we find that we cannot use the same scale for both relationships. For instance, we may wish to plot both the acceleration and velocity of a cam follower as a function of time but find that the values are so different that different scales are needed. The problem can be solved by creating two scales on the vertical

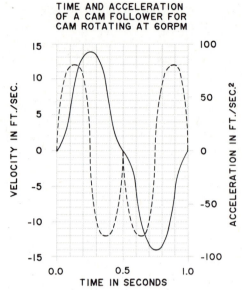

Figure 13.11 Multiple-scale graph.

axis (Figure 13.11). The two scales can be placed on the same side of the graph, but it is often more convenient to put one on the left side and the other on the right, as shown in Figure 13.11.

Pie Charts

Pie charts, as shown in Figure 13.5, show the relative size of each part of a whole. A pie chart is drawn as a circle that represents the entire item, such as a budget or the cost of a manufactured product. The circle is then divided into sectors (Figure 13.12) to show each component that makes up the total and how much each component contributes to the total. Finally, the chart and each of its sectors are labeled. A typical use for a pie chart would be to show what portion of the total weight of an automobile is steel,

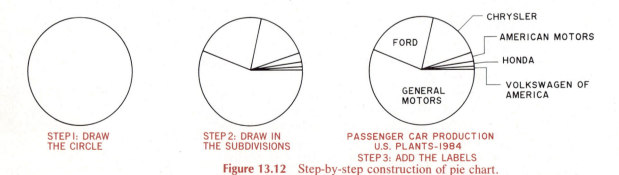

Figure 13.12 Step-by-step construction of pie chart.

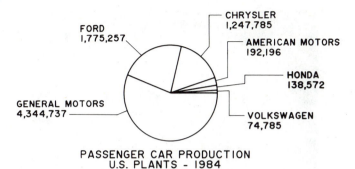

Figure 13.13   Pie chart that displays absolute amounts.

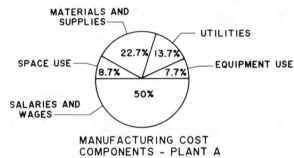

Figure 13.14   Pie chart that displays percentages.

plastic, rubber, glass, and other materials. Another typical use is to show how much of a budget is used for each of the spending categories in the total. Pie charts can display absolute amounts (Figure 13.13), percentages (Figure 13.14), or both.

Polar Charts

A type of line chart not often seen in presentations or reports but often encountered in equipment monitoring and control applications is circular and uses polar coordinates (Figure 13.15). This type of chart is often used when one of the variables being plotted is time. A recorder that operates similar to a clock turns the chart and one or more pens draw the line(s) (Figure 13.16). Another use for polar graphs is to represent intensity levels or patterns such as illumination patterns from a light source or radiation patterns from a radio or television broadcasting station.

Bar Charts

Bar charts can be classified as either vertical bar, often called column charts, or horizontal bar charts. Vertical bar charts are often used to present the magnitudes of quantities (Figure 13.17). Horizontal bars are more often used when periods of time are

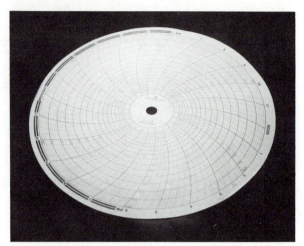

Figure 13.15 Example of chart that uses polar coordinates. (Courtesy of Ohio State Engineering Publications)

Figure 13.16 Recording instrument that draws polar charts. (Courtesy of Ohio State Engineering Publications)

being presented or when the lengths of the bars are too great to draw them vertically (Figure 13.18).

More than one relationship can be plotted on the same graph by shading or crosshatching the bars (Figure 13.19) or by drawing

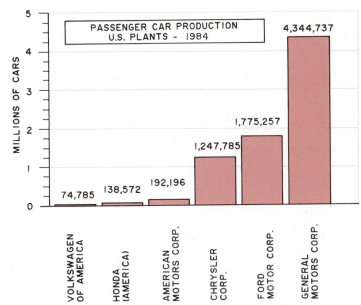

Figure 13.17 Vertical bar chart.

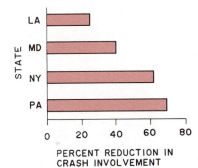

Figure 13.18 Horizontal bar chart. (Courtesy of the Insurance Institute for Highway Safety)

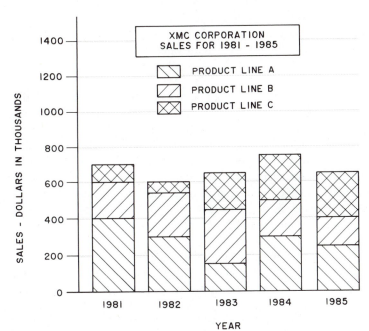

Figure 13.19 Vertical bar chart that shows more than one relationship using multiple shading of bars.

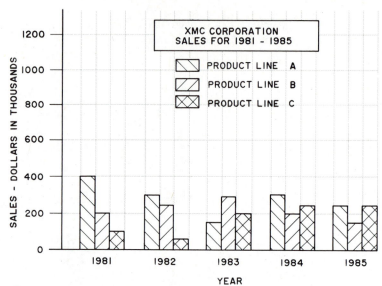

Figure 13.20 Vertical bar chart that shows more than one relationship using multiple bars and shading.

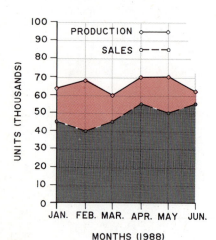

Figure 13.21 Surface chart.

multiple bars (Figure 13.20). Note that Figure 13.19 shows the total sales from all product lines as the total height of the bar. Figure 13.20 shows more clearly the relative contribution of each of the three product lines to total sales.

Surface Charts

Surface charts are drawn in a manner similar to line graphs. However, the area under the line is shaded to represent a surface or area. If the chart contains more than one line, the areas between each of the lines are shaded differently (Figure 13.21). The shading helps the viewer to determine the relative size of each shaded area. Since the exact area of each shaded surface is difficult to determine, surface charts are not generally intended to be used for careful analysis of data but rather to give an overall impression or feel for the information.

Three-Coordinate Charts

Three-coordinate charts, often called volume charts, are used when it is necessary to relate three variables on one chart. The third coordinate is plotted by adding a third axis (the *Z* axis) to the chart (Figure 13.22). Volume charts are generally difficult to construct by manual methods but are easy to do on a computer. As a result, most three-coordinate charts are computer generated

A THREE COORDINATE (X,Y,Z) PLOT

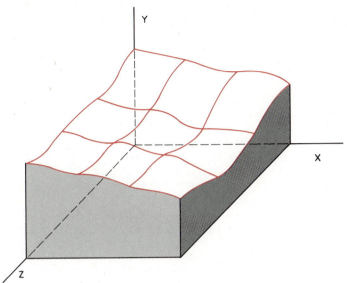

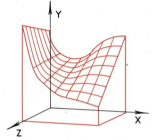

Figure 13.22 Three-coordinate chart or volume chart produced by manual drawing methods.

Figure 13.23 Three-coordinate chart or volume chart produced by computer. (*Computer-generated image*)

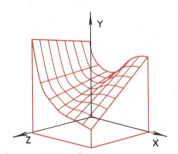

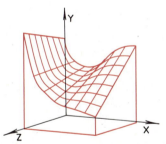

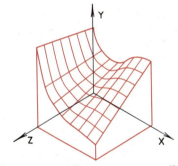

Figure 13.24 Computer-generated dimetric three-coordinate chart. (*Computer-generated image*)

Figure 13.25 Computer-generated trimetric three-coordinate chart. (*Computer-generated image*)

Figure 13.26 Computer-generated isometric three-coordinate chart. (*Computer-generated image*)

(Figure 13.23). Since each **data point** has X, Y, and Z coordinates, the chart can be displayed in any of several three-dimensional forms: oblique (Figure 13.23), dimetric (Figure 13.24), trimetric (Figure 13.25), or isometric (Figure 13.26).

Trilinear Charts

The trilinear chart is a valuable tool used in the fields of chemical engineering, metallurgy, and agriculture. It can be used to analyze

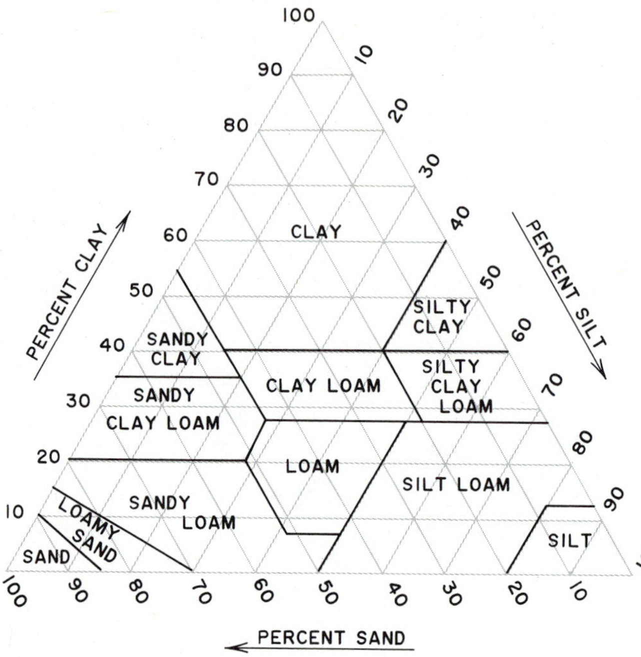

Figure 13.27 Trilinear chart. (Courtesy of The USDA-Soil Conservation Service)

mixtures containing three variables. For example, chemical compounds consisting of three elements can be studied using a trilinear chart.

The chart is formed by drawing an equilateral triangle. The altitude of this triangle represents 100% of each of the three variables that make up the compound, mixture, or alloy. Using this type of chart depends upon the geometric principle that the sum of the perpendiculars to the sides of an equilateral triangle from any point within the equilateral triangle is a constant and is equal to the altitude of the triangle. Figure 13.27 is an example of a trilinear chart.

13.2.4 Plotting a Graph

Selecting the Type of Graph

The function of a chart or graph is to display information in a clear and concise manner. This information may be entirely misleading if displayed using the wrong type of coordinates. For example, bearing wear with regard to the temperature of two

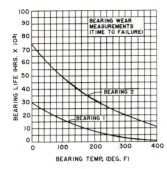
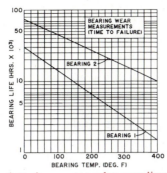

Figure 13.28 (a) Bearing wear data plotted on rectangular coordinates may be misleading. (b) Bearing wear data plotted on semilogarithmic coordinates shows true picture of the relationship.

different types of bearings is shown on a graph using rectangular coordinates (Figure 13.28a). The designer might be misled reading such a graph when the same data plotted on semilogarithmic coordinates would give the true picture of the relationship (Figure 13.28b). Therefore, it is important to know and understand something about the data that you wish to display. Only then can you make an intelligent decision as to how it should be displayed. The computer can be very helpful in this regard. It will allow you to see the data plotted in different forms very rapidly and then you can choose the best way it should be displayed.

Intentionally misleading charts and graphs are often used to distort the facts or to win over an opinion. Commonly, the vertical scale is exaggerated showing a trend that may not exist. In design work it is essential to present factual data in a clear and accurate manner.

Selecting a Size for a Graph

When preparing to plot a chart or graph, it is important to do some planning before putting any marks on the paper. First, the amount of space surrounding the plot itself should be considered. Most charts employ a horizontal axis (the abscissa) and a vertical axis (the ordinate). These axes will need to be labeled with the name of the quantity being recorded and the units being used. They will also need to be graduated and calibrated (Figure 13.29). It is important to reserve enough space for these purposes. A **graph title** will also be needed to identify the purpose of the graph and important information regarding it (Figure 13.29). If the chart title will be outside of the plotting area, space will need to be reserved for it also. Many times the graph title is placed in an unused part of the plotting area.

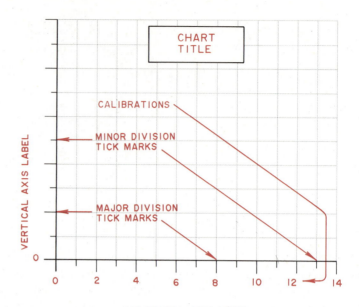

Figure 13.29 Important elements to consider when constructing graphs.

Selecting and Calibrating a Scale

When the approximate location and length of each axis line has been determined, it is necessary to work out a scale so that the axis line can be subdivided into increments. The first step is to determine the range of data values to be recorded for that axis. The scale is normally chosen so that the final plot uses nearly the entire plotting area. That is because it is easier to read a large plot than a small one and because the large plot presents a better appearance. To choose a scale, you must first determine the maximum and minimum values that you need to plot. It is important that the scale be such that the plotted points can all be contained within the plotting area.

After choosing the scale, it is necessary to decide what size increments to use, that is, how many subdivisions to have (Figure 13.30). The increments should be chosen for maximum readability. Too many increments will make the axis markings appear crowded and difficult to read. Too few will make it difficult to visually determine the values of the plotted points. The increments should be chosen so that the values at the increment points are even numbers. For instance, a scale of 0, . . . , 100 might be subdivided as 0, 20, 40, 60, 80, 100 or 0, 10, 20, 30, 40, 50, 60, 70, 80, 90, 100 but not 0, 12.5, 25, 37.5, 50, 67.5, 75, 87.5, 100

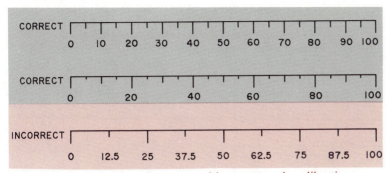

Figure 13.30 Examples of correct and incorrect scale calibrations.

(Figure 13.30). It is not necessary or even desirable to show the value of each subdivision on the axis. For example, a scale with subdivision markings every 5 units might have values shown every 20 units. As a rule, try to use increments that are some multiple of 2, 5, or 10.

Before we can place the tick marks, or **axis calibration** markings, on the scale, we must determine where each one is to go. To do this, we first calculate a scale modulus. The scale modulus is a factor that relates the length of the scale in inches or millimeters to the range of the data. It is calculated as the scale length divided by the scale range. The scale range is the difference between the maximum scale value and the minimum scale value:

$$m = (\text{scale length})/(\text{maximum} - \text{minimum})$$

If we think of the data as being represented by a function $y = f(v)$, then

$$m = (\text{scale length})/[f(v_2) - f(v_1)]$$

where $f(v_2)$ is the maximum scale value and $f(v_1)$ is the minimum scale value (Figure 13.31).

When the scale modulus has been calculated, the location of any data point v on the scale can be determined by rearranging the preceding equation to read

$$\text{Length} = m[f(v) - f(v_1)]$$

where length is the distance in inches or millimeters from the beginning of the scale to the location of v.

This method of calibrating a scale can be used for all scales, nonlinear as well as linear. The only requirements are that the function $f(v)$ is known and that it is continuous over the range of the scale. Figure 13.32 shows the use of the method to calibrate

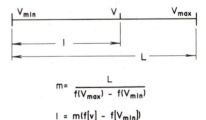

$$m = \frac{L}{f(V_{max}) - f(V_{min})}$$

$$l = m(f[v] - f[V_{min}])$$

Figure 13.31 Formulas for computing scale modulus and scale length when calibrating scales.

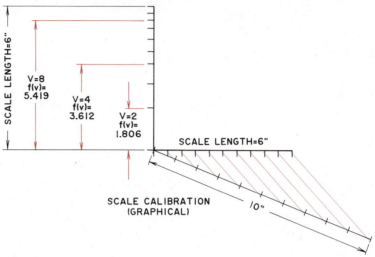

SCALE CALIBRATION - SCALE MODULUS
f(v)=log(l0)v , f(l)=0 , f(l0)=1
m=SCALE LENGTH/(l-0)=6/1=6" , f(v)=6x(logl0(v))

SCALE LENGTH=6"

V=8
f(v)=
5.419

V=4
f(v)=
3.612

V=2
f(v)=
1.806

SCALE LENGTH=6"

SCALE CALIBRATION
(GRAPHICAL)

10"

Figure 13.32   Using formulas to calibrate logarithmic scale for range of values from 1 to 10 along ordinate. Using graphical procedure to develop arithmetic scale for range of values from 1 to 10 along abscissa.

a logarithmic scale along the ordinate axis for a range of values from 1 to 10. In this case $f(v) = \log(10)v$. Note that $f(1) = 0$ and $f(10) = 1$. Therefore, $m = $ (scale length)$/(1 - 0)$. If the scale length is 6, then m is $6/(1 - 0) = 6$. For any value of v, the distance from the start of the scale is length $= 6[f(v)]$. For example, the distance to the calibration point for $v = 4$ is length $= 6[\log 10(4)] = 6(0.602) = 3.612$ in.

You can also calibrate a scale using graphical methods. Chapter 2 discusses a method in which you can divide a line into a required number of parts. The scale along the abscissa in Figure 13.32 was calibrated using this graphical procedure.

When more than one relationship is shown on a single chart, it is necessary to mark the points for each relationship differently so that there will be no question of which data points belong to which relationship. If the chart is to be done in color, different colors can be used. Otherwise different symbol types should be used. The symbol types should be as different as possible. A chart might use circles, squares, and triangles to plot three sets of relationships. Use of circles and polygons with more than five sides on the same chart should be avoided because they might look too much alike.

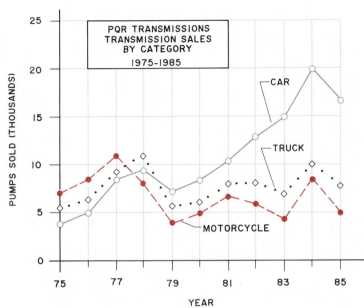

Figure 13.33 Multiple-line graph showing three relationships using different line styles to set them apart.

The lines or curves connecting the points should also be different for each relationship. Again, color can be used effectively to set them apart. Line style can do the same thing. A chart showing three relationships might use a solid line for one, a dashed line for the second, and a dotted line for the third (Figure 13.33).

A question that is frequently asked is whether the line connecting the plotted points should be a continuous smooth curve or a series of straight-line segments. The answer depends on the nature of the data that is being graphed. If it is known that the relationship would provide a smooth curve if you plotted enough points, then draw a smooth curve. However, if the data are only available as specific points and there is no information about the relationship between the plotted points, then straight lines should be used.

13.2.5 Adding Titles and Labels

Chart/Graph Title
In order for the plotted data to be read and understood, it is necessary to complete the chart with descriptive information. A chart title is needed. The chart title should be placed in a box along with notes stating what the graph is for, dates, locations,

and other identifying information. Other notes may be included to help explain the chart. The chart title should answer the questions who, what, where, when, how, and why.

The title can be placed on the chart by one of several different methods. For charts or graphs that are to be used as working charts, it may be only necessary to freehand letter the title and other notes. The letter heights in the title generally are .25 in. (6 mm); therefore, you should make your light guidelines accordingly. The letter heights may vary for some titles depending on space available.

Charts and graphs presented in formal reports, proposals, sales brochures, and so on, should have lettering in the title that is not hand lettered. This type of lettering can be done using simple lettering templates, an engraved lettering template set, automatic lettering machines, or artists' transfer type.

Simple lettering templates (Figure 13.34) can be used with either pencils or ink pens; however, ink pens are more commonly used. Plan your letter and word spacing in the title area with a light pencil. Then with the template resting against a straight edge such as a drafting machine blade or triangle, form the letters of your title with the pen tracing through the desired letters. Be sure that the pen remains vertical while you are lettering. This ensures that the letter line weights will remain uniform (Figure 13.34).

An engraved lettering template set is composed of technical pens of various line widths, a scribe, and several templates of varying letter sizes. Technical pens are inserted into a scribe. The size of the technical pen (line weight) is a function of the size of

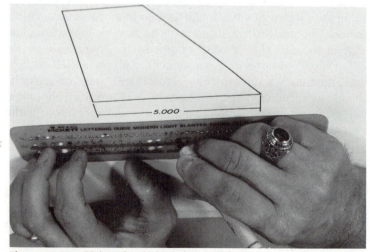

Figure 13.34 Using simple lettering template.

Figure 13.35 Engraved lettering template.

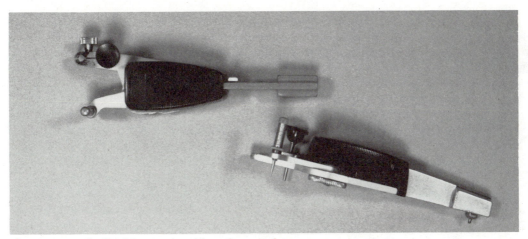

Figure 13.36 Scribe from engraved lettering template set.

the desired lettering. The template (the long scale with the engraved letters) tells the size pen to use for the proper line weight (Figure 13.35). The template is placed against a straight edge to ensure that the letters will be straight when traced. The scribe has three distinct points extending from its lower side (Figure 13.36). The point at the end of the handle (the smallest of the three) is inserted into the wide groove at the bottom of the lettering template. The thin point (called the stylus) is inserted into the groove representing the desired letter shown on the template. The point near the pen has a flat surface. This point acts as a support to ensure stability while lettering. Simply trace the stylus through the thin groove on the template to letter on the chart (Figure 13.37). Using this lettering device takes a little more practice than the simple lettering template; however, the results are generally of higher quality.

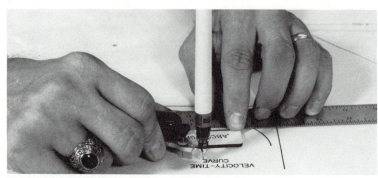

Figure 13.37 Using lettering set for chart title.

Figure 13.38 Automatic lettering device is used by professionals. (Courtesy of MUTOH America, Inc.)

An automatic lettering machine (Figure 13.38) can be very beneficial if you have to create a lot of lettering on charts and graphs. It has several features such as proportional spacing, degree of slant, memory, and predrawn symbols that can be used over and over again. The automatic lettering machine has a keyboard with which the operator types in the desired words to be lettered. After keying in these words, the machine will place the words in the desired position on the graph, automatically spacing the letters and words. These machines are expensive. However, if the user is in a business that has large volumes of lettering to create, then the expense can be justified.

Lastly, chart titles may be created using transfer type. Transfer type allows the user to rub off or burnish the desired letters onto the chart of graph so that the letters look like they have been printed (Figure 13.39). Transfer type is easy to use and can be purchased in many fonts or styles.

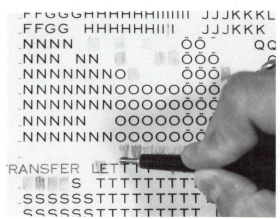

Figure 13.39 Using transfer type for charts and graphs.

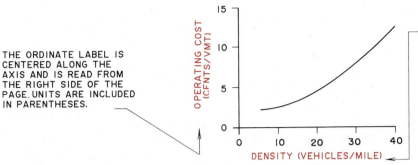

THE ORDINATE LABEL IS
CENTERED ALONG THE
AXIS AND IS READ FROM
THE RIGHT SIDE OF THE
PAGE. UNITS ARE INCLUDED
IN PARENTHESES.

THE ABSCISSA LABEL IS
CENTERED ALONG THE
AXIS AND IS READ FROM
THE BOTTOM OF THE PAGE.
UNITS ARE INCLUDED
IN PARENTHESES.

Figure 13.40 Axes labels show variable being plotted as well as units of measurement being used. Ordinate labeled sideways on sheet so it can be read from right side.

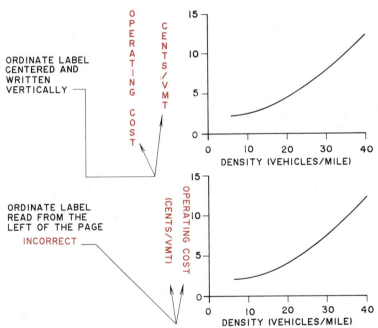

ORDINATE LABEL
CENTERED AND
WRITTEN
VERTICALLY

ORDINATE LABEL
READ FROM THE
LEFT OF THE PAGE

INCORRECT

Figure 13.41 Example of ordinate axis labeled so it is read vertically. Letters printed one below the other in upright position. Axis labels are never read from left of page.

Labeling the Axes

If the chart is a line or bar graph, then the axes need titles. Each **axis title,** or legend, should state the variable that is being plotted for that axis and the units of measurement being used (Figure 13.40). The vertical axis labels may be lettered sideways on the sheet to read from the right side or they may be computer plotted one below the other in an upright position (Figure 13.41). The methods described in lettering titles in this section can be used in lettering axis labels (Figure 13.41).

13.2.6 Modeling Empirical Data (Data Scatter)

Best-Fit Methods

Much of the engineering data that are plotted result from measurements taken in the laboratory or in the field. These data are called empirical data, meaning that they are measured rather than computed. Such data are often not precise due to inaccuracies in the measurement procedure or the equipment being used or due to uncontrolled changes taking place in the process being measured while the data are being taken. As a result, a plot of the data may show an erratic relationship known as data scatter (Figure 13.42). When it is known that the data should conform to some known relationship, then an attempt is usually made to fit a smooth curve describing that known relationship.

Many different methods have been developed to fit known curve forms to empirical data. One of the best known is the method of least squares. A curve of the known shape is drawn in a position that minimizes the sum of the distances or "errors" between the plotted points and the curve. Since all error values must be positive, the square of each error is used to avoid the problem of

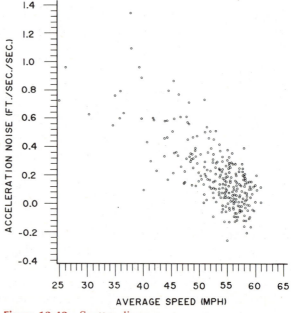

Figure 13.42 Scatter diagram.

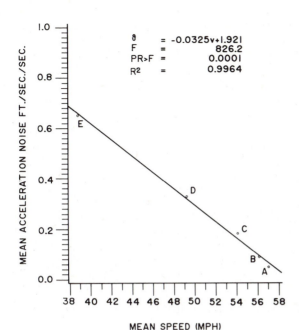

Figure 13.43 Least-squares fit of curve through experimental data.

"negative errors," that is, points where the mathematical difference between the plotted point and the curve is a negative value (Figure 13.43).

Types of Data

Empirical data may be taken from either a discrete process or a continuous process. Discrete processes are those that produce data at certain times. The Yellowstone National Park Geyser, Old Faithful, only erupts about once an hour, so data regarding the eruption can only be taken at those times. Electrical current, on the other hand, flows continuously, so data regarding voltage, current, waveform, and so on, can be taken on a continuous basis. Discrete data would not be plotted using a smooth curve. The multiple-line graph shown in Figure 13.33 is an example showing how discrete data should be plotted.

13.2.7 Computer-based Chart and Graph Programs

We have been looking at the techniques for producing charts and graphs by manual methods. However, an increasing percentage of chart and graph production is being done by computer. A large number of high-quality specialized software packages are being marketed for this purpose.

There are many advantages to using a computer-based graphing package rather than creating text and graphics manually. One advantage is speed. A knowledgeable user can produce a chart in minutes using the computer. A good chart created manually may take an hour or more. This situation is quite different from producing line drawings with CADD programs. A drawing produced with a CADD package might take as long as doing it manually. The difference is that with the CADD package you must create the drawing line by line. The graphing software draws the graph. You just enter the data and select the graph type and other options.

A second advantage of using a graphing package is that the results do not depend on one's manual drawing skills. A person with little or no drawing ability can produce good charts and graphs. Production of effective charts does require that the user select the correct chart type and that intelligent choices be made regarding scales, labels, colors, data organization, and similar factors.

Third, computer-generated charts and graphs are easy to modify. The data can be rearranged. Text font styles and lettering sizes can be changed. Scales can be changed, as can graduations, calibrations, and axis and chart labels. The modified chart can be redrawn effortlessly. The result is that the user is less likely to settle for a chart that isn't quite right.

Fourth, due to the ease of modifying the charts, several different chart types and layouts can be created, allowing the user to select from several alternatives. It is not unusual for a user to create and print three or four charts before selecting one as the best.

CADD packages can be used to draw charts and graphs. However, they provide little advantage over manual creation techniques. They do not provide any of the automatic scaling, labeling, plotting, or hatching that is common with graphing software. They do allow the chart to be modified and redrawn without having to start over.

Spreadsheet programs generally contain graphing routines. They allow the user to create a chart or graph from the spreadsheet data and dump it to a printer. They also provide a selection of charts along with scaling, labeling, hatching, and some choices of text. The charts they produce are generally good from the utilitarian data presentation standpoint. Their shortcoming is that they do not provide the "extras" that make a chart attractive. Figure 13.44 is a pie chart that was produced with a spreadsheet program.

Presentation graphics software, as the charting and graphing packages are called, provides many features that allow the user

COMPOSITION of FREE CUTTING BRASS

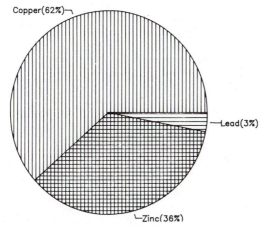

Figure 13.44 Pie diagram created with spreadsheet program. (*Computer-generated image*)

to transform a plain chart into a very attractive presentation quality document. Most graphing software supports the most popular types of charts and graphs. These include vertical and horizontal bar and line charts, and pie charts. Many also support surface area, Gantt, and high–low charts. Some support organization, bubble, and other specialized chart types.

Perhaps the major difference between chart and graph software included in other programs such as spreadsheets and stand-alone software is in the options available to the user. Graphing software in a spreadsheet generally expects its input to come from the spreadsheet. Stand-alone software will allow you to provide the data from a number of sources such as any of several spreadsheets, databases, or files as well as keyboard input. The stand-alone software also typically provides a much larger choice of options for text style (fonts) and for size and location and appearance of axis and chart labels.

Most stand-alone software provides a choice of automatic or manual scaling of the axes. A wider range of fills, chart orientations, three-dimensional effects, and other customization options are usually available in stand-alone packages. Template, macro, and symbol library support are also common.

Finally, the stand-alone software often supports a wider range of output choices. Output may be to a file in a specified format, to a clipboard, to a printer, to an imaging system for slides, or

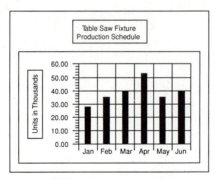

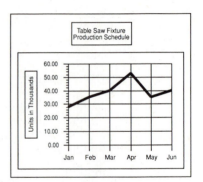

Figure 13.45 Vertical bar chart created with presentation graphics software. (*Computer-generated image*)

Figure 13.46 Line chart created with presentation graphics software using same data as Figure 13.45. (*Computer-generated image*)

other audiovisual equipment or to a large screen for direct viewing.

There are several advantages in being able to output a chart to a clipboard or to a file in a specified format. Many times the chart is to be used in a report or other publication. If it is ready to be used, then transferring it by using the clipboard is quick and easy. If it needs additional detailing, perhaps with an illustration package, then another file format may be better.

Each graphing package is somewhat different and requires some time to learn. Once learned, you can produce a wide range of charts and graphs in minutes. Figure 13.45 is a vertical bar chart that was produced using a presentation graphics package. Figure 13.46 is a line chart that was produced from the same data. The only step that was required to produce the line chart was to go to the menu and change the desired chart type from vertical bar to line. The graphing software did the rest.

13.3 *MERGING TEXT AND GRAPHICS BY COMPUTER*

Software to merge text and graphics to one page by computer and print the output, has become very popular. It is known by the popular term *desktop publishing*. So popular is desktop publishing that many users choose a computer primarily on the basis of the availability of such software.

Pick up any newspaper, product advertisement, magazine, or in fact, this textbook, and you will find that a high proportion of the pages contain both written material and pictorial information. The pictorial information may be logos, photographs, charts, illustrations, artist's renderings, or any of a wide range of non-verbal material. So accustomed are we to seeing graphic material interspersed with text that we may ignore advertisements that do not contain pictures. Likewise, textbooks without illustrations are judged to be "heavy" and difficult to read and understand.

Many times, small enterprises such as small retail businesses, religious, social, or professional organizations have difficulty producing attractive fliers, newsletters, and the like because they lack the facility for merging graphics with text. Equipment is expensive and limited to large enterprises. Page composition software is now available for microcomputers. The software goes by the general name "desktop publishing."

Of course, it has always been possible to merge text and graphics manually. The technique is often called "cut and paste" or "paste-up" since the various text and graphic items are cut from their respective sheets and reassembled on a new sheet. The reassembled document is then photographed to produce a plate for an offset press. Alternatively, the reassembled document may be reproduced on an office type copier. Desktop publishing software has many advantages over this manual method.

The basis of any publication that merges text and graphics is page composition or layout. Whether you are producing the page manually or by computer, you must decide what it should look like, where the text should go, where the graphics should go, style and size of text, and all of the details that go into production of a quality product. The desktop publishing packages are excellent tools that allow you to try several layouts, and they eliminate the requirement for manual dexterity with drawing and manual paste-up tools. They do not relieve you of the job of creating the page design.

When using a desktop publishing package, your first task is to create the design for the publication. You may start with a template and style sheet that have been created previously and stored or you may start from scratch. A page layout consists of areas called boxes that are to contain text or pictures and lines that are inserted to separate items. Gutters, which are spaces between columns, and other features are defined. If you are doing a repetitive operation where the same type of document is to be produced more than once (such as a weekly newsletter) the general overall design can be created once and saved as a template to be retrieved and modified as needed for each new production. If starting from scratch, you will typically start by defining the page size and the margins to be used (top, bottom, and sides).

Figure 13.47 Page layout for merging text and graphics. The template looks like this on the computer screen. (*Computer-generated image*)

If the page is to have multiple columns, you will set up the areas for each column. If the document is to have multiple pages, you should complete the design for one typical page and then insert the required number of these pages into the template for your document. Then each page can be modified as necessary to fit your specific needs.

You will also want to specify default parameters for type style, size, colors, and other variables. These can be changed also for specific needs on any page or in any text box since each text and picture box can have different parameters.

A good approach is to sketch your layout on a sheet of paper, planning what the page should look like. Sketch rectangular boxes on the sheet to show the location of each block of text and each graphic. Then lay out a template on the computer to match your sketch. Figure 13.47 is a template layout for a one-page document that will combine a drawing of the table saw fixture, a graph showing a six-month production schedule for the

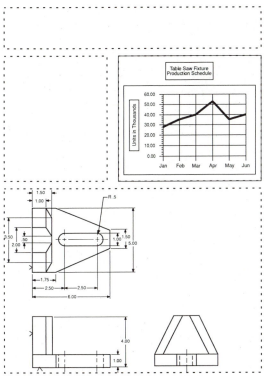

Figure 13.48 Template with graphic images inserted. (*Computer-generated image*)

fixture, and a paragraph describing the fixture. Dotted lines show the margins and the areas (boxes) for text and graphics. The graphics boxes have diagonal lines to distinguish them from text boxes. Of course, the lines will not appear on the finished document, but they do appear on the computer screen to guide you in assembling your material.

Figure 13.48 shows the template with drawing and graph added. The drawing was created using a CADD package. The graph was created using a presentation graphics package. Both were brought into the document by placing them on an electronic "clip board" and then "pasting" them from the clipboard to the document template.

The document was completed by creating the heading using the desktop publishing package's text editing feature and then reading the paragraph from a file. The paragraph had previously been written with a word processor. Figure 13.49 shows the completed document.

TABLE SAW FIXTURE
Model GCF–145A
DESIGN DRAWING AND PRODUCTION SCHEDULE

Table saw fixture, Model GCF–145A, is standard on Contractor Saw, Model GCTS–90357 and available for Models GCTS–90356 and GCTS–90357. Fixture production schedules for the first half-year are expected to increase inventory by 20,000 units in anticipation of construction upturns and resulting brisk mid-year demand.

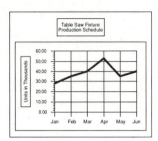

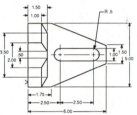

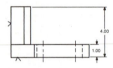

Figure 13.49 Completed page. Paragraph of text has been merged with the graphics and heading has been added. (*Computer-generated image*)

13.4 *COMMON PROBLEM*

A table saw fixture has been used as a common problem throughout this text. In this chapter, Figures 13.45 and 13.46 are graphs of a production schedule for the fixture. Figures 13.47–13.49 show the process of merging the drawing of the fixture (from Chapter 4) and a production schedule graph into a document.

13.5 *SUMMARY*

Various types of charts and graphs are used to present data in an understandable way. Pie charts show the relationship between the various parts of a whole. Line graphs show trends or the relationship between two variables. Bar charts are used to compare various quantities. Multiple-line graphs, surface charts, three-coordinate charts, map charts, and trilinear charts are other types of charts that can be used to display data.

When developing a chart or a graph, it is important to take into consideration chart size, scale calibrations, axis labels, and chart title. Scale calibration is a very important concern when designing a chart or graph. The primary consideration is to be sure that data can be interpolated or read from the chart without having to do time-consuming calculations.

Charts and graphs are constructed easily on computers using presentation graphics software, which will create the entire graph using data you provide. This software will create a wide range of chart and graph types and will generally provide a number of text sizes and styles, automatic or manual scaling options, and several other formatting and customizing options. Data for the graphs can be entered from the keyboard or input from a data file or other source.

Desktop publishing is the name given to the process of merging text and graphics by computer. Desktop publishing packages provide electronic "cut and paste" capability, thus eliminating the need to assemble your document manually. They provide a wide range of typesetting functions such as combining multiple type styles and sizes, tracking, kerning, scaling, leading, and many other editing and page layout options previously available only at a professional print shop.

PROBLEMS

Solve each problem either by manual or computer methods as defined by your instructor.

Problem 13.1
Draw and calibrate the axes for a line graph that will be used to plot data with abscissa (X axis) values ranging from 3 to 96 and ordinate (Y axis) values ranging from 270 to 9500.

Problem 13.2
Draw a complete pie chart showing the proportion of a student's budget that is spent in each of the following categories. Show the percentage for each item.

Tuition	$1800	Transportation	$400
Books	400	Clothing	600
Housing	1200	Entertainment	600
Food	2200	Personal and miscellaneous	400

Problem 13.3
Draw a pie chart similar to the one for problem 13.2 but use your own budget items and amounts.

Problem 13.4
Calculate the scale modulus m for the function $f(v) = v$. Use a 6-in. line and values of $v_{min} = 0$ and $v_{max} = 10$.

Problem 13.5
Calibrate a 6-in.-long scale using the data from problem 13.4 and your calculated value of m.

Problem 13.6
Calibrate a 4-in.-long scale for the following data: $f(v) = \sin(v)$, $v_{min} = 0°$, and $v_{max} = 90°$.

Problem 13.7
Calibrate a 3-in.-long scale for the following data: $f(v) = \log(10)v$, $v_{min} = 10$, and $v_{max} = 100$.

Problem 13.8
Calibrate a 5-in.-long scale for the following data: $f(v) = 2.5v$, $v_{min} = 0$, and $m = 0.2$.

Problem 13.9
ABLE Consulting Co. posted the following operating revenues and expenses (in thousands of dollars) for the years 1981–1985:

	1981	1982	1983	1984	1985
Revenue	1520	1716	2135	1987	2225
Expense	1215	1410	1530	1560	1923

Draw a surface chart for this data with the year on the abscissa (X axis). Calibrate and label the axes. Include a short title.

Problem 13.10
Draw a bar chart for the data of problem 13.9. Shade the bars to distinguish between revenue and expense.

Problem 13.11
Answer the following questions by examining your chart from problem 13.9 or 13.10:
(a) In what year was the most profit made?
(b) In what year was the highest percentage of profit made?
(c) In what year was total revenue the highest?
Confirm your answers mathematically.

Problem 13.12
The Clausius–Clapeyron equation relates the vapor pressure (p) of a pure substance with the absolute temperature (T).

$$\ln p = -\frac{\Delta H}{RT} + B$$

where

p = the vapor pressure in millimeters of mercury
T = the temperature in ° Kelvin
ΔH = the latent heat of vaporization
R = the universal gas constant
B = a constant

The vapor pressure of benzene is measured at two temperatures with the following results:

$$T_1 = 285.7\ °K \qquad p_1 = 52\ \text{mm Hg}$$
$$T_2 = 298.5\ °K \qquad p_2 = 97\ \text{mm Hg}$$

Estimate, using the graphic method, the vapor pressure of benzene at 320 K.

Problem 13.13
The rate at which heat is transferred from a hot wall to its surroundings is given by the formula:

$$\frac{Q}{A} = H(T - TO)$$

where

T = temperature of the wall (°C)
TO = temperature of the surroundings (usually a constant)
A = surface area of the wall (m^2)
Q = rate of transference of heat (kJ/h)
H = heat transfer coefficient (kJ/hm² °C)

The following information was obtained from measurements made in a heat transfer experiment:

T (°C)	$\frac{Q}{A}$ (kJ/hm²)
25	0
30	250
35	500
40	750
45	1000

What is the value of TO?

Using the appropriate coordinates, graphically determine the value of H for these conditions.

Problem 13.14

The equation proposed by Thiesen has proved to accurately follow the variation of the heat of vaporization (Q) with temperature.

$$Q = k(T_c - T)^n$$

The critical temperature for water (TC) is 647 K and the values of Q in calories per mole are 3250 and 6600 when temperatures (T) are 632 and 550 K respectively. What is the heat of vaporization for water at 347 K? k and n are constants. Solve this problem graphically using logarithmic graph paper.

Solve the following problems by writing the answer and the text reference from which you obtained the answer.

Problem 13.15

What is data scatter?

Problem 13.16

Describe the difference between a discrete process and a continuous process.

Problem 13.17

List four (4) advantages to using computer-based graphing software over manual methods.

Problem 13.18

You are to produce a graph based on some experimental data. You have a microcomputer that has CADD software, graphing software, and spreadsheet software available. You also have all the necessary tools to produce the graph using manual methods. Which method would you use to produce the graph? Justify your answer.

Problem 13.19

What is an electronic clipboard?

Problem 13.20

Describe the meaning of "cut and paste" with regard to merging text and graphics.

Problem 13.21

Define the term *gutter*.

Problem 13.22
Describe how desktop publishing can be used to produce a one-page document that includes both text and graphics.

Problem 13.23
Name an advantage of computer-based graphing packages over spreadsheet graphics.

Problem 13.24
Name three types of options available to users of computer-based graphing software.

Problem 13.25
What is the popular term used to describe merging text and graphics by computer?

Problem 13.26
Name three applications for merging text and graphics.

Problem 13.27
What basic step must be done when merging text and graphics either manually or by computer?

Problem 13.28
What is the term used to describe a skeleton page layout?

Problem 13.29
Name an advantage of using desktop publishing software over manual page paste-up methods.

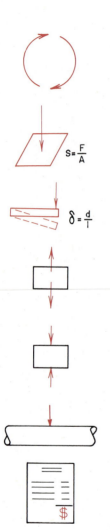

ITERATIVE   Involving repetition.

STRESS   Load applied to an object. The load is divided by area to obtain values that can be compared.

STRAIN   Deformation resulting from an applied load. The deformation is divided by the length over which the deformation occurs to obtain values that can be compared.

TENSILE STRENGTH   Resistance to being pulled apart.

COMPRESSIVE STRENGTH   Resistance to being crushed.

SHEAR STRENGTH   Resistance to being cut.

COST ANALYSIS   Detailed investigation of all factors affecting the total cost of a product or system.

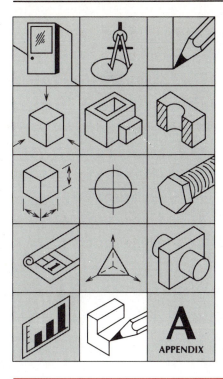

14

DESIGN PROCESS

14.1 INTRODUCTION

Design is the most creative part of the total engineering process. It includes all activities involved in creation from the original idea to the finished product or system (Figure 14.1). A designer may engage in many different activities during his or her career. Some of these may directly involve design; they may also involve research, development, testing, manufacturing, technical service, construction, sales, and

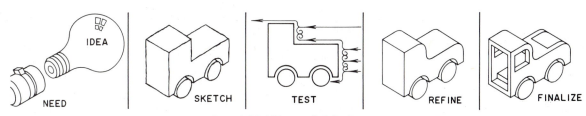

Figure 14.1 Design includes concept from initial idea to finished product.

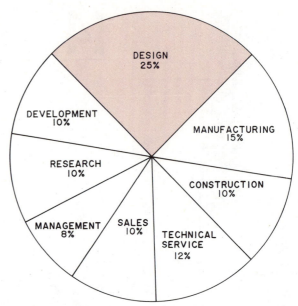

Figure 14.2 Engineering embraces many activities: *Design* is essence of engineering.

management (Figure 14.2). In design, all the technical knowledge, analysis, and creative thinking acquired in technical education and experience are focused on solving problems to meet the needs of humankind. Design is the process by which we create and modify products and systems.

Design is a creative process that requires interaction between the designer and other people or groups of people and also between the designer and machines. (A former chief engineer of a large manufacturing company used to tell his engineers, "Listen to the machines; they will tell you what they need." Learning to do this kind of listening is a skill that takes practice; it also produces results.) The designer uses interactive communications with the computer for technical data, analysis, the sharing of data, and computer graphics.

All design is an **iterative** process. We do not expect to achieve the best design on the first try. The steps of the design process are repeated until all concerned believe that an optimum solution has been reached within the time and money available.

Computer programs for design calculations and graphics have increased our ability to reach an optimum solution in less time. Computer graphics does not offer much of an advantage in speed over manual graphics when preparing a single drawing. However, once the data are entered, they can be modified and presented in many different ways very quickly. This is one of the advantages of computer graphics over manual graphics. In the "old days," when only manual graphics were available, we could seldom afford more than three trials to design a new part or machine. All the drawings were done by hand, and strength calculations were done by inputting the data separately to a calculator or computer. Today you can build a data base, display orthographic or pictorial drawings, calculate **stresses,** display stress patterns, estimate **strains,** calculate weights, and redesign using commercially available software packages. In many organizations, your data base can be sent directly to manufacturing to develop tools and to control the production machinery making the parts!

14.1.1 Product Design and Systems Design

Designers are concerned with product design and systems design. The product may be an original concept or, more commonly, a refinement of an existing product. The system may be a manufacturing process, a construction time schedule, a computer system, or the communication structure for a large group of people working toward a common objective.

Product design is the process of developing new products or modifying existing ones. This process starts when someone recognizes a need for a product. For example, Velcro fasteners were developed to meet the needs of astronaut clothing in our space program. Product redesign centers activities around the redesign of an existing product to update or improve the product for its intended use. This type of design activity may be functional or aesthetic. A lawn care company produces a new product line that includes a redesigned garden hose. This development makes it easier for the user to attach an accessory to the garden hose and also prevents the user from damaging the threads on the hose or the accessory (Figure 14.3).

Both product design and product redesign require the designer to have a firm knowledge of the use of the product, the market in which the product will be used, and technical data related to the product, its materials, and production. With product design or product redesign, the designer will have to interact with people. Regardless of the source of the idea, the designer must work with other creative people to further the scope of application of the product. These may include specialists in other technical disciplines, manufacturing people, marketing representatives, and consultants. These encounters may be on a one-to-one basis or in group discussions.

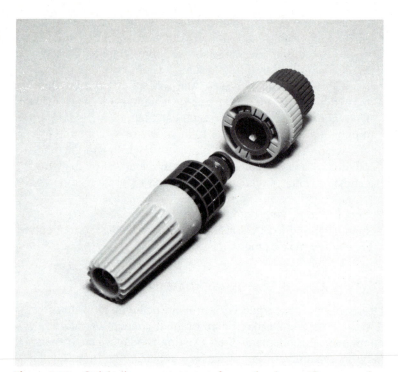

Figure 14.3 Quick disconnect system for garden hose. (Courtesy of OM Scott & Sons)

System design is the process of developing engineered systems that are used in manufacturing; establishing relationships among people, machines, and data; and developing complicated assemblies made of many parts. Chemical designers create chemical processing manufacturing facilities to alter and/or manufacture chemicals. Mechanical designers develop automated assembly lines that involve many mechanical processes. Auto manufacturers apply system design to build cars that contain simple and complex assemblies. Within an automobile there are many systems that require design such as the electrical system, the steering system, the drive train system, the heating and air-conditioning system, and the sound system.

Designers working on system designs must also work with people. They may work with fewer people related to the user and are more likely to work with technical people from many disciplines during the design process. It is necessary that all the systems must be integrated with each other in order that the goals of the organization be achieved.

I. Form a tentative hypothesis
II. Determine methods for testing the hypothesis
III. Test the hypothesis
IV. Analyze results of tests
V. Reformulate the hypothesis as a result of test reports
VI. Retest and refine hypothesis
VII. Announce results

Figure 14.4 One outline of the scientific method.

14.2 *SCIENTIFIC METHOD*

The scientific method can be described as a vertical process. It is not peculiar to engineers and designers; it is used by all scientists. It includes a logical sequence of steps from start through completion. Following this logical order assures that we do not leave anything out. It results in a solution that can be reproduced and defended (Figure 14.4). This method has been described in many different words, but it always includes the following steps:

Identify a problem.

Collect data.

Form a tentative hypothesis.

Test the hypothesis.

Accept, reject, or modify the hypothesis.

Announce the results.

When you consider the method for technical design, you may remember the steps outlined here:

D	Determine a need.
E	Establish known information.
S	Stimulate design concepts.
I	Investigate design concepts with analytical tools.
G	Gather investigation data and review.
N	Now *test* the design that appears best.
P	Propose to accept, modify, or reject the design.
R	Render a final decision.
O	Optimize design by iteration.

If you follow these steps, you will arrive at a useful design solution. This is not the only way to arrive at a solution: Some designers are intuitive, some use collective thinking, and some use the scientific method. Design solutions are never final: New information, new materials, and new tools mean that solutions can always be improved. So do not ignore the last step; you can continue to improve a good solution by repeating the process with new data. Keep it up and you will be a *design pro*.

14.2.1 Determine a Need

The process starts with the recognition of a need. The owner of a house overrun by mice might need a better mouse trap. A company that manufactures knives may have had complaints from people who have been cut. The need for safer packaging could result in fewer injuries to purchasers. The need for a car to appear different from last year's model is necessary in order that next year's model will keep up with the competition and sell well. A manufacturer of a high-quality item may identify the need to produce more products with lower labor cost in order to keep up with the competition. These are all examples of product and system designs that require change based on need.

At this stage of design, sketching will be one of your most valuable tools. As ideas and concepts for meeting the need are discussed, you and others can present sketches of possible ways to solve the problem. These may be pictorial or orthographic sketches of a product or a plan for a manufacturing process (Figures 14.5 and 14.6).

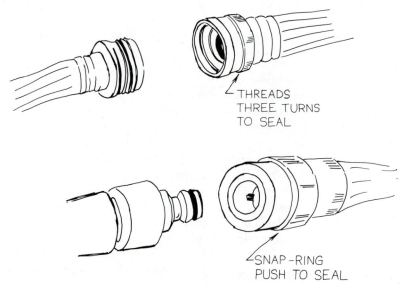

Figure 14.5 Preliminary sketch of idea for product.

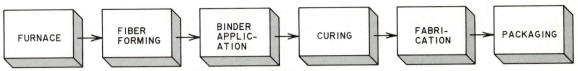

Figure 14.6 Preliminary flow chart for manufacturing facility.

14.2.2 Establish Known Information

Once the need has been identified, the designer must collect information. This information might include such items as sales history, availability of parts or materials (including new materials and new technology), market analysis, and financial and cost data (Figure 14.7). During this phase of the process, as much information as possible should be gathered so that later decisions will be based on the most complete collection of information possible.

At this step you will want to get as much expert opinion as possible. Do not attempt to design parts using new materials without advice from materials specialists. If your problem involves a new manufacturing method, get the manufacturing people involved; they know their existing process and may have many valuable ideas for changes. Interview computer systems experts

Figure 14.7 Design information needs to be collected from many sources.

before deciding that a new computer program will save time and money—a program to meet your needs may already exist.

14.2.3 Stimulate Design Concepts

At this stage you review your collected information and apply it to the need determined in step 1 (Section 14.2.1). Suppose it were a corrosion problem and the established information showed that you might:

Make the part of stainless steel.

Plate the part with cadmium.

Make the part of plastic material.

Change the corrosive environment.

A review of available data might show the following:

1. Cost information on the price of stainless steel and the cost of machining it might eliminate stainless steel as too expensive.
2. Cadmium vapors are hazardous to human health, and environmental protection requirements would dictate a separate plating facility or having the plating done by a qualified outside supplier.
3. There are many plastics, some are reinforced with fibers and some are useful as structural materials without reinforcement; you may need more information on which materials to use. You would need to examine the cost and performance of both plain and reinforced materials, noting that there are several types of reinforcements that may be used with each plastic and what type of reinforcements will meet your needs.
4. Changing the environment is a possibility many people overlook. It is sometimes the best and most attainable solution. Do not overlook it! In this problem we shall assume that we do not change the environment, so we shall concentrate on the part.

Now you would truly be involved in the iterative nature of design. As you learn more, you would modify your design concepts. The information may even lead you to go back to the source of the problem and redefine the need more precisely. When you are confident that you have adequate information (not perfect information—no one ever has perfect information), make some preliminary design sketches or drawings. These may be done on

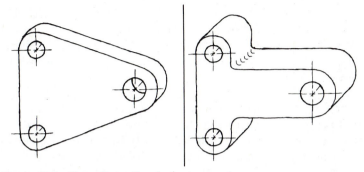

Figure 14.8 Two alternative design concepts.

the computer or manually. One thing you may be sure of: They will be changed. At this stage of design it is wise to consider several alternative designs. You have eliminated stainless steel, but how much would it cost to have a cast-iron part plated by an outside shop? Would polyester plastic reinforced with nylon be rigid enough? Would glass-fiber-reinforced polyester plastic be strong enough? During the designing can some material be taken out of the part or some manufacturing steps combined (instead of drilling holes, molding them in)? (See Figure 14.8).

14.2.4 Investigate Design Concepts with Analytical Tools

The power of your computer allows you to enter the design for one material and with slight revisions have the design in a second material. Call up a finite-element analysis program and you can check for stresses throughout the part (Figure 14.9). Is the bending stress too much for cast iron? If the plastic part is not strong enough, can you make it large enough to carry the load and still

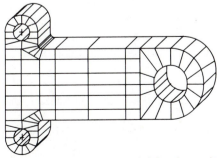

Figure 14.9 Finite-element analysis model for determining stresses in proposed design. (*Computer generated image.*)

fit in the assembly? Call up the assembly drawing into which this part must fit and check the space available. It is in this area of analysis that computer graphics and calculation programs give you power to do more iterations in less time. This means better design.

If you are not using analytical tools on the computer, you still need to apply tools such as calculations of **tensile strength, shear strength, compressive strength, cost analysis,** and producibility. Much valuable information on the design for efficient production has been developed in recent years by engineers, among them Boothroyd and Dewhurst.* You will need to consider their research in advanced design courses and in work projects.

14.2.5 Gather Investigation Data and Review

After you have investigated design concepts with various analytical tools, you need to review the results of the analysis. This process should include the input of representatives from every department or function concerned with the design. The results of the analysis should be given to these representatives before they are asked to comment on the designs.

Everyone should have had an opportunity to review the data before any meeting where the analysis is discussed (Figure 14.10). This is another example of where the designer must interact with many people during the design process. Review meetings should reduce the number of choices and result in a decision to proceed with a particular design or sets of designs.

14.2.6 Now Test the Design

Test the design. This testing may be narrowed down to one design or conducted on several designs to see which is the most appropriate solution to the problem. Many techniques may be used to test a design (Figure 14.11). These could include, but would not be limited to, the following:

Numerical methods are an extension of analysis procedures in which the performance of a design is analyzed under many

---

*Boothroyd, Geoffrey. *Automatic Assembly* (Mechanical Engineering Series: Vol 6). Dekken, 1982.
Boothroyd, Geoffrey. *Fundamentals of Metal Machining and Machine Tools.* Hemisphere, 1975.
Boothroyd, G. and Dewhurst P. Series of articles on design for assembly. *Machine Design* V 55 n 25 Nov. 10 '83 pp. 94–98; V 55 n 28 Dec. 8 '83 pp. 140–145; V 56 n 2 Jan. 28 '84 pp. 87–92; V 56 n 4 Feb. 23 '84 pp. 72–76.

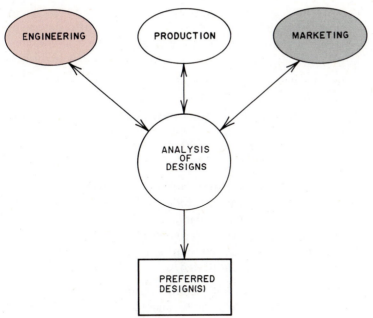

Figure 14.10 Everyone concerned reviews analysis of possible designs.

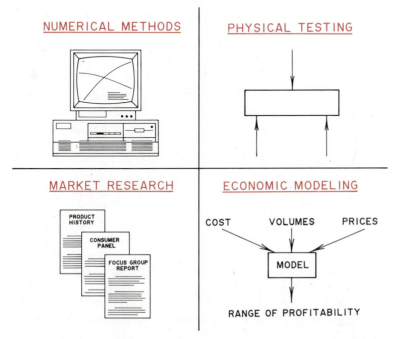

Figure 14.11 Many kinds of testing are needed to review a design.

conditions. Numerical answers are developed for the many conditions and statistical techniques applied to determine the most likely values and the probability of extreme values. A computer program is essential for a thorough analysis.

Physical testing may be either nondestructive or destructive. A product requiring a certain amount of bending strength may be bent and the strain measured (nondestructive) or simply bent until it fails (destructive). Models may be bent and stress lines located on these models to determine areas of maximum stress.

Market research will need to be conducted to estimate if there is a market for the product or system and how much of that market could be captured by the proposed product or system. If an existing product is being redesigned, you need to know if the redesign is as acceptable to the customers as the current design.

Economic analysis will need to be done to learn whether the product or system can be manufactured and distributed at a profit. An economic analysis is a mathematical model of the investment, the materials costs, the production costs, distribution costs, and overhead costs needed to bring the product to the customers. If done well, it considers various levels of production. In a capitalistic, free-enterprise economy, we do not produce goods at a loss (not for very long). An economic analysis may pinpoint certain areas of cost that need to be reviewed.

14.2.7 Propose to Accept, Modify, or Reject Design

The design team now proposes to the decision makers to accept, modify, or reject the design (Figure 14.12). The most common decision is to modify a design because we always learn more about a concept when we perform the testing. If the decision were to accept, then you could proceed to the next step. If the decision were to modify the design, then you go back to the appropriate design process level and work your way back to this level. Inexperienced designers who contribute to a project are often mystified when this occurs. It may be that their personal contribution was flawless. If some part of the project is unacceptable, then it may be necessary to review every step of the design process and consider alternatives. Your computer data base makes this task a lot easier than doing everything by hand. If the decision were to reject the design, then the project team must decide whether the design process should be started again from scratch or abandoned.

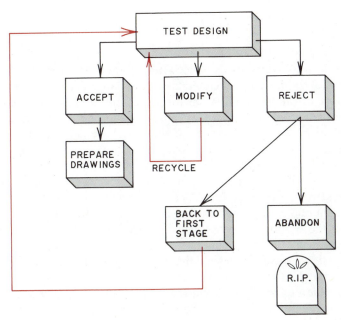

Figure 14.12 Three possible decisions for a design.

14.2.8 Render the Final Decision and Prepare Working Drawings

When the design is accepted, it must be communicated to those who will do the manufacturing or construction. Working drawings for producing the product or system must be prepared, checked, and published. The information needed and the steps in preparing it are outlined in Chapter 10.

14.2.9 Optimize Design by Iteration

Again, during this process, constant feedback must be maintained. The iterative process continues within the production drawing phase for many decisions will need to be made. Here there may be the opportunity to eliminate or combine parts to reduce the complexity and cost of an assembly. This will often simplify the production process. Some decisions will require going back to earlier decisions to get the proper information or to modify information based on new knowledge. Feedback should be continuous even after the manufacture of the product or system begins so that the information will be available for later modification or redesign of the product.

14.3 *LATERAL THINKING*

The scientific method was referred to as a vertical process with each step following in sequence after another. Another technique is useful for design; it is called lateral thinking. The characteristic feature of this method is that it reaches out broadly for input to a design problem from many possible sources. Some of these sources may seem odd or unconventional. If you are designing a bridge, you might consider asking an entomologist to join your design team. Why? An entomologist studies insects and some insects produce highly efficient structures. Can we as designers learn from other disciplines? Yes, of course, we can! The development of Velcro, noted earlier, borrowed the idea of little hooks from the burdock, a pesky weed whose burrs catch on clothing as you walk through fields.

First, we recognize a need that may arise from the market place or it may be the need to find a use for new technology. Then we assemble a team of people who can contribute to a design solution. People deeply involved in the technology and those from areas of expertise far from the problem are brought together. Ideas are generated to solve the problem. In the initial phase all ideas are acceptable, no matter how unrealistic they may seem. An odd idea from one person may inspire another to see a new way to solve the problem. After many ideas are presented, they are organized in patterns and resubmitted to the design team for elaboration and refinement. The process of iteration is used here also.

One method for lateral thinking is called brainstorming. People from diverse backgrounds get together, and the rules of the meeting are as follows:

1. Quantity of ideas is important, not quality.
2. No critical comments.
3. Build on other ideas.
4. The wilder the idea, the better.

Others may be called just interactive design meetings. In some cases, individuals may be supplied with unusual inputs to stimulate creativity. One such device is the "think tank," a ball with a window (Figure 14.13). Inside the ball are 20,000 little tags with words on them. The designer stirs the tags inside the ball and then tries to relate each word showing in the window to the design problem.

If you were an industrial designer working on a new automobile design, you might stir the tank and see the word *honey*. How could that relate to your design problem?

Figure 14.13   Think tank: Tool for lateral thinking.

Honey is a sticky substance. Could you make use of adhesives to join components of the auto?

Honey is often a transparent golden color. Would this color be appropriate for the design?

Honey is also a viscous liquid; it pours slowly when cool. Would a viscous liquid in the shock absorbers improve the ride?

Honey is stored in a structure of hexagonal cells that is efficient and strong. Would a structure of hexagonal cells strengthen the body panels?

These are some of the ideas a designer might consider when relating the word *honey* to a problem.

The lateral thinking process will not solve all design problems, but it is a spur to creativity. The design process always requires creativity from the individual and from the design team. Good design includes learning and practicing creative techniques, analyzing carefully, and being willing to learn wherever you can. Designers need to interact not only with technical people, but also with other creative people when searching for solutions to design problems.

14.4 *PRESENTATION*

Now that you have your design, how do you present it to your technical management, to general management, or to customers? No matter how good the design may be, it is worthless unless you are able to sell the design to others. If you consider what follows, you will be able to present your designs so that others can understand and accept them. The presentations may be oral, written, and/or graphic. A good overall presentation will probably include written reports, drawings, and oral reports to technical groups and management.

14.4.1 What Is Being Presented?

The designer must make clear whether the design involves a product or a system. Also, you must make clear what you are about to present; you may be able to do this in a sentence or two. A rambling presentation with no clear subject will lose your audience quickly. This would be the opening paragraph of a summary report.

14.4.2 Why Is the Presentation Being Made?

You should identify the objective for making the presentation. Are you trying to convince the audience of the need for the product? Are you selling the need for a better way to produce the product? Are you trying to obtain approval for a better redesign of the product? Again, this should be stated at the beginning of the presentation so the observer or reader will know exactly what you are trying to accomplish.

14.4.3 Who Is the Customer?

As the presenter you must first identify whether your audience will be composed of people within your own group, people within the same company, or consultants from outside your organization. Does your report go only to your supervisor or will it be sent to readers of diverse backgrounds? This could have a bearing on the understanding the audience may already have of the product or system. Also, the prior knowledge of the audience should be considered. Is the presentation to be made to production personnel, management level personnel, or the board of directors? Drawings in orthographic projection will be good for other designers; management and marketing personnel may understand pictorial drawings and exploded assemblies more readily. Each of these groups has different interests and brings a different background and point of view to the problem.

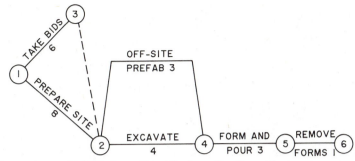

Figure 14.14 PERT chart: Graphic time plan for project.

14.4.4 When Will the Presentation Be Made?

Time constraints are important to consider for meetings. These include the time available to prepare the presentation and the time line needed for the design to be initiated. Graphic methods of presentation are useful here. The time span of a project might best be presented graphically using a PERT (Program Evaluation and Review Technique) chart that shows how the times for various parts of the project are related (Figure 14.14).

14.4.5 Where Will the Presentation Be Made?

Often you have the luxury of making the presentation in your conference room where you are very comfortable, where you know the availability of resources such as videotape or closed circuit television, and where you have unconditional access for setup time. Other times, you must take your presentation "on the road," where some of these factors may not be available. If this happens, you have much less control over the presentation environment and must be able to make quick decisions concerning the presentation. One other consideration when presenting at a location away from your facility is the transportation of the materials needed to make the presentation. It is important to be able to move the objects to the new location (size of the object vs. size and availability of the vehicle to move it) and to have the needed personnel to move it. Also, weather can play a role since rain and snow can ruin some well-prepared illustrations.

14.4.6 How Will the Information Be Presented?

To make the greatest impact, the presenter must decide the method of presentation and what visual aids are necessary. A written report should have a short (one- or two-page) summary in the beginning. An easily read drawing of the basic concept should be a part of this summary. Experimental data, analysis, estimates, and detailed drawings will follow as a major appendix.

Most readers will not examine the detailed information unless they are convinced by the summary that the concept is worthwhile. Many design presentations consist of verbal presentations. As you learned in school, a plain lecture is a poor way to transmit information. However, if aids are used well, the impact of the presentation can be greatly enhanced. Simple drawings, charts, graphs, or other large graphics are always effective. The charts and graphs should be simple and easy to comprehend. Color can be very effective. Three-dimensional scale models are sometimes used when the design has been refined enough and the cost of the system or product warrants the cost of having the model built. As is often said, a picture is worth a thousand words. If this is true, then a moving picture (videotape) is worth a thousand pictures and a three-dimensional model is worth even more.

14.4.7   How Much Will the Presentation Cost?

As with all phases of design, the cost of the presentation is important. The cost should be relative to the level to which the design has progressed, the level of the audience, and the importance of the decision resulting from your presentation. Remember, no matter how good the design is, it is worthless if you cannot convince others of its value.

14.4.8   Notes on Content of Presentations

Each of the preceding paragraphs describing presentations included a key word. These key words are questions that need to be answered in any report. Good technical reports always answer them; good news reports always answer them, and you will want to be sure that you do, too. The questions are:

What? Why? Who? When? Where? How? How much?

14.5   *PROTECTION OF DESIGNS*

A design is a valuable property. Much time and talent is required to develop a good product or a useful system. Those who create these things should have the right to enjoy the fruits of their labor. This can be accomplished in several ways.

Some people protect their ideas by not disclosing them and by safeguarding them from others who might benefit from detailed information about the invention. This method works best for systems that are used within a private organization. If you develop a new assembly method for a product or a new chemical process, competitors see the result but do not know how you achieved it. Nondisclosure can protect a design under these circumstances.

More commonly, new designs are protected by patents. A patent is a license granted by a government that gives the owners the exclusive right to use the design for their benefit. Others wishing to use the design must pay a license or royalty fee to use it. The United States has a patent office as do many foreign countries.

The application to patent an invention must be filed on special forms with the patent office and accompanied by drawings that describe the device. The requirements for these drawings are very strict. A sample drawing is shown in Figure 14.15. Both products and manufacturing processes can be patented. Once the application is filed, the inventor may mark the device as ''patent pending'' and offer it to the public. The patent office searches all its records to learn if the proposed design is similar to others or is truly different. Sometimes it may be found to be similar in some respects and different in others; then a patent is granted only for the areas in which it is different. The search and final granting of a patent may take several years. Once it is granted, the owner of a U.S. patent has the exclusive right to use or license the invention for 17 years and the possibility of another 17-year renewal.

Patents are costly. Much time and effort of designers, attorneys, and drafters is required to prepare an application. If the information is incomplete, the application will be rejected. One very important factor in patent records is time. If two people seek to patent similar devices at the same time, the one with the earliest record of conceiving the design will be granted the patent. For this reason written records, in ink and in bound notebooks, should be kept of creative discussions, meetings, research laboratory work, development trials, and tests. The notebooks should be checked weekly or monthly by a person not directly involved in the invention and witnessed. The reviewer should sign and date each page of notes to indicate that they existed at the time of the review.

Written material such as books, music, and computer software can be copyrighted. This is a form of license from the government that gives the creator the exclusive right to benefit from the written work. A copy of the work and an application must be submitted to the copyright office. The work can be labeled ''copyright'' or with the symbol ''©'' as soon as published and even before the application is submitted. A copyright prevents others from using

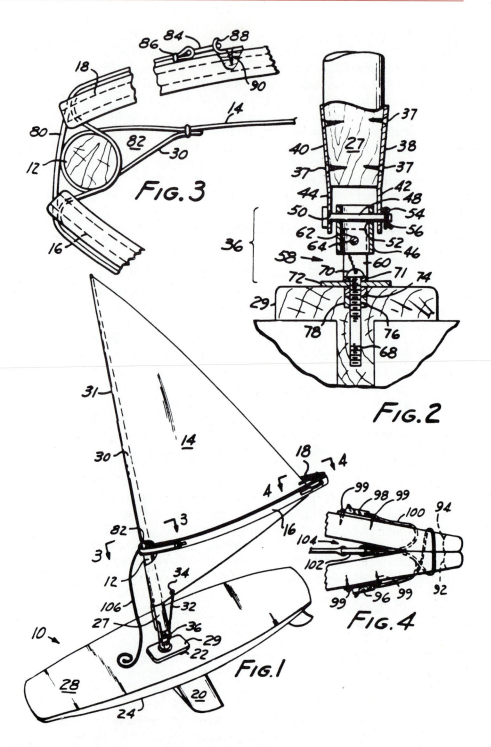

Figure 14.15 Example of patent drawing. (Courtesy of Roylance, Abrams, Berdo & Goodman)

the work for profit; there are regulations that allow copyrighted materials to be reproduced in small quantities for educational purposes without permission or payment of royalties.

14.6   *A DESIGN PROBLEM*

This design problem is not very difficult, yet it includes all the elements of the design process. What is it? To design an organizer for holding the supplies and materials needed by a student of graphics. As you follow the development of this project, you will see each step of the design process.

14.6.1   Determine a Need

Some of the problems that graphics students have include:

Forgotten or lost equipment.

Bent or wrinkled drawings.

Broken or crushed models.

Storage space not available in graphics laboratory.

Adequate drawing surface not available in residence.

Students are often seen scrounging in their backpacks for needed triangles, scales, or pencils; there is a need to keep graphics equipment so it can be easily found. Sometimes, laboratory drawings appear to have been through a war: wrinkled, wet, torn, or dirty. Finished work is valuable and needs protection. Students occasionally build models of a particular problem, but a crushed or broken model does not demonstrate a concept very well.

Some graphics laboratories are equipped with locked drawers in each work station so that students can leave drawings and equipment in the laboratory. Not all graphics laboratories have these facilities. Some students live in crowded quarters where they have no smooth surface available on which to solve graphics problems, and it would be nice to have a portable drawing surface. With these needs in mind we can pursue the next step in the design process.

14.6.2   Establish Known Information

The organizer should hold all the equipment that a student usually needs to solve graphics problems. Typical equipment was shown in Chapter 1 and is shown again in Figure 14.16. The sheets to be carried in the organizer would be those from the workbook or

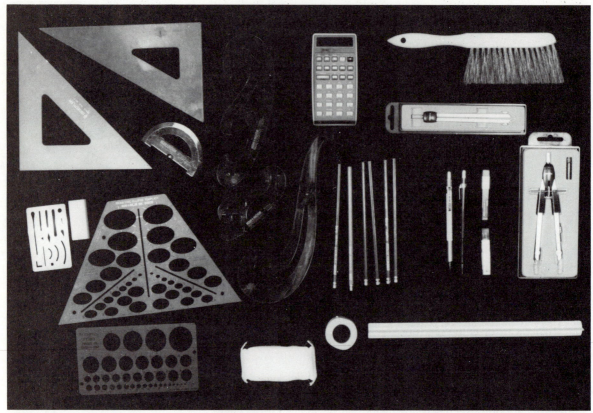

Figure 14.16 Tools to be carried in graphics organizer.

those from the textbook, usually $8\frac{1}{2} \times 11$ in. (Figure 14.17). A graphics model might be of foam, clay, or cardboard; the model of a cut block used in Chapter 11 is shown in Figure 14.18.

These materials establish the minimum dimensions of the organizer. The scale is the longest item; the model and cleaning pad have the greatest height. The flat pieces are the largest overall in combined width and depth. The minimum space requirements are sketched in Figure 14.19. You may have equipment that differs a little from these pieces, but these are typical. We should also allow a little extra space for tools a bit larger than those we have shown.

14.6.3 Stimulate Design Concepts

Here you can go into a "lateral thinking" mode. No ideas are unacceptable; unused ideas may stimulate a useful idea from another designer. Our design team may remember that some stu-

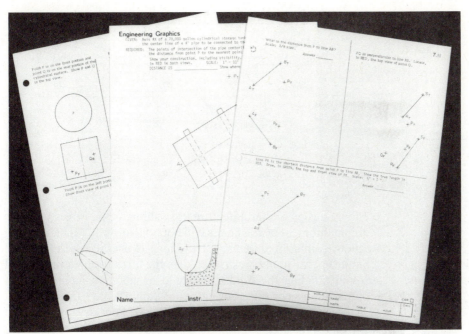

Figure 14.17 Workbook sheets that should be in organizer.

Figure 14.18 Three-dimensional model to carry in organizer.

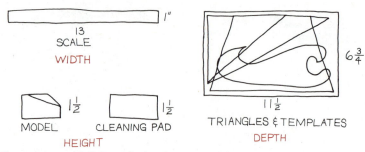

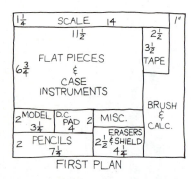

Figure 14.19 Quick analysis of space requirements to determine minimum width, height, and depth.

dents in design and architecture use fishing tackle boxes for their equipment; most tackle boxes will not hold an $8\frac{1}{2} \times 11$-in. drawing without rolling or folding, but the idea is useful: The individual compartments in the tackle box keep the lures from getting mixed up and tangling the hooks; it would be good to have different compartments for different pieces of equipment.

The drawings and equipment need to be kept dry, so we need a case that closes tight. Should it fit in the backpack? Some quick sketches of "floor plans" show that if it holds all tools on one level, it would be too large for most backpacks (Figure 14.20). Figure 14.21 gives the overall dimensions of a one-level organizer

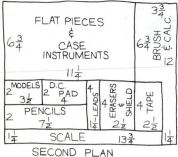

Figure 14.20 Two possible "floor plans" for organizer.

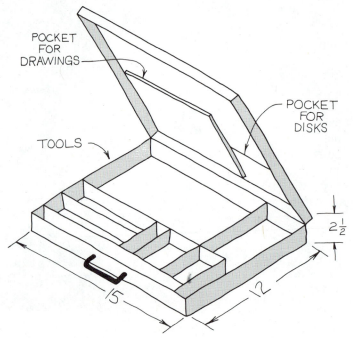

Figure 14.21 Concept sketch for organizer.

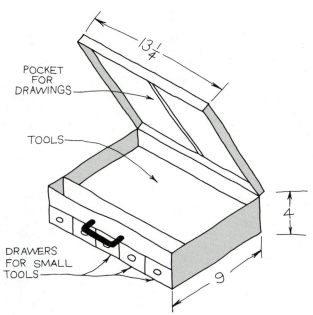

POCKET
FOR
DRAWINGS

$13\frac{1}{4}$

TOOLS

4

DRAWERS
FOR SMALL
TOOLS

9

Figure 14.22   Another concept sketch smaller in width and depth but larger in height than first concept.

with the second floor plan; note that the dusting brush handle would be toward the user and also that the scale would be handier than in the first plan. The lid has pockets for drawings in process or completed and a pocket for disks for computer graphics. Since this is too large for most packs, an alternative concept is developed in Figure 14.22. This model has a smaller floor plan accomplished by putting smaller parts in drawers under the large-area storage. The package is now 4 in. thick, perhaps too big for a backpack.

Since the organizer appears to have outgrown the backpack, it should be water resistant (rain resistant, not swimming pool resistant). At this point we may consider putting the workbook and the textbook into the organizer: It becomes much bigger and heavier. The team may decide to leave the books in the backpack.

Now we consider materials. Our model is of illustration board (Figure 14.23), but illustration board would not survive many rain storms. The participants in our concept session have steel and plastic tackle boxes. Both have survived several fishing expeditions, so plastics and steel are listed as candidate materials. One of the participants brings in a hand-made organizer made of walnut wood. It is beautiful, but walnut is too expensive (Figure 14.24). After considering some good ideas on the use of space in this example, wood is listed as a candidate material.

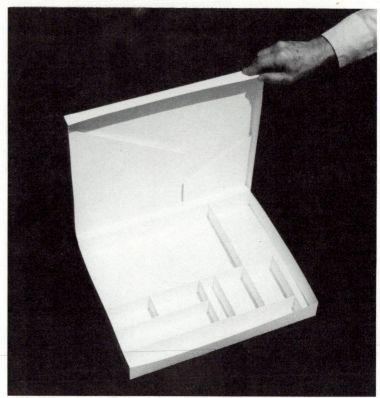

Figure 14.23   Three-dimensional model of a concept.

As the concept development progresses, we discuss the idea of a drawing surface as a clipboard with a rectilinear grid on one side and an isometric grid on the other to aid in sketching with translucent media.

This is not an exhaustive list of concepts. You and your fellow students could produce many more. Nor are any of the concepts an ultimate or optimum design. These are the concepts developed so far, and the next step is to investigate them.

14.6.4   Investigate Design Concepts with Analytical Tools

One of our first investigations is to determine if the designs will hold the needed tools. Figure 14.25 shows a model with the tools in place. Did they all fit? Is there enough space for each item? Could they be better arranged?

Another area would be strength calculations. How strong does the organizer need to be? What kind of strength: tensile, com-

Figure 14.24   Hand-made graphics kit box, fabric lined and beautifully finished.

pressive, shear, bending, or crushing? There are no large loads on the box, but it may have books stacked on top of it or some one might sit on it. So we should check crushing of the box, which would be related to the compressive strength of the outside walls and the interior dividers and the beam strength of the top and bottom. We could assume that the worst load would be a 200-lb person sitting on the organizer; if they sat down hard, the impact might be 300 lb for an instant. These calculations would be done for different materials to determine minimum thicknesses for walls and partitions. This leads us to materials analysis.

Different materials have different costs, different strengths, and different resistance to moisture. As an aid in eliminating poor choices for materials, we can make a simple manufacturing cost estimate. This estimate would take into account only the direct costs of production (Figure 14.26). Manufacturing overhead, distribution costs, and profit would be considered later in a total economic model. Strength of materials and resistance to moisture can be found in handbooks and current literature on materials. Tensile strength of three possible materials are shown here:

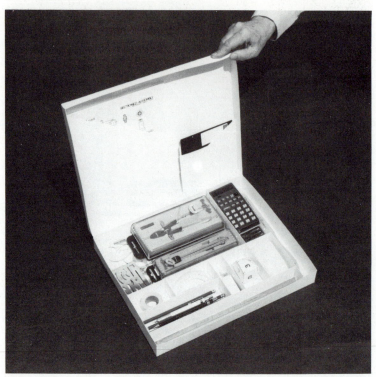

Figure 14.25 Three-dimensional model loaded with typical tools.

Material	Tensile Strength (psi)
Steel sheet	60,000
HD polyethylene	3100–5500
Polypropylene	5000

Steel has a much higher tensile strength than plastics, but it has a density of about eight times that of plastics. Considered as strength per unit weight, it is only one and a half times stronger than plastics.

The investigative steps described here are only a few of those necessary for a thorough analysis. The quality of the investigation is dependent on the value of the final result and the cost of incomplete information: If our organizer project is delayed, some people may not realize the benefits of this device; if a flood control project is delayed or includes errors, lives and much property may be lost. The flood control project would require a thorough and exacting analysis.

GraphiComp, Inc.

PRELIMINARY COST ESTIMATE

PROJECT GRAPHICS ORGANIZER

DATE 9/1/88   BY F.D.M.

DIRECT COSTS	$/Unit
Materials	.63
Labor	1.20
Sub-Total	1.83

SEMI-DIRECT COSTS	
Energy	.20
Tooling	.15
Supervision	.40
Sub-Total	.75

DISTRIBUTED FIXED COSTS	
Equipment	.13
Plant services	.40
Sub-total	.53

TOTAL UNIT COSTS   3.11

Figure 14.26 Brief manufacturing cost analysis.

14.6.5 Gather Investigation Data and Review

After the investigations are completed, the team would review the information. Suppose the cost analysis shows that the manufacturing cost of a steel box and a plastic box are about the same. Then we look at finishing costs. The steel box will need to be painted; a plastic box can be molded in color and will not rust. Latches for a steel box will be fabricated separately and probably spot welded to the basic box; latch parts can be molded directly into the plastic boxes. It appears that a plastic material is a better choice. Which one? This may be a time to recycle or iterate our analysis. Polypropylene has self-hinging properties (the top and bottom could be molded in one piece with a hinge strip in between). Exact material costs, mold costs, and production rates may need to be examined to make this decision.

The shallow design is larger in area but has no extra parts such as drawers. The drawers would require additional molds and some means of keeping them in place during transport. So our team may decide to pursue the shallow design.

After a review of the investigations, we decide on a shallow, injection-molded polyethylene container. Now we need to test the design. We could build one from polyethylene shapes or carve one from a block of polyethylene. In neither case would the physical properties be exactly like a molded shape. If this appears to be a good project, we might invest in a low-cost, temporary mold and mold a few boxes for testing.

14.6.6 Now Test the Design That Appears Best

Some of the tests that would need to be conducted include crushing strength, water resistance, and wear resistance (abrasion). (Figure 14.27).

A market test would also be a good idea. Some of the molded organizers could be finished and shown to graphics students and instructors to get their opinions about features, improvements, and willingness to buy such a device. Based on the willingness to buy the organizer at different selling prices, an economic model could be built on a computer to show profit or loss at various sales volumes, different prices, and different manufacturing concepts.

During our tests the molded box allows water to enter when exposed to a shower approximating a hard rain. A problem! A 200-lb weight dropped on the case 100 times from a height of 6 in. results in no noticeable damage. The prototype latches also stay closed during this test.

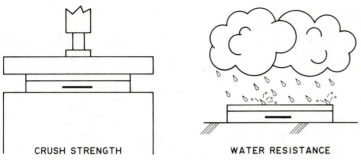

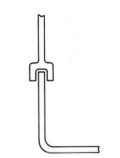

Figure 14.27 Three types of tests that need to be conducted for organizer.

Some comments from potential users are favorable. However, some want a provision to carry a T-square. Simulated use show that the dry cleaning pad sheds particles that find their way throughout the organizer, even to the computer disk stored in the lid. Another problem!

Physical and market testing disclose at least two problems. Now it is time to review the project to date.

14.6.7 Propose to Accept, Modify, or Reject the Design

The design as tested is unacceptable. There is considerable interest among potential users if the price could be held below $15. So our team decides to modify the design. The water problem is resolved by changing the match line joint from a simple mating of the two sides to a U-channel that provides a sort of labyrinth seal (Figure 14.28). Investigation of some quality luggage finds a similar joint used there. The dust from the dry cleaning pad could be contained if the pad were stored in a separate small container, such as a soap box sold for travel or a molded box such as the ones used to package screws. The internal corners of the organizer are also given generous fillets to improve cleanability (Figure 14.28).

If the modifications made are significant changes in terms of design or cost, we need to go back to test the design and then reconsider our decision to accept, modify, or reject. Discussion with a mold designer may indicate that the U-seal slightly increases costs; the separate container for the dry cleaning pad also increases costs. These are judged not to be large enough to reject the project or go back further in the iteration process.

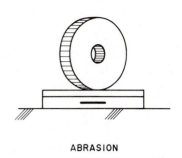

Figure 14.28 Improved design for match line and interior corners of organizer.

14.6.8 Render a Final Decision

This is it! Is the organizer as now conceived ready for production? A final review of the design, the market, and the economics will tell us if we should proceed.

Does the organizer meet the needs as determined in the beginning? Our sample of users says it would. Will enough people buy it to enable us to buy the tools to make it, buy the materials, pay the people, distribute the product, and make a profit? If our economic model says we have better than a 50% probability of doing this, our answer may be "go." (If we are very conservative, we may want a higher probability of success.)

This is the time to make the detail and assembly drawings. From the detail drawings, the molds will be designed to mold the two sides of the case (Figure 14.29). The portable drawing board will be purchased as sheets of pressed board, cut, sanded, and faced with grid surfaces on both sides. Pockets for drawings and disks will be cut and installed in the top according to the assembly drawings (Figure 14.30). The bands to hold the drawings and disks will be installed according to the assembly drawings and the two sides of the organizer joined as shown in the assembly. The completed organizer is shown in Figure 14.30. This final decision will also trigger a packaging study that becomes a miniproject within the larger project. Go!

14.6.9 Optimize Design by Iteration

Are we done? Never! As the detail design and assembly drawings are made, questions will arise about the parts. Decisions about exact contours and locations will need to be answered. We may need to review some steps of the design process to get answers. When production begins, new information and new questions will be found. And the final judges of our work, the *users,* will tell us if the design meets their needs. They will note the failures and the improvements needed.

If we intend to continue to produce graphics organizers, the input from the users must be continually fed back to the design process. The product must be continually reevaluated through the steps of the design process to meet the users' need and to lower costs. If we fail to do this, we shall join the list of organizations that are extinct because they failed to meet their customers' needs while maintaining their own economic health.

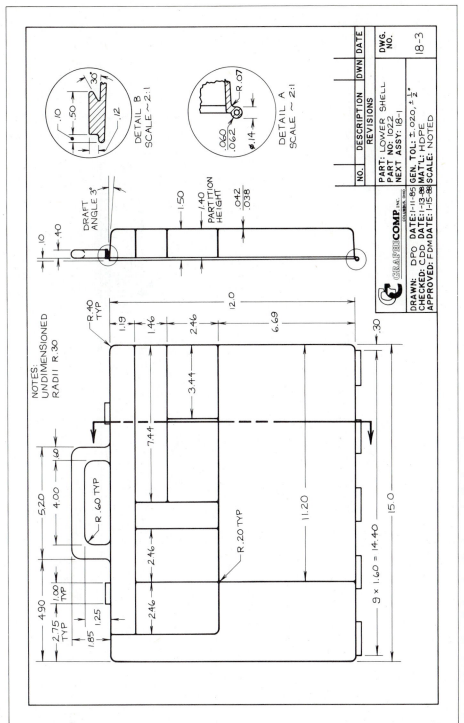

Figure 14.29 Detail drawing for lower half of organizer.

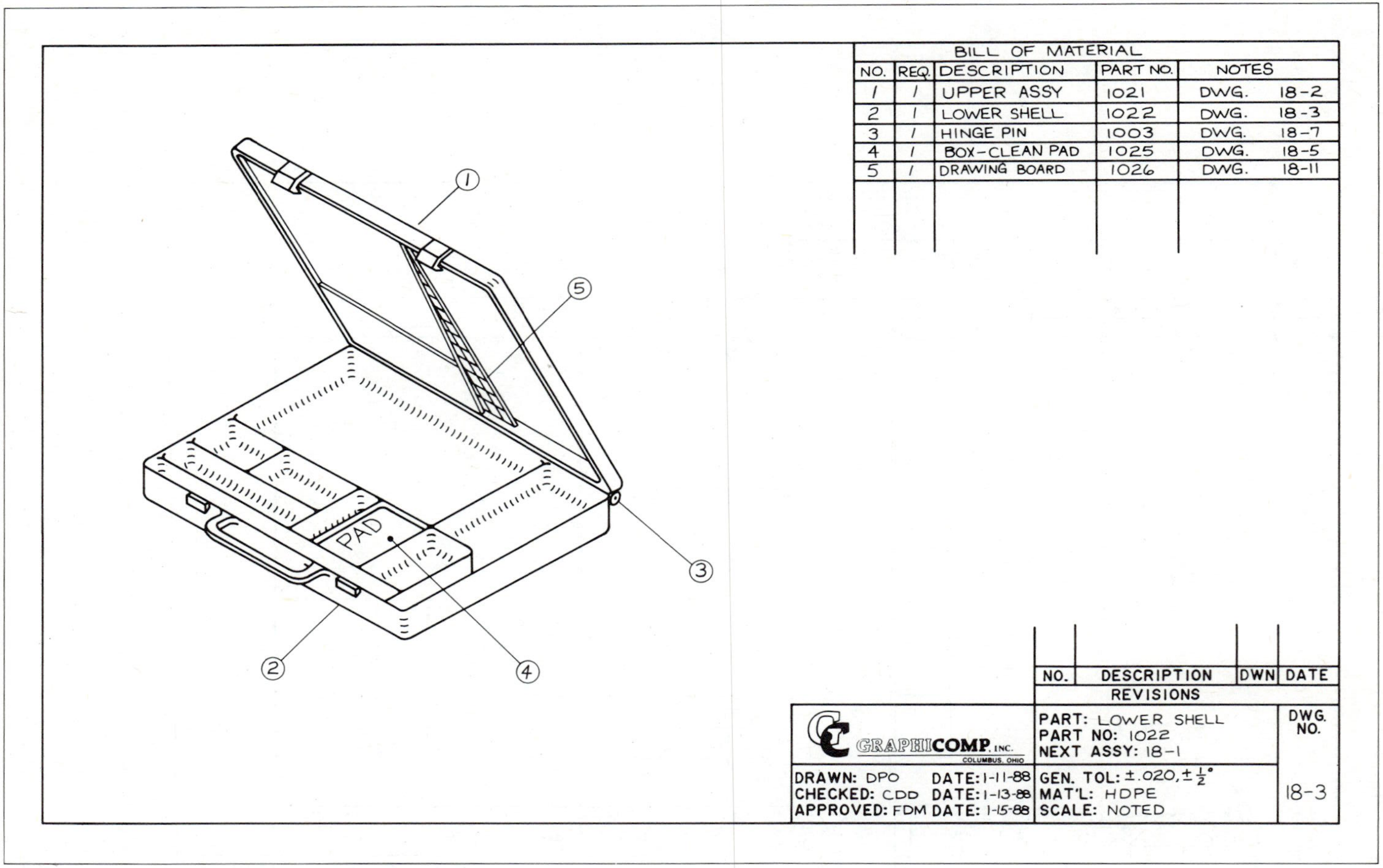

BILL OF MATERIAL				
NO.	REQ	DESCRIPTION	PART NO.	NOTES
1	1	UPPER ASSY	1021	DWG. 18-2
2	1	LOWER SHELL	1022	DWG. 18-3
3	1	HINGE PIN	1003	DWG. 18-7
4	1	BOX-CLEAN PAD	1025	DWG. 18-5
5	1	DRAWING BOARD	1026	DWG. 18-11

REVISIONS			
NO.	DESCRIPTION	DWN	DATE

GRAPHICOMP, INC.
COLUMBUS, OHIO

DRAWN: DPO   DATE: 1-11-88
CHECKED: CDD   DATE: 1-13-88
APPROVED: FDM   DATE: 1-15-88

PART: LOWER SHELL
PART NO: 1022   GEN. TOL: ±.020, ±$\frac{1}{2}$°
NEXT ASSY: 18-1   MATL: HDPE
SCALE: NOTED

DWG. NO. 18-3

Figure 14.30   Pictorial assembly of organizer.

14.7   *SUMMARY*

Design is the essence of engineering. It is a creative process and also an iterative process. The steps of the design process may be applied to any creative endeavor: a part, an assembled product, a process, or a system. Both vertical thinking and lateral thinking contribute to the design process. The steps may be remembered with the acronym DESIGN PRO:

D	Determine a need.
E	Establish known information.
S	Stimulate design concepts.
I	Investigate design concepts with analytical tools.
G	Gather investigation data and review.
N	Now test the design that appears best.
P	Propose to accept, modify, or reject the design.
R	Render a final decision.
O	Optimize design by iteration.

During the design process we need to communicate with others. Good communication requires that we answer basic questions in our written and oral presentations:

What?   Why?   Who?   When?   Where?   How?   How much?

A design is valuable property. Designs can be protected by patents and copyrights. If the design will be used in other countries, it must be protected under the laws of each country in which it will be used.

A design is never perfect. We go with the best we can conceive with the resources we have and then we iterate to improve it.

PROBLEMS

These problems are all presented verbally. You may receive problems in other situations as a combination of sketches and verbal requirements, or all sketches, or all verbal. Your instructor will specify how far you are to pursue the design process: ideas, concept sketches, plan for analysis, models, or working drawings.

Problem 14.1
Need: a portable device to hold an $8\frac{1}{2} \times 11$ sheet of paper for reference while using a computer. Many computer stations do not have table space for reference material. The device must be portable, low cost, lightweight, and inexpensive.

Problem 14.2
Need: a simple locking mechanism for automobile wheel covers. Thieves often steal wheel covers. Currently available locking devices require special key wrenches. The device must be low cost, and it must be easy for the owner to unlock to change a tire.

Problem 14.3
Need: a holder for eyeglasses that will protect them from damage while the owner works at a computer station. Lenses are easily damaged; many users remove their glasses or use a second pair for computer work. The user should not have to fold the side supports— this would require a two-handed operation.

Problem 14.4
Need: a stadium pack for sports events that would serve as a seat with backrest and provide storage for a vacuum bottle, umbrella or rain cape, and programs. It should be easy to carry and attractive.

Problem 14.5
Need: a portable bookshelf for student housing. The shelf should be collapsible for easy moving and should provide built-in bookends.

Problem 14.6
Need: a small buffer/grinder stand for a basement workshop. A small electric motor ($\frac{1}{2}$ horsepower) would drive the tool through a V-belt drive. The tool should accept different size pulleys, up to 8 in. diameter, on one end and different size wheel and brushes, up to 6 in. diameter, on the other end. The frame should be rigid (it might be a casting or a weldment).

Problem 14.7
Need: a candle holder that will immediately extinguish the flame if it is tipped over. A standard candle has a tapered base with a maximum diameter of $\frac{7}{8}$ in. The design should be aesthetically pleasing as well as safe.

Problem 14.8
Need: a single auto rack that can hold skis or a bicycle. Many people like to ski in cold weather and bike in warm. A single rack would reduce changeover time and costs. The rack must be attractive and it would be a plus if it could be removed without damage to the auto.

Problem 14.9
Need: a dispenser for toilet paper that will limit the user to removing not more than 6 sheets at a time. A common roll of toilet paper is about 5 in. diameter and $4\frac{1}{2}$ in. long. The core diameter is approximately $1\frac{5}{8}$ in. Each sheet is $4\frac{1}{2}$ in. square. Durability and ease of loading are prime considerations in this design.

Problem 14.10
Need: an entertainment center for a college room. It should be not more than 2 ft. square by 4 ft. high and should contain a television set, speakers, amplifier, AM/FM tuner, tape deck, record player (33 rpm or compact discs), and storage for tapes and records or discs. All wiring should be concealed and neatly run, and power surge protection should be provided. Security and portability are major considerations (somewhat in opposition).

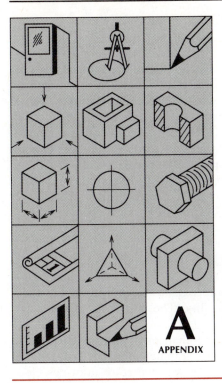

 A

ENGINEERING GRAPHICS: THE NEXT STEP

A.1 ENGINEERING GRAPHICS: THE FUTURE

A.1.1 Improvements in Graphics

The main purpose of graphics is to communicate ideas, designs, concepts, illustrations, and instructions. The ways that graphical features are constructed, whether manually drawn or produced with the aid of a computer system, are always evolving; they are always being improved in efficiency, clarity, and power. We would expect that the graphical methods described in this text will be improved during the future years. We will examine two important improvements that are beginning to be used in graphics for industrial and educational applications. We expect that 5–10 years from now these new methods, solid modeling and feature-based modeling, will form a significant portion of engineering graphics texts.

A.1.2 Line Drawings

Real objects are three dimensional (3D) in form. They have width, height, and depth. They can be represented on a three-axis coordinate system, using the X axis for width, the Y axis for height, and the Z axis for depth.

Traditionally designs and design geometries have been represented through drawings on a two-dimensional surface such as paper, which has only two axes. A two-dimensional computer-based drawing system (2D) uses (X, Y) coordinates to show views of an object on a computer screen in the same way as they would be drawn on paper. A system that goes beyond 2D will use another data set to draw views that represent an object in three dimensional appearance. Extensions to 2D drawings are called 2 1/2D. There is no precise definition of 2 1/2D. These systems approximate a true 3D view with added data that takes into account the depth of the object, but they do not use a single complete three-axis (X, Y, Z) coordinate system. There are many schemes that allow an object to be shown in 3D form despite being on a 2D surface.

Drawings that are limited to 2D or 2 1/2D require human assistance and interpretation to recognize design features and evaluate designs for manufacturability. The popularity of computer graphics systems has not changed 2D and 2 1/2D methods but has brought about two very positive changes in graphics processes. The use of computer systems as graphics tools has changed the way we create drawings and the way we store drawings electronically. They have provided the means to construct and use true 3D drawing systems. A 3D drawing system has the capability to draw any view of an object using a single data structure to draw all of the views. If the points and lines in any view are modified, the data structure will be changed so all other views will likewise be changed. These systems use points and edges to form line drawings, and since only point, edge, and face data are stored, only these parts of an object are known. Intersections of surfaces may not be known. Volume and weight of the objects are not known. If an object or its model is to be completely defined, more information must be known about it. This additional data can be found through the use of a solid modeling system.

A.2 SOLID MODELING SYSTEMS

A.2.1 Types of Solid Modeling Systems

A solid modeling system defines all points of an object, not just point, edge, and face features but interior points as well. This definition of interior points is accomplished through calculation and retention of more geometric features of the object. This gives better definition of the object, permits the calculation of physical features (weight, density, surface area) of the object, and allows program evaluation of the manufacturability of the part or assembly of parts. There are many algorithms for the definition of solid objects; however, the three methods discussed in the next sections are the most widely used. They are constructive solid geometry (CSG), boundary representation (BREP), and cell enumeration (CELL).

A.2.2   Constructive Solid Geometry

A CSG representation of an object uses a combination of primitive (basic building block) solid objects—block, cylinder, sphere, cone, wedge, torus, and cone—along with Boolean operations to form more complex objects.

Boolean operations permit areas and volumes to be added and subtracted using operations known as union, difference, and intersection. If two volumes or areas overlap, the union of them is all of the points in either of them, but the overlap area is not counted twice (Figure A.1a). The union of two objects is thus usually less than the sum of the two. The difference operation subtracts the overlap area (Figures A.1b,c). An example of difference is a drilling operation that subtracts a cylinder from a block, thus creating a block with a hole in it. The intersection operation is only the overlap area (Figure A.1d). The intersection of the drill with the block would be the hole the drill created.

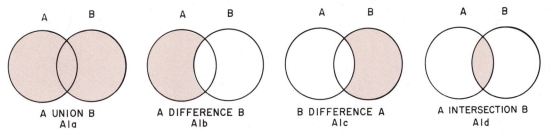

Figure A.1   Examples of Boolean operations.

Primitive objects are completely defined on the boundary and in the interior so that any objects formed from them are also completely defined. Four examples of primitive objects are shown in Figure A.2.

The operations used to combine the primitives are union, difference, and intersection, as described above. Boolean operations and the definition data are arranged in data structures resembling upside-down trees and are known as tree structures. Binary trees,

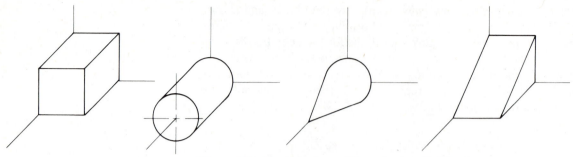

Figure A.2 Constructive solid geometry primitives.

which have only two branches from any intersection point (node), are normally used. This arrangement gives a logical structure to the data storage and object formation. The primitives are the leaves (the ends of the branches) of the binary tree and the Boolean operations are the nodes for the branches, as shown in Figure A.3.

The primitives used in model construction are valid, complete solids that are unambiguous (have a single clear meaning); therefore, the resulting solid part is valid, unambiguous, and complete. The Boolean operations union, difference, and intersection are defined so that they form a regular solid. A regular solid has no

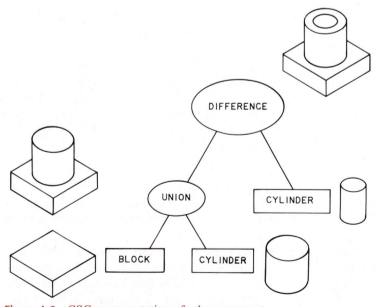

Figure A.3 CSG representation of a base support.

dangling edges, zero thickness faces, or undefined planes. Since all surface and interior points are defined, physical features such as weight, mass, moment of inertia, and volume can be calculated and the part can be drawn as a wire frame, shaded polygonal (surface) object or as a solid.

Most solid objects can be modeled through the use of a CSG system; however, objects with very complete, nonplanar boundaries and features can be modeled only through the use of a large number of small primitives. This means that a good representation of a complex object will require a large amount of memory and long program run time. Objects that have complex shapes are often modeled by other solid modeling methods.

A.2.3   Boundary Representation (BREP) Modeling

A BREP solid modeler has as its basis lists of all vertices, edges, faces, surface normals (vectors perpendicular to each surface), and surface definitions. These lists are linked to form a data structure that will represent an object, draw it as either a wire frame or a solid, define all surface and interior points, and calculate physical features. Figure A.4 shows how the boundaries are combined to form the model.

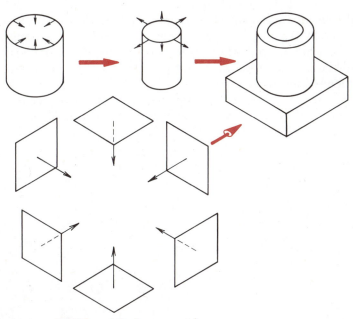

Figure A.4   BREP model of a support base.

The boundary and interior points of the model are defined from knowledge of the vertices and edges through the use of mathematically defined features such as parametric surfaces, Bezier surfaces, Coons patches, splines, and inward normal vectors and by use of Boolean operations. Validity, completeness, nonambiguity, and regularity are checked throughout part construction using Euler's law (which defines relationships among features of a solid), consistency checking, and conformity with regular Boolean operations. BREP modelers are able to efficiently model complex objects, provide values for the physical features of an object, and give an evaluation of the manufacturability of a part. It is quite easy to produce a drawing of the object using a BREP modeler, but the complete definition of the solid is less efficient than is the case with a CSG modeler.

A.2.4   Cell Enumeration Modeling

Definition of a solid has been described through the use of primitives and through the definition of vertices and surfaces (boundaries). Another method of modeling a solid, cell enumeration, is to divide a solid into many small cubes or cells (called regions) and then construct the model from this collection of cubes. The division of the object is done through a logical process by which the object is divided into equal parts and then each nonfull region (a cube that the object does not completely fill) is divided into equal parts and each of these smaller nonfull regions is again subdivided. This subdivision is continued to the level of accuracy desired by the program user. The size, number, and locations of the cells are determined by algorithms such as octree subdivision and binary subdivision and recorded in data structures called tree-structured linked lists.

Figure A.5 shows an object that has been divided into cells and enumerated by one of these algorithms. The solid circles are cells

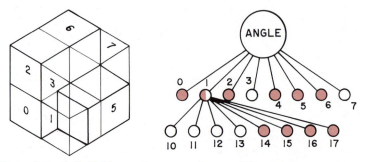

Figure A.5   Cell division and enumeration model.

that are full, the partially filled circles indicate a cell that must be subdivided, and the empty circles are the cells with no material in them. This very simple object shows only one level of division for all except cell 1.

Solids are the set of points on the boundary and interior of objects so the cells are combined to form a complete, valid, unambiguous solid. The physical properties of each cell are known so the properties of the resultant solid are known. Objects with complex geometry can be modeled through cell enumeration; the degree of object–model agreement is determined by the level of subdivision desired.

A.2.5   Combination Modelers

In actuality most solid modeling systems use a combination of the methods described. A CSG modeler will define an object through the primitive combinations but will need a conversion to a BREP model in order to display the object. Often the most efficient way for BREP modelers to determine physical characteristics of objects is to first conduct a cell division and then calculate the properties of the total cell structure.

A.3   FEATURE-BASED MODELING

The solid modeling methods previously described require considerable memory and execution time without greatly increasing the interface of design with manufacturing, assembly, inspection, installation, maintenance, and support/service efforts. Researchers and users seek to find methods for improving these interfaces and increasing function, shape, and tolerance data for the models and parts. Since designers, engineers, scientists, and technicians are better able to communicate in terms of existing forms, widely used features and typical operations of a solid modeling system are now being defined so that the solids are formed, manipulated, and evaluated through the use of these "features." Feature-based solid modeling may be defined through citing examples and listing the structure of a typical feature-based system.

A.3.1   Design-oriented Feature-based Modeler

A feature-based solid modeler that is oriented toward the design stage would use predefined functional parts such as angles, supports, mating surfaces, lightening (weight-reducing) holes, cooling holes, and typical beam cross sections. The part geometry would be defined using those features, and the part form, fit, and function would be evaluated and physical features calculated through use of the part features and geometry. The designer would select appropriate features in order to define the part and the way it functions.

A.3.2 Manufacturing-oriented Feature-based Modeler

A feature-based modeler oriented toward the manufacturing process would use features such as material shapes (blocks, angles, cross sections), machining tools (drill, lathe, mill), machining methods (drilling, broaching, grinding), tool selection (type, size, order), and joining operations (casting, forging, welding, adhesion) in order to design the model and check tolerances, fit, function, and manufacturability.

A.3.3 Quality-Control-oriented Feature-based Modeler

A feature-based modeler oriented toward quality control, inspection, and quality standards would use features that are related to quality assurance operations. The basic materials would be used along with features such as datum surfaces, geometric entities, quality standards, and inspection methods. The part would be designed and all information concerning it would be in consonance with the applicable quality assurance requirements for the particular part.

A.3.4 Elements of a Feature-based Modeler

These examples show that features are the elements that are important for a particular application; they are the entities that are familiar to the user. This allows the user to define a part in terms of familiar features rather than solely through the generic solid modeling of parts such as blocks, cylinders, edges, surfaces, and normals. There is a large overlap in the structure of the examples cited. This commonality enables us to construct a list of the steps or operations in a typical feature-based solid modeler.

Step 1. Design the model through the use of solid primitives, features, subfeatures, and operations.

Step 2. Modify the designed part as needed for the specific function.

Step 3. Form a representation of the part and use it to construct the data base and make a visual display of the model.

Step 4. Perform part evaluation: design, manufacturability, quality assurance, analysis, and assembly.

Step 5. Provide the drawings, data, and documentation for the end user.

A.4 *APPLICATIONS OF SOLID MODELING*

The formation of a solid model, whether by CSG, BREP, CELL, or features, permits the user to calculate physical properties of the model, evaluate the model for manufacturability, provide checking of assembly operations, and create geometry and data for succeeding programs. Solid modeling systems are often used to create the geometry needed for finite-element analysis programs; tolerance checking programs; error checking operations; fit and assembly checks; size, form, and shape determination and comparisons; geometric tolerancing and function checks; and tool- and pattern-making programs. These applications and capabilities greatly enhance the design process and make existing wire frame graphics systems only a part of the modeling process.

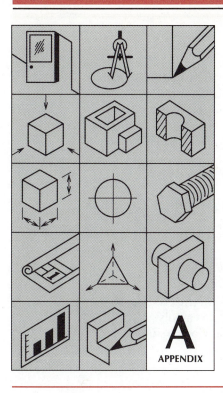

APPENDIX

B

TABLES

The following is a list of the tables in Appendix B. The numbers in the list are page numbers.

Table B.1   Preferred Limits and Fits (Inches)

Nominal Size Range Inches — Over	To	RC 1 Limits of Clearance	RC 1 Hole H5	RC 1 Shaft g4	RC 2 Limits of Clearance	RC 2 Hole H6	RC 2 Shaft g5	RC 3 Limits of Clearance	RC 3 Hole H7	RC 3 Shaft f6	RC 4 Limits of Clearance	RC 4 Hole H8	RC 4 Shaft f7
0	− 0.12	0.1	+ 0.2	− 0.1	0.1	+ 0.25	− 0.1	0.3	+ 0.4	− 0.3	0.3	+ 0.6	− 0.3
		0.45	0	− 0.25	0.55	0	− 0.3	0.95	0	− 0.55	1.3	0	− 0.7
0.12	− 0.24	0.15	+ 0.2	− 0.15	0.15	+ 0.3	− 0.15	0.4	+ 0.5	− 0.4	0.4	+ 0.7	− 0.4
		0.5	0	− 0.3	0.65	0	− 0.35	1.2	0	− 0.7	1.6	0	− 0.9
0.24	− 0.40	0.2	+ 0.25	− 0.2	0.2	+ 0.4	− 0.2	0.5	+ 0.6	− 0.5	0.5	+ 0.9	− 0.5
		0.6	0	− 0.35	0.85	0	− 0.45	1.5	0	− 0.9	2.0	0	− 1.1
0.40	− 0.71	0.25	+ 0.3	− 0.25	0.25	+ 0.4	− 0.25	0.6	+ 0.7	− 0.6	0.6	+ 1.0	− 0.6
		0.75	0	− 0.45	0.95	0	− 0.55	1.7	0	− 1.0	2.3	0	− 1.3
0.71	− 1.19	0.3	+ 0.4	− 0.3	0.3	+ 0.5	− 0.3	0.8	+ 0.8	− 0.8	0.8	+ 1.2	− 0.8
		0.95	0	− 0.55	1.2	0	− 0.7	2.1	0	− 1.3	2.8	0	− 1.6
1.19	− 1.97	0.4	+ 0.4	− 0.4	0.4	+ 0.6	− 0.4	1.0	+ 1.0	− 1.0	1.0	+ 1.6	− 1.0
		1.1	0	− 0.7	1.4	0	− 0.8	2.6	0	− 1.6	3.6	0	− 2.0
1.97	− 3.15	0.4	+ 0.5	− 0.4	0.4	+ 0.7	− 0.4	1.2	+ 1.2	− 1.2	1.2	+ 1.8	− 1.2
		1.2	0	− 0.7	1.6	0	− 0.9	3.1	0	− 1.9	4.2	0	− 2.4
3.15	− 4.73	0.5	+ 0.6	− 0.5	0.5	+ 0.9	− 0.5	1.4	+ 1.4	− 1.4	1.4	+ 2.2	− 1.4
		1.5	0	− 0.9	2.0	0	− 1.1	3.7	0	− 2.3	5.0	0	− 2.8
4.73	− 7.09	0.6	+ 0.7	− 0.6	0.6	+ 1.0	− 0.6	1.6	+ 1.6	− 1.6	1.6	+ 2.5	− 1.6
		1.8	0	− 1.1	2.3	0	− 1.3	4.2	0	− 2.6	5.7	0	− 3.2
7.09	− 9.85	0.6	+ 0.8	− 0.6	0.6	+ 1.2	− 0.6	2.0	+ 1.8	− 2.0	2.0	+ 2.8	− 2.0
		2.0	0	− 1.2	2.6	0	− 1.4	5.0	0	− 3.2	6.6	0	− 3.8
9.85	− 12.41	0.8	+ 0.9	− 0.8	0.7	+ 1.2	− 0.7	2.5	+ 2.0	− 2.5	2.2	+ 3.0	− 2.2
		2.3	0	− 1.4	2.8	0	− 1.6	5.7	0	− 3.7	7.2	0	− 4.2
12.41	− 15.75	1.0	+ 1.0	− 1.0	0.7	+ 1.4	− 0.7	3.0	+ 2.2	− 3.0	2.5	+ 3.5	− 2.5
		2.7	0	− 1.7	3.1	0	− 1.7	6.6	0	− 4.4	8.2	0	− 4.7
15.75	− 19.69	1.2	+ 1.0	− 1.2	0.8	+ 1.6	− 0.8	4.0	+ 2.5	− 4.0	2.8	+ 4.0	− 2.8
		3.0	0	− 2.0	3.4	0	− 1.8	8.1	0	− 5.6	9.3	0	− 5.3
19.69	− 30.09	1.6	+ 1.2	− 1.6	1.6	+ 2.0	− 1.6	5.0	+ 3.0	− 5.0	5.0	+ 5.0	− 5.0
		3.7	0	− 2.5	4.8	0	− 2.8	10.0	0	− 7.0	13.0	0	− 8.0
30.09	− 41.49	2.0	+ 1.6	− 2.0	2.0	+ 2.5	− 2.0	6.0	+ 4.0	− 6.0	6.0	+ 6.0	− 6.0
		4.6	0	− 3.0	6.1	0	− 3.6	12.5	0	− 8.5	16.0	0	−10.0
41.49	− 56.19	2.5	+ 2.0	− 2.5	2.5	+ 3.0	− 2.5	8.0	+ 5.0	− 8.0	8.0	+ 8.0	− 8.0
		5.7	0	− 3.7	7.5	0	− 4.5	16.0	0	−11.0	21.0	0	−13.0
56.19	− 76.39	3.0	+ 2.5	− 3.0	3.0	+ 4.0	− 3.0	10.0	+ 6.0	−10.0	10.0	+10.0	−10.0
		7.1	0	− 4.6	9.5	0	− 5.5	20.0	0	−14.0	26.0	0	−16.0
76.39	− 100.9	4.0	+ 3.0	− 4.0	4.0	+ 5.0	− 4.0	12.0	+ 8.0	−12.0	12.0	+12.0	−12.0
		9.0	0	− 6.0	12.0	0	− 7.0	25.0	0	−17.0	32.0	0	−20.0
100.9	− 131.9	5.0	+ 4.0	− 5.0	5.0	+ 6.0	− 5.0	16.0	+10.0	−16.0	16.0	+16.0	−16.0
		11.5	0	− 7.5	15.0	0	− 9.0	32.0	0	−22.0	42.0	0	−26.0
131.9	− 171.9	6.0	+ 5.0	− 6.0	6.0	+ 8.0	− 6.0	18.0	+12.0	−18.0	18.0	+20.0	−18.0
		14.0	0	− 9.0	19.0	0	−11.0	38.0	0	−26.0	50.0	0	−30.0
171.9	− 200	8.0	+ 6.0	− 8.0	8.0	+10.0	− 8.0	22.0	+16.0	−22.0	22.0	+25.0	−22.0
		18.0	0	−12.0	22.0	0	−12.0	48.0	0	−32.0	63.0	0	−38.0

Courtesy of USAS (B4.1)

Table B.1 (continued)

Class RC 5			Class RC 6			Class RC 7			Class RC 8			Class RC 9			Nominal Size Range Inches	
Limits of Clearance	Standard Limits		Limits of Clearance	Standard Limits		Limits of Clearance	Standard Limits		Limits of Clearance	Standard Limits		Limits of Clearance	Standard Limits			
	Hole H8	Shaft e7		Hole H9	Shaft e8		Hole H9	Shaft d8		Hole H10	Shaft c9		Hole H11	Shaft	Over	To
0.6 / 1.6	+0.6 / −0	−0.6 / −1.0	0.6 / 2.2	+1.0 / −0	−0.6 / −1.2	1.0 / 2.6	+1.0 / 0	−1.0 / −1.6	2.5 / 5.1	+1.6 / 0	−2.5 / −3.5	4.0 / 8.1	+2.5 / 0	−4.0 / −5.6	0 −	0.12
0.8 / 2.0	+0.7 / −0	−0.8 / −1.3	0.8 / 2.7	+1.2 / −0	−0.8 / −1.5	1.2 / 3.1	+1.2 / 0	−1.2 / −1.9	2.8 / 5.8	+1.8 / 0	−2.8 / −4.0	4.5 / 9.0	+3.0 / 0	−4.5 / −6.0	0.12−	0.24
1.0 / 2.5	+0.9 / −0	−1.0 / −1.6	1.0 / 3.3	+1.4 / −0	−1.0 / −1.9	1.6 / 3.9	+1.4 / 0	−1.6 / −2.5	3.0 / 6.6	+2.2 / 0	−3.0 / −4.4	5.0 / 10.7	+3.5 / 0	−5.0 / −7.2	0.24−	0.40
1.2 / 2.9	+1.0 / −0	−1.2 / −1.9	1.2 / 3.8	+1.6 / −0	−1.2 / −2.2	2.0 / 4.6	+1.6 / 0	−2.0 / −3.0	3.5 / 7.9	+2.8 / 0	−3.5 / −5.1	6.0 / 12.8	+4.0 / −0	−6.0 / −8.8	0.40−	0.71
1.6 / 3.6	+1.2 / −0	−1.6 / −2.4	1.6 / 4.8	+2.0 / −0	−1.6 / −2.8	2.5 / 5.7	+2.0 / 0	−2.5 / −3.7	4.5 / 10.0	+3.5 / 0	−4.5 / −6.5	7.0 / 15.5	+5.0 / 0	−7.0 / −10.5	0.71−	1.19
2.0 / 4.6	+1.6 / −0	−2.0 / −3.0	2.0 / 6.1	+2.5 / −0	−2.0 / −3.6	3.0 / 7.1	+2.5 / 0	−3.0 / −4.6	5.0 / 11.5	+4.0 / 0	−5.0 / −7.5	8.0 / 18.0	+6.0 / 0	−8.0 / −12.0	1.19−	1.97
2.5 / 5.5	+1.8 / −0	−2.5 / −3.7	2.5 / 7.3	+3.0 / −0	−2.5 / −4.3	4.0 / 8.8	+3.0 / 0	−4.0 / −5.8	6.0 / 13.5	+4.5 / 0	−6.0 / −9.0	9.0 / 20.5	+7.0 / 0	−9.0 / −13.5	1.97−	3.15
3.0 / 6.6	+2.2 / −0	−3.0 / −4.4	3.0 / 8.7	+3.5 / −0	−3.0 / −5.2	5.0 / 10.7	+3.5 / 0	−5.0 / −7.2	7.0 / 15.5	+5.0 / 0	−7.0 / −10.5	10.0 / 24.0	+9.0 / 0	−10.0 / −15.0	3.15−	4.73
3.5 / 7.6	+2.5 / −0	−3.5 / −5.1	3.5 / 10.0	+4.0 / −0	−3.5 / −6.0	6.0 / 12.5	+4.0 / 0	−6.0 / −8.5	8.0 / 18.0	+6.0 / 0	−8.0 / −12.0	12.0 / 28.0	+10.0 / 0	−12.0 / −18.0	4.73−	7.09
4.0 / 8.6	+2.8 / −0	−4.0 / −5.8	4.0 / 11.3	+4.5 / 0	−4.0 / −6.8	7.0 / 14.3	+4.5 / 0	−7.0 / −9.8	10.0 / 21.5	+7.0 / 0	−10.0 / −14.5	15.0 / 34.0	+12.0 / 0	−15.0 / −22.0	7.09−	9.85
5.0 / 10.0	+3.0 / 0	−5.0 / −7.0	5.0 / 13.0	+5.0 / 0	−5.0 / −8.0	8.0 / 16.0	+5.0 / 0	−8.0 / −11.0	12.0 / 25.0	+8.0 / 0	−12.0 / −17.0	18.0 / 38.0	+12.0 / 0	−18.0 / −26.0	9.85−	12.41
6.0 / 11.7	+3.5 / 0	−6.0 / −8.2	6.0 / 15.5	+6.0 / 0	−6.0 / −9.5	10.0 / 19.5	+6.0 / 0	−10.0 / −13.5	14.0 / 29.0	+9.0 / 0	−14.0 / −20.0	22.0 / 45.0	+14.0 / 0	−22.0 / −31.0	12.41−	15.75
8.0 / 14.5	+4.0 / 0	−8.0 / −10.5	8.0 / 18.0	+6.0 / 0	−8.0 / −12.0	12.0 / 22.0	+6.0 / 0	−12.0 / −16.0	16.0 / 32.0	+10.0 / 0	−16.0 / −22.0	25.0 / 51.0	+16.0 / 0	−25.0 / −35.0	15.75−	19.69
10.0 / 18.0	+5.0 / 0	−10.0 / −13.0	10.0 / 23.0	+8.0 / 0	−10.0 / −15.0	16.0 / 29.0	+8.0 / 0	−16.0 / −21.0	20.0 / 40.0	+12.0 / 0	−20.0 / −28.0	30.0 / 62.0	+20.0 / 0	−30.0 / −42.0	19.69−	30.09
12.0 / 22.0	+6.0 / 0	−12.0 / −16.0	12.0 / 28.0	+10.0 / 0	−12.0 / −18.0	20.0 / 36.0	+10.0 / 0	−20.0 / −26.0	25.0 / 51.0	+16.0 / 0	−25.0 / −35.0	40.0 / 81.0	+25.0 / 0	−40.0 / −56.0	30.09−	41.49
16.0 / 29.0	+8.0 / 0	−16.0 / −21.0	16.0 / 36.0	+12.0 / 0	−16.0 / −24.0	25.0 / 45.0	+12.0 / 0	−25.0 / −33.0	30.0 / 62.0	+20.0 / 0	−30.0 / −42.0	50.0 / 100	+30.0 / 0	−50.0 / −70.0	41.49−	56.19
20.0 / 36.0	+10.0 / 0	−20.0 / −26.0	20.0 / 46.0	+16.0 / 0	−20.0 / −30.0	30.0 / 56.0	+16.0 / 0	−30.0 / −40.0	40.0 / 81.0	+25.0 / 0	−40.0 / −56.0	60.0 / 125	+40.0 / 0	−60.0 / −85.0	56.19−	76.39
25.0 / 45.0	+12.0 / 0	−25.0 / −33.0	25.0 / 57.0	+20.0 / 0	−25.0 / −37.0	40.0 / 72.0	+20.0 / 0	−40.0 / −52.0	50.0 / 100	+30.0 / 0	−50.0 / −70.0	80.0 / 160	+50.0 / 0	−80.0 / −110	76.39−	100.9
30.0 / 56.0	+16.0 / 0	−30.0 / −40.0	30.0 / 71.0	+25.0 / 0	−30.0 / −46.0	50.0 / 91.0	+25.0 / 0	−50.0 / −66.0	60.0 / 125	+40.0 / 0	−60.0 / −85.0	100 / 200	+60.0 / 0	−100 / −140	100.9 −131.9	
35.0 / 67.0	+20.0 / 0	−35.0 / −47.0	35.0 / 85.0	+30.0 / 0	−35.0 / −55.0	60.0 / 110.0	+30.0 / 0	−60.0 / −80.0	80.0 / 160	+50.0 / 0	−80.0 / −110	130 / 260	+80.0 / 0	−130 / −180	131.9 −171.9	
45.0 / 86.0	+25.0 / 0	−45.0 / −61.0	45.0 / 110.0	+40.0 / 0	−45.0 / −70.0	80.0 / 145.0	+40.0 / 0	−80.0 / −105.0	100 / 200	+60.0 / 0	−100 / −140	150 / 310	+100 / 0	−150 / −210	171.9 −200	

Courtesy of USAS (B4.1)

Nominal Size Range Inches Over — To	Class LC 1 Limits of Clearance	Class LC 1 Standard Limits Hole H6	Class LC 1 Standard Limits Shaft h5	Class LC 2 Limits of Clearance	Class LC 2 Standard Limits Hole H7	Class LC 2 Standard Limits Shaft h6	Class LC 3 Limits of Clearance	Class LC 3 Standard Limits Hole H8	Class LC 3 Standard Limits Shaft h7	Class LC 4 Limits of Clearance	Class LC 4 Standard Limits Hole H10	Class LC 4 Standard Limits Shaft h9	Class LC 5 Limits of Clearance	Class LC 5 Standard Limits Hole H7	Class LC 5 Standard Limits Shaft g6
0 — 0.12	0 / 0.45	+0.25 / −0	+0 / −0.2	0 / 0.65	+0.4 / −0	+0 / −0.25	0 / 1	+0.6 / −0	+0 / −0.4	0 / 2.6	+1.6 / −0	+0 / −1.0	0.1 / 0.75	+0.4 / −0	−0.1 / −0.35
0.12— 0.24	0 / 0.5	+0.3 / −0	+0 / −0.2	0 / 0.8	+0.5 / −0	+0 / −0.3	0 / 1.2	+0.7 / −0	+0 / −0.5	0 / 3.0	+1.8 / −0	+0 / −1.2	0.15 / 0.95	+0.5 / −0	−0.15 / −0.45
0.24— 0.40	0 / 0.65	+0.4 / −0	+0 / −0.25	0 / 1.0	+0.6 / −0	+0 / −0.4	0 / 1.5	+0.9 / −0	+0 / −0.6	0 / 3.6	+2.2 / −0	+0 / −1.4	0.2 / 1.2	+0.6 / −0	−0.2 / −0.6
0.40— 0.71	0 / 0.7	+0.4 / −0	+0 / −0.3	0 / 1.1	+0.7 / −0	+0 / −0.4	0 / 1.7	+1.0 / −0	+0 / −0.7	0 / 4.4	+2.8 / −0	+0 / −1.6	0.25 / 1.35	+0.7 / −0	−0.25 / −0.65
0.71— 1.19	0 / 0.9	+0.5 / −0	+0 / −0.4	0 / 1.3	+0.8 / −0	+0 / −0.5	0 / 2	+1.2 / −0	+0 / −0.8	0 / 5.5	+3.5 / −0	+0 / −2.0	0.3 / 1.6	+0.8 / −0	−0.3 / −0.8
1.19— 1.97	0 / 1.0	+0.6 / −0	+0 / −0.4	0 / 1.6	+1.0 / −0	+0 / −0.6	0 / 2.6	+1.6 / −0	+0 / −1	0 / 6.5	+4.0 / −0	+0 / −2.5	0.4 / 2.0	+1.0 / −0	−0.4 / −1.0
1.97— 3.15	0 / 1.2	+0.7 / −0	+0 / −0.5	0 / 1.9	+1.2 / −0	+0 / −0.7	0 / 3	+1.8 / −0	+0 / −1.2	0 / 7.5	+4.5 / −0	+0 / −3	0.4 / 2.3	+1.2 / −0	−0.4 / −1.1
3.15— 4.73	0 / 1.5	+0.9 / −0	+0 / −0.6	0 / 2.3	+1.4 / −0	+0 / −0.9	0 / 3.6	+2.2 / −0	+0 / −1.4	0 / 8.5	+5.0 / −0	+0 / −3.5	0.5 / 2.8	+1.4 / −0	−0.5 / −1.4
4.73— 7.09	0 / 1.7	+1.0 / −0	+0 / −0.7	0 / 2.6	+1.6 / −0	+0 / −1.0	0 / 4.1	+2.5 / −0	+0 / −1.6	0 / 10	+6.0 / −0	+0 / −4	0.6 / 3.2	+1.6 / −0	−0.6 / −1.6
7.09— 9.85	0 / 2.0	+1.2 / −0	+0 / −0.8	0 / 3.0	+1.8 / −0	+0 / −1.2	0 / 4.6	+2.8 / −0	+0 / −1.8	0 / 11.5	+7.0 / −0	+0 / −4.5	0.6 / 3.6	+1.8 / −0	−0.6 / −1.8
9.85— 12.41	0 / 2.1	+1.2 / −0	+0 / −0.9	0 / 3.2	+2.0 / −0	+0 / −1.2	0 / 5	+3.0 / −0	+0 / −2.0	0 / 13	+8.0 / −0	+0 / −5	0.7 / 3.9	+2.0 / −0	−0.7 / −1.9
12.41— 15.75	0 / 2.4	+1.4 / −0	+0 / −1.0	0 / 3.6	+2.2 / −0	+0 / −1.4	0 / 5.7	+3.5 / −0	+0 / −2.2	0 / 15	+9.0 / −0	+0 / −6	0.7 / 4.3	+2.2 / −0	−0.7 / −2.1
15.75— 19.69	0 / 2.6	+1.6 / −0	+0 / −1.0	0 / 4.1	+2.5 / −0	+0 / −1.6	0 / 6.5	+4 / −0	+0 / −2.5	0 / 16	+10.0 / −0	+0 / −6	0.8 / 4.9	+2.5 / −0	−0.8 / −2.4
19.69— 30.09	0 / 3.2	+2.0 / −0	+0 / −1.2	0 / 5.0	+3 / −0	+0 / −2	0 / 8	+5 / −0	+0 / −3	0 / 20	+12.0 / −0	+0 / −8	0.9 / 5.9	+3.0 / −0	−0.9 / −2.9
30.09— 41.49	0 / 4.1	+2.5 / −0	+0 / −1.6	0 / 6.5	+4 / −0	+0 / −2.5	0 / 10	+6 / −0	+0 / −4	0 / 26	+16.0 / −0	+0 / −10	1.0 / 7.5	+4.0 / −0	−1.0 / −3.5
41.49— 56.19	0 / 5.0	+3.0 / −0	+0 / −2.0	0 / 8.0	+5 / −0	+0 / −3	0 / 13	+8 / −0	+0 / −5	0 / 32	+20.0 / −0	+0 / −12	1.2 / 9.2	+5.0 / −0	−1.2 / −4.2
56.19— 76.39	0 / 6.5	+4.0 / −0	+0 / −2.5	0 / 10	+6 / −0	+0 / −4	0 / 16	+10 / −0	+0 / −6	0 / 41	+25.0 / −0	+0 / −16	1.2 / 11.2	+6.0 / −0	−1.2 / −5.2
76.39— 100.9	0 / 8.0	+5.0 / −0	+0 / −3.0	0 / 13	+8 / −0	+0 / −5	0 / 20	+12 / −0	+0 / −8	0 / 50	+30.0 / −0	+0 / −20	1.4 / 14.4	+8.0 / −0	−1.4 / −6.4
100.9 — 131.9	0 / 10.0	+6.0 / −0	+0 / −4.0	0 / 16	+10 / −0	+0 / −6	0 / 26	+16 / −0	+0 / −10	0 / 65	+40.0 / −0	+0 / −25	1.6 / 17.6	+10.0 / −0	−1.6 / −7.6
131.9 — 171.9	0 / 13.0	+8.0 / −0	+0 / −5.0	0 / 20	+12 / −0	+0 / −8	0 / 32	+20 / −0	+0 / −12	0 / 80	+50.0 / −0	+0 / −30	1.8 / 21.8	+12.0 / −0	−1.8 / −9.8
171.9 — 200	0 / 16.0	+10.0 / −0	+0 / −6.0	0 / 26	+16 / −0	+0 / −10	0 / 41	+25 / −0	+0 / −16	0 / 100	+60.0 / −0	+0 / −40	1.8 / 27.8	+16.0 / −0	−1.8 / −11.8

Courtesy of USAS (B4.1)

Table B.1   (continued)

Class LC 6			Class LC 7			Class LC 8			Class LC 9			Class LC 10			Class LC 11			Nominal Size Range Inches	
Limits of Clearance	Standard Limits Hole H9	Standard Limits Shaft f8	Limits of Clearance	Standard Limits Hole H10	Standard Limits Shaft e9	Limits of Clearance	Standard Limits Hole H10	Standard Limits Shaft d9	Limits of Clearance	Standard Limits Hole H11	Standard Limits Shaft c10	Limits of Clearance	Standard Limits Hole H12	Standard Limits Shaft	Limits of Clearance	Standard Limits Hole H13	Standard Limits Shaft	Over	To
0.3 / 1.9	+1.0 / 0	−0.3 / −0.9	0.6 / 3.2	+1.6 / 0	−0.6 / −1.6	1.0 / 3.6	+1.6 / −0	−1.0 / −2.0	2.5 / 6.6	+2.5 / −0	−2.5 / −4.1	4 / 12	+4 / −0	−4 / −8	5 / 17	+6 / −0	−5 / −11	0	0.12
0.4 / 2.3	+1.2 / 0	−0.4 / −1.1	0.8 / 3.8	+1.8 / 0	−0.8 / −2.0	1.2 / 4.2	+1.8 / −0	−1.2 / −2.4	2.8 / 7.6	+3.0 / −0	−2.8 / −4.6	4.5 / 14.5	+5 / −0	−4.5 / −9.5	6 / 20	+7 / −0	−6 / −13	0.12	0.24
0.5 / 2.8	+1.4 / 0	−0.5 / −1.4	1.0 / 4.6	+2.2 / 0	−1.0 / −2.4	1.6 / 5.2	+2.2 / −0	−1.6 / −3.0	3.0 / 8.7	+3.5 / −0	−3.0 / −5.2	5 / 17	+6 / −0	−5 / −11	7 / 25	+9 / −0	−7 / −16	0.24	0.40
0.6 / 3.2	+1.6 / 0	−0.6 / −1.6	1.2 / 5.6	+2.8 / 0	−1.2 / −2.8	2.0 / 6.4	+2.8 / −0	−2.0 / −3.6	3.5 / 10.3	+4.0 / −0	−3.5 / −6.3	6 / 20	+7 / −0	−6 / −13	8 / 28	+10 / −0	−8 / −18	0.40	0.71
0.8 / 4.0	+2.0 / 0	−0.8 / −2.0	1.6 / 7.1	+3.5 / 0	−1.6 / −3.6	2.5 / 8.0	+3.5 / −0	−2.5 / −4.5	4.5 / 13.0	+5.0 / −0	−4.5 / −8.0	7 / 23	+8 / −0	−7 / −15	10 / 34	+12 / −0	−10 / −22	0.71	1.19
1.0 / 5.1	+2.5 / 0	−1.0 / −2.6	2.0 / 8.5	+4.0 / 0	−2.0 / −4.5	3.0 / 9.5	+4.0 / −0	−3.0 / −5.5	5 / 15	+6 / −0	−5 / −9	8 / 28	+10 / −0	−8 / −18	12 / 44	+16 / −0	−12 / −28	1.19	1.97
1.2 / 6.0	+3.0 / 0	−1.2 / −3.0	2.5 / 10.0	+4.5 / 0	−2.5 / −5.5	4.0 / 11.5	+4.5 / −0	−4.0 / −7.0	6 / 17.5	+7 / −0	−6 / −10.5	10 / 34	+12 / −0	−10 / −22	14 / 50	+18 / −0	−14 / −32	1.97	3.15
1.4 / 7.1	+3.5 / 0	−1.4 / −3.6	3.0 / 11.5	+5.0 / 0	−3.0 / −6.5	5.0 / 13.5	+5.0 / −0	−5.0 / −8.5	7 / 21	+9 / −0	−7 / −12	11 / 39	+14 / −0	−11 / −25	16 / 60	+22 / −0	−16 / −38	3.15	4.73
1.6 / 8.1	+4.0 / 0	−1.6 / −4.1	3.5 / 13.5	+6.0 / 0	−3.5 / −7.5	6 / 16	+6 / −0	−6 / −10	8 / 24	+10 / −0	−8 / −14	12 / 44	+16 / −0	−12 / −28	18 / 68	+25 / −0	−18 / −43	4.73	7.09
2.0 / 9.3	+4.5 / 0	−2.0 / −4.8	4.0 / 15.5	+7.0 / 0	−4.0 / −8.5	7 / 18.5	+7 / −0	−7 / −11.5	10 / 29	+12 / −0	−10 / −17	16 / 52	+18 / −0	−16 / −34	22 / 78	+28 / −0	−22 / −50	7.09	9.85
2.2 / 10.2	+5.0 / 0	−2.2 / −5.2	4.5 / 17.5	+8.0 / 0	−4.5 / −9.5	7 / 20	+8 / −0	−7 / −12	12 / 32	+12 / −0	−12 / −20	20 / 60	+20 / −0	−20 / −40	28 / 88	+30 / −0	−28 / −58	9.85	12.41
2.5 / 12.0	+6.0 / 0	−2.5 / −6.0	5.0 / 20.0	+9.0 / 0	−5 / −11	8 / 23	+9 / −0	−8 / −14	14 / 37	+14 / −0	−14 / −23	22 / 66	+22 / −0	−22 / −44	30 / 100	+35 / −0	−30 / −65	12.41	15.75
2.8 / 12.8	+6.0 / 0	−2.8 / −6.8	5.0 / 21.0	+10.0 / 0	−5 / −11	9 / 25	+10 / −0	−9 / −15	16 / 42	+16 / −0	−16 / −26	25 / 75	+25 / −0	−25 / −50	35 / 115	+40 / −0	−35 / −75	15.75	19.69
3.0 / 16.0	+8.0 / 0	−3.0 / −8.0	6.0 / 26.0	+12.0 / −0	−6 / −14	10 / 30	+12 / −0	−10 / −18	18 / 50	+20 / −0	−18 / −30	28 / 88	+30 / −0	−28 / −58	40 / 140	+50 / −0	−40 / −90	19.69	30.09
3.5 / 19.5	+10.0 / 0	−3.5 / −9.5	7.0 / 33.0	+16.0 / −0	−7 / −17	12 / 38	+16 / −0	−12 / −22	20 / 61	+25 / −0	−20 / −36	30 / 110	+40 / −0	−30 / −70	45 / 165	+60 / −0	−45 / −105	30.09	41.49
4.0 / 24.0	+12.0 / 0	−4.0 / −12.0	8.0 / 40.0	+20.0 / −0	−8 / −20	14 / 46	+20 / −0	−14 / −26	25 / 75	+30 / −0	−25 / −45	40 / 140	+50 / −0	−40 / −90	60 / 220	+80 / −0	−60 / −140	41.49	56.19
4.5 / 30.5	+16.0 / 0	−4.5 / −14.5	9.0 / 50.0	+25.0 / −0	−9 / −25	16 / 57	+25 / −0	−16 / −32	30 / 95	+40 / −0	−30 / −55	50 / 170	+60 / −0	−50 / −110	70 / 270	+100 / −0	−70 / −170	56.19	76.39
5.0 / 37.0	+20.0 / 0	−5 / −17	10.0 / 60.0	+30.0 / −0	−10 / −30	18 / 68	+30 / −0	−18 / −38	35 / 115	+50 / −0	−35 / −65	50 / 210	+80 / −0	−50 / −130	80 / 330	+125 / −0	−80 / −205	76.39	100.9
6.0 / 47.0	+25.0 / 0	−6 / −22	12.0 / 67.0	+40.0 / −0	−12 / −27	20 / 85	+40 / −0	−20 / −45	40 / 140	+60 / −0	−40 / −80	60 / 260	+100 / −0	−60 / −160	90 / 410	+160 / −0	−90 / −250	100.9	131.9
7.0 / 57.0	+30.0 / 0	−7 / −27	14.0 / 94.0	+50.0 / −0	−14 / −44	25 / 105	+50 / −0	−25 / −55	50 / 180	+80 / −0	−50 / −100	80 / 330	+125 / −0	−80 / −205	100 / 500	+200 / −0	−100 / −300	131.9	171.9
7.0 / 72.0	+40.0 / 0	−7 / −32	14.0 / 114.0	+60.0 / −0	−14 / −54	25 / 125	+60 / −0	−25 / −65	50 / 210	+100 / −0	−50 / −110	90 / 410	+160 / −0	−90 / −250	125 / 625	+250 / −0	−125 / −375	171.9	200

Courtesy of USAS (B4.1)

Nominal Size Range, Inches Over	To	Class LT 1 Fit	Class LT 1 Hole H7	Class LT 1 Shaft js6	Class LT 2 Fit	Class LT 2 Hole H8	Class LT 2 Shaft js7	Class LT 3 Fit	Class LT 3 Hole H7	Class LT 3 Shaft k6	Class LT 4 Fit	Class LT 4 Hole H8	Class LT 4 Shaft k7	Class LT 5 Fit	Class LT 5 Hole H7	Class LT 5 Shaft n6	Class LT 6 Fit	Class LT 6 Hole H7	Class LT 6 Shaft n7
0	0.12	−0.10 / +0.50	+0.4 / −0	+0.10 / −0.10	−0.2 / +0.8	+0.6 / −0	+0.2 / −0.2							−0.5 / +0.15	+0.4 / −0	+0.5 / +0.25	−0.65 / +0.15	+0.4 / −0	+0.65 / +0.25
0.12	0.24	−0.15 / +0.65	+0.5 / −0	+0.15 / −0.15	−0.25 / +0.95	+0.7 / −0	+0.25 / −0.25							−0.6 / +0.2	+0.5 / −0	+0.6 / +0.3	−0.8 / +0.2	+0.5 / −0	+0.8 / +0.3
0.24	0.40	−0.2 / +0.8	+0.6 / −0	+0.2 / −0.2	−0.3 / +1.2	+0.9 / −0	+0.3 / −0.3	−0.5 / +0.5	+0.6 / −0	+0.5 / +0.1	−0.7 / +0.8	+0.9 / −0	+0.7 / +0.1	−0.8 / +0.2	+0.6 / −0	+0.8 / +0.4	−1.0 / +0.2	+0.6 / −0	+1.0 / +0.4
0.40	0.71	−0.2 / +0.9	+0.7 / −0	+0.2 / −0.2	−0.35 / +1.35	+1.0 / −0	+0.35 / −0.35	−0.5 / +0.6	+0.7 / −0	+0.5 / +0.1	−0.8 / +0.9	+1.0 / −0	+0.8 / +0.1	−0.9 / +0.2	+0.7 / −0	+0.9 / +0.5	−1.2 / +0.2	+0.7 / −0	+1.2 / +0.5
0.71	1.19	−0.25 / +1.05	+0.8 / −0	+0.25 / −0.25	−0.4 / +1.6	+1.2 / −0	+0.4 / −0.4	−0.6 / +0.7	+0.8 / −0	+0.6 / +0.1	−0.9 / +1.1	+1.2 / −0	+0.9 / +0.1	−1.1 / +0.2	+0.8 / −0	+1.1 / +0.6	−1.4 / +0.2	+0.8 / −0	+1.4 / +0.6
1.19	1.97	−0.3 / +1.3	+1.0 / −0	+0.3 / −0.3	−0.5 / +2.1	+1.6 / −0	+0.5 / −0.5	−0.7 / +0.9	+1.0 / −0	+0.7 / +0.1	−1.1 / +1.5	+1.6 / −0	+1.1 / +0.1	−1.3 / +0.3	+1.0 / −0	+1.3 / +0.7	−1.7 / +0.3	+1.0 / −0	+1.7 / +0.7
1.97	3.15	−0.3 / +1.5	+1.2 / −0	+0.3 / −0.3	−0.6 / +2.4	+1.8 / −0	+0.6 / −0.6	−0.8 / +1.1	+1.2 / −0	+0.8 / +0.1	−1.3 / +1.7	+1.8 / −0	+1.3 / +0.1	−1.5 / +0.4	+1.2 / −0	+1.5 / +0.8	−2.0 / +0.4	+1.2 / −0	+2.0 / +0.8
3.15	4.73	−0.4 / +1.8	+1.4 / −0	+0.4 / −0.4	−0.7 / +2.9	+2.2 / −0	+0.7 / −0.7	−1.0 / +1.3	+1.4 / −0	+1.0 / +0.1	−1.5 / +2.1	+2.2 / −0	+1.5 / +0.1	−1.9 / +0.4	+1.4 / −0	+1.9 / +1.0	−2.4 / +0.4	+1.4 / −0	+2.4 / +1.0
4.73	7.09	−0.5 / +2.1	+1.6 / −0	+0.5 / −0.5	−0.8 / +3.3	+2.5 / −0	+0.8 / −0.8	−1.1 / +1.5	+1.6 / −0	+1.1 / +0.1	−1.7 / +2.4	+2.5 / −0	+1.7 / +0.1	−2.2 / +0.4	+1.6 / −0	+2.2 / +1.2	−2.8 / +0.4	+1.6 / −0	+2.8 / +1.2
7.09	9.85	−0.6 / +2.4	+1.8 / −0	+0.6 / −0.6	−0.9 / +3.7	+2.8 / −0	+0.9 / −0.9	−1.4 / +1.6	+1.8 / −0	+1.4 / +0.2	−2.0 / +2.6	+2.8 / −0	+2.0 / +0.2	−2.6 / +0.4	+1.8 / −0	+2.6 / +1.4	−3.2 / +0.4	+1.8 / −0	+3.2 / +1.4
9.85	12.41	−0.6 / +2.6	+2.0 / −0	+0.6 / −0.6	−1.0 / +4.0	+3.0 / −0	+1.0 / −1.0	−1.4 / +1.8	+2.0 / −0	+1.4 / +0.2	−2.2 / +2.8	+3.0 / −0	+2.2 / +0.2	−2.6 / +0.6	+2.0 / −0	+2.6 / +1.4	−3.4 / +0.6	+2.0 / −0	+3.4 / +1.4
12.41	15.75	−0.7 / +2.9	+2.2 / −0	+0.7 / −0.7	−1.0 / +4.5	+3.5 / −0	+1.0 / −1.0	−1.6 / +2.0	+2.2 / −0	+1.6 / +0.2	−2.4 / +3.3	+3.5 / −0	+2.4 / +0.2	−3.0 / +0.6	+2.2 / −0	+3.0 / +1.6	−3.8 / +0.6	+2.2 / −0	+3.8 / +1.6
15.75	19.69	−0.8 / +3.3	+2.5 / −0	+0.8 / −0.8	−1.2 / +5.2	+4.0 / −0	+1.2 / −1.2	−1.8 / +2.3	+2.5 / −0	+1.8 / +0.2	−2.7 / +3.8	+4.0 / −0	+2.7 / +0.2	−3.4 / +0.7	+2.5 / −0	+3.4 / +1.8	−4.3 / +0.7	+2.5 / −0	+4.3 / +1.8

Courtesy of USAS (B4.1)

Table B.1 (continued)

Nominal Size Range Inches		Class LN 1			Class LN 2			Class LN 3		
		Limits of Interference	Standard Limits		Limits of Interference	Standard Limits		Limits of Interference	Standard Limits	
Over	To		Hole H6	Shaft n5		Hole H7	Shaft p6		Hole H7	Shaft r6
0 — 0.12		0 / 0.45	+ 0.25 / − 0	+0.45 / +0.25	0 / 0.65	+ 0.4 / − 0	+ 0.65 / + 0.4	0.1 / 0.75	+ 0.4 / − 0	+ 0.75 / + 0.5
0.12 — 0.24		0 / 0.5	+ 0.3 / − 0	+0.5 / +0.3	0 / 0.8	+ 0.5 / − 0	+ 0.8 / + 0.5	0.1 / 0.9	+ 0.5 / 0	+ 0.9 / + 0.6
0.24 — 0.40		0 / 0.65	+ 0.4 / − 0	+0.65 / +0.4	0 / 1.0	+ 0.6 / − 0	+ 1.0 / + 0.6	0.2 / 1.2	+ 0.6 / − 0	+ 1.2 / + 0.8
0.40 — 0.71		0 / 0.8	+ 0.4 / − 0	+0.8 / +0.4	0 / 1.1	+ 0.7 / − 0	+ 1.1 / + 0.7	0.3 / 1.4	+ 0.7 / − 0	+ 1.4 / + 1.0
0.71 — 1.19		0 / 1.0	+ 0.5 / − 0	+1.0 / +0.5	0 / 1.3	+ 0.8 / − 0	+ 1.3 / + 0.8	0.4 / 1.7	+ 0.8 / − 0	+ 1.7 / + 1.2
1.19 — 1.97		0 / 1.1	+ 0.6 / − 0	+1.1 / +0.6	0 / 1.6	+ 1.0 / − 0	+ 1.6 / + 1.0	0.4 / 2.0	+ 1.0 / − 0	+ 2.0 / + 1.4
1.97 — 3.15		0.1 / 1.3	+0.8 / −0	+1.3 / + 0.7	0.2 / 2.1	+ 1.2 / − 0	+ 2.1 / + 1.4	0.4 / 2.3	+ 1.2 / − 0	+ 2.3 / + 1.6
3.15 — 4.73		0.1 / 1.6	+ 0.9 / − 0	+1.6 / +1.0	0.2 / 2.5	+ 1.4 / − 0	+ 2.5 / + 1.6	0.6 / 2.9	+ 1.4 / − 0	+ 2.9 / + 2.0
4.73 — 7.09		0.2 / 1.9	+ 1.0 / − 0	+1.9 / +1.2	0.2 / 2.8	+ 1.6 / − 0	+ 2.8 / + 1.8	0.9 / 3.5	+ 1.6 / − 0	+ 3.5 / + 2.5
7.09 — 9.85		0.2 / 2.2	+ 1.2 / − 0	+2.2 / +1.4	0.2 / 3.2	+ 1.8 / − 0	+ 3.2 / + 2.0	1.2 / 4.2	+ 1.8 / − 0	+ 4.2 / + 3.0
9.85 — 12.41		0.2 / 2.3	+ 1.2 / − 0	+2.3 / +1.4	0.2 / 3.4	+ 2.0 / − 0	+ 3.4 / + 2.2	1.5 / 4.7	+ 2.0 / − 0	+ 4.7 / + 3.5
12.41 — 15.75		0.2 / 2.6	+ 1.4 / − 0	+2.6 / +1.6	0.3 / 3.9	+ 2.2 / − 0	+ 3.9 / + 2.5	2.3 / 5.9	+ 2.2 / − 0	+ 5.9 / + 4.5
15.75 — 19.69		0.2 / 2.8	+ 1.6 / − 0	+2.8 / +1.8	0.3 / 4.4	+ 2.5 / − 0	+ 4.4 / + 2.8	2.5 / 6.6	+ 2.5 / − 0	+ 6.6 / + 5.0
19.69 — 30.09			+ 2.0 / − 0		0.5 / 5.5	+ 3 / − 0	+ 5.5 / + 3.5	4 / 9	+ 3 / − 0	+ 9 / + 7
30.09 — 41.49			+ 2.5 / − 0		0.5 / 7.0	+ 4 / − 0	+ 7.0 / + 4.5	5 / 11.5	+ 4 / − 0	+11.5 / + 9
41.49 — 56.19			+ 3.0 / − 0		1 / 9	+ 5 / − 0	+ 9 / + 6	7 / 15	+ 5 / − 0	+15 / +12
56.19 — 76.39			+ 4.0 / − 0		1 / 11	+ 6 / − 0	+11 / + 7	10 / 20	+ 6 / − 0	+20 / +16
76.39 — 100.9			+ 5.0 / − 0		1 / 14	+ 8 / − 0	+14 / + 9	12 / 25	+ 8 / − 0	+25 / +20
100.9 — 131.9			+ 6.0 / − 0		2 / 18	+10 / − 0	+18 / +12	15 / 31	+10 / − 0	+31 / +25
131.9 — 171.9			+ 8.0 / − 0		4 / 24	+12 / − 0	+24 / +16	18 / 38	+12 / − 0	+38 / +30
171.9 — 200			+10.0 / − 0		4 / 30	+16 / − 0	+30 / +20	24 / 50	+16 / − 0	+50 / +40

Courtesy of USAS (B4.1)

A APPENDIX

Nominal Size Range Inches		Class FN 1			Class FN 2			Class FN 3			Class FN 4			Class FN 5		
		Limits of Interference	Standard Limits		Limits of Interference	Standard Limits		Limits of Interference	Standard Limits		Limits of Interference	Standard Limits		Limits of Interference	Standard Limits	
Over	To		Hole H6	Shaft		Hole H7	Shaft s6		Hole H7	Shaft t6		Hole H7	Shaft u6		Hole H8	Shaft x7
0	0.12	0.05 / 0.5	+0.25 / − 0	+ 0.5 / + 0.3	0.2 / 0.85	+0.4 / − 0	+ 0.85 / + 0.6				0.3 / 0.95	+ 0.4 / − 0	+ 0.95 / + 0.7	0.3 / 1.3	+ 0.6 / − 0	+ 1.3 / + 0.9
0.12	0.24	0.1 / 0.6	+0.3 / − 0	+ 0.6 / + 0.4	0.2 / 1.0	+0.5 / − 0	+ 1.0 / + 0.7				0.4 / 1.2	+0.5 / − 0	+ 1.2 / + 0.9	0.5 / 1.7	+ 0.7 / − 0	+ 1.7 / + 1.2
0.24	0.40	0.1 / 0.75	+0.4 / − 0	+ 0.75 / + 0.5	0.4 / 1.4	+0.6 / − 0	+ 1.4 / + 1.0				0.6 / 1.6	+0.6 / − 0	+ 1.6 / + 1.2	0.5 / 2.0	+ 0.9 / − 0	+ 2.0 / + 1.4
0.40	0.56	0.1 / 0.8	+0.4 / − 0	+ 0.8 / + 0.5	0.5 / 1.6	+0.7 / − 0	+ 1.6 / + 1.2				0.7 / 1.8	+ 0.7 / − 0	+ 1.8 / + 1.4	0.6 / 2.3	+ 1.0 / − 0	+ 2.3 / + 1.6
0.56	0.71	0.2 / 0.9	+0.4 / − 0	+ 0.9 / + 0.6	0.5 / 1.6	+0.7 / − 0	+ 1.6 / + 1.2				0.7 / 1.8	+ 0.7 / − 0	+ 1.8 / + 1.4	0.8 / 2.5	+ 1.0 / − 0	+ 2.5 / + 1.8
0.71	0.95	0.2 / 1.1	+0.5 / − 0	+ 1.1 / + 0.7	0.6 / 1.9	+0.8 / − 0	+ 1.9 / + 1.4				0.8 / 2.1	+ 0.8 / − 0	+ 2.1 / + 1.6	1.0 / 3.0	+ 1.2 / − 0	+ 3.0 / + 2.2
0.95	1.19	0.3 / 1.2	+0.5 / − 0	+ 1.2 / + 0.8	0.6 / 1.9	+0.8 / − 0	+ 1.9 / + 1.4	0.8 / 2.1	+0.8 / − 0	+ 2.1 / + 1.6	1.0 / 2.3	+ 0·8 / − 0	+ 2.3 / + 1.8	1.3 / 3.3	+ 1.2 / − 0	+ 3.3 / + 2.5
1.19	1.58	0.3 / 1.3	+0.6 / − 0	+ 1.3 / + 0.9	0.8 / 2.4	+1.0 / − 0	+ 2.4 / + 1.8	1.0 / 2.6	+1.0 / − 0	+ 2.6 / + 2.0	1.5 / 3.1	+1.0 / − 0	+ 3.1 / + 2.5	1.4 / 4.0	+ 1.6 / − 0	+ 4.0 / + 3.0
1.58	1.97	0.4 / 1.4	+0.6 / − 0	+ 1.4 / + 1.0	0.8 / 2.4	+1.0 / − 0	+ 2.4 / + 1.8	1.2 / 2.8	+1.0 / − 0	+ 2.8 / + 2.2	1.8 / 3.4	+1.0 / − 0	+ 3.4 / + 2.8	2.4 / 5.0	+ 1.6 / − 0	+ 5.0 / + 4.0
1.97	2.56	0.6 / 1.8	+0.7 / − 0	+ 1.8 / + 1.3	0.8 / 2.7	+1.2 / − 0	+ 2.7 / + 2.0	1.3 / 3.2	+1.2 / − 0	+ 3.2 / + 2.5	2.3 / 4.2	+1.2 / − 0	+ 4.2 / + 3.5	3.2 / 6.2	+ 1.8 / − 0	+ 6.2 / + 5.0
2.56	3.15	0.7 / 1.9	+0.7 / − 0	+ 1.9 / + 1.4	1.0 / 2.9	+1.2 / − 0	+ 2.9 / + 2.2	1.8 / 3.7	+1.2 / − 0	+ 3.7 / + 3.0	2.8 / 4.7	+1.2 / − 0	+ 4.7 / + 4.0	4.2 / 7.2	+ 1.8 / − 0	+ 7.2 / + 6.0
3.15	3.94	0.9 / 2.4	+0.9 / − 0	+ 2.4 / + 1.8	1.4 / 3.7	+1.4 / − 0	+ 3.7 / + 2.8	2.1 / 4.4	+1.4 / − 0	+ 4.4 / + 3.5	3.6 / 5.9	+1.4 / − 0	+ 5.9 / + 5.0	4.8 / 8.4	+ 2.2 / − 0	+ 8.4 / + 7.0
3.94	4.73	1.1 / 2.6	+0.9 / − 0	+ 2.6 / + 2.0	1.6 / 3.9	+1.4 / − 0	+ 3.9 / + 3.0	2.6 / 4.9	+1.4 / − 0	+ 4.9 / + 4.0	4.6 / 6.9	+1.4 / − 0	+ 6.9 / + 6.0	5.8 / 9.4	+ 2.2 / − 0	+ 9.4 / + 8.0
4.73	5.52	1.2 / 2.9	+1.0 / − 0	+ 2.9 / + 2.2	1.9 / 4.5	+1.6 / − 0	+ 4.5 / + 3.5	3.4 / 6.0	+1.6 / − 0	+ 6.0 / + 5.0	5.4 / 8.0	+1.6 / − 0	+ 8.0 / + 7.0	7.5 / 11.6	+ 2.5 / − 0	+11.6 / +10.0
5.52	6.30	1.5 / 3.2	+1.0 / − 0	+ 3.2 / + 2.5	2.4 / 5.0	+1.6 / − 0	+ 5.0 / + 4.0	3.4 / 6.0	+1.6 / − 0	+ 6.0 / + 5.0	5.4 / 8.0	+1.6 / − 0	+ 8.0 / + 7.0	9.5 / 13.6	+ 2.5 / − 0	+13.6 / +12.0
6.30	7.09	1.8 / 3.5	+1.0 / − 0	+ 3.5 / + 2.8	2.9 / 5.5	+1.6 / − 0	+ 5.5 / + 4.5	4.4 / 7.0	+1.6 / − 0	+ 7.0 / + 6.0	6.4 / 9.0	+1.6 / − 0	+ 9.0 / + 8.0	9.5 / 13.6	+ 2.5 / − 0	+13.6 / +12.0
7.09	7.88	1.8 / 3.8	+1.2 / − 0	+ 3.8 / + 3.0	3.2 / 6.2	+1.8 / − 0	+ 6.2 / + 5.0	5.2 / 8.2	+1.8 / − 0	+ 8.2 / + 7.0	7.2 / 10.2	+1.8 / − 0	+10.2 / + 9.0	11.2 / 15.8	+ 2.8 / − 0	+15.8 / +14.0
7.88	8.86	2.3 / 4.3	+1.2 / − 0	+ 4.3 / + 3.5	3.2 / 6.2	+1.8 / − 0	+ 6.2 / + 5.0	5.2 / 8.2	+1.8 / − 0	+ 8.2 / + 7.0	8.2 / 11.2	+1.8 / − 0	+11.2 / +10.0	13.2 / 17.8	+ 2.8 / − 0	+17.8 / +16.0
8.86	9.85	2.3 / 4.3	+1.2 / − 0	+ 4.3 / + 3.5	4.2 / 7.2	+1.8 / − 0	+ 7.2 / + 6.0	6.2 / 9.2	+1.8 / − 0	+ 9.2 / + 8.0	10.2 / 13.2	+1.8 / − 0	+13.2 / +12.0	13.2 / 17.8	+ 2.8 / − 0	+17.8 / +16.0
9.85	11.03	2.8 / 4.9	+1.2 / − 0	+ 4.9 / + 4.0	4.0 / 7.2	+2.0 / − 0	+ 7.2 / + 6.0	7.0 / 10.2	+2.0 / − 0	+10.2 / + 9.0	10.0 / 13.2	+2.0 / − 0	+13.2 / +12.0	15.0 / 20.0	+ 3.0 / − 0	+20.0 / +18.0
11.03	12.41	2.8 / 4.9	+1.2 / − 0	+ 4.9 / + 4.0	5.0 / 8.2	+2.0 / − 0	+ 8.2 / + 7.0	7.0 / 10.2	+2.0 / − 0	+10.2 / + 9.0	12.0 / 15.2	+2.0 / − 0	+15.2 / +14.0	17.0 / 22.0	+ 3.0 / − 0	+22.0 / +20.0
12.41	13.98	3.1 / 5.5	+1.4 / − 0	+ 5.5 / + 4.5	5.8 / 9.4	+2.2 / − 0	+ 9.4 / + 8.0	7.8 / 11.4	+2.2 / − 0	+11.4 / +10.0	13.8 / 17.4	+2.2 / − 0	+17.4 / +16.0	18.5 / 24.2	+ 3.5 / + 0	+24.2 / +22.0
13.98	15.75	3.6 / 6.1	+1.4 / − 0	+ 6.1 / + 5.0	5.8 / 9.4	+2.2 / − 0	+ 9.4 / + 8.0	9.8 / 13.4	+2.2 / − 0	+13.4 / +12.0	15.8 / 19.4	+2.2 / − 0	+19.4 / +18.0	21.5 / 27.2	+ 3.5 / − 0	+27.2 / +25.0
15.75	17.72	4.4 / 7.0	+1.6 / − 0	+ 7.0 / + 6.0	6.5 / 10.6	+2.5 / − 0	+10.6 / + 9.0	9.5 / 13.6	+2.5 / − 0	+13.6 / +12.0	17.5 / 21.6	+2.5 / − 0	+21.6 / +20.0	24.0 / 30.5	+ 4.0 / − 0	+30.5 / +28.0
17.72	19.69	4.4 / 7.0	+1.6 / − 0	+ 7.0 / + 6.0	7.5 / 11.6	+2.5 / − 0	+11.6 / +10.0	11.5 / 15.6	+2.5 / − 0	+15.6 / +14.0	19.5 / 23.6	+2.5 / − 0	+23.6 / +22.0	26.0 / 32.5	+ 4.0 / − 0	+32.5 / +30.0

Courtesy of USAS (B4.1)

Table B.2 Preferred Limits and Fits (Metric)

Dimensions in mm.

BASIC SIZE		LOOSE RUNNING Hole H11	Shaft c11	Fit	FREE RUNNING Hole H9	Shaft d9	Fit	CLOSE RUNNING Hole H8	Shaft f7	Fit	SLIDING Hole H7	Shaft g6	Fit	LOCATIONAL CLEARANCE Hole H7	Shaft h6	Fit
1	MAX	1.060	0.940	0.180	1.025	0.980	0.070	1.014	0.994	0.030	1.010	0.998	0.018	1.010	1.000	0.016
	MIN	1.000	0.880	0.060	1.000	0.955	0.020	1.000	0.984	0.006	1.000	0.992	0.002	1.000	0.994	0.000
1.2	MAX	1.260	1.140	0.180	1.225	1.180	0.070	1.214	1.194	0.030	1.210	1.198	0.018	1.210	1.200	0.016
	MIN	1.200	1.080	0.060	1.200	1.155	0.020	1.200	1.184	0.006	1.200	1.192	0.002	1.200	1.194	0.000
1.6	MAX	1.660	1.540	0.180	1.625	1.580	0.070	1.614	1.594	0.030	1.610	1.598	0.018	1.610	1.600	0.016
	MIN	1.600	1.480	0.060	1.600	1.555	0.020	1.600	1.584	0.006	1.600	1.592	0.002	1.600	1.594	0.000
2	MAX	2.060	1.940	0.180	2.025	1.980	0.070	2.014	1.994	0.030	2.010	1.998	0.018	2.010	2.000	0.016
	MIN	2.000	1.880	0.060	2.000	1.955	0.020	2.000	1.984	0.006	2.000	1.992	0.002	2.000	1.994	0.000
2.5	MAX	2.560	2.440	0.180	2.525	2.480	0.070	2.514	2.494	0.030	2.510	2.498	0.018	2.510	2.500	0.016
	MIN	2.500	2.380	0.060	2.500	2.455	0.020	2.500	2.484	0.006	2.500	2.492	0.002	2.500	2.494	0.000
3	MAX	3.060	2.940	0.180	3.025	2.980	0.070	3.014	2.994	0.030	3.010	2.998	0.018	3.010	3.000	0.016
	MIN	3.000	2.880	0.060	3.000	2.955	0.020	3.000	2.984	0.006	3.000	2.992	0.002	3.000	2.994	0.000
4	MAX	4.075	3.930	0.220	4.030	3.970	0.090	4.018	3.990	0.040	4.012	3.996	0.024	4.012	4.000	0.020
	MIN	4.000	3.855	0.070	4.000	3.940	0.030	4.000	3.978	0.010	4.000	3.988	0.004	4.000	3.992	0.000
5	MAX	5.075	4.930	0.220	5.030	4.970	0.090	5.018	4.990	0.040	5.012	4.996	0.024	5.012	5.000	0.020
	MIN	5.000	4.855	0.070	5.000	4.940	0.030	5.000	4.978	0.010	5.000	4.988	0.004	5.000	4.992	0.000
6	MAX	6.075	5.930	0.220	6.030	5.970	0.090	6.018	5.990	0.040	6.012	5.996	0.024	6.012	6.000	0.020
	MIN	6.000	5.855	0.070	6.000	5.940	0.030	6.000	5.978	0.010	6.000	5.988	0.004	6.000	5.992	0.000
8	MAX	8.090	7.920	0.260	8.036	7.960	0.112	8.022	7.987	0.050	8.015	7.995	0.029	8.015	8.000	0.024
	MIN	8.000	7.830	0.080	8.000	7.924	0.040	8.000	7.972	0.013	8.000	7.986	0.005	8.000	7.991	0.000
10	MAX	10.090	9.920	0.260	10.036	9.960	0.112	10.022	9.987	0.050	10.015	9.995	0.029	10.015	10.000	0.024
	MIN	10.000	9.830	0.080	10.000	9.924	0.040	10.000	9.972	0.013	10.000	9.986	0.005	10.000	9.991	0.000
12	MAX	12.110	11.905	0.315	12.043	11.950	0.136	12.027	11.984	0.061	12.018	11.994	0.035	12.018	12.000	0.029
	MIN	12.000	11.795	0.095	12.000	11.907	0.050	12.000	11.966	0.016	12.000	11.983	0.006	12.000	11.989	0.000
16	MAX	16.110	15.905	0.315	16.043	15.950	0.136	16.027	15.984	0.061	16.018	15.994	0.035	16.018	16.000	0.029
	MIN	16.000	15.795	0.095	16.000	15.907	0.050	16.000	15.966	0.016	16.000	15.983	0.006	16.000	15.989	0.000
20	MAX	20.130	19.890	0.370	20.052	19.935	0.169	20.033	19.980	0.074	20.021	19.993	0.041	20.021	20.000	0.034
	MIN	20.000	19.760	0.110	20.000	19.883	0.065	20.000	19.959	0.020	20.000	19.980	0.007	20.000	19.987	0.000
25	MAX	25.130	24.890	0.370	25.052	24.935	0.169	25.033	24.980	0.074	25.021	24.993	0.041	25.021	25.000	0.034
	MIN	25.000	24.760	0.110	25.000	24.883	0.065	25.000	24.959	0.020	25.000	24.980	0.007	25.000	24.987	0.000
30	MAX	30.130	29.890	0.370	30.052	29.935	0.169	30.033	29.980	0.074	30.021	29.993	0.041	30.021	30.000	0.034
	MIN	30.000	29.760	0.110	30.000	29.883	0.065	30.000	29.959	0.020	30.000	29.980	0.007	30.000	29.987	0.000

Courtesy of ANSI (Reproduced from B4.2)

Dimensions in mm.

BASIC SIZE		LOOSE RUNNING			FREE RUNNING			CLOSE RUNNING			SLIDING			LOCATIONAL CLEARANCE		
		Hole H11	Shaft c11	Fit	Hole H9	Shaft d9	Fit	Hole H8	Shaft f7	Fit	Hole H7	Shaft g6	Fit	Hole H7	Shaft h6	Fit
40	MAX	40.160	39.880	0.440	40.062	39.920	0.204	40.039	39.975	0.089	40.025	39.991	0.050	40.025	40.000	0.041
	MIN	40.000	39.720	0.120	40.000	39.858	0.080	40.000	39.950	0.025	40.000	39.975	0.009	40.000	39.984	0.000
50	MAX	50.160	49.870	0.450	50.062	49.920	0.204	50.039	49.975	0.089	50.025	49.991	0.050	50.025	50.000	0.041
	MIN	50.000	49.710	0.130	50.000	49.858	0.080	50.000	49.950	0.025	50.000	49.975	0.009	50.000	49.984	0.000
60	MAX	60.190	59.860	0.520	60.074	59.900	0.248	60.046	59.970	0.106	60.030	59.990	0.059	60.030	60.000	0.049
	MIN	60.000	59.670	0.140	60.000	59.826	0.100	60.000	59.940	0.030	60.000	59.971	0.010	60.000	59.981	0.000
80	MAX	80.190	79.850	0.530	80.074	79.900	0.248	80.046	79.970	0.106	80.030	79.990	0.059	80.030	80.000	0.049
	MIN	80.000	79.660	0.150	80.000	79.826	0.100	80.000	79.940	0.030	80.000	79.971	0.010	80.000	79.981	0.000
100	MAX	100.220	99.830	0.610	100.087	99.880	0.294	100.054	99.964	0.125	100.035	99.988	0.069	100.035	100.000	0.057
	MIN	100.000	99.610	0.170	100.000	99.793	0.120	100.000	99.929	0.036	100.000	99.966	0.012	100.000	99.978	0.000
120	MAX	120.220	119.820	0.620	120.087	119.880	0.294	120.054	119.964	0.125	120.035	119.988	0.069	120.035	120.000	0.057
	MIN	120.000	119.600	0.180	120.000	119.793	0.120	120.000	119.929	0.036	120.000	119.966	0.012	120.000	119.978	0.000
160	MAX	160.250	159.790	0.710	160.100	159.855	0.345	160.063	159.957	0.146	160.040	159.986	0.079	160.040	160.000	0.065
	MIN	160.000	159.540	0.210	160.000	159.755	0.145	160.000	159.917	0.043	160.000	159.961	0.014	160.000	159.975	0.000
200	MAX	200.290	199.760	0.820	200.115	199.830	0.400	200.072	199.950	0.168	200.046	199.985	0.090	200.046	200.000	0.075
	MIN	200.000	199.470	0.240	200.000	199.715	0.170	200.000	199.904	0.050	200.000	199.956	0.015	200.000	199.971	0.000
250	MAX	250.290	249.720	0.860	250.115	249.830	0.400	250.072	249.950	0.168	250.046	249.985	0.090	250.046	250.000	0.075
	MIN	250.000	249.430	0.280	250.000	249.715	0.170	250.000	249.904	0.050	250.000	249.956	0.015	250.000	249.971	0.000
300	MAX	300.320	299.670	0.970	300.130	299.810	0.450	300.081	299.944	0.189	300.052	299.983	0.101	300.052	300.000	0.084
	MIN	300.000	299.350	0.330	300.000	299.680	0.190	300.000	299.892	0.056	300.000	299.951	0.017	300.000	299.968	0.000
400	MAX	400.360	399.600	1.120	400.140	399.790	0.490	400.089	399.938	0.208	400.057	399.982	0.111	400.057	400.000	0.093
	MIN	400.000	399.240	0.400	400.000	399.650	0.210	400.000	399.881	0.062	400.000	399.946	0.018	400.000	399.964	0.000
500	MAX	500.400	499.520	1.280	500.155	499.770	0.540	500.097	499.932	0.228	500.063	499.980	0.123	500.063	500.000	0.103
	MIN	500.000	499.120	0.480	500.000	499.615	0.230	500.000	499.869	0.068	500.000	499.940	0.020	500.000	499.960	0.000

Courtesy of ANSI (Reproduced from B4.2)

Table B.2 *(continued)*   Dimensions in mm.

BASIC SIZE		LOCATIONAL TRANSN.			LOCATIONAL TRANSN.			LOCATIONAL INTERF.			MEDIUM DRIVE			FORCE		
		Hole H 7	Shaft k6	Fit	Hole H7	Shaft n6	Fit	Hole H7	Shaft p6	Fit	Hole H7	Shaft s6	Fit	Hole H7	Shaft u6	Fit
1	MAX	1.010	1.006	0.010	1.010	1.010	0.006	1.010	1.012	0.004	1.010	1.020	-0.004	1.010	1.024	-0.008
	MIN	1.000	1.000	-0.006	1.000	1.004	-0.010	1.000	1.006	-0.012	1.000	1.014	-0.020	1.000	1.018	-0.024
1.2	MAX	1.210	1.206	0.010	1.210	1.210	0.006	1.210	1.212	0.004	1.210	1.220	-0.004	1.210	1.224	-0.008
	MIN	1.200	1.200	-0.006	1.200	1.204	-0.010	1.200	1.206	-0.012	1.200	1.214	-0.020	1.200	1.218	-0.024
1.6	MAX	1.610	1.606	0.010	1.610	1.610	0.006	1.610	1.612	0.004	1.610	1.620	-0.004	1.610	1.624	-0.008
	MIN	1.600	1.600	-0.006	1.600	1.604	-0.010	1.600	1.606	-0.012	1.600	1.614	-0.020	1.600	1.618	-0.024
2	MAX	2.010	2.006	0.010	2.010	2.010	0.006	2.010	2.012	0.004	2.010	2.020	-0.004	2.010	2.024	-0.008
	MIN	2.000	2.000	-0.006	2.000	2.004	-0.010	2.000	2.006	-0.012	2.000	2.014	-0.020	2.000	2.018	-0.024
2.5	MAX	2.510	2.506	0.010	2.510	2.510	0.006	2.510	2.512	0.004	2.510	2.520	-0.004	2.510	2.524	-0.008
	MIN	2.500	2.500	-0.006	2.500	2.504	-0.010	2.500	2.506	-0.012	2.500	2.514	-0.020	2.500	2.518	-0.024
3	MAX	3.010	3.006	0.010	3.010	3.010	0.006	3.010	3.012	0.004	3.010	3.020	-0.004	3.010	3.024	-0.008
	MIN	3.000	3.000	-0.006	3.000	3.004	-0.010	3.000	3.006	-0.012	3.000	3.014	-0.020	3.000	3.018	-0.024
4	MAX	4.012	4.009	0.011	4.012	4.016	0.004	4.012	4.020	0.000	4.012	4.027	-0.007	4.012	4.031	-0.011
	MIN	4.000	4.001	-0.009	4.000	4.008	-0.016	4.000	4.012	-0.020	4.000	4.019	-0.027	4.000	4.023	-0.031
5	MAX	5.012	5.009	0.011	5.012	5.016	0.004	5.012	5.020	0.000	5.012	5.027	-0.007	5.012	5.031	-0.011
	MIN	5.000	5.001	-0.009	5.000	5.008	-0.016	5.000	5.012	-0.020	5.000	5.019	-0.027	5.000	5.023	-0.031
6	MAX	6.012	6.009	0.011	6.012	6.016	0.004	6.012	6.020	0.000	6.012	6.027	-0.007	6.012	6.031	-0.011
	MIN	6.000	6.001	-0.009	6.000	6.008	-0.016	6.000	6.012	-0.020	6.000	6.019	-0.027	6.000	6.023	-0.031
8	MAX	8.015	8.010	0.014	8.015	8.019	0.005	8.015	8.024	0.000	8.015	8.032	-0.008	8.015	8.037	-0.013
	MIN	8.000	8.001	-0.010	8.000	8.010	-0.019	8.000	8.015	-0.024	8.000	8.023	-0.032	8.000	8.028	-0.037
10	MAX	10.015	10.010	0.014	10.015	10.019	0.005	10.015	10.024	0.000	10.015	10.032	-0.008	10.015	10.037	-0.013
	MIN	10.000	10.001	-0.010	10.000	10.010	-0.019	10.000	10.015	-0.024	10.000	10.023	-0.032	10.000	10.028	-0.037
12	MAX	12.018	12.012	0.017	12.018	12.023	0.006	12.018	12.029	0.000	12.018	12.039	-0.010	12.018	12.044	-0.015
	MIN	12.000	12.001	-0.012	12.000	12.012	-0.023	12.000	12.018	-0.029	12.000	12.028	-0.039	12.000	12.033	-0.044
16	MAX	16.018	16.012	0.017	16.018	16.023	0.006	16.018	16.029	0.000	16.018	16.039	-0.010	16.018	16.044	-0.015
	MIN	16.000	16.001	-0.012	16.000	16.012	-0.023	16.000	16.018	-0.029	16.000	16.028	-0.039	16.000	16.033	-0.044
20	MAX	20.021	20.015	0.019	20.021	20.028	0.006	20.021	20.035	-0.001	20.021	20.048	-0.014	20.021	20.054	-0.020
	MIN	20.000	20.002	-0.015	20.000	20.015	-0.028	20.000	20.022	-0.035	20.000	20.035	-0.048	20.000	20.041	-0.054
25	MAX	25.021	25.015	0.019	25.021	25.028	0.006	25.021	25.035	-0.001	25.021	25.048	-0.014	25.021	25.061	-0.027
	MIN	25.000	25.002	-0.015	25.000	25.015	-0.028	25.000	25.022	-0.035	25.000	25.035	-0.048	25.000	25.048	-0.061
30	MAX	30.021	30.015	0.019	30.021	30.028	0.006	30.021	30.035	-0.001	30.021	30.048	-0.014	30.021	30.061	-0.027
	MIN	30.000	30.002	-0.015	30.000	30.015	-0.028	30.000	30.022	-0.035	30.000	30.035	-0.048	30.000	30.048	-0.061

Courtesy of ANSI (Reproduced from B4.2)

Dimensions in mm

BASIC SIZE		LOCATIONAL TRANSN. Hole H7	Shaft k6	Fit	LOCATIONAL TRANSN. Hole H7	Shaft n6	Fit	LOCATIONAL INTERF. Hole H7	Shaft p6	Fit	MEDIUM DRIVE Hole H7	Shaft s6	Fit	FORCE Hole H7	Shaft u6	Fit
40	MAX	40.025	40.018	0.023	40.025	40.033	0.008	40.025	40.042	-0.001	40.025	40.059	-0.018	40.025	40.076	-0.035
	MIN	40.000	40.002	-0.018	40.000	40.017	-0.033	40.000	40.026	-0.042	40.000	40.043	-0.059	40.000	40.060	-0.076
50	MAX	50.025	50.018	0.023	50.025	50.033	0.008	50.025	50.042	-0.001	50.025	50.059	-0.018	50.025	50.086	-0.045
	MIN	50.000	50.002	-0.018	50.000	50.017	-0.033	50.000	50.026	-0.042	50.000	50.043	-0.059	50.000	50.070	-0.086
60	MAX	60.030	60.021	0.028	60.030	60.039	0.010	60.030	60.051	-0.002	60.030	60.072	-0.023	60.030	60.106	-0.057
	MIN	60.000	60.002	-0.021	60.000	60.020	-0.039	60.000	60.032	-0.051	60.000	60.053	-0.072	60.000	60.087	-0.106
80	MAX	80.030	80.021	0.028	80.030	80.039	0.010	80.030	80.051	-0.002	80.030	80.078	-0.029	80.030	80.121	-0.072
	MIN	80.000	80.002	-0.021	80.000	80.020	-0.039	80.000	80.032	-0.051	80.000	80.059	-0.078	80.000	80.102	-0.121
100	MAX	100.035	100.025	0.032	100.035	100.045	0.012	100.035	100.059	-0.002	100.035	100.093	-0.036	100.035	100.146	-0.089
	MIN	100.000	100.003	-0.025	100.000	100.023	-0.045	100.000	100.037	-0.059	100.000	100.071	-0.093	100.000	100.124	-0.146
120	MAX	120.035	120.025	0.032	120.035	120.045	0.012	120.035	120.059	-0.002	120.035	120.101	-0.044	120.035	120.166	-0.109
	MIN	120.000	120.003	-0.025	120.000	120.023	-0.045	120.000	120.037	-0.059	120.000	120.079	-0.101	120.000	120.144	-0.166
160	MAX	160.040	160.028	0.037	160.040	160.052	0.013	160.040	160.068	-0.003	160.040	160.125	-0.060	160.040	160.215	-0.150
	MIN	160.000	160.003	-0.028	160.000	160.027	-0.052	160.000	160.043	-0.068	160.000	160.100	-0.125	160.000	160.190	-0.215
200	MAX	200.046	200.033	0.042	200.046	200.060	0.015	200.046	200.079	-0.004	200.046	200.151	-0.076	200.046	200.265	-0.190
	MIN	200.000	200.004	-0.033	200.000	200.031	-0.060	200.000	200.050	-0.079	200.000	200.122	-0.151	200.000	200.236	-0.265
250	MAX	250.046	250.033	0.042	250.046	250.060	0.015	250.046	250.079	-0.004	250.046	250.169	-0.094	250.046	250.313	-0.238
	MIN	250.000	250.004	-0.033	250.000	250.031	-0.060	250.000	250.050	-0.079	250.000	250.140	-0.169	250.000	250.284	-0.313
300	MAX	300.052	300.036	0.048	300.052	300.066	0.018	300.052	300.088	-0.004	300.052	300.202	-0.118	300.052	300.382	-0.298
	MIN	300.000	300.004	-0.036	300.000	300.034	-0.066	300.000	300.056	-0.088	300.000	300.170	-0.202	300.000	300.350	-0.382
400	MAX	400.057	400.040	0.053	400.057	400.073	0.020	400.057	400.098	-0.005	400.057	400.244	-0.151	400.057	400.471	-0.378
	MIN	400.000	400.004	-0.040	400.000	400.037	-0.073	400.000	400.062	-0.098	400.000	400.208	-0.244	400.000	400.435	-0.471
500	MAX	500.063	500.045	0.058	500.063	500.080	0.023	500.063	500.108	-0.005	500.063	500.292	-0.189	500.063	500.580	-0.477
	MIN	500.000	500.005	-0.045	500.000	500.040	-0.080	500.000	500.068	-0.108	500.000	500.252	-0.292	500.000	500.540	-0.580

Courtesy of ANSI (Reproduced from B4.2)

Table B.3 American National Standard Unified Inch Screw Threads

Nominal Size		Basic Major Diameter	Threads/in.											Nominal Size
			Series With Graded Pitches			Series With Constant Pitches								
Primary	Secondary		Coarse UNC	Fine UNF	Extra-Fine UNEF	4UN	6UN	8UN	12UN	16UN	20UN	28UN	32UN	
0		0.0600	—	80	—	—	—	—	—	—	—	—	—	0
	1	0.0730	64	72	—	—	—	—	—	—	—	—	—	1
2		0.0860	56	64	—	—	—	—	—	—	—	—	—	2
	3	0.0990	48	56	—	—	—	—	—	—	—	—	—	3
4		0.1120	40	48	—	—	—	—	—	—	—	—	—	4
5		0.1250	40	44	—	—	—	—	—	—	—	—	—	5
6		0.1380	32	40	—	—	—	—	—	—	—	—	UNC	6
8		0.1640	32	36	—	—	—	—	—	—	—	—	UNC	8
10		0.1900	24	32	—	—	—	—	—	—	—	—	UNF	10
	12	0.2160	24	28	32	—	—	—	—	—	—	UNF	UNEF	12
1/4		0.2500	20	28	32	—	—	—	—	—	UNC	UNF	UNEF	1/4
5/16		0.3125	18	24	32	—	—	—	—	—	20	28	UNEF	5/16
3/8		0.3750	16	24	32	—	—	—	—	UNC	20	28	UNEF	3/8
7/16		0.4375	14	20	28	—	—	—	—	16	UNF	UNEF	32	7/16
1/2		0.5000	13	20	28	—	—	—	—	16	UNF	UNEF	32	1/2
9/16		0.5625	12	18	24	—	—	—	UNC	16	20	28	32	9/16
5/8		0.6250	11	18	24	—	—	—	12	16	20	28	32	5/8
	11/16	0.6875	—	—	24	—	—	—	12	16	20	28	32	11/16
3/4		0.7500	10	16	20	—	—	—	12	UNF	UNEF	28	32	3/4
	13/16	0.8125	—	—	20	—	—	—	12	16	UNEF	28	32	13/16
7/8		0.8750	9	14	20	—	—	—	12	16	UNEF	28	32	7/8
	15/16	0.9375	—	—	20	—	—	—	12	16	UNEF	28	32	15/16
1		1.0000	8	12	20	—	—	UNC	UNF	16	UNEF	28	32	1
	1 1/16	1.0625	—	—	18	—	—	8	12	16	20	28	—	1 1/16
1 1/8		1.1250	7	12	18	—	—	8	UNF	16	20	28	—	1 1/8
	1 3/16	1.1875	—	—	18	—	—	8	12	16	20	28	—	1 3/16
1 1/4		1.2500	7	12	18	—	—	8	UNF	16	20	28	—	1 1/4
	1 5/16	1.3125	—	—	18	—	—	8	12	16	20	28	—	1 5/16
1 3/8		1.3750	6	12	18	—	UNC	8	UNF	16	20	28	—	1 3/8
	1 7/16	1.4375	—	—	18	—	6	8	12	16	20	28	—	1 7/16
1 1/2		1.5000	6	12	18	—	UNC	8	UNF	16	20	28	—	1 1/2
	1 9/16	1.5625	—	—	18	—	6	8	12	16	20	—	—	1 9/16
1 5/8		1.6250	—	—	18	—	6	8	12	16	20	—	—	1 5/8
	1 11/16	1.6875	—	—	18	—	6	8	12	16	20	—	—	1 11/16
1 3/4		1.7500	5	—	—	—	6	8	12	16	20	—	—	1 3/4
	1 13/16	1.8125	—	—	—	—	6	8	12	16	20	—	—	1 13/16
1 7/8		1.8750	—	—	—	—	6	8	12	16	20	—	—	1 7/8
	1 15/16	1.9375	—	—	—	—	6	8	12	16	20	—	—	1 15/16
2		2.0000	4 1/2	—	—	—	6	8	12	16	20	—	—	2
	2 1/8	2.1250	—	—	—	—	6	8	12	16	20	—	—	2 1/8
2 1/4		2.2500	4 1/2	—	—	—	6	8	12	16	20	—	—	2 1/4
	2 3/8	2.3750	—	—	—	—	6	8	12	16	20	—	—	2 3/8
2 1/2		2.5000	4	—	—	UNC	6	8	12	16	20	—	—	2 1/2
	2 5/8	2.6250	—	—	—	4	6	8	12	16	20	—	—	2 5/8
2 3/4		2.7500	4	—	—	UNC	6	8	12	16	20	—	—	2 3/4
	2 7/8	2.8750	—	—	—	4	6	8	12	16	20	—	—	2 7/8
3		3.0000	4	—	—	UNC	6	8	12	16	20	—	—	3
	3 1/8	3.1250	—	—	—	4	6	8	12	16	—	—	—	3 1/8
3 1/4		3.2500	4	—	—	UNC	6	8	12	16	—	—	—	3 1/4
	3 3/8	3.3750	—	—	—	4	6	8	12	16	—	—	—	3 3/8
3 1/2		3.5000	4	—	—	UNC	6	8	12	16	—	—	—	3 1/2
	3 5/8	3.6250	—	—	—	4	6	8	12	16	—	—	—	3 5/8
3 3/4		3.7500	4	—	—	UNC	6	8	12	16	—	—	—	3 3/4
	3 7/8	3.8750	—	—	—	4	6	8	12	16	—	—	—	3 7/8
4		4.0000	4	—	—	UNC	6	8	12	16	—	—	—	4
	4 1/8	4.1250	—	—	—	4	6	8	12	16	—	—	—	4 1/8
4 1/4		4.2500	—	—	—	4	6	8	12	16	—	—	—	4 1/4
	4 3/8	4.3750	—	—	—	4	6	8	12	16	—	—	—	4 3/8
4 1/2		4.5000	—	—	—	4	6	8	12	16	—	—	—	4 1/2
	4 5/8	4.6250	—	—	—	4	6	8	12	16	—	—	—	4 5/8
4 3/4		4.7500	—	—	—	4	6	8	12	16	—	—	—	4 3/4
	4 7/8	4.8750	—	—	—	4	6	8	12	16	—	—	—	4 7/8
5		5.0000	—	—	—	4	6	8	12	16	—	—	—	5
	5 1/8	5.1250	—	—	—	4	6	8	12	16	—	—	—	5 1/8
5 1/4		5.2500	—	—	—	4	6	8	12	16	—	—	—	5 1/4
	5 3/8	5.3750	—	—	—	4	6	8	12	16	—	—	—	5 3/8
5 1/2		5.5000	—	—	—	4	6	8	12	16	—	—	—	5 1/2
	5 5/8	5.6250	—	—	—	4	6	8	12	16	—	—	—	5 5/8
5 3/4		5.7500	—	—	—	4	6	8	12	16	—	—	—	5 3/4
	5 7/8	5.8750	—	—	—	4	6	8	12	16	—	—	—	5 7/8
6		6.0000	—	—	—	4	6	8	12	16	—	—	—	6

NOTE:

(1) Series designation shown indicates the UN thread form; however, the UNR thread form may be specified by substituting UNR in place of UN in all designations for external use only.

Courtesy of ANSI (B1.1)

A

APPENDIX

Table B.4   Standard Coarse Pitch Metric Threads

Nominal Size	Pitch	Nominal Size	Pitch
		20	2.5
1.6	0.35	22	2.5[1]
2	0.4	24	3
2.5	0.45	27	3[1]
		30	3.5
3	0.5	36	4
3.5	0.6	42	4.5
4	0.7	48	5
5	0.8	56	5.5
6	1	64	6
8	1.25	72	6
10	1.5	80	6
12	1.75	90	6
14	2	100	6
16	2		

NOTE:
(1) For high strength structural steel fasteners only.
Courtesy of ANSI (B1.13M)

Table B.5   Standard Fine Pitch Metric Threads

Nominal Size	Pitch		Nominal Size	Pitch	
8	1	—	55	1.5	—
10	0.75	1.25	56	—	2
12	1 1.5[1]	1.25	60	1.5	—
14		1.5	64	—	2
			65	1.5	—
15	1	—	70	1.5	—
16	—	1.5	72	—	2
17	1	—	75	1.5	—
18	—	1.5	80	1.5	2
20	1	1.5	85	—	2
22	—	1.5	90	—	2
24	—	2	95	—	2
25	1.5	—	100	—	2
27	—	2	105	—	2
30	1.5	2	110	—	2
33	—	2	120	—	2
35	1.5	—	130	—	2
36	—	2	140	—	2
39	—	2	150	—	2
40	1.5	—	160	—	3
42	—	2	170	—	3
45	1.5	—	180	—	3
48	—	2	190	—	3
50	1.5	—	200	—	3

NOTE:
(1) Only for wheel studs and nuts.
Courtesy of ANSI (B1.13M)

Table B.6   Dimensions of Hex Bolts

Nominal Size or Basic Product Dia (17)		E Body Dia (7) Max	F Width Across Flats (4) Basic	F Max	F Min	G Width Across Corners Max	G Min	H Height Basic	H Max	H Min	R Radius of Fillet Max	R Min	L_T Thread Length 6 in. and shorter Basic	L_T Thread Length over 6 in. Basic
1/4	0.2500	0.260	7/16	0.438	0.425	0.505	0.484	11/64	0.188	0.150	0.03	0.01	0.750	1.000
5/16	0.3125	0.324	1/2	0.500	0.484	0.577	0.552	7/32	0.235	0.195	0.03	0.01	0.875	1.125
3/8	0.3750	0.388	9/16	0.562	0.544	0.650	0.620	1/4	0.268	0.226	0.03	0.01	1.000	1.250
7/16	0.4375	0.452	5/8	0.625	0.603	0.722	0.687	19/64	0.316	0.272	0.03	0.01	1.125	1.375
1/2	0.5000	0.515	3/4	0.750	0.725	0.866	0.826	11/32	0.364	0.302	0.03	0.01	1.250	1.500
5/8	0.6250	0.642	15/16	0.928	0.906	1.083	1.033	27/64	0.444	0.378	0.06	0.02	1.500	1.750
3/4	0.7500	0.768	1 1/8	1.125	1.088	1.299	1.240	1/2	0.524	0.455	0.06	0.02	1.750	2.000
7/8	0.8750	0.895	1 5/16	1.312	1.269	1.516	1.447	37/64	0.604	0.531	0.06	0.02	2.000	2.250
1	1.0000	1.022	1 1/2	1.500	1.450	1.732	1.653	43/64	0.700	0.591	0.09	0.03	2.250	2.500
1 1/8	1.1250	1.149	1 11/16	1.688	1.631	1.949	1.859	3/4	0.780	0.658	0.09	0.03	2.500	2.750
1 1/4	1.2500	1.277	1 7/8	1.875	1.812	2.165	2.066	27/32	0.876	0.749	0.09	0.03	2.750	3.000
1 3/8	1.3750	1.404	2 1/16	2.062	1.994	2.382	2.273	29/32	0.940	0.810	0.09	0.03	3.000	3.250
1 1/2	1.5000	1.531	2 1/4	2.250	2.175	2.598	2.480	1	1.036	0.902	0.09	0.03	3.250	3.500
1 3/4	1.7500	1.785	2 5/8	2.625	2.538	3.031	2.893	1 5/32	1.196	1.054	0.12	0.04	3.750	4.000
2	2.0000	2.039	3	3.000	2.900	3.464	3.306	1 11/32	1.388	1.175	0.12	0.04	4.250	4.500
2 1/4	2.2500	2.305	3 3/8	3.375	3.262	3.897	3.719	1 1/2	1.548	1.327	0.19	0.06	4.750	5.000
2 1/2	2.5000	2.559	3 3/4	3.750	3.625	4.330	4.133	1 21/32	1.708	1.479	0.19	0.06	5.250	5.500
2 3/4	2.7500	2.827	4 1/8	4.125	3.988	4.763	4.546	1 13/16	1.869	1.632	0.19	0.06	5.750	6.000
3	3.0000	3.081	4 1/2	4.500	4.350	5.196	4.959	2	2.060	1.815	0.19	0.06	6.250	6.500
3 1/4	3.2500	3.335	4 7/8	4.875	4.712	5.629	5.372	2 3/16	2.251	1.936	0.19	0.06	6.750	7.000
3 1/2	3.5000	3.589	5 1/4	5.250	5.075	6.062	5.786	2 5/16	2.380	2.057	0.19	0.06	7.250	7.500
3 3/4	3.7500	3.858	5 5/8	5.625	5.437	6.495	6.198	2 1/2	2.572	2.241	0.19	0.06	7.750	8.000
4	4.0000	4.111	6	6.000	5.800	6.928	6.612	2 11/16	2.764	2.424	0.19	0.06	8.250	8.500

Courtesy of ANSI (B18.2.1)

A
APPENDIX

Table B.7 Dimensions of Square Bolts

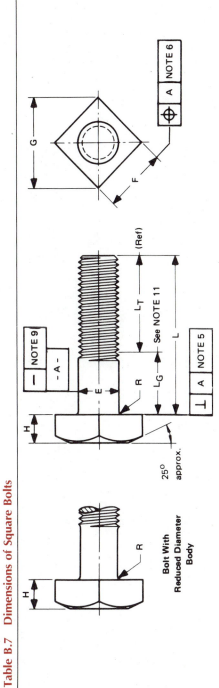

Bolt With
Reduced Diameter
Body

25° approx.

Bolt With Reduced Diameter Body

NOTE 9

NOTE 6

(Ref)

See NOTE 11

NOTE 5

Nominal Size or Basic Product Dia (17)		E Body Dia (7), (14) Max	F Width Across Flats (4)			G Width Across Corners		H Height				R Radius of Fillet			L_T Thread Length For Bolt Lengths (11)		
			Basic	Max	Min	Max	Min	Basic	Max	Min	Max	Min	6 in. and shorter Basic	over 6 in. Basic		
1/4	0.2500	0.260	3/8	0.375	0.362	0.530	0.498	11/64	0.188	0.156	0.03	0.01	0.750	1.000		
5/16	0.3125	0.324	1/2	0.500	0.484	0.707	0.665	13/64	0.220	0.186	0.03	0.01	0.875	1.125		
3/8	0.3750	0.388	9/16	0.562	0.544	0.795	0.747	1/4	0.268	0.232	0.03	0.01	1.000	1.250		
7/16	0.4375	0.452	5/8	0.625	0.603	0.884	0.828	19/64	0.316	0.278	0.03	0.01	1.125	1.375		
1/2	0.5000	0.515	3/4	0.750	0.725	1.061	0.995	21/64	0.348	0.308	0.03	0.01	1.250	1.500		
5/8	0.6250	0.642	15/16	0.938	0.906	1.326	1.244	27/64	0.444	0.400	0.06	0.02	1.500	1.750		
3/4	0.7500	0.768	1 1/8	1.125	1.088	1.591	1.494	1/2	0.524	0.476	0.06	0.02	1.750	2.000		
7/8	0.8750	0.895	1 5/16	1.312	1.269	1.856	1.742	19/32	0.620	0.568	0.06	0.02	2.000	2.250		
1	1.0000	1.022	1 1/2	1.500	1.450	2.121	1.991	21/32	0.684	0.628	0.09	0.03	2.250	2.500		
1 1/8	1.1250	1.149	1 11/16	1.688	1.631	2.386	2.239	3/4	0.780	0.720	0.09	0.03	2.500	2.750		
1 1/4	1.2500	1.277	1 7/8	1.875	1.812	2.652	2.489	27/32	0.876	0.812	0.09	0.03	2.750	3.000		
1 3/8	1.3750	1.404	2 1/16	2.062	1.994	2.917	2.738	29/32	0.940	0.872	0.09	0.03	3.000	3.250		
1 1/2	1.5000	1.531	2 1/4	2.250	2.175	3.182	2.986	1	1.036	0.964	0.09	0.03	3.250	3.500		

Courtesy of ANSI (B18.2.1)

A-28 APPENDIX B TABLES

Table B.8 Dimensions of Hex Nuts and Hex Jam Nuts

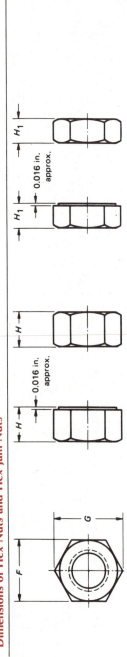

Nominal Size or Basic Major Diam. of Thread		F Width Across Flats			G Width Across Corners		H Thickness Hex Nuts			H₁ Thickness Hex Jam Nuts			Runout of Bearing Face FIM Hex Nuts — Specified Proof Load		Hex Jam Nuts All Strength Levels
		Basic	Max.	Min.	Max.	Min.	Basic	Max.	Min.	Basic	Max.	Min.	Up to 150,000 psi (Max.)	150,000 psi and Greater (Max.)	Max.
1/4	0.2500	7/16	0.438	0.428	0.505	0.488	7/32	0.226	0.212	5/32	0.163	0.150	0.015	0.010	0.015
5/16	0.3125	1/2	0.500	0.489	0.577	0.557	17/64	0.273	0.258	3/16	0.195	0.180	0.016	0.011	0.016
3/8	0.3750	9/16	0.562	0.551	0.650	0.628	21/64	0.337	0.320	7/32	0.227	0.210	0.017	0.012	0.017
7/16	0.4375	11/16	0.688	0.675	0.794	0.768	3/8	0.385	0.365	1/4	0.260	0.240	0.018	0.013	0.018
1/2	0.5000	3/4	0.750	0.736	0.866	0.840	7/16	0.448	0.427	5/16	0.323	0.302	0.019	0.014	0.019
9/16	0.5625	7/8	0.875	0.861	1.010	0.982	31/64	0.496	0.473	5/16	0.324	0.301	0.020	0.015	0.020
5/8	0.6250	15/16	0.938	0.922	1.083	1.051	35/64	0.559	0.535	3/8	0.387	0.363	0.021	0.016	0.021
3/4	0.7500	1 1/8	1.125	1.088	1.299	1.240	41/64	0.665	0.617	27/64	0.446	0.398	0.023	0.018	0.023
7/8	0.8750	1 5/16	1.312	1.269	1.516	1.447	3/4	0.776	0.724	31/64	0.510	0.458	0.025	0.020	0.025
1	1.0000	1 1/2	1.500	1.450	1.732	1.653	55/64	0.887	0.831	35/64	0.575	0.519	0.027	0.022	0.027
1 1/8	1.1250	1 11/16	1.688	1.631	1.949	1.859	31/32	0.999	0.939	39/64	0.639	0.579	0.030	0.025	0.030
1 1/4	1.2500	1 7/8	1.875	1.812	2.165	2.066	1 1/16	1.094	1.030	23/32	0.751	0.687	0.033	0.028	0.033
1 3/8	1.3750	2 1/16	2.062	1.994	2.382	2.273	1 11/64	1.206	1.138	25/32	0.815	0.747	0.036	0.031	0.036
1 1/2	1.5000	2 1/4	2.250	2.175	2.598	2.480	1 9/32	1.317	1.245	27/32	0.880	0.808	0.039	0.034	0.039
See Notes		1			2								3		

Courtesy of ANSI (B18.2.2)

Table B.9   Dimensions of Square Nuts

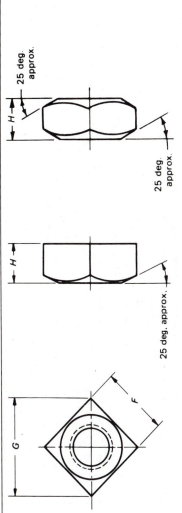

25 deg. approx.

Optional

Nominal Size or Basic Major Diam. of Thread		F Width Across Flats			G Width Across Corners		H Thickness		
		Basic	Max.	Min.	Max.	Min.	Basic	Max.	Min.
1/4	0.2500	7/16	0.438	0.425	0.619	0.554	7/32	0.235	0.203
5/16	0.3125	9/16	0.562	0.547	0.795	0.721	17/64	0.283	0.249
3/8	0.3750	5/8	0.625	0.606	0.884	0.802	21/64	0.346	0.310
7/16	0.4375	3/4	0.750	0.728	1.061	0.970	3/8	0.394	0.356
1/2	0.5000	13/16	0.812	0.788	1.149	1.052	7/16	0.458	0.418
5/8	0.6250	1	1.000	0.969	1.414	1.300	35/64	0.569	0.525
3/4	0.7500	1 1/8	1.125	1.088	1.591	1.464	21/32	0.680	0.632
7/8	0.8750	1 5/16	1.312	1.269	1.856	1.712	49/64	0.792	0.740
1	1.0000	1 1/2	1.500	1.450	2.121	1.961	7/8	0.903	0.847
1 1/8	1.1250	1 11/16	1.688	1.631	2.386	2.209	1	1.030	0.970
1 1/4	1.2500	1 7/8	1.875	1.812	2.652	2.458	1 3/32	1.126	1.062
1 3/8	1.3750	2 1/16	2.062	1.994	2.917	2.708	1 13/64	1.237	1.169
1 1/2	1.5000	2 1/4	2.250	2.175	3.182	2.956	1 5/16	1.348	1.276
See Note		1							

Courtesy of ANSI (B18.2.2)

Table B.10 Dimensions of Hex Cap Screws

Nominal Size or Basic Product Diameter (18)	E Body Diameter (8)		F Width Across Flats			G Width Across Corners (4)		H Height			J Wrenching Height (4)	L_T Thread Length For Screw Lengths (10)		Y Transition Thread Length (10)	Runout of Bearing Surface FIM (5)
	Max	Min	Basic	Max	Min	Max	Min	Basic	Max	Min	Min	6 in. and Shorter Basic	Over 6 in. Basic	Max	Max
1/4 0.2500	0.2500	0.2450	7/16	0.438	0.428	0.505	0.488	5/32	0.163	0.150	0.106	0.750	1.000	0.250	0.010
5/16 0.3125	0.3125	0.3065	1/2	0.500	0.489	0.577	0.557	13/64	0.211	0.195	0.140	0.875	1.125	0.278	0.011
3/8 0.3750	0.3750	0.3690	9/16	0.562	0.551	0.650	0.628	15/64	0.243	0.226	0.160	1.000	1.250	0.312	0.012
7/16 0.4375	0.4375	0.4305	5/8	0.625	0.612	0.722	0.698	9/32	0.291	0.272	0.195	1.125	1.375	0.357	0.013
1/2 0.5000	0.5000	0.4930	3/4	0.750	0.736	0.866	0.840	5/16	0.323	0.302	0.215	1.250	1.500	0.385	0.014
9/16 0.5625	0.5625	0.5545	13/16	0.812	0.798	0.938	0.910	23/64	0.371	0.348	0.250	1.375	1.625	0.417	0.015
5/8 0.6250	0.6250	0.6170	15/16	0.938	0.922	1.083	1.051	25/64	0.403	0.378	0.269	1.500	1.750	0.455	0.017
3/4 0.7500	0.7500	0.7410	1 1/8	1.125	1.100	1.299	1.254	15/32	0.483	0.455	0.324	1.750	2.000	0.500	0.020
7/8 0.8750	0.8750	0.8660	1 5/16	1.312	1.285	1.516	1.465	35/64	0.563	0.531	0.378	2.000	2.250	0.556	0.023
1 1.0000	1.0000	0.9900	1 1/2	1.500	1.469	1.732	1.675	39/64	0.627	0.591	0.416	2.250	2.500	0.625	0.026
1 1/8 1.1250	1.1250	1.1140	1 11/16	1.688	1.631	1.949	1.859	11/16	0.718	0.658	0.461	2.500	2.750	0.714	0.029
1 1/4 1.2500	1.2500	1.2390	1 7/8	1.875	1.812	2.165	2.066	25/32	0.813	0.749	0.530	2.750	3.000	0.714	0.033
1 3/8 1.3750	1.3750	1.3630	2 1/16	2.062	1.994	2.382	2.273	27/32	0.878	0.810	0.569	3.000	3.250	0.833	0.036
1 1/2 1.5000	1.5000	1.4880	2 1/4	2.230	2.175	2.598	2.480	15/16	0.974	0.902	0.640	3.250	3.500	0.833	0.039
1 3/4 1.7500	1.7500	1.7380	2 5/8	2.625	2.538	3.031	2.893	1 3/32	1.134	1.054	0.748	3.750	4.000	1.000	0.046
2 2.0000	2.0000	1.9880	3	3.000	2.900	3.464	3.306	1 7/32	1.263	1.175	0.825	4.250	4.500	1.111	0.052
2 1/4 2.2500	2.2500	2.2380	3 3/8	3.375	3.262	3.897	3.719	1 3/8	1.423	1.327	0.933	4.750	5.000	1.111	0.059
2 1/2 2.5000	2.5000	2.4880	3 3/4	3.750	3.625	4.330	4.133	1 17/32	1.583	1.479	1.042	5.250	5.500	1.250	0.065
2 3/4 2.7500	2.7500	2.7380	4 1/8	4.125	3.988	4.763	4.546	1 11/16	1.744	1.632	1.151	5.750	6.000	1.250	0.072
3 3.0000	3.0000	2.9880	4 1/2	4.500	4.350	5.196	4.959	1 7/8	1.935	1.815	1.290	6.250	6.500	1.250	0.079

Courtesy of ANSI (B18.2.1)

Table B.11   Dimensions of Slotted Flat Countersunk Head Cap Screws

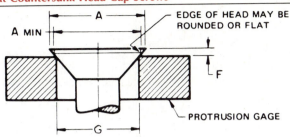

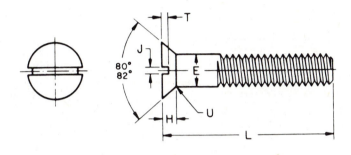

Nominal Size[1] or Basic Screw Diameter		E Body Diameter		A Head Diameter		H Head Height	J Slot Width		T Slot Depth		U Fillet Radius	F Protrusion Above Gaging Diameter		G Gaging Diameter
		Max	Min	Max, Edge Sharp	Min, Edge Rounded or Flat	Ref	Max	Min	Max	Min	Max	Max	Min	
1/4	0.2500	0.2500	0.2450	0.500	0.452	0.140	0.075	0.064	0.068	0.045	0.100	0.046	0.030	0.424
5/16	0.3125	0.3125	0.3070	0.625	0.567	0.177	0.084	0.072	0.086	0.057	0.125	0.053	0.035	0.538
3/8	0.3750	0.3750	0.3690	0.750	0.682	0.210	0.094	0.081	0.103	0.068	0.150	0.060	0.040	0.651
7/16	0.4375	0.4375	0.4310	0.812	0.736	0.210	0.094	0.081	0.103	0.068	0.175	0.065	0.044	0.703
1/2	0.5000	0.5000	0.4930	0.875	0.791	0.210	0.106	0.091	0.103	0.068	0.200	0.071	0.049	0.756
9/16	0.5625	0.5625	0.5550	1.000	0.906	0.244	0.118	0.102	0.120	0.080	0.225	0.078	0.054	0.869
5/8	0.6250	0.6250	0.6170	1.125	1.020	0.281	0.133	0.116	0.137	0.091	0.250	0.085	0.058	0.982
3/4	0.7500	0.7500	0.7420	1.375	1.251	0.352	0.149	0.131	0.171	0.115	0.300	0.099	0.068	1.208
7/8	0.8750	0.8750	0.8660	1.625	1.480	0.423	0.167	0.147	0.206	0.138	0.350	0.113	0.077	1.435
1	1.0000	1.0000	0.9900	1.875	1.711	0.494	0.188	0.166	0.240	0.162	0.400	0.127	0.087	1.661
1 1/8	1.1250	1.1250	1.1140	2.062	1.880	0.529	0.196	0.178	0.257	0.173	0.450	0.141	0.096	1.826
1 1/4	1.2500	1.2500	1.2390	2.312	2.110	0.600	0.211	0.193	0.291	0.197	0.500	0.155	0.105	2.052
1 3/8	1.3750	1.3750	1.3630	2.562	2.340	0.665	0.226	0.208	0.326	0.220	0.550	0.169	0.115	2.279
1 1/2	1.5000	1.5000	1.4880	2.812	2.570	0.742	0.258	0.240	0.360	0.244	0.600	0.183	0.124	2.505

[1] Where specifying nominal size in decimals, zeros preceding decimal and in the fourth decimal place shall be omitted.
Courtesy of ANSI (B18.6.2)

Table B.12 Dimensions of Slotted Round Head Cap Screws

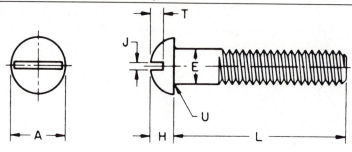

Nominal Size[1] or Basic Screw Diameter		E		A		H		J		T		U	
		Body Diameter		Head Diameter		Head Height		Slot Width		Slot Depth		Fillet Radius	
		Max	Min	Max	Min	Max	Min	Max	Min	Max	Min	Max	Min
1/4	0.2500	0.2500	0.2450	0.437	0.418	0.191	0.175	0.075	0.064	0.117	0.097	0.031	0.016
5/16	0.3125	0.3125	0.3070	0.562	0.540	0.245	0.226	0.084	0.072	0.151	0.126	0.031	0.016
3/8	0.3750	0.3750	0.3690	0.625	0.603	0.273	0.252	0.094	0.081	0.168	0.138	0.031	0.016
7/16	0.4375	0.4375	0.4310	0.750	0.725	0.328	0.302	0.094	0.081	0.202	0.167	0.047	0.016
1/2	0.5000	0.5000	0.4930	0.812	0.786	0.354	0.327	0.106	0.091	0.218	0.178	0.047	0.016
9/16	0.5625	0.5625	0.5550	0.937	0.909	0.409	0.378	0.118	0.102	0.252	0.207	0.047	0.016
5/8	0.6250	0.6250	0.6170	1.000	0.970	0.437	0.405	0.133	0.116	0.270	0.220	0.062	0.031
3/4	0.7500	0.7500	0.7420	1.250	1.215	0.546	0.507	0.149	0.131	0.338	0.278	0.062	0.031

[1] Where specifying nominal size in decimals, zeros preceding decimal and in the fourth decimal place shall be omitted.
Courtesy of ANSI (B18.6.2)

Table B.13   Dimensions of Slotted Fillister Head Cap Screws

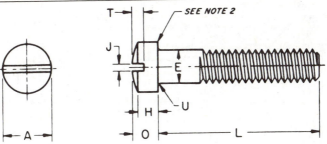

Nominal Size[1] or Basic Screw Diameter		E Body Diameter		A Head Diameter		H Head Side Height		O Total Head Height		J Slot Width		T Slot Depth		U Fillet Radius	
		Max	Min	Max	Min	Max	Min	Max	Min	Max	Min	Max	Min	Max	Min
1/4	0.2500	0.2500	0.2450	0.375	0.363	0.172	0.157	0.216	0.194	0.075	0.064	0.097	0.077	0.031	0.016
5/16	0.3125	0.3125	0.3070	0.437	0.424	0.203	0.186	0.253	0.230	0.084	0.072	0.115	0.090	0.031	0.016
3/8	0.3750	0.3750	0.3690	0.562	0.547	0.250	0.229	0.314	0.284	0.094	0.081	0.142	0.112	0.031	0.016
7/16	0.4375	0.4375	0.4310	0.625	0.608	0.297	0.274	0.368	0.336	0.094	0.081	0.168	0.133	0.047	0.016
1/2	0.5000	0.5000	0.4930	0.750	0.731	0.328	0.301	0.413	0.376	0.106	0.091	0.193	0.153	0.047	0.016
9/16	0.5625	0.5625	0.5550	0.812	0.792	0.375	0.346	0.467	0.427	0.118	0.102	0.213	0.168	0.047	0.016
5/8	0.6250	0.6250	0.6170	0.875	0.853	0.422	0.391	0.521	0.478	0.133	0.116	0.239	0.189	0.062	0.031
3/4	0.7500	0.7500	0.7420	1.000	0.976	0.500	0.466	0.612	0.566	0.149	0.131	0.283	0.223	0.062	0.031
7/8	0.8750	0.8750	0.8660	1.125	1.098	0.594	0.556	0.720	0.668	0.167	0.147	0.334	0.264	0.062	0.031
1	1.0000	1.0000	0.9900	1.312	1.282	0.656	0.612	0.803	0.743	0.188	0.166	0.371	0.291	0.062	0.031

[1] Where specifying nominal size in decimals, zeros preceding decimal and in the fourth decimal place shall be omitted.
[2] A slight rounding of the edges at periphery of head shall be permissible provided the diameter of the bearing circle is equal to no less than 90 per cent of the specified minimum head diameter.
Courtesy of ANSI (B18.6.2)

Table B.14 Dimensions of Hex Head Machine Screws

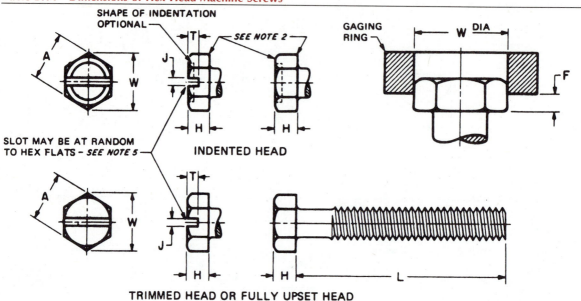

Nominal Size[1] or Basic Screw Diameter	A Regular Head[7] Width Across Flats		W Regular Head[7] Across Corners	A Large Head[7,8] Width Across Flats		W Large Head[7,8] Across Corners	H Head Height		J Slot Width		T Slot Depth		F Protrusion Beyond Gaging Ring
	Max	Min	Min	Max	Min	Min	Max	Min	Max	Min	Max	Min	Min
1 0.0730	0.125	0.120	0.134	—	—	—	0.044	0.036	—	—	—	—	0.022
2 0.0860	0.125	0.120	0.134	—	—	—	0.050	0.040	—	—	—	—	0.024
3 0.0990	0.188	0.181	0.202	—	—	—	0.055	0.044	—	—	—	—	0.026
4 0.1120	0.188	0.181	0.202	0.219	0.213	0.238	0.060	0.049	0.039	0.031	0.036	0.025	0.029
5 0.1250	0.188	0.181	0.202	0.250	0.244	0.272	0.070	0.058	0.043	0.035	0.042	0.030	0.035
6 0.1380	0.250	0.244	0.272	—	—	—	0.093	0.080	0.048	0.039	0.046	0.033	0.048
8 0.1640	0.250	0.244	0.272	0.312	0.305	0.340	0.110	0.096	0.054	0.045	0.066	0.052	0.058
10 0.1900	0.312	0.305	0.340	—	—	—	0.120	0.105	0.060	0.050	0.072	0.057	0.063
12 0.2160	0.312	0.305	0.340	0.375	0.367	0.409	0.155	0.139	0.067	0.056	0.093	0.077	0.083
1/4 0.2500	0.375	0.367	0.409	0.438	0.428	0.477	0.190	0.172	0.075	0.064	0.101	0.083	0.103
5/16 0.3125	0.500	0.489	0.545	—	—	—	0.230	0.208	0.084	0.072	0.122	0.100	0.125
3/8 0.3750	0.562	0.551	0.614	—	—	—	0.295	0.270	0.094	0.081	0.156	0.131	0.162

Courtesy of ANSI (B18.6.3)

Table B.15   Dimensions of Slotted Flat Countersunk Head Machine Screws

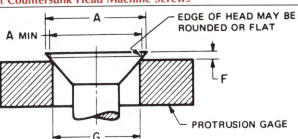

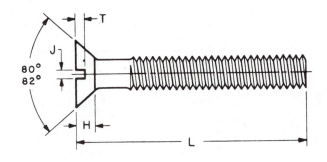

Nominal Size[1] or Basic Screw Diameter		L — These Lengths or Shorter are Undercut.	A — Head Diameter		H — Head Height	J — Slot Width		T — Slot Depth		F — Protrusion Above Gaging Diameter		G — Gaging Diameter
			Max, Edge Sharp	Min, Edge Rounded or Flat	Ref	Max	Min	Max	Min	Max	Min	
0000	0.0210	—	0.043	0.037	0.011	0.008	0.004	0.007	0.003	*	*	*
000	0.0340	—	0.064	0.058	0.016	0.011	0.007	0.009	0.005	*	*	*
00	0.0470	—	0.093	0.085	0.028	0.017	0.010	0.014	0.009	*	*	*
0	0.0600	1/8	0.119	0.099	0.035	0.023	0.016	0.015	0.010	0.026	0.016	0.078
1	0.0730	1/8	0.146	0.123	0.043	0.026	0.019	0.019	0.012	0.028	0.016	0.101
2	0.0860	1/8	0.172	0.147	0.051	0.031	0.023	0.023	0.015	0.029	0.017	0.124
3	0.0990	1/8	0.199	0.171	0.059	0.035	0.027	0.027	0.017	0.031	0.018	0.148
4	0.1120	3/16	0.225	0.195	0.067	0.039	0.031	0.030	0.020	0.032	0.019	0.172
5	0.1250	3/16	0.252	0.220	0.075	0.043	0.035	0.034	0.022	0.034	0.020	0.196
6	0.1380	3/16	0.279	0.244	0.083	0.048	0.039	0.038	0.024	0.036	0.021	0.220
8	0.1640	1/4	0.332	0.292	0.100	0.054	0.045	0.045	0.029	0.039	0.023	0.267
10	0.1900	5/16	0.385	0.340	0.116	0.060	0.050	0.053	0.034	0.042	0.025	0.313
12	0.2160	3/8	0.438	0.389	0.132	0.067	0.056	0.060	0.039	0.045	0.027	0.362
1/4	0.2500	7/16	0.507	0.452	0.153	0.075	0.064	0.070	0.046	0.050	0.029	0.424
5/16	0.3125	1/2	0.635	0.568	0.191	0.084	0.072	0.088	0.058	0.057	0.034	0.539
3/8	0.3750	9/16	0.762	0.685	0.230	0.094	0.081	0.106	0.070	0.065	0.039	0.653
7/16	0.4375	5/8	0.812	0.723	0.223	0.094	0.081	0.103	0.066	0.073	0.044	0.690
1/2	0.5000	3/4	0.875	0.775	0.223	0.106	0.091	0.103	0.065	0.081	0.049	0.739
9/16	0.5625	—	1.000	0.889	0.260	0.118	0.102	0.120	0.077	0.089	0.053	0.851
5/8	0.6250	—	1.125	1.002	0.298	0.133	0.116	0.137	0.088	0.097	0.058	0.962
3/4	0.7500	—	1.375	1.230	0.372	0.149	0.131	0.171	0.111	0.112	0.067	1.186

[1] Where specifying nominal size in decimals, zeros preceding decimal and in the fourth decimal place shall be omitted.

*Not practical to gage.

Courtesy of ANSI (B18.6.3)

Table B.16   Dimensions of Slotted Fillister Head Machine Screws

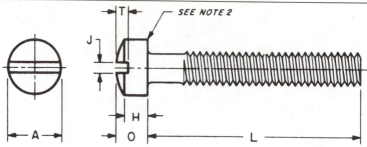

Nominal Size[1] or Basic Screw Diameter		A Head Diameter		H Head Side Height		O Total Head Height		J Slot Width		T Slot Depth	
		Max	Min	Max	Min	Max	Min	Max	Min	Max	Min
0000	0.0210	0.038	0.032	0.019	0.011	0.025	0.015	0.008	0.004	0.012	0.006
000	0.0340	0.059	0.053	0.029	0.021	0.035	0.027	0.012	0.006	0.017	0.011
00	0.0470	0.082	0.072	0.037	0.028	0.047	0.039	0.017	0.010	0.022	0.015
0	0.0600	0.096	0.083	0.043	0.038	0.055	0.047	0.023	0.016	0.025	0.015
1	0.0730	0.118	0.104	0.053	0.045	0.066	0.058	0.026	0.019	0.031	0.020
2	0.0860	0.140	0.124	0.062	0.053	0.083	0.066	0.031	0.023	0.037	0.025
3	0.0990	0.161	0.145	0.070	0.061	0.095	0.077	0.035	0.027	0.043	0.030
4	0.1120	0.183	0.166	0.079	0.069	0.107	0.088	0.039	0.031	0.048	0.035
5	0.1250	0.205	0.187	0.088	0.078	0.120	0.100	0.043	0.035	0.054	0.040
6	0.1380	0.226	0.208	0.096	0.086	0.132	0.111	0.048	0.039	0.060	0.045
8	0.1640	0.270	0.250	0.113	0.102	0.156	0.133	0.054	0.045	0.071	0.054
10	0.1900	0.313	0.292	0.130	0.118	0.180	0.156	0.060	0.050	0.083	0.064
12	0.2160	0.357	0.334	0.148	0.134	0.205	0.178	0.067	0.056	0.094	0.074
1/4	0.2500	0.414	0.389	0.170	0.155	0.237	0.207	0.075	0.064	0.109	0.087
5/16	0.3125	0.518	0.490	0.211	0.194	0.295	0.262	0.084	0.072	0.137	0.110
3/8	0.3750	0.622	0.590	0.253	0.233	0.355	0.315	0.094	0.081	0.164	0.133
7/16	0.4375	0.625	0.589	0.265	0.242	0.368	0.321	0.094	0.081	0.170	0.135
1/2	0.5000	0.750	0.710	0.297	0.273	0.412	0.362	0.106	0.091	0.190	0.151
9/16	0.5625	0.812	0.768	0.336	0.308	0.466	0.410	0.118	0.102	0.214	0.172
5/8	0.6250	0.875	0.827	0.375	0.345	0.521	0.461	0.133	0.116	0.240	0.193
3/4	0.7500	1.000	0.945	0.441	0.406	0.612	0.542	0.149	0.131	0.281	0.226

[1] Where specifying nominal size in decimals, zeros preceding decimal and in the fourth decimal place shall be omitted.

[2] A slight rounding of the edges at periphery of head shall be permissible provided the diameter of the bearing circle is equal to no less than 90 per cent of the specified minimum head diameter.

Courtesy of ANSI (B18.6.3)

Table B.17   Dimensions of Square and Hex Machine Screw Nuts

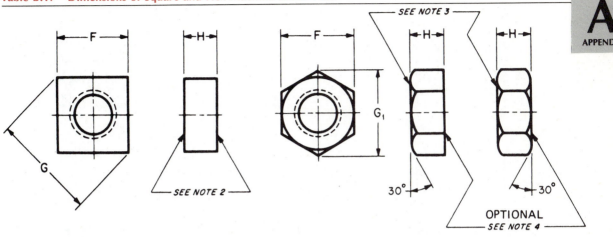

Nominal Size[1] or Basic Thread Diameter		F Width Across Flats			G Width Across Corners Square		G₁ Width Across Corners Hex		H Thickness	
		Basic	Max	Min	Max	Min	Max	Min	Max	Min
0	0.0600	5/32	0.156	0.150	0.221	0.206	0.180	0.171	0.050	0.043
1	0.0730	5/32	0.156	0.150	0.221	0.206	0.180	0.171	0.050	0.043
2	0.0860	3/16	0.188	0.180	0.265	0.247	0.217	0.205	0.066	0.057
3	0.0990	3/16	0.188	0.180	0.265	0.247	0.217	0.205	0.066	0.057
4	0.1120	1/4	0.250	0.241	0.354	0.331	0.289	0.275	0.098	0.087
5	0.1250	5/16	0.312	0.302	0.442	0.415	0.361	0.344	0.114	0.102
6	0.1380	5/16	0.312	0.302	0.442	0.415	0.361	0.344	0.114	0.102
8	0.1640	11/32	0.344	0.332	0.486	0.456	0.397	0.378	0.130	0.117
10	0.1900	3/8	0.375	0.362	0.530	0.497	0.433	0.413	0.130	0.117
12	0.2160	7/16	0.438	0.423	0.619	0.581	0.505	0.482	0.161	0.148
1/4	0.2500	7/16	0.438	0.423	0.619	0.581	0.505	0.482	0.193	0.178
5/16	0.3125	9/16	0.562	0.545	0.795	0.748	0.650	0.621	0.225	0.208
3/8	0.3750	5/8	0.625	0.607	0.884	0.833	0.722	0.692	0.257	0.239

[1] Where specifying nominal size in decimals, zeros preceding decimal and in the fourth decimal place shall be omitted.

[2] Square machine screw nuts shall have tops and bottoms flat, without chamfer. The bearing surface shall be perpendicular to the axis of the threaded hole pitch cylinder within a tolerance of 4 deg.

[3] Hexagon machine screw nuts shall have tops flat with chamfered corners. Diameter of top circle shall be equal to the maximum width across flats within a tolerance of minus 15 per cent. The bearing surface shall be perpendicular to the axis of the threaded hole pitch cylinder within a tolerance of 4 deg.

[4] Bottoms of hexagon machine screw nuts are normally flat, but for special purposes may be chamfered, if so specified by purchaser.

Courtesy of ANSI (B18.6.3)

Table B.18   Dimensions of Slotted Headless Set Screws

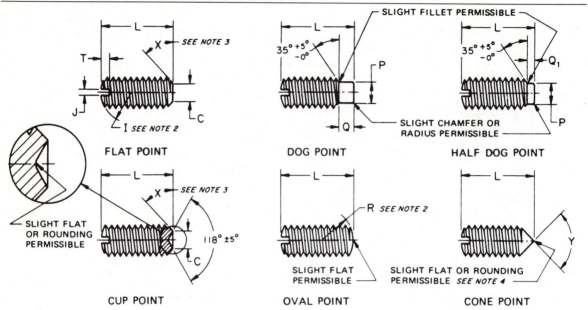

FLAT POINT · DOG POINT · HALF DOG POINT · CUP POINT · OVAL POINT · CONE POINT

Nominal Size[1] or Basic Screw Diameter		I Crown Radius	J Slot Width		T Slot Depth		C Cup and Flat Point Diameters		P Dog Point Diameters		Point Length				R Oval Point Radius	Y Cone Point Angle 90° ±2° For These Nominal Lengths or Longer; 118° ±2° For Shorter Screws
											Dog		Half Dog			
		Basic	Max	Min	Max	Min	Max	Min	Max	Min	Max	Min	Max	Min	Basic	
0	0.0600	0.060	0.014	0.010	0.020	0.016	0.033	0.027	0.040	0.037	0.032	0.028	0.017	0.013	0.045	5/64
1	0.0730	0.073	0.016	0.012	0.020	0.016	0.040	0.033	0.049	0.045	0.040	0.036	0.021	0.017	0.055	3/32
2	0.0860	0.086	0.018	0.014	0.025	0.019	0.047	0.039	0.057	0.053	0.046	0.042	0.024	0.020	0.064	7/64
3	0.0990	0.099	0.020	0.016	0.028	0.022	0.054	0.045	0.066	0.062	0.052	0.048	0.027	0.023	0.074	1/8
4	0.1120	0.112	0.024	0.018	0.031	0.025	0.061	0.051	0.075	0.070	0.058	0.054	0.030	0.026	0.084	5/32
5	0.1250	0.125	0.026	0.020	0.036	0.026	0.067	0.057	0.083	0.078	0.063	0.057	0.033	0.027	0.094	3/16
6	0.1380	0.138	0.028	0.022	0.040	0.030	0.074	0.064	0.092	0.087	0.073	0.067	0.038	0.032	0.104	3/16
8	0.1640	0.164	0.032	0.026	0.046	0.036	0.087	0.076	0.109	0.103	0.083	0.077	0.043	0.037	0.123	1/4
10	0.1900	0.190	0.035	0.029	0.053	0.043	0.102	0.088	0.127	0.120	0.095	0.085	0.050	0.040	0.142	1/4
12	0.2160	0.216	0.042	0.035	0.061	0.051	0.115	0.101	0.144	0.137	0.115	0.105	0.060	0.050	0.162	5/16
1/4	0.2500	0.250	0.049	0.041	0.068	0.058	0.132	0.118	0.156	0.149	0.130	0.120	0.068	0.058	0.188	5/16
5/16	0.3125	0.312	0.055	0.047	0.083	0.073	0.172	0.156	0.203	0.195	0.161	0.151	0.083	0.073	0.234	3/8
3/8	0.3750	0.375	0.068	0.060	0.099	0.089	0.212	0.194	0.250	0.241	0.193	0.183	0.099	0.089	0.281	7/16
7/16	0.4375	0.438	0.076	0.068	0.114	0.104	0.252	0.232	0.297	0.287	0.224	0.214	0.114	0.104	0.328	1/2
1/2	0.5000	0.500	0.086	0.076	0.130	0.120	0.291	0.270	0.344	0.334	0.255	0.245	0.130	0.120	0.375	9/16
9/16	0.5625	0.562	0.096	0.086	0.146	0.136	0.332	0.309	0.391	0.379	0.287	0.275	0.146	0.134	0.422	5/8
5/8	0.6250	0.625	0.107	0.097	0.161	0.151	0.371	0.347	0.469	0.456	0.321	0.305	0.164	0.148	0.469	3/4
3/4	0.7500	0.750	0.134	0.124	0.193	0.183	0.450	0.425	0.562	0.549	0.383	0.367	0.196	0.180	0.562	7/8

[1] Where specifying nominal size in decimals, zeros preceding decimal and in the fourth decimal place shall be omitted.

[2] Tolerance on radius for nominal sizes up to and including 5 (0.125 in.) shall be plus 0.015 in. and minus 0.000, and for larger sizes, plus 0.031 in. and minus 0.000. Slotted ends on screws may be flat at option of manufacturer.

[3] Point angle X shall be 45° plus 5°, minus 0°, for screws of nominal lengths equal to or longer than those listed in Column Y, and 30° minimum for screws of shorter nominal lengths.

[4] The extent of rounding or flat at apex of cone point shall not exceed an amount equivalent to 10 per cent of the basic screw diameter.

Courtesy of ANSI (B18.6.2)

Table B.19   Dimensions of Square Head Set Screws

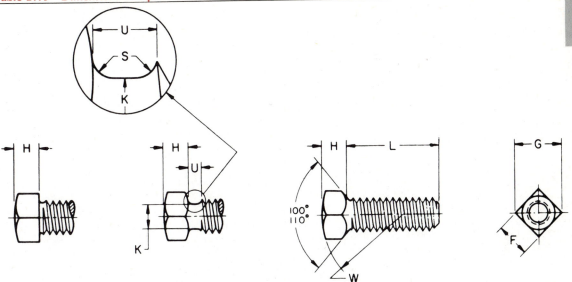

OPTIONAL HEAD CONSTRUCTIONS

Nominal Size[1] or Basic Screw Diameter		F Width Across Flats		G Width Across Corners		H Head Height		K Neck Relief Diameter		S Neck Relief Fillet Radius	U Neck Relief Width	W Head Radius
		Max	Min	Max	Min	Max	Min	Max	Min	Max	Min	Min
10	0.1900	0.188	0.180	0.265	0.247	0.148	0.134	0.145	0.140	0.027	0.083	0.48
1/4	0.2500	0.250	0.241	0.354	0.331	0.196	0.178	0.185	0.170	0.032	0.100	0.62
5/16	0.3125	0.312	0.302	0.442	0.415	0.245	0.224	0.240	0.225	0.036	0.111	0.78
3/8	0.3750	0.375	0.362	0.530	0.497	0.293	0.270	0.294	0.279	0.041	0.125	0.94
7/16	0.4375	0.438	0.423	0.619	0.581	0.341	0.315	0.345	0.330	0.046	0.143	1.09
1/2	0.5000	0.500	0.484	0.707	0.665	0.389	0.361	0.400	0.385	0.050	0.154	1.25
9/16	0.5625	0.562	0.545	0.795	0.748	0.437	0.407	0.454	0.439	0.054	0.167	1.41
5/8	0.6250	0.625	0.606	0.884	0.833	0.485	0.452	0.507	0.492	0.059	0.182	1.56
3/4	0.7500	0.750	0.729	1.060	1.001	0.582	0.544	0.620	0.605	0.065	0.200	1.88
7/8	0.8750	0.875	0.852	1.237	1.170	0.678	0.635	0.731	0.716	0.072	0.222	2.19
1	1.0000	1.000	0.974	1.414	1.337	0.774	0.726	0.838	0.823	0.081	0.250	2.50
1 1/8	1.1250	1.125	1.096	1.591	1.505	0.870	0.817	0.939	0.914	0.092	0.283	2.81
1 1/4	1.2500	1.250	1.219	1.768	1.674	0.966	0.908	1.064	1.039	0.092	0.283	3.12
1 3/8	1.3750	1.375	1.342	1.945	1.843	1.063	1.000	1.159	1.134	0.109	0.333	3.44
1 1/2	1.5000	1.500	1.464	2.121	2.010	1.159	1.091	1.284	1.259	0.109	0.333	3.75

[1] Where specifying nominal size in decimals, zeros preceding decimal and in the fourth decimal place shall be omitted.
Courtesy of ANSI (B18.6.2)

Table B.20   Shoulder Screws

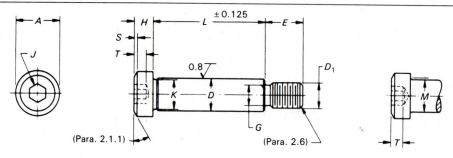

(Para. 2.1.1)   (Para. 2.6)

Nominal Screw Size or Basic Shoulder Diameter	D Shoulder Diameter		A Head Diameter		H Head Height		S Chamfer or Radius	J Hexagon Socket Size	T Key Engagement	M Head Fillet Extension Diameter	R Shoulder Neck Fillet Radius
	Max.	Min.	Max.	Min.	Max.	Min.	Max.	Nom.	Min.	Max.	Min.
6.5	6.487	6.451	10.00	9.78	4.50	4.32	0.6	3	2.4	7.5	.2
8.0	7.987	7.951	13.00	12.73	5.50	5.32	0.8	4	3.3	9.2	.4
10.0	9.987	9.951	16.00	15.73	7.00	6.78	1.0	5	4.2	11.2	.4
13.0	12.984	12.941	18.00	17.73	9.00	8.78	1.2	6	4.9	15.2	.6
16.0	15.984	15.941	24.00	23.67	11.00	10.73	1.6	8	6.6	18.2	.6
20.0	19.980	19.928	30.00	29.67	14.00	13.73	2.0	10	8.8	22.4	.8
25.0	24.980	24.928	36.00	35.61	16.00	15.73	2.4	12	10.0	27.4	.8
See Para.	2.3.1		2.1.2				2.1.1	2.2.1	2.2.2	2.1.5	2.1.5

Courtesy of ASME/ANSI (B18.3.3M)

Table B.21   Dimensions of Preferred Sizes of Type A Plain Washers

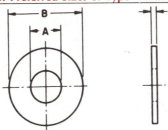

Nominal Washer Size			Inside Diameter A			Outside Diameter B			Thickness C		
			Basic	Tolerance Plus	Minus	Basic	Tolerance Plus	Minus	Basic	Max	Min
—	—		0.078	0.000	0.005	0.188	0.000	0.005	0.020	0.025	0.016
—	—		0.094	0.000	0.005	0.250	0.000	0.005	0.020	0.025	0.016
—	—		0.125	0.008	0.005	0.312	0.008	0.005	0.032	0.040	0.025
No. 6	0.138		0.156	0.008	0.005	0.375	0.015	0.005	0.049	0.065	0.036
No. 8	0.164		0.188	0.008	0.005	0.438	0.015	0.005	0.049	0.065	0.036
No. 10	0.190		0.219	0.008	0.005	0.500	0.015	0.005	0.049	0.065	0.036
3/16	0.188		0.250	0.015	0.005	0.562	0.015	0.005	0.049	0.065	0.036
No. 12	0.216		0.250	0.015	0.005	0.562	0.015	0.005	0.065	0.080	0.051
1/4	0.250	N	0.281	0.015	0.005	0.625	0.015	0.005	0.065	0.080	0.051
1/4	0.250	W	0.312	0.015	0.005	0.734	0.015	0.007	0.065	0.080	0.051
5/16	0.312	N	0.344	0.015	0.005	0.688	0.015	0.007	0.065	0.080	0.051
5/16	0.312	W	0.375	0.015	0.005	0.875	0.030	0.007	0.083	0.104	0.064
3/8	0.375	N	0.406	0.015	0.005	0.812	0.015	0.007	0.065	0.080	0.051
3/8	0.375	W	0.438	0.015	0.005	1.000	0.030	0.007	0.083	0.104	0.064
7/16	0.438	N	0.469	0.015	0.005	0.922	0.015	0.007	0.065	0.080	0.051
7/16	0.438	W	0.500	0.015	0.005	1.250	0.030	0.007	0.083	0.104	0.064
1/2	0.500	N	0.531	0.015	0.005	1.062	0.030	0.007	0.095	0.121	0.074
1/2	0.500	W	0.562	0.015	0.005	1.375	0.030	0.007	0.109	0.132	0.086
9/16	0.562	N	0.594	0.015	0.005	1.156	0.030	0.007	0.095	0.121	0.074
9/16	0.562	W	0.625	0.015	0.005	1.469	0.030	0.007	0.109	0.132	0.086
5/8	0.625	N	0.656	0.030	0.007	1.312	0.030	0.007	0.095	0.121	0.074
5/8	0.625	W	0.688	0.030	0.007	1.750	0.030	0.007	0.134	0.160	0.108
3/4	0.750	N	0.812	0.030	0.007	1.469	0.030	0.007	0.134	0.160	0.108
3/4	0.750	W	0.812	0.030	0.007	2.000	0.030	0.007	0.148	0.177	0.122
7/8	0.875	N	0.938	0.030	0.007	1.750	0.030	0.007	0.134	0.160	0.108
7/8	0.875	W	0.938	0.030	0.007	2.250	0.030	0.007	0.165	0.192	0.136
1	1.000	N	1.062	0.030	0.007	2.000	0.030	0.007	0.134	0.160	0.108
1	1.000	W	1.062	0.030	0.007	2.500	0.030	0.007	0.165	0.192	0.136
1 1/8	1.125	N	1.250	0.030	0.007	2.250	0.030	0.007	0.134	0.160	0.108
1 1/8	1.125	W	1.250	0.030	0.007	2.750	0.030	0.007	0.165	0.192	0.136
1 1/4	1.250	N	1.375	0.030	0.007	2.500	0.030	0.007	0.165	0.192	0.136
1 1/4	1.250	W	1.375	0.030	0.007	3.000	0.030	0.007	0.165	0.192	0.136
1 3/8	1.375	N	1.500	0.030	0.007	2.750	0.030	0.007	0.165	0.192	0.136
1 3/8	1.375	W	1.500	0.045	0.010	3.250	0.045	0.010	0.180	0.213	0.153
1 1/2	1.500	N	1.625	0.030	0.007	3.000	0.030	0.007	0.165	0.192	0.136
1 1/2	1.500	W	1.625	0.045	0.010	3.500	0.045	0.010	0.180	0.213	0.153
1 5/8	1.625		1.750	0.045	0.010	3.750	0.045	0.010	0.180	0.213	0.153
1 3/4	1.750		1.875	0.045	0.010	4.000	0.045	0.010	0.180	0.213	0.153
1 7/8	1.875		2.000	0.045	0.010	4.250	0.045	0.010	0.180	0.213	0.153
2	2.000		2.125	0.045	0.010	4.500	0.045	0.010	0.180	0.213	0.153
2 1/4	2.250		2.375	0.045	0.010	4.750	0.045	0.010	0.220	0.248	0.193
2 1/2	2.500		2.625	0.045	0.010	5.000	0.045	0.010	0.238	0.280	0.210
2 3/4	2.750		2.875	0.065	0.010	5.250	0.065	0.010	0.259	0.310	0.228
3	3.000		3.125	0.065	0.010	5.500	0.065	0.010	0.284	0.327	0.249

Courtesy of ANSI (B18.22.1)

Table B.22   Dimensions of Woodruff Keys

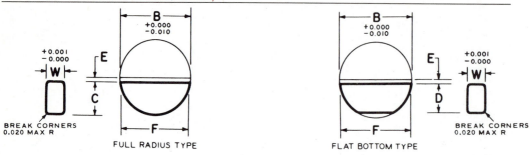

FULL RADIUS TYPE   FLAT BOTTOM TYPE

Key No.	Nominal Key Size W × B	Actual Length F +0.000-0.010	Height of Key				Distance Below Center E
			C		D		
			Max	Min	Max	Min	
202	1/16 × 1/4	0.248	0.109	0.104	0.109	0.104	1/64
202.5	1/16 × 5/16	0.311	0.140	0.135	0.140	0.135	1/64
302.5	3/32 × 5/16	0.311	0.140	0.135	0.140	0.135	1/64
203	1/16 × 3/8	0.374	0.172	0.167	0.172	0.167	1/64
303	3/32 × 3/8	0.374	0.172	0.167	0.172	0.167	1/64
403	1/8 × 3/8	0.374	0.172	0.167	0.172	0.167	1/64
204	1/16 × 1/2	0.491	0.203	0.198	0.194	0.188	3/64
304	3/32 × 1/2	0.491	0.203	0.198	0.194	0.188	3/64
404	1/8 × 1/2	0.491	0.203	0.198	0.194	0.188	3/64
305	3/32 × 5/8	0.612	0.250	0.245	0.240	0.234	1/16
405	1/8 × 5/8	0.612	0.250	0.245	0.240	0.234	1/16
505	5/32 × 5/8	0.612	0.250	0.245	0.240	0.234	1/16
605	3/16 × 5/8	0.612	0.250	0.245	0.240	0.234	1/16
406	1/8 × 3/4	0.740	0.313	0.308	0.303	0.297	1/16
506	5/32 × 3/4	0.740	0.313	0.308	0.303	0.297	1/16
606	3/16 × 3/4	0.740	0.313	0.308	0.303	0.297	1/16
806	1/4 × 3/4	0.740	0.313	0.308	0.303	0.297	1/16
507	5/32 × 7/8	0.866	0.375	0.370	0.365	0.359	1/16
607	3/16 × 7/8	0.866	0.375	0.370	0.365	0.359	1/16
707	7/32 × 7/8	0.866	0.375	0.370	0.365	0.359	1/16
807	1/4 × 7/8	0.866	0.375	0.370	0.365	0.359	1/16
608	3/16 × 1	0.992	0.438	0.433	0.428	0.422	1/16
708	7/32 × 1	0.992	0.438	0.433	0.428	0.422	1/16
808	1/4 × 1	0.992	0.438	0.433	0.428	0.422	1/16
1008	5/16 × 1	0.992	0.438	0.433	0.428	0.422	1/16
1208	3/8 × 1	0.992	0.438	0.433	0.428	0.422	1/16
609	3/16 × 1 1/8	1.114	0.484	0.479	0.475	0.469	5/64
709	7/32 × 1 1/8	1.114	0.484	0.479	0.475	0.469	5/64
809	1/4 × 1 1/8	1.114	0.484	0.479	0.475	0.469	5/64
1009	5/16 × 1 1/8	1.114	0.484	0.479	0.475	0.469	5/64

Courtesy of USAS (B17.2)

Table B.23   Tap Drill Sizes—Inches

American National Standard Unified Threads

Nominal Diameter	UNC Thds per in.	UNC Tap Drill	UNF Thds per in.	UNF Tap Drill	UNEF Thds per in.	UNEF Tap Drill
# 5(.125)	40	#38	44	#37	—	—
# 6(.138)	32	#36	40	#33	—	—
# 8(.164)	32	#29	36	#29	—	—
#10(.190)	24	#25	32	#21	—	—
#12(.216)	24	#16	28	#14	32	#13
1/4 (.25)	20	# 7	28	# 3	32	7/32
5/16(.31)	18	F	24	I	32	9/32
3/8 (.38)	16	5/16	24	Q	32	11/32
7/16(.44)	14	U	20	25/64	28	13/32
1/2 (.50)	13	27/64	20	29/64	28	15/32
9/16(.56)	12	31/64	18	33/64	24	33/64
5/8 (.63)	11	17/32	18	37/64	24	37/64
3/4 (.75)	10	21/32	16	11/16	20	45/64
7/8 (.88)	9	49/64	14	13/16	20	53/64
1″ (1.00)	8	7/8	12	59/64	20	61/64
1 1/8 (1.13)	7	63/64	12	1 3/64	18	1 5/64
1 1/4 (1.25)	7	1 7/64	12	1 11/64	18	1 3/16
1 3/8 (1.38)	6	1 7/32	12	1 19/64	18	1 5/16
1 1/2 (1.50)	6	1 11/32	12	1 27/64	18	1 7/16
1 3/4 (1.75)	5	1 9/16	—	—	—	—
2 (2.00)	4 1/2	1 25/32	—	—	—	—

Table B.24   Tap Drill Sizes—Metric

Nominal Diameter (mm)	Coarse Pitch	Coarse Tap Drill	Fine Pitch	Fine Tap Drill
2	0.4	1.6	—	—
2.5	0.45	2.05	—	—
3	0.5	2.50	—	—
3.5	0.6	2.90	—	—
5	0.8	4.20	—	—
6.3	1.0	5.30	—	—
8	1.25	6.80	1.0	7.00
10	1.5	8.50	1.25	8.75
12	1.75	10.30	1.25	10.50
14	2.0	12.00	1.5	12.50
16	2.0	14.00	1.5	14.50
20	2.5	17.50	1.5	18.50
24	3.0	21.00	2.0	22.00
30	3.5	26.50	2.0	28.00
36	4.0	32.00	3.0	33.00
42	4.5	37.50	3.0	39.00
48	5.0	43.00	3.0	45.00

Table B.25   Standard Welding Symbols

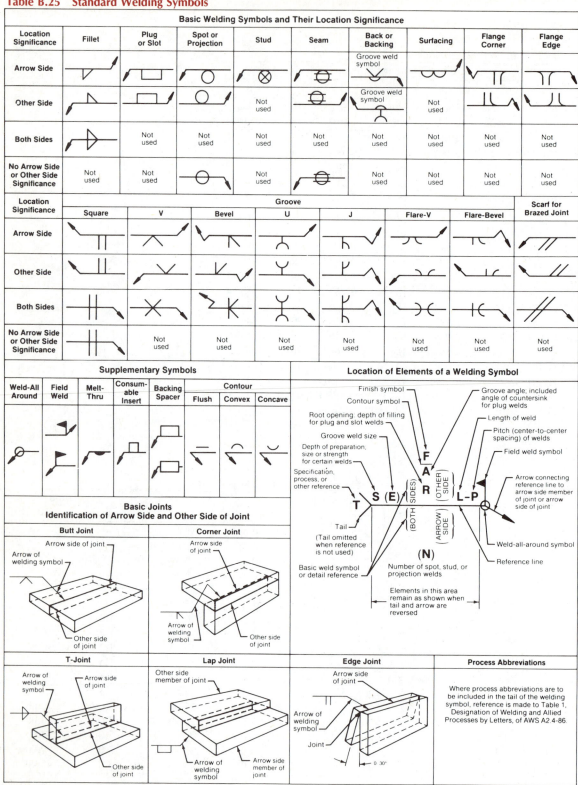

Courtesy of American Welding Society (A2.1-86)

Table B.25 (*continued*)

Typical Welding Symbols

Double-Fillet Welding Symbol	Chain Intermittent Fillet Welding Symbol	Staggered Intermittent Fillet Welding Symbol
Plug Welding Symbol	Back Welding Symbol	Backing Welding Symbol
Spot Welding Symbol	Stud Welding Symbol	Seam Welding Symbol
Square-Groove Welding Symbol	Single-V Groove Welding Symbol	Double-Bevel-Groove Welding Symbol
Symbol with Back Gouging	Flare-V Groove Welding Symbol	Flare-Bevel-Groove Welding Symbol
Multiple Reference Lines	Complete Penetration	Edge Flange Welding Symbol
Flash or Upset Welding Symbol	Melt-Thru Symbol	Joint with Backing
Joint with Spacer	Flush Contour Symbol	Convex Contour Symbol

*It should be understood that these charts are intended only as shop aids. The only complete and official presentation of the standard welding symbols is in A2 4.

Table B.26   Fraction–Decimal Conversion Table

Fraction			2-Place Decimal	3-Place Decimal
		1/32	.03	.031
	1/16		.06	.062
		3/32	.09	.094
1/8			.13	.125
		5/32	.16	.156
	3/16		.19	.188
		7/32	.22	.219
1/4			.25	.250
		9/32	.28	.281
	5/16		.31	.312
		11/32	.34	.344
3/8			.38	.375
		13/32	.41	.406
	7/16		.44	.438
		15/32	.47	.469
1/2			.50	.500
		17/32	.53	.531
	9/16		.56	.562
		19/32	.59	.594
5/8			.63	.625
		21/32	.66	.656
	11/16		.69	.688
		23/32	.72	.719
3/4			.75	.750
		25/32	.78	.781
	13/16		.81	.812
		27/32	.84	.844
7/8			.88	.875
		29/32	.91	.906
	15/16		.94	.938
		31/32	.97	.969
1			1.00	1.000

Data assembled by authors

INDEX

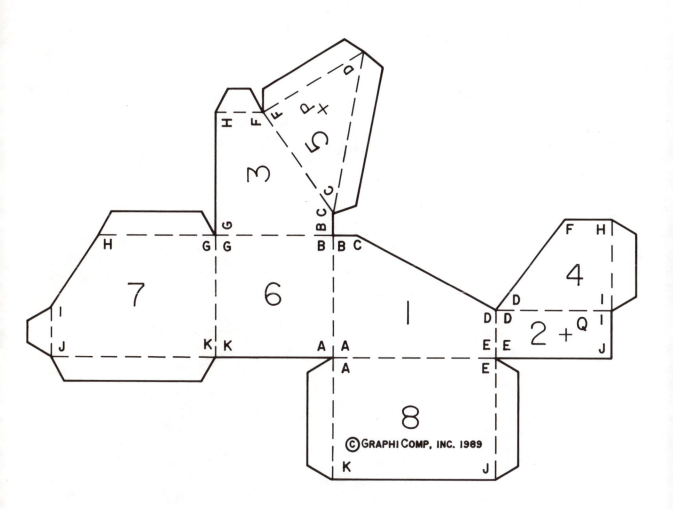